BEGINNING ALGEBRA

BEGINNING ALGEBRA

Dennis Weltman
North Harris County College

Gilbert Perez
San Antonio College

Wadsworth Publishing Company
Belmont, California
A Division of Wadsworth, Inc.

Mathematics Editor: **Anne Scanlan-Rohrer**
Assistant Editor: **Tamiko Verkler**
Editorial Assistant: **Leslie With**
Production Editor: **Sandra Craig**
Print Buyer: **Karen Hunt**
Managing Designer: **Andrew H. Ogus**
Cover and Text Designer: **Adriane Bosworth**
Cover Photo: **Detail of Acoma pot, New Mexico; The Oakland Museum History Department**
Art Editor: **Adriane Bosworth**
Technical Illustrators: **Judith Ogus/Random Arts** (text), **Guy Magallanes** (answer section)
Copy Editor: **Mary Roybal**
Compositor: **Jonathan Peck Typographers, Santa Cruz**

Printed in the United States of America   34

1 2 3 4 5 6 7 8 9 10—94 93 92 91 90

Library of Congress Cataloging-in-Publication Data

Weltman, Dennis.
    Beginning algebra / Dennis Weltman, Gilbert Perez.
    p.   cm.
    ISBN 0-534-11778-3
    1. Algebra.   I. Perez, Gilbert.   II. Title.
QA152.2.W438   1990   89-38613
512.9—dc20   CIP

To our parents,
who have always encouraged
and supported us

To Beth and Dottie,
thanks for putting
up with us

About the Authors

Dennis Weltman has a B.S. and an M.S. in mathematics from Southern Methodist University, and he teaches at North Harris County College.

Gilbert Perez has a B.A. in mathematics from Rice University and an M.S. from the University of Houston, and he teaches at San Antonio College.

They are also the authors of *Intermediate Algebra*, Second Edition, published by Wadsworth Publishing Company.

Contents

CHAPTER 4

Factoring Polynomials 169

CHAPTER 5

Rational Expressions 226

To the Instructor

Since beginning algebra is a developmental course, this textbook has two fundamental goals. The first goal is to introduce students to the elementary concepts of algebra. The second goal is to prepare students for a rigorous course in intermediate algebra and college algebra. It should be clear from the table of contents that topics required to meet the first goal are included. In addition, the book has the following features:

Features ▶ **Conversational writing style** Material is presented in an easy-to-understand, sometimes humorous style.

Examples Many detailed examples are included.

Exercises A large number of routine exercises help students build skill and confidence.

Reviews Common errors to avoid, key words and properties, chapter reviews, chapter tests, and cumulative review exercises are also featured.

The following features will help students attain the second goal:

Exercises Each exercise set contains a full range of problems extending from the routine to the challenging. Many problems involve noninteger coefficients and solutions.

Advanced topics Some intermediate algebra topics are introduced at the beginning algebra level—for example, absolute value equations, factoring by grouping, and factoring the sum and difference of two cubes.

Applications An abundance of word problems is provided. In addition, elementary motion and mixture problems will prepare students for the more complicated motion and mixture problems of intermediate algebra.

This text is aimed at students who have had little or no experience or success in algebra. These students are often weak in arithmetic. A complete review of the arithmetic of rational numbers, order of operations, and properties of real numbers is included in Chapter 1. You may choose not to spend class time on all of this material, but students will be able to review the topics they feel less secure about.

Ancillaries ▶ **Instructor's Manual**   An Instructor's Manual, written by Jim Polito of North Harris County College, contains four forms of tests for each chapter, two midterm tests, and two final examinations. Answers to the tests are provided in an answer key at the back of the manual. The Instructor's Manual also provides answers to all even-numbered exercises in the text.

Student Solutions Manual   A Student Solutions Manual, written by Richard Spangler of Tacoma Community College, contains worked-out solutions and hints to all odd-numbered exercises.

EXP Test   This full-featured test-generating system for the IBM PC and compatible microcomputers is available to adopters of the text.

JEWEL Test   Also available to adopters of the text is JEWEL Test, a random number generator testing system for the IBM PC and compatibles and for Apple II systems.

Videotapes   A series of videotapes by John Jobe, Oklahoma State University, reviews key topics in the text.

Acknowledgments ▶ As with any book, there are many people to thank. We are grateful for the professional assistance and encouragement received from Anne Scanlan-Rohrer, Sandra Craig, Andrew Ogus, Adriane Bosworth, and Ragu Raghavan. Of course, we are also grateful to all the other people at Wadsworth who helped make this book a reality. The following people reviewed the manuscript, in part or in its entirety: Mary Anne Anthony, Rancho Santiago College; Judy Cain, Tompkins Cortland Community College; Joseph De Blassio, Community College of Allegheny County; Eunice F. Everett, Seminole Community College; Jaclyn Le Febvre, Illinois Central College; Gael Mericle, Mankato State University; Thomas J. Ribley, Valencia Community College; Sharon Sledge, San Jacinto College; Penny Slingerland, Mount Hood Community College; and Michael White, Jefferson Community College. We would also like to thank Nancy Adcox, Mount San Antonio College; Ray Byrum, North Harris County College; Charles Peveto, North Harris County College; Jim Polito, North Harris County College; Dick Spangler, Tacoma Community College; and Joe Stien, Illinois Central College, for checking the text and the Instructor's Manual for accuracy. We would like to thank our colleagues at North Harris County College and San Antonio College for their assistance and encouragement. Most of all, our families and friends have offered much support from the beginning.

To the Student

Have you ever watched a tennis match on television and marveled at how the pros made it look so easy? You might have noticed that your math instructor makes algebra look easy. And yet in both cases, when you try to play tennis, or work some algebra problems, you find it is not so easy. What's the reason? A tennis pro practices for hours each day. He or she played tennis for years before becoming good enough to play professionally. Your math instructor has been working with math for *years*, spending many hours a day in preparation for class. This is not to imply that you should strive for, or expect to achieve, the level of proficiency of your math instructor; but it does imply that if you want to do well in a math course you have to practice it for hours, and that the more you practice, the better you should be. If you don't want to waste your time and money registering for the same algebra course semester after semester, here's what you should do:

1. *Attend class regularly and ask questions.* There are almost no students who are capable of learning algebra without going to class. You need to see and hear the instructor's explanations. While you can't expect to understand everything, you should have an idea of how to work the problems. If not, *ask!*

2. *Do your homework soon after class.* You will forget most of what was discussed in class within a few hours unless you do your homework during this time. Waiting until the last minute before the next class meeting to do your homework is the worst thing you can do.

3. *Read the book carefully.*
   a. It is best to read each section before the material is discussed in class. Try to follow the examples and work some of the exercises.
   b. After class, reread the section carefully. Read the statement of the problem in each example and try to solve the problem before reading the solution given in the example.

c. It is unlikely that your instructor will assign all of the exercises for homework. Some time after you have done your homework, try to work some of the unassigned exercises. Exercises requiring a **calculator** are identified by colored exercise numbers.

d. We have boxed some **common mistakes** that many students make. Make a special note of these when you work the exercises. They are marked by this symbol:

e. The **answers** to the odd-numbered exercises are in the back of the book. You should not look at the answers until you have finished working each problem. Remember, on a quiz or test you don't have the answers to refer to.

f. When you have completed a chapter, use the **chapter reviews** to go back over the important properties, formulas, and so on, that were introduced in that chapter. Exercises from each section are provided for you to get more practice. When you think you have reviewed the chapter sufficiently, take the timed **chapter test** provided. The **Test Your Memory** exercises at the end of each chapter will help to keep "old" topics fresh in your mind.

4. If you need additional help, a Student Solutions Manual, by Richard Spangler (Tacoma Community College), contains worked-out solutions and hints to all odd-numbered exercises.

CHAPTER 1

The Real Number System and Its Properties

1.1 Sets of Real Numbers

One of the most basic concepts in any mathematics course is that of a **set**. A set is defined to be a collection of objects. For example, the set that contains the **digits** of the decimal number system is written

$$\{0, 1, 2, 3, 4, 5, 6, 7, 8, 9\}$$

The numbers within the set are called **elements**, or **members**, of the set, and they are separated by commas and written within braces, { }, which is called **set notation**. We often name a set by using a capital letter; for example,

$$D = \{0, 1, 2, 3, 4, 5, 6, 7, 8, 9\}$$

When we mention set D, we know that we are referring to the set of digits. If we want to indicate that 3 is an element of set D, we write

$$3 \in D$$

If we want to indicate that $\frac{2}{3}$ is not an element of set D, we write

$$\frac{2}{3} \notin D$$

Given two sets X and Y, if each element in set X is contained in set Y, then we say X is a **subset** of Y and write

$$X \subseteq Y$$

Consider the set $E = \{0, 2, 4, 6, 8\}$. Then $E \subseteq D$, since each element of set E is also an element of set D. On the other hand, consider set $C = \{-5, \frac{2}{3},$

3, 9}. Since $-5 \in C$ but $-5 \notin D$, set C is not a subset of set D, which is written

$$C \nsubseteq D$$

An interesting set is the **empty set**, or **null set**, which contains no elements. It is denoted by either { } or $\varnothing$. An example of the empty set is the set of numbers that are both less than 2 and greater than 5.

NOTE ▶

1. *Every set is a subset of itself.* $D \subseteq D$ since every element of set D is also an element of set D.

2. *The empty set is a subset of any set.* $\varnothing \subseteq D$ because if $\varnothing$ were not a subset of D, then there would have to be an element of $\varnothing$ that is not in D; but there aren't any elements in $\varnothing$!

Sometimes we are interested in the set of elements that two sets P and W have in common. This set is called the **intersection** of P and W and is denoted by

$$P \cap W$$

Other times we might be interested in putting the elements of sets P and W together into one set. This set is called the **union** of P and W and is denoted by

$$P \cup W$$

EXAMPLE 1.1.1

Consider the sets $E = \{0, 2, 4, 6, 8\}$, $F = \{1, 2, 3, 4, 5\}$, and $O = \{1, 3, 5, 7, 9\}$. Find the following.

1. $E \cap F$
   $E \cap F = \{2, 4\}$, since the elements 2 and 4 appear in both set E and set F.

2. $E \cup F$
   $E \cup F = \{0, 1, 2, 3, 4, 5, 6, 8\}$, since each element appears in either set E or set F.

3. $F \cap \varnothing, F \cup \varnothing$
   $F \cap \varnothing = \varnothing, F \cup \varnothing = F$

4. $E \cap O$
   $E \cap O = \varnothing$, since E and O have no elements in common.

NOTE ▶ When two sets have no elements in common, they are said to be **disjoint**.

Let us now consider some specific sets—the sets of numbers we will be working with throughout this book.

Natural numbers: $N = \{1, 2, 3, 4, \ldots\}$

Whole numbers: $W = \{0, 1, 2, 3, 4, \ldots\}$

Integers: $I = \{\ldots, -3, -2, -1, 0, 1, 2, 3, \ldots\}$

Rational numbers: Q = set of all fractions, p/q, where p and q are integers and q is not zero.

or Q = set of all terminating or repeating decimals.

Irrational numbers: H = set of all nonterminating and nonrepeating decimals.

Real numbers: R = set of all decimal numbers.

or R = set of all numbers that are rational or irrational.

The first four sets build on the preceding sets and are related as follows:

$$N \subseteq W \subseteq I \subseteq Q$$

The last statement, $I \subseteq Q$, follows from the fact that any integer can be written as a fraction by writing it over 1. For example, $6 = \frac{6}{1}$, $-2 = \frac{-2}{1}$.

We can describe the rational numbers in two ways because a fraction can be changed to a decimal by dividing.

EXAMPLE 1.1.2

Change the following fractions to decimals.

1. $\dfrac{5}{8}$

$$\begin{array}{r} 0.625 \\ 8\overline{)5.000} \\ \underline{4\ 8} \\ 20 \\ \underline{16} \\ 40 \\ \underline{40} \\ 0 \end{array}$$

$\dfrac{5}{8} = 0.625$

2. $\dfrac{17}{11}$

$$
\begin{array}{r}
1.5454\ldots \\
11\overline{)17.00000} \\
\underline{11} \\
6\,0 \\
\underline{5\,5} \\
50 \\
\underline{44} \\
60 \\
\underline{55} \\
50 \\
\underline{44} \\
60
\end{array}
$$

$\dfrac{17}{11} = 1.5454\ldots$

NOTE ▶

1. $1.5454\ldots$ can be written as $1.\overline{54}$ where the bar indicates that the numbers under it are repeated.

2. Each time a fraction is converted to a decimal, the decimal terminates, as in part 1 of Example 1.1.2, or repeats, as in part 2.

We define the irrational numbers as being the set of decimal numbers that are *not* rational. The two sets are disjoint, that is,

$$Q \cap H = \varnothing$$

Some examples of irrational numbers are $0.2020020002\ldots$ and π, the ratio of the circumference to the diameter of a circle:

$$\pi = 3.1415926536\ldots$$

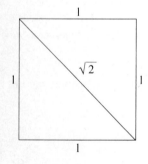

More common examples of irrational numbers are square roots. Each side of the square in Figure 1.1.1 has a length of one unit. The diagonal has a length of the square root of 2, written $\sqrt{2}$. (See Appendix II.) $\sqrt{2}$ is some number that when multiplied by itself is 2:

$$\sqrt{2} = 1.4142135623\ldots$$

Most of the square roots of the natural numbers are irrational.

Together, the rational numbers and irrational numbers compose the real numbers, that is,

$$Q \cup H = R$$

The relationship of these sets of numbers to each other is illustrated in Figure 1.1.2.

Figure 1.1.1

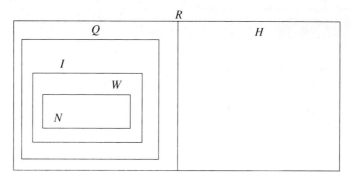

Figure 1.1.2

EXAMPLE 1.1.3

Identify the sets to which the numbers $\frac{3}{4}$, -7, $\sqrt{5}$, and 0 belong.

$\frac{3}{4}$ is a rational number and therefore also a real number, that is, $\frac{3}{4} \in Q$ and $\frac{3}{4} \in R$.

-7 is an integer and therefore also a rational number and a real number, that is, $-7 \in I$, $-7 \in Q$, and $-7 \in R$.

$\sqrt{5}$ is an irrational number and therefore also a real number, that is, $\sqrt{5} \in H$ and $\sqrt{5} \in R$.

0 is a whole number, integer, rational number, and real number, that is, $0 \in W$, $0 \in I$, $0 \in Q$, and $0 \in R$.

EXAMPLE 1.1.4

Determine whether the following statements are true or false.

1. $W \cap H = \varnothing$

   True. All whole numbers are rational numbers and therefore are not irrational.

2. $\{6, -\frac{11}{4}, \frac{0}{3}, 4.5\} \subseteq Q$

   True. Each of the four numbers is a fraction, can be written as a fraction, or is a terminating decimal.

3. $\frac{8}{0} \in R$

   False. $\frac{8}{0}$ is undefined because the denominator of a fraction cannot be zero.

EXERCISES 1.1

Consider the sets $A = \{4, 8, 12, 16, 20\}$, $B = \{6, 7, 8, 9, 10\}$, $T = \{0, 3, 6, 9, 12, \ldots\}$, $S = \{\ldots, -5, -3, -1, 1, 3, 5, 7, \ldots\}$, and N. Find the following.

1. $A \cap B$ **2.** $A \cap T$ **3.** $B \cap T$ **4.** $T \cap N$

5. $A \cap S$ **6.** $T \cap S$ **7.** $S \cap N$ **8.** $S \cap B$

9. $A \cup B$ **10.** $A \cup T$ **11.** $B \cup T$ **12.** $N \cup T$

13. $S \cup N$ **14.** $B \cup S$ **15.** $A \cap \varnothing$ **16.** $A \cup \varnothing$

17. $A \cup (B \cap T)$ **18.** $A \cap (B \cup T)$ **19.** $B \cap (T \cup S)$ **20.** $(B \cap T) \cup (B \cap S)$

Change the following fractions to decimals.

21. $\dfrac{2}{3}$ **22.** $\dfrac{1}{3}$ **23.** $\dfrac{4}{9}$ **24.** $\dfrac{7}{9}$ **25.** $\dfrac{2}{11}$ **26.** $\dfrac{7}{11}$

27. $\dfrac{15}{11}$ **28.** $\dfrac{30}{11}$ **29.** $\dfrac{3}{4}$ **30.** $\dfrac{5}{4}$ **31.** $\dfrac{11}{8}$ **32.** $\dfrac{7}{8}$

33. $\dfrac{5}{16}$ **34.** $\dfrac{13}{64}$ **35.** $\dfrac{3}{7}$ **36.** $\dfrac{5}{7}$

Identify the sets to which each of the following numbers belongs.

37. 6   **38.** -13 **39.** $\dfrac{5}{-11}$ **40.** $\dfrac{3}{8}$

41. $\sqrt{3}$ **42.** $\sqrt{7}$ **43.** 3.5   **44.** $6.2\overline{3}$

45. $8.535535553\ldots$ **46.** $\dfrac{4}{1}$ **47.** $\dfrac{5}{0}$ **48.** $\dfrac{\sqrt{2}}{9}$

Determine whether the following statements are true or false.

49. $8.0 \in I$ **50.** $0 \in N$ **51.** $\dfrac{6}{7} \in I$ **52.** $3.4444\ldots \in H$

53. $N \cap H = \varnothing$ **54.** $N \cup I = I$ **55.** $\varnothing \subseteq R$ **56.** $W \subseteq W$

1.2 The Real Number Line and Absolute Value

An important concept in mathematics is that of the **real number line**. A ruler is a common example of part of a number line, in that numbers are assigned to specific points on the ruler. In a similar fashion, we can assign every real number to a point on a line and every point on the line to a real number. We commonly draw the real number line with tic marks on it, like a ruler, labeling the locations of some of the integers, as in Figure 1.2.1. When we locate the point assigned to a specific number and draw a dot there, we say that we are

graphing that number. To **graph a set** means to graph each of the numbers in the set.

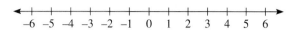

Figure 1.2.1

EXAMPLE 1.2.1

Graph the following sets.

1. $\{\frac{3}{4}, -\frac{7}{2}, 4.3, 2\}$

$\frac{3}{4}$ is three-fourths of the way from 0 to 1. $-\frac{7}{2} = -3.5$ is one-half of the way from -3 to -4. 4.3 is three-tenths of the way from 4 to 5.

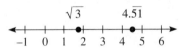

2. $\{\sqrt{3}, 4.\overline{51}\}$

We will use approximations to graph these numbers. $\sqrt{3} \doteq 1.7$, from Table 2 in Appendix III, and $4.\overline{51} = 4.515151 \ldots \doteq 4.5$. ($\doteq$ means approximately equal to.)

The real number line gives us a pictorial way of understanding the concept of the **order of the real numbers**. Given two distinct real numbers, we see that one is smaller than the other. This relationship is noted with one of the **inequality symbols** given here.

Symbol	Stands for	Examples
$<$	is less than	$3 < 8,\ -7 < -2$
$\leq$	is less than or equal to	$-5 \leq 2,\ -4 \leq -4$
$>$	is greater than	$6 > 0,\ 0 > -3$
$\geq$	is greater than or equal to	$\frac{5}{2} \geq 1,\ \frac{9}{1} \geq 9$

NOTE ▶

1. On the real number line the numbers get larger to the right and smaller to the left. Therefore, given two real numbers, we note that the graph of the smaller one lies to the left of the graph of the larger one.

2. A number is positive if it is greater than zero, and it is negative if it is less than zero. Zero is neither positive nor negative.

3. The two symbols $\leq$ and $\geq$ state that one of two possible relationships holds. For example, $3 \leq 5$ says that either 3 is less than 5 or 3 equals 5.

4. The inequality symbols $<$ and $>$ always point toward the smaller number.

EXAMPLE 1.2.2

Place the appropriate inequality symbol, $<$ or $>$, between each pair of numbers.

1. $\frac{5}{3}$ _____ 5

   $\frac{5}{3} = 1.\overline{6}$, so $\frac{5}{3} < 5$.

2. $\sqrt{6}$ _____ 3

   From Table 2 in Appendix III, $\sqrt{6} \doteq 2.449$; so $\sqrt{6} < 3$.

3. -2 _____ -8

   The graph of -2 lies to the right of the graph of -8; therefore, $-2 > -8$.

A very important concept in mathematics is that of absolute value.

Absolute Value

The **absolute value** of a number is the distance from zero to that number on the real number line. The absolute value of a number x is denoted by $|x|$.

EXAMPLE 1.2.3

Evaluate each of the following.

1. $|3|$

   Since 3 is three units to the right of zero, $|3| = 3$.

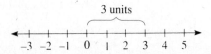

2. $|-4|$

Since -4 is four units to the left of zero, $|-4| = 4$.

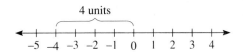

3. $|0|$

Since 0 is zero units from zero, $|0| = 0$.

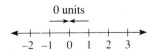

Note that the absolute value of a number is never negative, since distance is never negative. It is always positive or zero. Absolute value measures distance, not direction.

EXAMPLE 1.2.4

Place the appropriate symbol, $<$, $>$, or $=$, between each pair of numbers.

1. $|-5|$ _____ 3

Since $|-5| = 5$, $|-5| > 3$.

2. $|-8|$ _____ $|8|$

Since $|-8| = 8$ and $|8| = 8$, $|-8| = |8|$.

EXERCISES 1.2

Graph the following sets of numbers on the real number line.

1. $\{5, 0, -3, -6\}$

2. $\{-4, -1, 2, 7\}$

3. $\{\frac{2}{3}, -\frac{5}{2}, \frac{7}{5}\}$

4. $\{-\frac{3}{4}, \frac{4}{3}, \frac{11}{2}\}$

5. $\{2.7, -3.1, 5.\overline{3}\}$

6. $\{-1.6, -4.2, 2.\overline{8}\}$

7. $\{\sqrt{8}, 8, \sqrt{2}, 2\}$

8. $\{\sqrt{5}, 5, \sqrt{10}, 10\}$

9. $\{-2, \frac{5}{3}, -4.5, \sqrt{7}\}$

10. $\{-6, -\frac{7}{4}, 1.4, \sqrt{12}\}$

11. $\{|-3|, 6.\overline{37}, |\frac{5}{6}|\}$

12. $\{|-0.5|, |4|, 1.\overline{06}\}$

Evaluate the following.

13. $|6|$

14. $\left|-\frac{1}{5}\right|$

15. $|\pi|$

16. $|\sqrt{21}|$

17. $3|-7|$

18. $9|-2|$

19. $|-8| + 3$

20. $|-4| + |-1|$

Place the appropriate symbol, <, >, or =, between the following pairs of numbers.

21. -4 _____ -11

22. -7 _____ -3

23. $\frac{5}{2}$ _____ 3

24. $\frac{4}{7}$ _____ 2

25. -8 _____ 2

26. -10 _____ 1

27. 0 _____ -5

28. 0 _____ -9

29. 2 _____ $\sqrt{2}$

30. 3 _____ $\sqrt{3}$

31. $\frac{2}{3}$ _____ 0.6

32. 0.75 _____ $\frac{7}{9}$

33. $|-15|$ _____ $|-8|$

34. $|-5|$ _____ $|-13|$

35. $|-20|$ _____ 6

36. 14 _____ $|-7|$

37. $|-24|$ _____ $|24|$

38. $\left|\frac{3}{2}\right|$ _____ $\left|-\frac{3}{2}\right|$

39. $|-12| + 5$ _____ $|1| + |16|$

40. $|18| + |-9|$ _____ $|-18| - 9$

1.3 Addition of Rational Numbers

In this section we study the addition of integers and rational numbers, assuming the reader is adept at the arithmetic of whole numbers. The other arithmetic operations on the integers and rational numbers are covered in Sections 1.4, 1.5, and 1.6. The arithmetic of irrational numbers will be dealt with in Chapter 8.

The operation of addition has certain properties that are useful in algebra. For instance, we know that $7 + 3 = 10$ and $3 + 7 = 10$, so $7 + 3 = 3 + 7$. This illustrates the commutative property of addition.

Commutative Property of Addition
If a and b are real numbers, then $a + b = b + a$

The commutative property allows us to change the order of numbers that we are adding. We also know that

$$2 + (5 + 13) \quad \text{and} \quad (2 + 5) + 13$$
$$= 2 + 18 \qquad\qquad\qquad = 7 + 13$$
$$= 20 \qquad\qquad\qquad\quad\; = 20$$

Thus, $2 + (5 + 13) = (2 + 5) + 13$. This illustrates the associative property of addition.

Associative Property of Addition

If a, b, and c are real numbers, then

$$a + (b + c) = (a + b) + c$$

When we are performing two or more addition operations, this property allows us to change the grouping of the operations. Together, these two properties allow us to add a string of numbers in any convenient order.

EXAMPLE 1.3.1

State the property that is illustrated in each step.

$17 + (99 + 3) + 1 = 17 + (3 + 99) + 1$ Commutative property of addition

$= (17 + 3) + (99 + 1)$ Associative property of addition

$= 20 + 100$

$= 120$

Another important property is that the sum (the answer when numbers are added) of a number and zero is the same number. For example, $7 + 0 = 7$, $0 + 13 = 13$. This property is called the identity property of addition, and 0 is called the **additive identity** element.

Identity Property of Addition

For any real number a,

$$a + 0 = a \qquad \text{and} \qquad 0 + a = a$$

If you have a debt of $350, this amount is expressed as a negative number, $-\$350$. If you have debts of $350, $200, and $75, your total indebtedness can be calculated by adding these positive numbers and then making the answer negative. Since $350 + 200 + 75 = 625$, then $(-\$350) + (-\$200) + (-\$75) = -\$625$. This reasoning motivates the following rule.

Rule 1	Examples
To add two or more negative numbers, add the absolute values of the numbers and make the answer negative.	$(-7) + (-8) = -15$ $(-4) + (-4) + (-4) = -12$

Returning to the example of indebtedness, we can motivate a rule for adding positive and negative numbers. If you paid $300 on your debt of $625, you still would have a debt of $325, or

$$(-\$625) + \$300 = -\$325$$

If you now paid $350 on this debt of $325, you would have a credit of $25, or

$$(-\$325) + \$350 = \$25$$

In each case, the sum is found by subtracting positive numbers (absolute values) and using the sign of the number with the larger absolute value to determine the sign of the answer.

Rule 2	Example
To add two numbers with opposite signs, subtract the smaller absolute value from the larger absolute value. The answer will have the same sign as the number with the larger absolute value.	$-15 + 5$ $\lvert -15 \rvert = 15,\ \lvert 5 \rvert = 5$ $15 - 5 = 10$ Since -15 has the larger absolute value, $-15 + 5 = -10$

EXAMPLE 1.3.2

Add the following numbers.

1. $-12 + 28$

Since $28 - 12 = 16$ and 28 has the larger absolute value, $-12 + 28 = 16$.

2. $-45 + (-15)$

Since both numbers are negative and $45 + 15 = 60$, $-45 + (-15) = -60$.

3. $-30 + 11 + (-8) + 5$

$= -30 + (-8) + 11 + 5$	Commutative property
$= [-30 + (-8)] + [11 + 5]$	Associative property
$= [-38] + 16$	Rule 1
$= -22$	Rule 2

4. $-2 + 2$

Since $2 - 2 = 0$, $-2 + 2 = 0$.

Whenever the sum of two numbers is zero, as in part 4 above, the numbers are said to be **additive inverses**, or **opposites**. They have the same absolute value (they are the same distance from zero) but are on opposite sides of zero.

Inverse Property of Addition
For any number a, $a + (-a) = 0$ and $-a + a = 0$.

Number	Additive Inverse	Sum
-2	2	$-2 + 2 = 0$
5	-5	$5 + (-5) = 0$
0	0	$0 + 0 = 0$

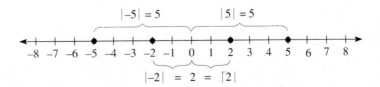

Placing a negative sign in front of a number changes it to its additive inverse. Thus,

$$-(5) \text{ is the additive inverse of } 5, \text{ which is } -5$$
$$-(-2) \text{ is the additive inverse of } -2, \text{ which is } 2$$

The second statement illustrates the double negative property.

Double Negative Property
For any number a, $-(-a) = a$.

Before studying the addition of fractions, we need to define several terms. We call 3 and 5 **factors** of 15 since $3 \cdot 5 = 15$. 15 also has factors of 15 and 1. 7 has factors of only 7 and 1. (In general we consider only integer factors.) When a natural number other than 1 has only itself and 1 as factors, we say that number is **prime**. Some prime numbers are 2, 3, 5, 7, 11, 13, 17, 19, 23, ... If a natural number other than 1 is not prime, we say it is **composite**. In a fraction, the top number is called the **numerator** and the bottom number is called the **denominator**. Since $\frac{2}{4} = 0.5 = \frac{1}{2}$, the fractions $\frac{2}{4}$ and $\frac{1}{2}$ represent the same rational number and are said to be **equivalent fractions**. $\frac{1}{2}$ is a ratio of smaller integers and is usually simpler to work with. "Simplifying" a fraction from $\frac{2}{4}$ to $\frac{1}{2}$ is called **reducing** the fraction. This is accomplished by finding the prime factors of the numerator and denominator and then dividing out the factors that the numerator and denominator have in common. In our example,

$$\frac{2}{4} = \frac{\overset{1}{\cancel{2}}}{\underset{1}{\cancel{2} \cdot 2}} = \frac{1}{2}$$

Each time a common factor is divided out, a factor of 1 is left. When all the common factors have been divided out, the fraction is said to be in **lowest terms**.

EXAMPLE 1.3.3

Reduce the following fractions to lowest terms.

1. $\dfrac{18}{42}$ $\dfrac{18}{42} = \dfrac{2 \cdot 3 \cdot 3}{2 \cdot 3 \cdot 7} = \dfrac{3}{7}$

2. $\dfrac{7}{63}$ $\qquad \dfrac{7}{63} = \dfrac{\cancel{7}^{1}}{3 \cdot 3 \cdot \cancel{7}} = \dfrac{1}{9}$ Don't forget the 1 on top!

3. $-\dfrac{14}{7}$ $\qquad -\dfrac{14}{7} = -\dfrac{\cancel{7} \cdot 2}{\cancel{7}_{1}} = -\dfrac{2}{1} = -2$

REMARK After performing an operation involving fractions, we always want the answer in lowest terms.

If you were to cut a cake into nine equal pieces, then each piece would be $\frac{1}{9}$ of the original cake. If two pieces are eaten, then $\frac{2}{9}$ of the cake is eaten. If another five pieces are eaten, then altogether

$$\frac{2}{9} + \frac{5}{9} = \frac{2+5}{9} = \frac{7}{9}$$

of the cake is gone. We add only the numerators and use the denominator they have in common.

Rule 3	Examples
To add fractions that have the same denominator, add the numerators and write this sum over the common denominator.	$\dfrac{2}{5} + \dfrac{4}{5} = \dfrac{2+4}{5} = \dfrac{6}{5}$ $\dfrac{5}{18} + \dfrac{1}{18} + \dfrac{7}{18} = \dfrac{5+1+7}{18}$ $= \dfrac{13}{18}$

EXAMPLE 1.3.4

Perform the indicated additions.

1. $\dfrac{7}{12} + \dfrac{1}{12}$ $\qquad \dfrac{7}{12} + \dfrac{1}{12} = \dfrac{7+1}{12}$

$\qquad\qquad\qquad = \dfrac{8}{12}$ This is not in lowest terms.

$\qquad\qquad\qquad = \dfrac{\cancel{2} \cdot \cancel{2} \cdot 2}{\cancel{2} \cdot \cancel{2} \cdot 3}$ Factor and reduce.

$\qquad\qquad\qquad = \dfrac{2}{3}$

2. $\dfrac{-9}{10} + \dfrac{3}{10}$

$$\dfrac{-9}{10} + \dfrac{3}{10} = \dfrac{-9 + 3}{10}$$

$$= \dfrac{-6}{10} \qquad \text{Rule 2}$$

$$= \dfrac{-2 \cdot 3}{2 \cdot 5} \qquad \text{Reduce to lowest terms.}$$

$$= \dfrac{-3}{5}$$

3. $2\frac{5}{8} + 5\frac{7}{8}$ $\qquad\qquad\qquad\qquad\qquad$ Convert to fractions.

$$2\frac{5}{8} + 5\frac{7}{8} = \dfrac{21}{8} + \dfrac{47}{8} = \dfrac{21 + 47}{8}$$

$$= \dfrac{68}{8} = \dfrac{2 \cdot 2 \cdot 17}{2 \cdot 2 \cdot 2} = \dfrac{17}{2} \qquad \text{Leave as a fraction.}$$

To add fractions with different denominators, such as $\frac{5}{12} + \frac{7}{30}$, we must first change the fractions to equivalent fractions with a common denominator. 360, 180, and 60 are all common denominators, each of which can be divided evenly by 12 and 30. We would like to find the smallest, or least, common denominator. The **least common denominator**, which is abbreviated **LCD**, can be found in the following manner.

How to Find the LCD	Example
1. Factor each of the denominators into products of prime numbers.	$12 = 2 \cdot 2 \cdot 3$ $30 = 2 \cdot 3 \cdot 5$
2. Form the LCD by taking the product of all the different factors. Use each factor the same number of times as in the denominator where it appears the most times.	There are different factors of 2, 3, and 5. 2 appears more times in 12, twice, so we need two factors of 2. We need only one 3 and one 5. LCD $= 2 \cdot 2 \cdot 3 \cdot 5 = 60$

Once the LCD has been found, we divide the denominator of each fraction into the LCD and multiply the numerator and denominator by this result. For example, $60 \div 12 = 5$, so we multiply $\frac{5}{12}$ on both top and bottom by 5:

$$\dfrac{5 \cdot 5}{12 \cdot 5} = \dfrac{25}{60}$$

Similarly, $60 \div 30 = 2$, so we multiply $\frac{7}{30}$ on both top and bottom by 2:

$$\frac{7 \cdot 2}{30 \cdot 2} = \frac{14}{60}$$

Thus, $\quad \dfrac{5}{12} + \dfrac{7}{30} = \dfrac{5 \cdot 5}{12 \cdot 5} + \dfrac{7 \cdot 2}{30 \cdot 2}$

$$= \frac{25}{60} + \frac{14}{60} = \frac{39}{60}$$

$$= \frac{3 \cdot 13}{2 \cdot 2 \cdot 3 \cdot 5} = \frac{13}{20}$$

EXAMPLE 1.3.5

Perform the indicated additions.

1. $\dfrac{5}{14} + \dfrac{1}{6} + \dfrac{3}{4}$ $\qquad 14 = 2 \cdot 7$

$\qquad\qquad\qquad\qquad\quad 6 = 2 \cdot 3 \qquad \text{LCD} = 2 \cdot 2 \cdot 3 \cdot 7 = 84$

$\qquad\qquad\qquad\qquad\quad 4 = 2 \cdot 2$

$$\frac{5}{14} + \frac{1}{6} + \frac{3}{4} = \frac{5 \cdot 6}{14 \cdot 6} + \frac{1 \cdot 14}{6 \cdot 14} + \frac{3 \cdot 21}{4 \cdot 21}$$

$$= \frac{30}{84} + \frac{14}{84} + \frac{63}{84} = \frac{107}{84}$$

2. $3 + \dfrac{-5}{4}$

$$3 + \frac{-5}{4} = \frac{3}{1} + \frac{-5}{4} \qquad\qquad\qquad\qquad \text{Write 3 as a fraction, } \tfrac{3}{1}.$$

$$= \frac{3 \cdot 4}{1 \cdot 4} + \frac{-5}{4} = \frac{12}{4} + \frac{-5}{4} = \frac{7}{4}$$

3. $5\frac{2}{3} + 2\frac{7}{9}$

$$5\tfrac{2}{3} + 2\tfrac{7}{9} = \frac{17}{3} + \frac{25}{9} = \frac{17 \cdot 3}{3 \cdot 3} + \frac{25}{9} \qquad \text{Convert to fractions.}$$

$$= \frac{51}{9} + \frac{25}{9} = \frac{76}{9}$$

When rational numbers are written in decimal form, we use the following rule to add them.

Rule 4	Example
1. Write the decimals vertically, lining up the decimal points. **2.** Use the rules for addition of integers. **3.** Place the decimal point in the answer directly beneath the other decimal points.	$-15.83 + 6.914$ Subtract the absolute values. 15.83 $\underline{-6.914}$ 8.916 The answer is -8.916.

EXAMPLE 1.3.6

Perform the indicated additions.

$$-4.78 + (-11.5) + (-0.436)$$

Add the absolute values:

$$\begin{array}{r} 4.78 \\ 11.5 \\ \underline{0.436} \\ 16.716 \end{array}$$

The answer is -16.716.

The distance around a plane figure is called the **perimeter**. For a triangle it is found by adding the lengths of its three sides [see Figure 1.3.1(a)]. For a rectangle the perimeter is the sum of all four sides [see Figure 1.3.1(b)].

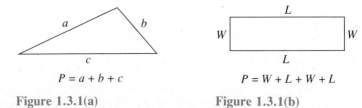

$P = a + b + c$

Figure 1.3.1(a)

$P = W + L + W + L$

Figure 1.3.1(b)

EXAMPLE 1.3.7   Find the perimeter of the triangle shown in Figure 1.3.2.

15.78 cm

6.24 cm

10.9 cm

Figure 1.3.2

$$P = 15.78 \text{ cm} + 6.24 \text{ cm} + 10.9 \text{ cm}$$
$$= 32.92 \text{ cm}$$

$$\begin{array}{r} 15.78 \\ 6.24 \\ 10.9 \\ \hline 32.92 \end{array}$$

EXERCISES 1.3

State the property or rule that is illustrated in each step, except those marked with an asterisk, which are simple operations.

1. $-7 + (4 + (-3)) = -7 + (-3 + 4)$
$\qquad = (-7 + (-3)) + 4$
$\qquad \overset{*}{=} (-10) + 4$
$\qquad \overset{*}{=} -6$

2. $(-8 + 13) + (-2) = -2 + (-8 + 13)$
$\qquad = (-2 + (-8)) + 13$
$\qquad \overset{*}{=} (-10) + 13$
$\qquad \overset{*}{=} 3$

3. $-11 + (6 + (-7)) + 12 = -11 + (-7 + 6) + 12$
$\qquad = (-11 + (-7)) + (6 + 12)$
$\qquad \overset{*}{=} -18 + 18$
$\qquad = 0$

4. $-19 + 25 \overset{*}{=} -19 + (19 + 6)$
$\qquad = (-19 + 19) + 6$
$\qquad = 0 + 6$
$\qquad = 6$

Reduce the following fractions to lowest terms.

5. $\dfrac{10}{12}$

6. $\dfrac{15}{18}$

7. $\dfrac{24}{30}$

8. $\dfrac{32}{40}$

9. $\dfrac{15}{45}$

10. $\dfrac{12}{60}$

11. $\dfrac{75}{25}$

12. $\dfrac{39}{13}$

13. $-\dfrac{70}{21}$

14. $-\dfrac{66}{75}$

15. $\dfrac{84}{63}$

16. $\dfrac{120}{96}$

Perform the indicated additions.

17. $-32 + 29$

18. $-60 + 41$

19. $-43 + 70$

20. $-57 + 82$

21. $-14 + (-75)$

22. $-22 + (-61)$

23. $-11 + (-8) + (-14)$

24. $-17 + (-6) + (-20)$

25. $-51 + 29 + (-38)$

26. $41 + (-16) + 24$

27. $15 + (-21) + 40 + (-28)$

28. $(-33) + 18 + (-9) + 62$

29. $\dfrac{3}{11} + \dfrac{5}{11}$

30. $\dfrac{2}{13} + \dfrac{7}{13}$

31. $\dfrac{3}{20} + \dfrac{9}{20} + \dfrac{13}{20}$

32. $\dfrac{5}{36} + \dfrac{11}{36} + \dfrac{17}{36}$

33. $-\dfrac{5}{4} + \dfrac{11}{4}$

34. $-\dfrac{7}{6} + \dfrac{13}{6}$

35. $-\dfrac{13}{45} + \left(-\dfrac{4}{45}\right)$

36. $-\dfrac{3}{28} + \left(-\dfrac{19}{28}\right)$

37. $\dfrac{5}{3} + \dfrac{7}{8}$

38. $\dfrac{3}{7} + \dfrac{5}{2}$

39. $\dfrac{11}{6} + \dfrac{3}{4} + 2$

40. $\dfrac{7}{10} + \dfrac{7}{4} + 1$

41. $\dfrac{11}{12} + \dfrac{7}{20}$

42. $\dfrac{5}{18} + \dfrac{11}{42}$

43. $-\dfrac{7}{9} + \dfrac{1}{6}$

44. $-\dfrac{4}{15} + \dfrac{2}{25}$

45. $\dfrac{9}{22} + \left(-\dfrac{3}{4}\right) + \left(\dfrac{20}{33}\right)$

46. $\dfrac{13}{21} + \left(-\dfrac{5}{6}\right) + \left(\dfrac{4}{35}\right)$

47. $-\dfrac{2}{3} + \dfrac{5}{6} + \left(-\dfrac{11}{18}\right)$

48. $-\dfrac{3}{4} + \dfrac{7}{12} + \left(-\dfrac{1}{24}\right)$

49. $2\dfrac{3}{5} + 4\dfrac{2}{3}$

50. $3\dfrac{1}{4} + 1\dfrac{5}{7}$

51. $3\dfrac{2}{9} + 2\dfrac{5}{21}$

52. $6\dfrac{1}{8} + 2\dfrac{5}{12}$

53. $-5.41 + (-2.793)$

54. $-11.5 + (-9.47)$

55. $62.08 + (-4.93)$

56. $6.004 + (-2.356)$

57. $-2.5 + (-20.46) + (-0.903)$

58. $-4.08 + (-0.462) + (-81.7)$

Find the perimeters of the following figures.

59. Triangle with sides 6.85 cm, 8.13 cm, and 12.46 cm

60. Triangle with sides $3\dfrac{1}{4}$ in., $4\dfrac{1}{2}$ in., and $6\dfrac{3}{16}$ in.

61. Rectangle with length $25\dfrac{2}{3}$ ft and width $20\dfrac{1}{6}$ ft

62. Rectangle with length 43.8 m and width 19.05 m

63. Square with each of its four sides being $3\dfrac{5}{6}$ yd

64. Suzie owes $864.00 to her credit card company. Over the next month, she makes charges of $59.17 and $250.68. She then pays $300 on the account and is charged $15.24 for finance charges. Express this information as a sum of positive and negative numbers, and find her account balance at the end of the month.

1.4 Subtraction of Rational Numbers

When the air temperature rises from 65°F to 72°F, the temperature change is 7°F, which is found by subtracting $72 - 65 = 7$. What if the temperature drops from 72°F to 65°F? The temperature change can be represented by -7°F. If we subtract the beginning temperature from the ending temperature, as before, then $65 - 72$ must be -7. We can arrive at this answer by changing $65 - 72$ to $65 + (-72)$. When the temperature rises from 5°F below zero to 5°F above zero, we know that the temperature change is 10°F. A temperature below zero is represented by a negative number. So when we subtract $5 - (-5)$, the answer must be 10. Again we can arrive at this answer by changing $5 - (-5)$ to $5 + 5$.

Rule 5	**Examples**
To subtract a number, add its additive inverse (opposite). Hence, if a and b represent numbers, then $a - b = a + (-b)$.	$-5 - 23 = -5 + (-23)$ $= -28$ $\dfrac{2}{7} - \left(-\dfrac{3}{7}\right) = \dfrac{2}{7} + \dfrac{3}{7}$ $= \dfrac{5}{7}$

It is important to note that subtraction is not commutative like addition. $8 - 2 = 6$, but $2 - 8 = 2 + (-8) = -6$. So we must be careful not to change the order in which numbers are subtracted.

EXAMPLE 1.4.1

Perform the indicated subtractions.

1. $-15 - (-48) = -15 + 48 = 33$

2. $-4.059 - 19.68 = -4.059 + (-19.68)$
$= -23.739$

$$\begin{array}{r} 4.059 \\ +19.68 \\ \hline 23.739 \end{array}$$

3. $\dfrac{3}{14} - \dfrac{7}{8} = \dfrac{3}{14} + \left(-\dfrac{7}{8}\right)$

$\qquad = \dfrac{3 \cdot 4}{14 \cdot 4} + \left(-\dfrac{7 \cdot 7}{8 \cdot 7}\right)$

$\qquad = \dfrac{12}{56} + \left(-\dfrac{49}{56}\right) = \dfrac{12 + (-49)}{56} = -\dfrac{37}{56}$

When a problem involves more than one subtraction operation or involves subtraction and addition, it would be unclear if there were no rules governing the order in which the operations are to be performed. For example, $8 - 5 - 2$ might be attempted in two ways.

Correct Solution	Incorrect Solution
$8 - 5 - 2$	$8 - 5 - 2$
$= 3 - 2$	$= 8 - 3$
$= 1$	$= 5$

The first solution is correct, because of the following rules.

Order of Operations	
Rule	**Examples**
A. Perform all operations within grouping symbols—that is, parentheses, brackets, or braces—first.	$21 - (35 + 16)$ $= 21 - 51$ $= 21 + (-51)$ $= -30$
B. If there are no grouping symbols, perform all additions and subtractions in order from left to right.	$21 - 35 + 16$ $= 21 + (-35) + 16$ $= -14 + 16$ $= 2$

NOTE ▶ Rule A implies that when grouping symbols are included within other grouping symbols, perform the operations within the innermost symbols and work outward.

EXAMPLE 1.4.2

Perform the indicated operations following the rules governing the order of operations.

1. $\dfrac{3}{5} - \dfrac{2}{3} - \left(-\dfrac{1}{6}\right)$ The parentheses here are used as separators, keeping the minus sign apart from the negative sign of $\frac{1}{6}$.

$$\frac{3}{5} - \frac{2}{3} - \left(-\frac{1}{6}\right) = \frac{3}{5} + \left(-\frac{2}{3}\right) + \frac{1}{6} \qquad \text{Rule 5}$$

$$= \frac{3 \cdot 6}{5 \cdot 6} + \left(-\frac{2 \cdot 10}{3 \cdot 10}\right) + \frac{1 \cdot 5}{6 \cdot 5} \qquad \text{LCD} = 30$$

$$= \frac{18}{30} + \left(-\frac{20}{30}\right) + \frac{5}{30}$$

$$= \frac{18 + (-20) + 5}{30} = \frac{23 + (-20)}{30} = \frac{3}{30}$$

$$= \frac{\overset{1}{\cancel{3}}}{2 \cdot \cancel{3} \cdot 5} = \frac{1}{10}$$

REMARK   After using Rule 5 to change all subtractions to additions, we can add the numbers in any order.

2. $5.08 - (-2.493 - 0.305) + 1.6$

$= 5.08 - (-2.493 + (-0.305)) + 1.6$

$= 5.08 - (-2.798) + 1.6$

$= 5.08 + 2.798 + 1.6$

$= 9.478$

```
  2.493
 +0.305
  2.798

  5.08
  2.798
 +1.6
  9.478
```

3. $-24 - [15 - 42 - (-6 + 30)]$

$= -24 - [15 - 42 - (24)]$ Start with innermost parentheses.

$= -24 - [15 + (-42) + (-24)]$ Rule 5

$= -24 - [15 + (-66)]$

$= -24 - [-51]$

$= -24 + 51$

$= 27$

4. $-\frac{3}{8} - \left|\frac{1}{4} - 2\right|$

NOTE ▶   The absolute value symbols act like grouping symbols. Therefore, the operation within the absolute value symbols is performed first, then the absolute value of the result is found.

$$-\frac{3}{8} - \left|\frac{1}{4} - 2\right| = -\frac{3}{8} - \left|\frac{1}{4} + \left(-\frac{2}{1}\right)\right|$$

$$= -\frac{3}{8} - \left|\frac{1}{4} + \left(-\frac{2 \cdot 4}{1 \cdot 4}\right)\right|$$

$$= -\frac{3}{8} - \left|\frac{1}{4} + \left(-\frac{8}{4}\right)\right|$$

$$= -\frac{3}{8} - \left|-\frac{7}{4}\right|$$

$$= -\frac{3}{8} - \frac{7}{4}$$

$$= -\frac{3}{8} + \left(-\frac{7}{4}\right)$$

$$= -\frac{3}{8} + \left(-\frac{7 \cdot 2}{4 \cdot 2}\right)$$

$$= -\frac{3}{8} + \left(-\frac{14}{8}\right)$$

$$= -\frac{17}{8}$$

In any triangle, the sum of the measures of the three angles is always 180°. If we know the measure of two angles, we can find the third by adding the two known angles and subtracting the result from 180°.

EXAMPLE 1.4.3

Two angles of a triangle measure 34.3° and 101.8°. Find the measure of the third angle.

The measure of the third angle is

$$180° - (34.3° + 101.8°)$$

$$= 180° - 136.1°$$

$$= 43.9°$$

EXERCISES 1.4

Perform the indicated operations.

1. $-18 - 35$

2. $-7 - 49$

3. $26 - 48$

4. $31 - 54$

5. $-12 - (-29)$

6. $-20 - (-41)$

7. $4.39 - (-12.8)$

8. $16.5 - (-8.12)$

9. $-94.3 - (-62.08)$

10. $-81.17 - (-33.5)$

11. $-\frac{3}{7} - \frac{5}{14}$

12. $-\frac{5}{6} - \frac{1}{12}$

13. $\frac{4}{25} - \frac{7}{15}$

14. $\frac{8}{35} - \frac{11}{21}$

15. $-6 - (-14) + (-27)$

16. $11 - 42 + (-8)$

17. $19 - (-4) + (-12) - 38$

18. $-21 + (-8) - 15 - (-43)$

19. $\frac{7}{3} - \frac{2}{9} - \left(-\frac{5}{12}\right)$

20. $-\frac{5}{8} - \frac{11}{4} - \left(-\frac{3}{10}\right)$

21. $-2\frac{3}{5} - \left(1\frac{1}{2}\right) + 4\frac{2}{3}$

22. $3\frac{1}{3} - 2\frac{3}{4} + \left(-1\frac{7}{8}\right)$

23. $-\frac{3}{8} + 2 - \frac{5}{7} - \left(-\frac{11}{14}\right)$

24. $-\frac{7}{11} - \left(-\frac{7}{3}\right) - 1 + \frac{5}{2}$

25. $14 - (-12 - 25)$

26. $-3 - (6 - 49)$

27. $-4.15 + 2.99 - (-16.8 + 9)$

28. $11.08 - 24 - (6.72 - 16.081)$

29. $\left(\frac{2}{9} + \frac{5}{6}\right) - \left(\frac{3}{2} - \frac{11}{3}\right)$

30. $\left(\frac{3}{5} - \frac{7}{15}\right) - \left(-\frac{8}{3} + \frac{13}{25}\right)$

31. $6\frac{1}{2} - \left(2\frac{5}{8} + 5\frac{1}{4}\right) - 1\frac{3}{4}$

32. $3\frac{7}{10} - \left(-1\frac{3}{5} + 3\frac{1}{2}\right) + 1\frac{8}{15}$

33. $-15 - |5 - (-5 - 5)|$

34. $-18 - |-6 - (6 - (-6))|$

35. $-\frac{4}{17} - \left|\frac{1}{2} - 3\right|$

36. $\frac{2}{21} - \left|-\frac{3}{2} - 1\right|$

37. $-\frac{5}{13} + \left|\frac{5}{13} - \frac{5}{3}\right| + \frac{5}{3}$

38. $\frac{7}{12} - \left|\frac{7}{12} - \frac{3}{5}\right| + \frac{3}{5}$

39. $-18 - [28 - (4 - 19)]$

40. $-3 - [-51 - (-11 + 32)]$

41. $6.1 - [(3.92 - 13.5) - 8.14]$

42. $-2.08 - [(5.6 - 2.98) - 17.52]$

43. $-\frac{9}{5} + \left[\frac{3}{10} - \left(\frac{3}{2} + 2\right)\right]$

44. $-\frac{7}{6} + \left[-\frac{9}{2} - \left(3 - \frac{1}{3}\right)\right]$

Find the measure of the third angle of each triangle with the given angles.

45. $14.9°, 61.8°$

46. $25.4°, 98.7°$

47. $42.18°, 108.9°$

48. $52.06°, 121.3°$

49. Mary has a checking account balance of $-\$85.31$ (she is overdrawn). The bank charges her \$30 for two returned checks. She makes a deposit of \$426.50 and then writes checks for \$75.00, \$16.38, and \$295.99. Express this information as a sum and/or difference, then find her account balance.

50. At noon a meteorologist in Verhoyansk, Siberia, recorded a temperature of $-18°$F. He continues making observations at 4-hour intervals and records the temperature rise $4°$, fall $12°$, fall $9°$, fall $6°$, and fall $3°$. Express this information as a sum and/or difference, then find the final temperature.

1.5 Multiplication of Rational Numbers

The operation of multiplication has the same properties as addition. For instance, we know that $6 \cdot 2 = 12$ and $2 \cdot 6 = 12$, so $6 \cdot 2 = 2 \cdot 6$. This illustrates the commutative property of multiplication.

Commutative Property of Multiplication
If a and b are real numbers, then
$a \cdot b = b \cdot a$

The commutative property allows us to change the order of numbers that we are multiplying.

We also know that

$$2 \cdot (5 \cdot 7) \qquad \text{and} \qquad (2 \cdot 5) \cdot 7$$
$$= 2 \cdot 35 \qquad\qquad\qquad = 10 \cdot 7$$
$$= 70 \qquad\qquad\qquad\qquad = 70$$

Thus, $2 \cdot (5 \cdot 7) = (2 \cdot 5) \cdot 7$. This illustrates the associative property of multiplication.

Associative Property of Multiplication

If a, b, and c are real numbers, then

$$a \cdot (b \cdot c) = (a \cdot b) \cdot c$$

When we are performing two or more multiplication operations, this property allows us to change the grouping of the operations. Together, the commutative and associative properties allow us to multiply a string of numbers in any convenient order.

Another important property is that the product (the answer when numbers are multiplied) of a number and 1 is the same number. For example, $-8 \cdot 1 = -8$, $1 \cdot 45 = 45$. This property is called the identity property of multiplication, and 1 is called the **multiplicative identity** element.

Identity Property of Multiplication

For any real number a,

$$a \cdot 1 = a \qquad \text{and} \qquad 1 \cdot a = a$$

EXAMPLE 1.5.1

Name the property illustrated in each step.

$$5 \cdot (13 \cdot 2) = 5 \cdot (2 \cdot 13) \qquad \text{Commutative property of multiplication}$$
$$= (5 \cdot 2) \cdot 13 \qquad \text{Associative property of multiplication}$$
$$= 10 \cdot 13 = 130$$

If you owe $5 to each of four people, your total debt can be computed as follows:

$$(-5) + (-5) + (-5) + (-5) = -20$$

or
$$4 \cdot (-5) = -20$$

This illustrates the following rule.

Rule 6	Examples
The product of a positive and a negative number is a negative number.	$5(-2) = -10$ $(-9)(12) = -108$

By Rule 6, $-1 \cdot 3 = -3$. From Section 1.3, $-(3) = -3$. Thus, $-1 \cdot 3 = -(3)$, or in words, negative one times a number is the same as its additive inverse. Now let us use this reasoning to motivate the rule for multiplying two negative numbers:

$$-3 \cdot (-4) = (-1 \cdot 3) \cdot (-4)$$

$$= -1(3 \cdot (-4)) \qquad \text{Associative property of multiplication}$$

$$= -1(-12) \qquad \text{Rule 6}$$

$$= -(-12)$$

$$= 12 \qquad \text{Double negative property}$$

Rule 7	Example
The product of two negative numbers is positive.	$(-8)(-7) = 56$

EXAMPLE 1.5.2

Perform the indicated multiplications.

1. $\underbrace{(-10)(-3)}(-6)$

$$= \quad 30 \quad \cdot (-6) \qquad \text{Rule 7}$$

$$= \quad -180 \qquad \text{Rule 6}$$

2. $\underbrace{(-2)(2)}(-2)(-2)(-2)$

$= \underbrace{-4 \ \cdot \ (-2)}(-2)(-2)$

$= \underbrace{8 \ \cdot \ (-2)}(-2)$

$= \underbrace{-16 \qquad \cdot (-2)}$

$= 32$

REMARK By the associative property of multiplication, we could have multiplied starting from the right.

From these two examples we can summarize Rules 6 and 7 into one rule.

Generalization of Rules 6 and 7	Examples
If the number of negative factors is odd, the product is negative.	$(2)(5)(-2)(5) = -100$ $(-1)(-2)(-3)(-4)(-5) = -120$
If the number of negative factors is even, the product is positive.	$(-8)(-2)(5) = 80$ $(-5)(-2)(-1)(-2)(-5)(-1)$ $= 100$

Note that the number of positive factors does not affect the sign of the product.

We will now apply the generalization of Rules 6 and 7 to rational numbers, both in fraction and decimal form.

Rule 8	Example
To multiply fractions: **1.** Write the product of the numerators over the product of the denominators. **2.** Factor the numerators and denominators and reduce the fraction. **3.** Multiply any remaining factors in the numerator and denominator.	$\dfrac{9}{8} \cdot \dfrac{10}{21} = \dfrac{9 \cdot 10}{8 \cdot 21}$ $= \dfrac{3 \cdot 3 \cdot 2 \cdot 5}{2 \cdot 2 \cdot 2 \cdot 3 \cdot 7}$ $= \dfrac{15}{28}$

EXAMPLE 1.5.3

Perform the indicated multiplications.

1. $\dfrac{11}{12} \cdot \dfrac{4}{15} \cdot \dfrac{5}{33} = \dfrac{11 \cdot 4 \cdot 5}{12 \cdot 15 \cdot 33}$

$= \dfrac{\cancel{11} \cdot \cancel{2} \cdot \cancel{2} \cdot \cancel{5}}{2 \cdot 2 \cdot 3 \cdot 3 \cdot \cancel{5} \cdot 3 \cdot \cancel{11}} = \dfrac{1}{27}$

2. $\left(-\dfrac{20}{3}\right)\left(-\dfrac{9}{25}\right) = \dfrac{20 \cdot 9}{3 \cdot 25}$ The product is positive because there is an even number of negative factors.

$= \dfrac{2 \cdot 2 \cdot \cancel{5} \cdot 3 \cdot 3}{\cancel{3} \cdot \cancel{5} \cdot 5}$

$= \dfrac{12}{5}$

3. $\left(-2\dfrac{3}{5}\right)\left(-1\dfrac{1}{2}\right)\left(-3\dfrac{1}{3}\right) = \left(-\dfrac{13}{5}\right)\left(-\dfrac{3}{2}\right)\left(-\dfrac{10}{3}\right)$ Change to fractions.

$= -\dfrac{13 \cdot 3 \cdot 10}{5 \cdot 2 \cdot 3}$ The product is negative because there is an odd number of negative factors.

$= -\dfrac{13 \cdot \cancel{3} \cdot \cancel{2} \cdot \cancel{5}}{\cancel{5} \cdot \cancel{2} \cdot \cancel{3}}$

$= -\dfrac{13}{1} = -13$

4. $\dfrac{2}{5} \cdot \dfrac{5}{2} = \dfrac{2 \cdot 5}{5 \cdot 2} = \dfrac{1}{1} = 1$

Whenever the product of two numbers is one, as in part 4 above, the numbers are said to be **multiplicative inverses**, or **reciprocals**.

Inverse Property of Multiplication
For any number a, except zero,
$a \cdot \dfrac{1}{a} = 1$ and $\dfrac{1}{a} \cdot a = 1$

Number	**Multiplicative inverse**	**Product**
-3	$-\dfrac{1}{3}$	$(-3)\left(-\dfrac{1}{3}\right) = 1$
$\dfrac{4}{7}$	$\dfrac{7}{4}$	$\dfrac{4}{7} \cdot \dfrac{7}{4} = 1$
0	No multiplicative inverse	

Zero does not have a multiplicative inverse because there is no number that can be multiplied by zero to get 1. In fact, the product of zero and any other number is zero.

Zero Product Property
For any number a, $a \cdot 0 = 0$.

When a problem contains the operations of multiplication and addition and/or subtraction, we must again use the rules governing the order of operations. For convenience, we repeat the previous rules here and add the new rules concerning multiplication.

Order of Operations	
Rule	**Examples**
A. Perform all operations within grouping symbols first.	$6(5 + 2)$ $= 6(7) = 42$
B. If there are no grouping symbols, perform the operations in the following order: **1.** multiplications **2.** additions and subtractions, in order from left to right	$15 - 4 \cdot 3 + 8$ $= 15 - 12 + 8$ $= 3 + 8$ $= 11$

EXAMPLE 1.5.4

Perform the indicated operations following the rules governing the order of operations.

1. $\dfrac{3}{4} - \dfrac{4}{5} + \dfrac{7}{6} \cdot \dfrac{3}{5} = \dfrac{3}{4} - \dfrac{4}{5} + \dfrac{7 \cdot 3}{6 \cdot 5}$ Multiply first.

   $= \dfrac{3}{4} - \dfrac{4}{5} + \dfrac{7 \cdot 3}{2 \cdot 3 \cdot 5}$ Reduce.

$$= \frac{3}{4} - \frac{4}{5} + \frac{7}{10}$$

$$= \frac{3 \cdot 5}{4 \cdot 5} - \frac{4 \cdot 4}{5 \cdot 4} + \frac{7 \cdot 2}{10 \cdot 2} \qquad \text{LCD} = 20.$$

$$= \frac{15}{20} - \frac{16}{20} + \frac{14}{20}$$

$$= \frac{15 - 16 + 14}{20} = \frac{15 + (-16) + 14}{20}$$

$$= \frac{-1 + 14}{20} = \frac{13}{20}$$

2. $5[-2.8 + 3.2(-0.25)]$ Perform the operations within the brackets first.

$= 5[-2.8 + (-0.8)]$ Multiply before you add.

$= 5[-3.6]$

$= -18$

3. $6 - 2|-5 - 11|$ Perform the subtraction within the absolute value symbols first.

$= 6 - 2|-5 + (-11)|$

$= 6 - 2|-16|$ Evaluate the absolute value.

$= 6 - 2 \cdot 16$ Multiply before you subtract.

$= 6 - 32 = 6 + (-32) = -26$

The last property we will consider in this section is one of the most important in algebra. The distributive property, which involves multiplication and addition, allows us to work a problem such as $4(5 + 3)$ two different ways. Following Rule A for the order of operations,

$4(5 + 3)$

$= 4(8)$

$= 32$

We also can distribute 4 across the sum, yielding

$4(5 + 3) = 4 \cdot 5 + 4 \cdot 3$

         $= 20 + 12$ Rule B of the order of operations

         $= 32$

$4(5 + 3)$ is a *product* of the two *factors* 4 and $(5 + 3)$, whereas $4 \cdot 5 + 4 \cdot 3$ is a *sum* of the two *terms* $4 \cdot 5$ and $4 \cdot 3$. Thus, the distributive property allows us to rewrite a product as a sum or a sum as a product.

<table>
<tr><td align="center">**Distributive Property**</td></tr>
<tr><td align="center">If a, b, and c are real numbers, then

$a(b + c) = a \cdot b + a \cdot c$</td></tr>
</table>

EXAMPLE 1.5.5

Use the distributive property to rewrite each expression.

1. $6(8 - 3) = 6 \cdot 8 - 6 \cdot 3 = 48 - 18 = 30$

   Since subtraction is defined as addition of the additive inverse, we also can distribute over subtraction.

2. $7(5 + 6 - 3) = 7 \cdot 5 + 7 \cdot 6 - 7 \cdot 3$
   $$= 35 + 42 - 21 = 56$$

   Multiplication distributes over two or more terms that are being added or subtracted.

3. $9 \cdot 18 + 9 \cdot 2 = 9(18 + 2)$
   $$= 9(20) = 180$$

Parts 2 and 3 of Example 1.5.5 illustrate that the distributive property has two "directions" in which it is used.

Some formulas in geometry involve both multiplication and addition. The perimeter of a rectangle with a length of L and a width of W can be written as $P = 2L + 2W$.

NOTE ▶ In algebra, when no operation is indicated between two quantities, the operation is understood to be multiplication. For example,

$$P = 2L + 2W$$

means

$$P = 2 \cdot L + 2 \cdot W$$

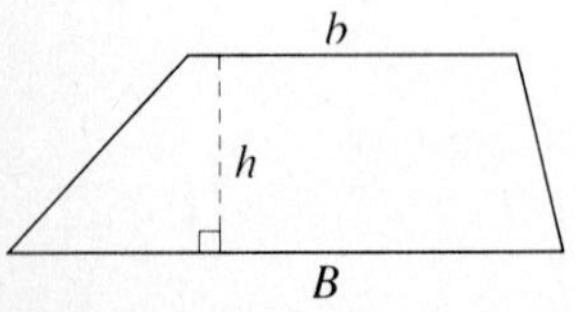

The area of a trapezoid (see Figure 1.5.1) with bases b and B and height h is

$$A = \tfrac{1}{2}(b + B)h$$

Figure 1.5.1

EXAMPLE 1.5.6

Find the area of a trapezoid with bases $b = 1.8$ cm and $B = 3.4$ cm and height $h = 2.75$ cm.

$A = \frac{1}{2}(b + B)h = \frac{1}{2}(1.8 + 3.4)(2.75)$ or $0.5(1.8 + 3.4)(2.75)$

$= 0.5(5.2)(2.75) = 2.6(2.75)$

$= 7.15$ cm^2

EXERCISES 1.5

Name the property illustrated in each step, except those marked with an asterisk, which are simple operations.

1. $\frac{2}{3}\left(\frac{3}{2} \cdot 8\right) = \left(\frac{2}{3} \cdot \frac{3}{2}\right) \cdot 8$

$= 1 \cdot 8$

$= 8$

2. $\frac{1}{4}(4 \cdot 7) = \left(\frac{1}{4} \cdot 4\right) \cdot 7$

$= 1 \cdot 7$

$= 7$

3. $5(-6 + 6) = 5(0)$

$= 0$

4. $9(-2 + 2) = 9(0)$

$= 0$

5. $\quad 3[6 + (4 + 13)]$

$= 3[(6 + 4) + 13]$

$\overset{*}{=} 3[10 + 13]$

$= 3 \cdot 10 + 3 \cdot 13$

$\overset{*}{=} 30 + 39 \overset{*}{=} 69$

6. $\quad 2\left[\frac{1}{2} + \left(-\frac{4}{3}\right) + \frac{5}{6}\right]$

$= 2 \cdot \frac{1}{2} + 2\left(-\frac{4}{3}\right) + 2\left(\frac{5}{6}\right)$

$= 1 + \left(-\frac{8}{3}\right) + \frac{5}{3}$

$\overset{*}{=} -\frac{5}{3} + \frac{5}{3}$

$= 0$

Use the distributive property to rewrite each expression.

7. $5(4 + 7)$

8. $11(9 + 2)$

9. $3(12 - 4 - 2)$

10. $8(5 - 3 - 1)$

11. $2 \cdot 7 + 2 \cdot 3$

12. $6 \cdot 4 + 6 \cdot 12$

13. $4 \cdot 3 + 4 \cdot 5 - 4 \cdot 2$

14. $10 \cdot 9 + 10 \cdot 3 - 10 \cdot 5$

15. $7 \cdot 11 + 7 \cdot 4 + 7$

16. $9 \cdot 5 + 9 \cdot 2 - 9$

Perform the indicated operations. When necessary, use the rules governing the order of operations.

17. $(-5)(16)$

18. $(9)(-12)$

19. $(-4)(11)(-5)$

20. $(-2)(15)(-6)$

21. $(-3)(3)(-3)(-3)$

22. $(-4)(-4)(4)(-4)$

23. $(-18)(-1000)$

24. $(-235)(10)$

25. $\left(-\frac{11}{4}\right)\left(-\frac{18}{5}\right)$

26. $\left(-\frac{15}{14}\right)\left(-\frac{21}{25}\right)$

27. $\left(\frac{1}{8}\right)\left(-\frac{2}{13}\right)\left(\frac{26}{3}\right)$

28. $\left(\frac{16}{9}\right)\left(-\frac{5}{12}\right)\left(\frac{33}{10}\right)$

29. $\left(-\frac{4}{7}\right)\left(-\frac{5}{6}\right)(-3)$

30. $\left(-\frac{6}{11}\right)(-4)\left(-\frac{3}{8}\right)$

31. $\left(-2\frac{3}{8}\right)\left(1\frac{3}{5}\right)$

32. $\left(-3\frac{3}{4}\right)\left(5\frac{1}{5}\right)$

33. $\left(-4\frac{2}{3}\right)\left(-3\frac{1}{7}\right)\left(1\frac{4}{11}\right)$

34. $\left(-2\frac{4}{9}\right)\left(-4\frac{1}{5}\right)\left(4\frac{1}{6}\right)$

35. $(-2.54)(15.5)$

36. $(-11.8)(3.25)$

37. $(-6.71)(2.4)(-0.75)$

38. $(-3.08)(6.2)(-0.15)$

39. $(-72.6)(10,000)$

40. $(-0.0502)(-100,000)$

41. $-6 - 11(4) + 18$

42. $-15 + 5(-8) + 43$

43. $3 \cdot 8 - 8 \cdot 5$

44. $9 \cdot 2 - 2 \cdot 7$

45. $\frac{2}{9} - \frac{4}{15} \cdot \frac{5}{8}$

46. $\frac{11}{12} - \frac{3}{8} \cdot \frac{2}{5}$

47. $2\left(-\frac{11}{6}\right) + \frac{2}{21} \cdot \frac{5}{4}$

48. $3\left(-\frac{13}{9}\right) + \left(-\frac{4}{15}\right)\left(\frac{7}{2}\right)$

49. $\frac{5}{14} + \left(-\frac{6}{17}\right)\left(\frac{4}{21}\right)\left(\frac{34}{5}\right)$

50. $\frac{16}{7} + \left(\frac{4}{11}\right)\left(-\frac{33}{10}\right)(7)$

51. $4\frac{1}{2} - \left(2\frac{1}{3}\right)\left(2\frac{1}{4}\right)$

52. $3\frac{2}{3} - \left(1\frac{5}{6}\right)\left(2\frac{5}{11}\right)$

53. $-3[5 - 2(4 - 11)]$

54. $7[-3 - 4(6 - 15)]$

55. $8 - 5(-2 - 14)$

56. $9 - 4(3 - 12)$

57. $(-6 - 7)(-17 + 8 \cdot 3)$

58. $(5 - 11)(6 - 4 \cdot 9)$

59. $-6.5 + 4.82(3.13 - 11.08)$

60. $-3.49 + 14.5(8.41 - 12.15)$

61. $[1.57 - (4.6)(-2.15)](8.5)$

62. $[16.3 - 16.3(0.4)](-6.5)$

63. $-\frac{2}{3} + \frac{4}{5}\left(\frac{7}{6} - 2\right)$

64. $-\frac{11}{5} + \frac{3}{10}\left(\frac{4}{9} - 1\right)$

65. $\frac{3}{4}\left[\frac{8}{5} + \frac{14}{9}\left(-\frac{6}{7}\right)\right]$

66. $\frac{17}{3}\left[-\frac{9}{4} - \left(\frac{3}{10}\right)\left(-\frac{7}{2}\right)\right]$

67. $\left[\left(2\frac{1}{7}\right)\left(2\frac{4}{5}\right) - 8\frac{1}{3}\right]\left(2\frac{1}{4}\right)$

68. $\left[3\frac{1}{3} - \left(-1\frac{2}{3}\right)\left(-2\frac{1}{10}\right)\right]\left(4\frac{1}{2}\right)$

69. $-2|-18 + 7| - 7$

70. $-3|-4 + 3(-8)| + 15$

71. $\frac{5}{6} - \frac{3}{8}\left|\frac{1}{6} - \frac{2}{9}\right|$

72. $\frac{3}{4} - \frac{5}{9}\left|\frac{13}{10} - 2\right|$

Use the formulas given earlier in this section to find the following.

73. The perimeter of a rectangle with length $8\frac{1}{3}$ ft and width $5\frac{5}{6}$ ft

74. The perimeter of a rectangle with length 4.61 m and width 3.5 m

75. The area of a trapezoid with bases 3.8 mm and 6.23 mm and height 4.5 mm

76. The area of a trapezoid with bases $2\frac{1}{4}$ in. and $3\frac{7}{8}$ in. and height $2\frac{2}{7}$ in.

77. Gustavo invested \$4,500 in bonds paying 9.75% simple interest. How much interest does he earn in $2\frac{1}{2}$ years? Use the formula $I = Prt$ where $I =$ interest, $P =$ principal (amount invested), $r =$ interest rate (in decimal form), and $t =$ time (in years).

1.6 Division of Rational Numbers

If you owe a total of \$28 to four people and you owe the same amount to each person, then the amount owed each person can be determined by dividing the total debt by the number of people:

$$-28 \div 4 = -7$$

Your debt to each person is $-\$7$. In this division problem, -28 is called the **dividend**, 4 is the **divisor**, and -7 is the **quotient**. We can check a division problem (when the remainder is zero) by determining whether the quotient times the divisor equals the dividend:

$$-7 \cdot 4 = -28$$

In this way we can determine the sign of the quotient when dividing a positive number by a negative number,

$$12 \div (-6) = -2 \qquad \text{since } (-2)(-6) = 12$$

and when dividing a negative number by a negative number,

$$(-20) \div (-4) = 5 \qquad \text{since } 5(-4) = -20$$

The above results are summarized in the following rule.

Rule 9	Examples
The quotient of two numbers of unlike sign is negative.	$-16 \div 8 = -2$
	$35 \div (-7) = -5$
The quotient of two negative numbers is positive.	$(-39) \div (-3) = 13$

Division problems involving zero need special attention. We can say that $0 \div 8 = 0$ since $0 \cdot 8 = 0$. But $5 \div 0$ presents a problem. If we say that the quotient is some number x, then $5 \div 0 = x$ implies $x \cdot 0 = 5$. But we know that $x \cdot 0 = 0$ for any number x. So $5 \div 0$ does not have an answer and we say it is **undefined**. We have a different problem trying to obtain an answer for $0 \div 0$. If we say that the quotient is some number x, then $0 \div 0 = x$ implies $x \cdot 0 = 0$. But this statement is true for any number x, which implies there is no unique answer. Therefore, we say $0 \div 0$ is **indeterminate**. Let us summarize these results.

Rule 10	Examples
Zero divided by any *nonzero* number is zero.	$0 \div (-15) = 0$
You cannot divide by zero.	$11 \div 0$ and $0 \div 0$ cannot be done.

Division problems can be written as fractions. For example,

$$12 \div 3 = \frac{12}{3} = 4$$

Using fractions, we note the following results.

$$-\frac{18}{3} = -6 \qquad \frac{18}{-3} = -6 \qquad -\left(\frac{18}{3}\right) = -6$$

Also,

$$\frac{-30}{-3} = 10 \qquad -\left(-\frac{30}{3}\right) = -(-10) = 10 \qquad -\left(\frac{30}{-3}\right) = -(-10) = 10$$

Generalizing the previous concepts, if $b \neq 0$, then

> **1.** $\dfrac{-a}{b} = \dfrac{a}{-b} = -\dfrac{a}{b}$
>
> **2.** $\dfrac{-a}{-b} = -\left(\dfrac{-a}{b}\right) = -\left(\dfrac{a}{-b}\right) = \dfrac{a}{b}$
>
> **3.** $\dfrac{0}{b} = 0$
>
> **4.** $\dfrac{a}{0}$ cannot be done; $\dfrac{0}{0}$ cannot be done.

When a number is divided by 2, the answer is the same as when that number is multiplied by $\frac{1}{2}$, the multiplicative inverse, or reciprocal, of 2. For example,

$$6 \div 2 = 3 \quad \text{and} \quad 6 \cdot \frac{1}{2} = 3$$

In fact, we can generalize this result to obtain a rule for dividing by any fraction.

Rule 11	Example
To divide by a fraction, multiply by its multiplicative inverse: $$x \div \frac{a}{b} = x \cdot \frac{b}{a}$$	$$\frac{3}{5} \div \frac{9}{4} = \frac{3}{5} \cdot \frac{4}{9}$$ $$= \frac{3 \cdot 2 \cdot 2}{5 \cdot 3 \cdot 3}$$ $$= \frac{4}{15}$$

EXAMPLE 1.6.1

Perform the indicated divisions.

1. $-\dfrac{7}{4} \div \dfrac{21}{10} = -\dfrac{7}{4} \cdot \dfrac{10}{21}$ Rule 11

$\qquad = -\dfrac{7 \cdot 2 \cdot 5}{2 \cdot 2 \cdot 3 \cdot 7}$ Rules 6 and 8

$\qquad = -\dfrac{5}{6}$

2. $-\dfrac{26}{25} \div (-39) = -\dfrac{26}{25} \cdot \left(-\dfrac{1}{39}\right)$ Rule 11

$\qquad = \dfrac{2 \cdot \cancel{13}}{5 \cdot 5 \cdot 3 \cdot \cancel{13}}$ Rules 7 and 8

$\qquad = \dfrac{2}{75}$

3. $15 \div \left(-3\tfrac{1}{3}\right) = 15 \div \left(-\dfrac{10}{3}\right)$ Convert to a fraction.

$\qquad = 15 \cdot \left(-\dfrac{3}{10}\right)$

$\qquad = -\dfrac{3 \cdot \cancel{5} \cdot 3}{2 \cdot \cancel{5}}$

$\qquad = -\dfrac{9}{2}$

EXAMPLE 1.6.2

Ragu drove 200 miles from Houston to San Antonio and used $9\tfrac{3}{5}$ gallons of gasoline. What was his gasoline mileage for this trip?

SOLUTION

Gasoline mileage is a certain number of miles per gallon. The word *per* tells us to *divide*. So we need to divide 200 miles by $9\tfrac{3}{5}$ gallons:

$200 \div 9\tfrac{3}{5} = 200 \div \dfrac{48}{5}$

$\qquad = 200 \cdot \dfrac{5}{48}$

$\qquad = \dfrac{\cancel{2} \cdot \cancel{2} \cdot 2 \cdot 5 \cdot 5 \cdot 5}{\cancel{2} \cdot \cancel{2} \cdot 2 \cdot 2 \cdot 3}$

$\qquad = \dfrac{125}{6}$, or $20\tfrac{5}{6}$ miles per gallon

When a problem contains more than one division operation, or when it contains division and multiplication, addition, and/or subtraction, we must again use the rules governing the order of operations. We now amend these rules as follows:

Order of Operations	
Rule	**Examples**
A. Perform all operations within grouping symbols first.	$(11 - 3) \div 4$ $= 8 \div 4 = 2$
B. If there are no grouping symbols, perform the operations in the following order: **1.** multiplications *and divisions, in order from left to right* **2.** additions and subtractions, in order from left to right	$20 - \underbrace{24 \div 6} \cdot 2$ $= 20 - \underbrace{4 \cdot 2}$ $= 20 - \quad 8$ $= 12$

EXAMPLE 1.6.3

Perform the indicated operations following the rules governing the order of operations.

1. $-60 \div 10 \div (-2) = -6 \div (-2)$ Rule B, 1
$$= 3$$

2. $5 + 30 \div (-5)(4 - 7) = 5 + 30 \div (-5)(4 + (-7))$
$$= 5 + \underbrace{30 \div (-5)} \cdot (-3)$$
$$= 5 + \underbrace{(-6) \cdot (-3)}$$
$$= 5 + \quad 18$$
$$= 23$$

3. $8[24 \div (3 - 9)]$
$$= 8[24 \div (-6)]$$
$$= 8[-4]$$
$$= -32$$

4. $\dfrac{\frac{15}{14}}{\frac{2}{3} + \frac{5}{6}}$

A fraction that contains fractions in its numerator and/or denominator is called a **complex fraction**. The division bar acts like grouping symbols, causing us to perform any operations in the numerator separately from any operations in the denominator. Then the division is performed last. In this example,

$$\frac{\frac{15}{14}}{\frac{2}{3} + \frac{5}{6}} = \frac{\frac{15}{14}}{\frac{4}{6} + \frac{5}{6}}$$

$$= \frac{\frac{15}{14}}{\frac{9}{6}} = \frac{\frac{15}{14}}{\frac{3}{2}} = \frac{15}{14} \div \frac{3}{2}$$

$$= \frac{15}{14} \cdot \frac{2}{3}$$

$$= \frac{3 \cdot 5 \cdot 2}{2 \cdot 7 \cdot 3}$$

$$= \frac{5}{7}$$

5. $\dfrac{\frac{3}{8} - 5\left(\frac{7}{10}\right)}{\frac{3}{4} + \frac{1}{12}} = \dfrac{\frac{3}{8} - \frac{5 \cdot 7}{5 \cdot 2}}{\frac{3}{4} + \frac{1}{12}}$

$$= \frac{\frac{3}{8} - \frac{7}{2}}{\frac{3}{4} + \frac{1}{12}}$$

LCD on top is 8.
LCD on bottom is 12.

$$= \frac{\frac{3}{8} - \frac{28}{8}}{\frac{9}{12} + \frac{1}{12}}$$

$$= \frac{\frac{3}{8} + \left(-\frac{28}{8}\right)}{\frac{9}{12} + \frac{1}{12}}$$

$$= \frac{-\frac{25}{8}}{\frac{10}{12}} = \frac{-\frac{25}{8}}{\frac{5}{6}} = -\frac{25}{8} \div \frac{5}{6}$$

$$= -\frac{25}{8} \cdot \frac{6}{5}$$

$$= -\frac{5 \cdot 5 \cdot 2 \cdot 3}{2 \cdot 2 \cdot 2 \cdot 5}$$

$$= -\frac{15}{4}$$

EXERCISES 1.6

Perform the indicated operations. When necessary, use the rules governing the order of operations.

1. $-42 \div 6$

2. $32 \div (-8)$

3. $-56 \div (-7)$

4. $\dfrac{-40}{-5}$

5. $\dfrac{64}{-4}$

6. $\dfrac{-120}{8}$

7. $\dfrac{96}{-16}$

8. $\dfrac{-91}{7}$

9. $\dfrac{-184}{-8}$

10. $\dfrac{-114}{-6}$

11. $\dfrac{0}{11}$

12. $\dfrac{-6}{0}$

13. $\dfrac{0}{0}$

14. $\dfrac{0}{-3}$

15. $\frac{4}{5} \div \left(-\frac{12}{35}\right)$

16. $-\frac{11}{7} \div \left(-\frac{9}{14}\right)$

17. $-\frac{16}{25} \div \frac{28}{15}$

18. $-4\frac{1}{2} \div 2\frac{1}{4}$

19. $-6\frac{1}{2} \div \left(-3\frac{9}{10}\right)$

20. $6 \div \left(-2\frac{2}{3}\right)$

21. $-8 \div 3\frac{1}{5}$

22. $\frac{4}{21} \div 14$

23. $-\frac{17}{9} \div 4$

24. $\dfrac{\frac{15}{8}}{-\frac{27}{4}}$

25. $\dfrac{-\frac{18}{7}}{-\frac{6}{35}}$

26. $\dfrac{-\frac{13}{10}}{\frac{26}{25}}$

27. $\dfrac{\frac{12}{5}}{\frac{32}{15}}$

28. $\dfrac{9}{-\frac{3}{4}}$

29. $\dfrac{-\frac{10}{3}}{6}$

30. $\dfrac{-\frac{12}{5}}{-4}$

31. $\dfrac{-8}{-\frac{6}{5}}$

32. $\dfrac{-11}{\frac{1}{6}}$

33. $-48 \div 8 \div (-2)$

34. $64 \div (-4)(2)$

35. $2 + 12 \div 3 - 9 \cdot 4$

36. $3 \cdot 6 \div 2 - 12 \div 3$

37. $15 + 22 \div (-11) - 8$

38. $-14 - 39 \div 3 + 10$

39. $75 - 16 \div 4(-2)$

40. $63 + 27 \div (-9)(3)$

41. $-6 \div \frac{9}{2} - \frac{4}{7} \div \frac{10}{21}$

42. $-\frac{3}{8} \div \frac{33}{10} + \left(-\frac{17}{30}\right)\left(\frac{45}{51}\right)$

43. $(4 + 16) \div (4 \div 2 \cdot 2)$

44. $(6 + 12) \div (9 \div 3 \cdot 3)$

45. $(-19 - 44) \div 7 \div (-3)$

46. $(29 - 110) \div (-3) \div (-3)$

47. $15 \cdot 14 \div 35 - 6$

48. $24 \cdot 25 \div 15 - 39$

49. $\frac{3}{4} \div \left(\frac{5}{8} - \frac{5}{2}\right)$

50. $\frac{25}{12} \div \left(-\frac{2}{9} - 2\right)$

51. $\frac{6}{11} - \frac{5}{6}\left(9 \div \frac{15}{7}\right)$

52. $\frac{7}{16} + \frac{13}{20}\left(\frac{35}{12} \div \frac{7}{4}\right)$

53. $[4 + (12 + 8 \div 2 - 6)] \div 2$

54. $[56 \div (-3 \cdot 2 + 13)] \div (-8)$

55. $28 \div [13 - 72 \div (21 - 9)]$

56. $75 \div [29 + 42 \div (18 - 21)]$

57. $14 - |32 \div (-8)|5$

58. $30 \div |26 - 4(5)|$

59. $\dfrac{\frac{2}{5} + \frac{7}{10}}{\frac{4}{3} - 5}$

60. $\dfrac{\frac{3}{7} - 3}{\frac{5}{4} + 1}$

61. $\dfrac{\left(\frac{4}{15}\right)\left(\frac{10}{3}\right) - \frac{7}{6}}{\frac{9}{2} - \frac{1}{3}}$

62. $\dfrac{\frac{9}{8} - \frac{11}{6}}{3\left(\frac{7}{4}\right) - \frac{29}{12}}$

63. $\dfrac{6\frac{3}{4} - 3\frac{5}{6}}{4\frac{1}{2} + 2\frac{3}{8}}$

64. $\dfrac{3\frac{2}{5} - 5\frac{1}{2}}{1\frac{3}{4} - 1\frac{3}{10}}$

65. Al drove 320 miles and used $12\frac{1}{2}$ gallons of gasoline. What was his gasoline mileage for the trip, that is, how many miles per gallon did he get?

66. Ray drove 450 miles and used $14\frac{2}{3}$ gallons of gasoline. What was his gasoline mileage for the trip?

67. Steve drove 240 miles in $4\frac{2}{5}$ hours. What was his average speed for the trip, that is, how many miles per hour did he average?

68. Glenda drove 560 miles in $8\frac{2}{5}$ hours. What was her average speed for the trip?

69. Dottie ran for $69\frac{3}{10}$ minutes, covering $8\frac{2}{5}$ miles. What was her average pace, that is, how many minutes per mile did she average?

70. Mary ran for $33\frac{1}{4}$ minutes, covering $3\frac{1}{2}$ miles. What was her average pace?

1.7 Exponents and Roots

When all the factors in a product are the same number, we have a shorter way of writing it, which we call **power notation**, or **exponential notation**. For example, $3 \cdot 3 \cdot 3 \cdot 3$ can be written as 3^4, where 3, the factor that is being multiplied, is called the **base**, and 4, the number of factors, is called the **power**, or **exponent**. The exponential expression 3^4 is read "three to the fourth power." Other examples are $7 \cdot 7 \cdot 7 \cdot 7 \cdot 7 = 7^5$, read "seven to the fifth power"; $6 \cdot 6 = 6^2$, read "six to the second power," or "six **squared**"; and $4 \cdot 4 \cdot 4 = 4^3$, read "four to the third power," or "four **cubed**." In general,

$$a^n = \underbrace{a \cdot a \cdot a \cdot \ldots \cdot a}_{n \text{ factors}} \qquad \text{where } a \text{ is any number and } n \text{ is a natural number}$$

When we multiply negative numbers, we usually write them in parentheses, such as $(-2)(-2)(-2)(-2) = 16$. In the same way, when an exponential expression has a negative base, we enclose the base in parentheses. For example, $(-2)^4 = (-2)(-2)(-2)(-2) = 16$. Therefore, if an exponential expression is preceded by a negative sign that is not in parentheses, the base is positive:

$$-2^4 = -(2^4) = -(2 \cdot 2 \cdot 2 \cdot 2) = -16$$

EXAMPLE 1.7.1

Multiply the following.

1. 5^3 $5^3 = 5 \cdot 5 \cdot 5 = 125$

2. $(-8)^2$ $(-8)^2 = (-8)(-8) = 64$

3. $(-3)^3$ $(-3)^3 = (-3)(-3)(-3) = -27$

4. -5^2 $-5^2 = -(5 \cdot 5) = -25$

5. 1^6 $1^6 = 1 \cdot 1 \cdot 1 \cdot 1 \cdot 1 \cdot 1 = 1$ One to any power is one.

6. 0^4 $0^4 = 0 \cdot 0 \cdot 0 \cdot 0 = 0$ Zero to any power is zero.

We are often interested in a process that is the opposite of raising a number to a power. This process is called finding a **root** of a number. We will work with roots in more detail in Chapter 8. In this section we only briefly consider square roots. For example, since $3^2 = 9$, then the principal **square root** of 9 is 3, which is indicated by

$$\sqrt{9} = 3$$

EXAMPLE 1.7.2

Find the following square roots.

1. $\sqrt{169}$ $\sqrt{169} = 13$, since $13^2 = 169$.

2. $\sqrt{5}$

Unlike the first example, we cannot find a rational number that is the square root of 5. As we discussed in Sections 1.1 and 1.2, $\sqrt{5}$ is an irrational number that we can approximate using Table 2 in Appendix III or a hand-held calculator:

$$\sqrt{5} \doteq 2.236$$

When an expression contains powers, roots, and any of the four arithmetic operations, we must use the rules governing the order of operations. We list the rules below in the final form as far as this textbook is concerned.

Order of Operations	
Rule	**Examples**
A. Perform all operations within grouping symbols first.	$(7 - 2)^3$ $= 5^3$ $= 125$
B. If there are no grouping symbols, perform the operations in the following order: **1.** powers and roots **2.** multiplications and divisions, in order from left to right **3.** additions and subtractions, in order from left to right	$5 + 6 \cdot 12 \div 3^2 - 3$ $= 5 + 6 \cdot 12 \div 9 - 3$ $= 5 + 72 \div 9 - 3$ $= 5 + 8 - 3$ $= 13 - 3$ $= 10$

EXAMPLE 1.7.3

Perform the indicated operations following the rules governing the order of operations.

1. $3 \cdot 2^2 - 5 \cdot 2 - 9 = 3 \cdot 4 - 5 \cdot 2 - 9$
$$= 12 - 10 - 9$$
$$= 2 - 9 = -7$$

2. $\dfrac{1 - \sqrt{(-1)^2 - 4(-6)}}{2} = \dfrac{1 - \sqrt{1 - 4(-6)}}{2}$
$$= \dfrac{1 - \sqrt{1 - (-24)}}{2}$$
$$= \dfrac{1 - \sqrt{25}}{2}$$
$$= \dfrac{1 - 5}{2}$$
$$= \dfrac{-4}{2} = -2$$

EXERCISES 1.7

Evaluate the following powers and roots.

1. 6^3 2. 11^2 3. 2^5 4. 5^4 5. $(-7)^2$ 6. $(-4)^4$

7. -8^2 8. -3^4 9. $\left(\dfrac{2}{5}\right)^2$ 10. $\left(\dfrac{3}{4}\right)^3$ 11. $(\sqrt{9})^2$ 12. $(\sqrt{4})^4$

13. $\sqrt{25}$ 14. $\sqrt{16}$ 15. $\sqrt{1}$ 16. $\sqrt{0}$ 17. $\sqrt{100}$ 18. $\sqrt{144}$

19. $\sqrt{36}$ 20. $\sqrt{4}$ 21. $\sqrt{49}$ 22. $\sqrt{64}$ 23. $\sqrt{11^2}$ 24. $\sqrt{9^2}$

Perform the indicated operations following the rules governing the order of operations.

25. $7 + 2\sqrt{16}$ 26. $5 - 3\sqrt{4}$ 27. $-14 + 30 \div \sqrt{25}$

28. $-24 + 42 \div \sqrt{49}$ 29. $-2 \cdot 3^3 + 7 \cdot 3^2 - 5 \cdot 3$ 30. $5 \cdot 4^3 - 3 \cdot 4^2 - 8 \cdot 4$

31. $6(-2)^2 + 13(-2) + 2$ 32. $3(-5)^2 + 7(-5) - 38$ 33. $\dfrac{4}{3}\left(\dfrac{5}{6}\right)^2 - \dfrac{11}{12}$

34. $\dfrac{21}{10} \div \left(\dfrac{7}{6}\right)^2 + \dfrac{6}{7}$ 35. $35 \div (5 - 7)^2$ 36. $-75 \div (3 - 8)^2$

37. $[6 - 2(2^3 - 5)] \div 13$ 38. $[9 + 3(1 - 4^2)] \div 9$ 39. $5\sqrt{15 + 7^2}$

40. $12\sqrt{11 + 5^2}$ 41. $\dfrac{-4 - \sqrt{4^2 - 4(-12)}}{2}$ 42. $\dfrac{-2 - \sqrt{2^2 - 4(-15)}}{2}$

43. $\dfrac{5 + \sqrt{(-5)^2 - 4 \cdot 2(-3)}}{2 \cdot 2}$

44. $\dfrac{-7 - \sqrt{7^2 - 4 \cdot 3 \cdot 2}}{2 \cdot 3}$

45. $\dfrac{3.6 - 4(1.05)}{(0.5)^2 - 0.1}$

46. $\dfrac{-5.24 + 2.6(8.5)}{(1.6)^2 - 2.32}$

47. $\dfrac{2(-2)^2 + 3(-2)(3) - 20 \cdot 3^2}{8(-2) - 20 \cdot 3}$

48. $\dfrac{3(-4)^2 + 7(-4) + 2}{9(-4)^2 - 1}$

49. $\dfrac{2\left(\frac{5}{2}\right)^2 - 5\left(\frac{5}{2}\right) - 3}{2\left(\frac{5}{2}\right)^2 + 5\left(\frac{5}{2}\right) + 2}$

50. $\dfrac{3\left(\frac{4}{3}\right)^2 - 13\left(\frac{4}{3}\right) + 12}{2\left(\frac{4}{3}\right)^2 - 3\left(\frac{4}{3}\right) - 9}$

51. A right circular cylinder has a radius of $r = 1.5$ inches and a height of $h = 6.8$ inches. Find the surface area S of the cylinder using the formula $S = 2\pi r^2 + 2\pi rh$. Leave your answer in terms of π.

52. A pyramid has a square base that measures $b = 2\frac{1}{4}$ feet on a side and a slant height of $s = 3\frac{1}{3}$ feet. Find the surface area S of the pyramid using the formula $S = b^2 + 4\left(\frac{1}{2}bs\right)$.

Chapter 1 Review

Terms to Remember

Notation

Set notation	$\{\ .\ .\ .\}$
Element of	$\in$
Subset of	$\subseteq$
Empty set	$\emptyset$ or $\{\ \ \}$
Intersection	$\cap$
Union	$\cup$
Natural numbers	N
Whole numbers	W
Integers	I
Rational numbers	Q
Irrational numbers	H
Real numbers	R
Less than	$<$
Less than or equal to	$\leq$
Greater than	$>$
Greater than or equal to	$\geq$
Absolute value of a number x	$|x|$
Least common denominator	LCD
Exponential, or power, notation	a^n
Square root	$\sqrt{\ \ }$
Approximately equal to	$\doteq$

Order of Operations

A. Perform all operations within grouping symbols first.

B. If there are no grouping symbols, perform the operations in the following order:

1. powers and roots
2. multiplications and divisions, in order from left to right
3. additions and subtractions, in order from left to right

Sets of Numbers

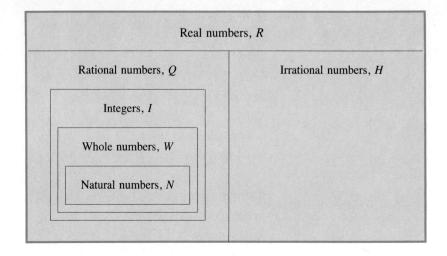

Properties of Real Numbers

Let a, b, and c represent any real numbers.

Property	Addition	Multiplication
Commutative	$a + b = b + a$	$a \cdot b = b \cdot a$
Associative	$(a + b) + c = a + (b + c)$	$(a \cdot b) \cdot c = a \cdot (b \cdot c)$
Identity	$a + 0 = a$ and $0 + a = a$	$a \cdot 1 = a$ and $1 \cdot a = a$
Inverse	$a + (-a) = 0$ and $-a + a = 0$	$a \cdot \dfrac{1}{a} = 1$ and $\dfrac{1}{a} \cdot a = 1, a \neq 0$
Distributive	$a \cdot (b + c) = a \cdot b + a \cdot c$	

Review Exercises

1.1 Consider the sets $A = \{3, 6, 9, 12, 15\}$, $B = \{3, 4, 5, 6, 7, 8, 9\}$, $C = \{4, 7, 10, 13, 16\}$, $D = \{\ldots, -6, -4, -2, 0, 2, 4, 6, \ldots\}$, and N. Find the following.

1. $A \cup B$ **2.** $A \cap C$ **3.** $A \cap B$ **4.** $B \cap D$ **5.** $A \cup C$ **6.** $D \cap N$

Change the following fractions to decimals.

7. $\frac{4}{3}$ **8.** $\frac{5}{9}$ **9.** $\frac{13}{8}$ **10.** $\frac{4}{7}$

Identify the sets to which the following numbers belong.

11. $-\frac{3}{4}$ **12.** $\sqrt{48}$ **13.** $7.070070007 \ldots$ **14.** -4

Determine whether the following statements are true or false.

15. $\frac{5}{1} \in I$ **16.** $H \subseteq \emptyset$ **17.** $H \cup Q = R$ **18.** $1.3333 \ldots \in Q$

1.2   Graph the following sets of numbers on a real number line.

19. $\{3, 5, 0, -1\}$ **20.** $\{\frac{1}{4}, -\frac{7}{3}, \frac{9}{5}\}$ **21.** $\{\sqrt{3}, 3, \sqrt{6}, 6\}$

22. $\{4.2, 3.\overline{5}, -1.4\}$ **23.** $\{|-2|, |2|, |-\frac{5}{2}|\}$ **24.** $\{|-1.7|, \sqrt{8}, 3.14\}$

Evaluate the following.

25. $|-15|$ **26.** $5|-6|$ **27.** $4 + |-4|$ **28.** $|5| + |-14|$

Place the appropriate symbol, $<$, $>$, or $=$, between the following pairs of numbers.

29. -18 _____ -5 **30.** $\frac{7}{3}$ _____ 2.7 **31.** 5 _____ $\sqrt{5}$

32. 7 _____ $|7|$ **33.** 0 _____ $|-5|$ **34.** $|-4| + 5$ _____ $|3| + |-2|$

1.3

35. State the property or rule that is illustrated in each step, except those marked with an asterisk.

$$1 + (-4 + 8) + (-5) = 1 + (8 + (-4)) + (-5)$$
$$= (1 + 8) + (-4 + (-5))$$
$$\overset{*}{=} 9 + (-9)$$
$$= 0$$

Reduce the following fractions to lowest terms.

36. $\dfrac{48}{36}$ **37.** $\dfrac{30}{75}$ **38.** $-\dfrac{24}{56}$ **39.** $\dfrac{14}{84}$

Perform the indicated additions.

40. $-48 + 27$ **41.** $-21 + (-14) + (-46)$ **42.** $7 + (-31) + 12$

43. $\frac{4}{13} + \frac{6}{13}$ **44.** $\frac{4}{15} + \left(-\frac{13}{15}\right)$ **45.** $\frac{3}{5} + \frac{2}{3}$

46. $\frac{5}{16} + \frac{7}{20}$ **47.** $-\frac{3}{8} + \frac{5}{9} + \frac{11}{36}$ **48.** $-\frac{1}{2} + \left(-\frac{5}{18}\right) + \frac{7}{6}$

49. $3\frac{4}{5} + 2\frac{3}{10}$ **50.** $-4.91 + 11.4 + (-8.05)$

Find the perimeters of the following figures.

51. Triangle with sides $5\frac{1}{3}$ in., $4\frac{5}{6}$ in., and $2\frac{7}{9}$ in.

52. Rectangle with length 21.9 m and width 16.2 m

1.4 Perform the indicated operations.

53. $-22 - 14$

54. $-43 - (-29) + (-12)$

55. $14.79 - (-75.6)$

56. $\frac{7}{18} - \frac{7}{6}$

57. $-\frac{4}{5} + \frac{2}{15} - \frac{9}{10}$

58. $4\frac{1}{2} - 3\frac{2}{3} + \left(-5\frac{1}{6}\right)$

59. $-27 - (6 - 52)$

60. $58 - (-15 + 13)$

61. $-\frac{8}{7} - \left(\frac{3}{14} - \frac{1}{2}\right)$

62. $\frac{2}{3} - \left(\frac{7}{13} - \frac{14}{39}\right)$

63. $-\frac{3}{8} - \left|\frac{1}{4} - 7\right|$

64. $\frac{5}{16} - \left|2 - \frac{21}{8}\right| - \frac{3}{4}$

65. $-7 - [63 - (5 - 21)]$

66. $\frac{5}{3} - \left[-\frac{11}{12} - \left(\frac{3}{4} + 3\right)\right]$

1.5 Name the property illustrated in each step.

67.
$$5\left[-2 + 2\left(1 + \tfrac{1}{2}\right)\right] = 5\left[-2 + \left(2 \cdot 1 + 2 \cdot \tfrac{1}{2}\right)\right]$$
$$= 5\left[-2 + \left(2 + 2 \cdot \tfrac{1}{2}\right)\right]$$
$$= 5[-2 + (2 + 1)]$$
$$= 5[(-2 + 2) + 1]$$
$$= 5[0 + 1]$$
$$= 5 \cdot 1$$
$$= 5$$

Use the distributive property to rewrite each expression.

68. $-3(4 + 9)$

69. $5(6 - 8 + 3)$

70. $5 \cdot 9 + 5 \cdot 12$

71. $7 \cdot 2 + 7 \cdot 4 - 7 \cdot 11$

Perform the indicated operations. When necessary, use the rules governing the order of operations.

72. $(-14)(2)(-5)$

73. $(-5)(-5)(-5)$

74. $\left(-\frac{15}{8}\right)\left(\frac{6}{35}\right)$

75. $(-12)\left(-\frac{15}{16}\right)$

76. $\left(5\frac{2}{5}\right)\left(-\frac{10}{9}\right)(-6)$

77. $\left(\frac{1}{9}\right)\left(-\frac{6}{7}\right)\left(-\frac{21}{4}\right)$

78. $(-3.47)(5.09)$

79. $16 - 6 \cdot 5 - 5$

80. $\frac{5}{3} \cdot \frac{1}{2} - \frac{1}{2} \cdot \frac{7}{6}$

81. $2\left(1\frac{3}{4}\right) + \frac{11}{8}\left(-\frac{2}{5}\right)$

82. $7 - 2[-3 + 5 \cdot 7]$

83. $8[3(7 - 4) + 2]$

84. $9[6(5 - 3) + 4]$

85. $-\frac{3}{5} + \frac{1}{2}\left(\frac{2}{3} - 4\right)$

86. $2.4[3.15 - (14.2 - 2.05)]$

87. $\frac{7}{20} - \left[\left(\frac{11}{24} - \frac{7}{15}\right) \cdot 12 + 2\right]$

88. $-4|-11 + 5| - 16$

89. $2\frac{1}{4}\left|\frac{11}{3} - \frac{29}{5}\right| - 3$

Use the formulas given earlier to find the following.

90. The perimeter of a rectangle with length $5\frac{3}{4}$ ft and width $2\frac{1}{2}$ ft

91. The area of a trapezoid with bases 6.032 m and 12.768 m and height 4.25 m

1.6 Perform the indicated operations. When necessary, use the rules governing the order of operations.

92. $-48 \div 3$

93. $-70 \div (-5)$

94. $\frac{6}{7} \div \left(-\frac{15}{49}\right)$

95. $-\frac{16}{15} \div 10$

96. $\left(-2\frac{4}{7}\right) \div \left(-\frac{4}{21}\right)$

97. $\dfrac{-40}{\frac{5}{8}}$

98. $\dfrac{\frac{26}{11}}{-\frac{39}{44}}$

99. $\dfrac{\frac{56}{45}}{20}$

100. $16 - 48 \div 6 + 10$

101. $6 \cdot 40 \div 5 + 3$

102. $8 + 12 \div (7 - 3)$

103. $-32 - 6 + 99 \div (-11)$

104. $\frac{4}{9} \div \frac{10}{21} \cdot \frac{25}{7} - \frac{1}{3}$

105. $4\frac{1}{2} \div \left(-3\frac{3}{4}\right) \div 2\frac{1}{10}$

106. $[5 - 11(28 \div 7)] \div (-13)$

107. $-27 \div |6 - 5 \cdot 3|$

108. $\left(-4\frac{6}{7}\right) \div \left(\frac{2}{7} - \frac{3}{2}\right)$

109. $\dfrac{\frac{5}{12} - \frac{3}{4}}{2 - \frac{1}{6}}$

110. $\dfrac{2\left(\frac{4}{5}\right) + \frac{1}{2}}{\frac{5}{4} - 2}$

111. Clara drove 280 miles and used $7\frac{7}{9}$ gallons of gasoline. What was her gasoline mileage for the trip, that is, how many miles per gallon did she get?

112. Rosie drove 84 miles in $1\frac{2}{5}$ hours. What was her average speed for the trip, that is, how many miles per hour did she average?

113. Miguel ran for 36 minutes, covering $4\frac{1}{2}$ miles. What was his average pace, that is, how many minutes per mile did he average?

1.7 Evaluate the following powers and roots.

114. 15^2

115. 10^3

116. $(-12)^2$

117. -1^6

118. $\left(\frac{5}{7}\right)^2$

119. $\sqrt{169}$

120. $\sqrt{121}$

121. $\sqrt{225}$

Perform the indicated operations following the rules governing the order of operations.

122. $6 - 11\sqrt{49}$

123. $-20 - 72 \div \sqrt{36}$

124. $3 \cdot 5^3 - 8 \cdot 5^2 + 4 \cdot 5$

125. $\frac{15}{14} \div \left(\frac{3}{4}\right)^2 - \frac{5}{6}$

126. $56 \div (3 - 5)^2$

127. $[4 - 2(2 - 5^2)] \div (-10)$

128. $6\sqrt{7^2 - 24}$

129. $\dfrac{14 + \sqrt{(-14)^2 - 4 \cdot 3 \cdot 8}}{2 \cdot 3}$

130. $\dfrac{6(-2)^2 - (-2)}{3(-2)^2 - (-2)}$

131. $\dfrac{2(-3)^2 - (-3) - 15}{3(-3)^2 - 11(-3) + 6}$

Chapter 1 Test

(You should be able to complete this test in 60 minutes.)

Consider the sets $A = \{-4, -2, 0, 2, 4\}$, $B = \{2, 4, 6, 8, \ldots\}$, and N.

1. Find $A \cap B$.

2. Find $A \cup B$.

3. True or false? $B \subseteq N$

4. Change $\frac{15}{4}$ to a decimal.

5. Identify the sets to which -8 belongs.

6. Graph the set $\{-\frac{5}{3}, 2.3, |-3|\}$ on a real number line.

7. Evaluate $3 + |-8|$.

Place the appropriate symbol, $<$, $>$, or $=$, between each pair of numbers.

8. -13 _____ -6

9. $\frac{9}{4}$ _____ $|-2|$

10. State the property or rule that is illustrated in each step, except those marked with an asterisk.

$$3[-8 + 2(1 + 3)] = 3[-8 + 2 \cdot 1 + 2 \cdot 3]$$
$$= 3[-8 + 2 + 6]$$
$$\overset{*}{=} 3[-8 + 8]$$
$$= 3[0]$$
$$= 0$$

11. Use the distributive property to rewrite $4 \cdot 9 + 4 \cdot 5$.

12. Reduce $\frac{45}{72}$ to lowest terms.

13. Find the perimeter of a rectangle with length $6\frac{1}{4}$ ft and width $4\frac{1}{2}$ ft.

14. Find the area of a trapezoid with bases 13 ft and 5 ft and height 7 ft.

15. Steve ran for $40\frac{3}{10}$ minutes, covering $6\frac{1}{5}$ miles. What was his average pace, that is, how many minutes per mile did he average?

Evaluate the following powers and roots.

16. -10^4

17. $\sqrt{64}$

18. $-\sqrt{144}$

Perform the indicated operations. When necessary, use the rules governing the order of operations.

19. $-63 + 17$

20. $-\frac{2}{9} + \left(-\frac{5}{6}\right)$

21. $4.87 - 12.09$

22. $-3\frac{2}{3} - 2\frac{4}{15}$

23. $\frac{7}{10} - \frac{3}{4} - \frac{6}{5}$

24. $-14 - 28 + 54$

25. $\left(-\frac{8}{9}\right)\left(-\frac{15}{22}\right)$

26. $(-5)(-6)(2)(-2)$

27. $52 \div (-13)$

28. $-72 \div (-12)$

29. $\frac{5}{12} \div \left(-\frac{10}{27}\right) \cdot \frac{7}{6}$

30. $\frac{4}{11} - \frac{13}{24} \cdot 6$

31. $5 \cdot 2^3 - 6 \cdot 7$

32. $-16 - 96 \div (-8)$

33. $5.4(-0.62) - 3.208$

34. $-14 - 5\sqrt{49}$

35. $-4 - \left(\frac{11}{4} - \frac{2}{3}\right)$

36. $-60 - (15 + 29)$

37. $7 - 3[-12 - (-20)(4)]$

38. $\frac{7}{3}\left(\frac{21}{4} - \frac{5}{6} \cdot \frac{9}{35}\right)$

39. $3 + (5^2 + 3 \cdot 8) \div (-7)$

40. $6\sqrt{(5 - 1)^2 + 9}$

41. $\dfrac{9 - \sqrt{(-9)^2 - 4 \cdot 20}}{2}$

42. $\dfrac{\frac{5}{7} + \frac{5}{2}}{\frac{9}{4} - \frac{11}{6}}$

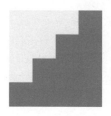

CHAPTER 2

Linear Equations and Inequalities in One Variable

2.1 Algebraic Expressions and Combining Like Terms

The mathematical expression used to calculate the circumference of a circle is $2\pi r$. In this expression, r represents the length of the radius of a circle. The letter r, which represents a number, is called a variable. A **variable** is a character used to represent any number from a specified set of numbers. A character that represents a particular number is called a **constant**. In the expression $2\pi r$, both 2 and π are constants. In the expression $7x + \frac{1}{3}y + 10$, the variables are x and y, whereas the constants are 7, $\frac{1}{3}$, and 10. The expressions $2\pi r$ and $7x + \frac{1}{3}y + 10$ are examples of algebraic expressions.

Algebraic Expression

An **algebraic expression** is any meaningful combination of constants and variables that involves powers, roots, grouping symbols, or the four operations of addition, subtraction, multiplication, and division.

REMARK   The following are algebraic expressions.

$$8x - \frac{4y}{z} + 1.2, \qquad \pi r^2, \qquad \sqrt{100 - x^2}, \qquad 5(2x + 4y)$$

Frequently we want to substitute numbers for the variables in an algebraic expression and then determine the value of the resulting expression. This process is called **evaluating an algebraic expression** for the given value(s) of the variable(s).

EXAMPLE 2.1.1

Evaluate the following algebraic expressions for the given values of the variables.

1. $3y^2 + 5y - 11$ for $y = -1$

Replacing each y by -1 we obtain

$$3(-1)^2 + 5(-1) - 11 = 3 \cdot 1 - 5 - 11$$
$$= 3 - 5 - 11$$
$$= -2 - 11$$
$$= -13$$

2. $9xy - 2y^2$ for $x = 2$ and $y = 5$

$$9 \cdot 2 \cdot 5 - 2 \cdot 5^2 = 90 - 2 \cdot 25$$
$$= 90 - 50$$
$$= 40$$

3. $\frac{1}{3}x^2 - \frac{1}{4}y^2 + 1$ for $x = -2$ and $y = 3$

$\frac{1}{3}(-2)^2 - \frac{1}{4}(3)^2 + 1$

$= \frac{1}{3}(4) - \frac{1}{4}(9) + 1$

$= \frac{4}{3} - \frac{9}{4} + 1$ Obtain the LCD 12.

$= \frac{16}{12} - \frac{27}{12} + \frac{12}{12}$

$= -\frac{11}{12} + \frac{12}{12}$

$= \frac{1}{12}$

4. $\sqrt{38 - x^2 + y^2}$ for $x = 5$ and $y = 6$

$\sqrt{38 - 5^2 + 6^2} = \sqrt{38 - 25 + 36}$
$= \sqrt{13 + 36}$
$= \sqrt{49}$
$= 7$

5. $5(2x + 4y)$ for $x = -2$ and $y = -4$

$5[2(-2) + 4(-4)] = 5[-4 + (-16)]$
$= 5(-20)$
$= -100$

6. $\dfrac{x - 5y}{\sqrt{3z}}$ for $x = 1$, $y = -3$, and $z = 12$

$\dfrac{1 - 5(-3)}{\sqrt{3 \cdot 12}} = \dfrac{1 - (-15)}{\sqrt{36}}$

$= \dfrac{16}{6}$ Reduce your answer.

$$= \frac{8}{3}$$

The previous example illustrates that some algebraic expressions might be very complicated. Mathematicians have defined some key words to help describe these expressions.

Algebraic expressions that are being added are called **terms**. The algebraic expression $7x + \frac{1}{3}y + 10$ has three terms: $7x$, $\frac{1}{3}y$, and 10. The terms of the algebraic expression $9xy - 2y^2$ are $9xy$ and $-2y^2$, since $9xy - 2y^2 = 9xy + (-2y^2)$.

Algebraic expressions that are being multiplied are called **factors**. The algebraic expression $2\pi r$ has three factors: 2, π, and r, since $2\pi r = 2 \cdot \pi \cdot r$.

Two additional concepts need to be mentioned before we can simplify algebraic expressions.

Coefficient

The numerical factor of a term is called the **numerical coefficient**, or just **coefficient**, of that term.

In the algebraic expression $5x - 2y + z$, the coefficient of the first term is 5, the coefficient of the second term is -2, and the coefficient of the last term is 1 since $z = 1 \cdot z$.

Like Terms

Like terms (or similar terms) are terms whose variable factors and exponents are *exactly* the same.

In the algebraic expression $2x + 5x + 3y$, the terms $2x$ and $5x$ are like terms. The algebraic expression $6x^2 - 7x + 4y + 1$ has no like terms since the variable factors and exponents on any of the terms are not exactly the same.

We can add or subtract like terms by applying the distributive property. Consider the following example.

EXAMPLE 2.1.2

Simplify the following algebraic expressions by combining like terms.

1. $3x + 5x = (3 + 5)x$ Distributive property

$\qquad\quad = 8x$

NOTE ▶ When combining like terms we simply add or subtract the coefficients; the variable factors remain exactly the same.

2. $2x + 5x + 3y = (2x + 5x) + 3y$ Associative property

$\qquad\qquad\quad = (2 + 5)x + 3y$ Distributive property

$\qquad\qquad\quad = 7x + 3y$

3. $5x + 4y - x + 10y = 5x - x + 4y + 10y$ Commutative property

$\qquad\qquad\qquad\quad = (5x - x) + (4y + 10y)$ Associative property

$\qquad\qquad\qquad\quad = (5 - 1)x + (4 + 10)y$ Distributive property

$\qquad\qquad\qquad\quad = 4x + 14y$

4. $6x^2 - 7x + 4y + 1$ This expression cannot be simplified since it has no like terms.

5. $x + \frac{1}{2} - \frac{5}{3} + \frac{7}{2}x = x + \frac{7}{2}x + \frac{1}{2} - \frac{5}{3}$ Commutative property

$\qquad\qquad\qquad = \frac{9}{2}x - \frac{7}{6}$ Combine like terms.

In the next example we will see how some algebraic expressions with grouping symbols can be simplified.

EXAMPLE 2.1.3

Simplify the following algebraic expressions by combining like terms.

1. $5x + 3(4x - 7) = 5x + 3(4x) - 3 \cdot 7$ Distributive property

$\qquad\qquad\quad = 5x + 12x - 21$

$\qquad\qquad\quad = 17x - 21$ Combine like terms.

2. $2(x - 4) - 5(x + 8) = 2x - 2 \cdot 4 - 5x - 5 \cdot 8$ Distributive property

$\qquad\qquad\qquad = 2x - 8 - 5x - 40$

$\qquad\qquad\qquad = 2x - 5x - 8 - 40$ Commutative property

$\qquad\qquad\qquad = -3x - 48$ Combine like terms.

3. $8(2x + y) - (7x + 3) = 8(2x + y) - 1(7x + 3)$

$\qquad\qquad\qquad\quad = 8 \cdot 2x + 8y - 1(7x) - 1 \cdot (3)$ Distributive property

$\qquad\qquad\qquad\quad = 16x + 8y - 7x - 3$

$\qquad\qquad\qquad\quad = 16x - 7x + 8y - 3$ Commutative property

$\qquad\qquad\qquad\quad = 9x + 8y - 3$ Combine like terms.

NOTE ▶ $-(7x + 3) = -1(7x + 3)$

Try to avoid this mistake:

Incorrect	Correct
$8(2x + y) - (7x + 3)$ $= 16x + 8y - 7x + 3$	$8(2x + y) - (7x + 3)$ $= 8(2x + y) - 1 \cdot (7x + 3)$ $= 16x + 8y - 7x - 3$
Here the sign of 3 was not changed.	When subtracting an expression within parentheses, we must change the signs of *all* of the terms inside the parentheses.

In Sections 1.3 and 1.5 we found the perimeters of triangles and rectangles by adding the lengths of all of their sides. If the lengths of the sides are unknown, we often use algebraic expressions to represent the sides and the perimeter.

EXAMPLE 2.1.4

The length of a rectangle is 3 feet more than the width. Find an expression for the perimeter.

SOLUTION

If we let x represent the width, then the length is $x + 3$ since it is 3 feet more than the width (see Figure 2.1.1).

Since there are two widths and two lengths, an expression for the perimeter is

$P = 2W + 2L$

$\quad = 2x + 2(x + 3)$

$\quad = 2x + 2x + 6$

$\quad = 4x + 6$

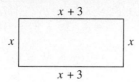

Figure 2.1.1

EXERCISES 2.1

Evaluate the following algebraic expressions for the given values of the variables.

1. $5x + 8$; $x = 3$

2. $2x - 1$; $x = -1$

3. $4 - 6x$; $x = 4$

4. $-8 - 8x$; $x = 5$

5. $x^2 + 7x - 4$; $x = 2$

6. $2x^2 - 3x - 11$; $x = 6$

7. $8 - x - x^2$; $x = -7$

8. $12 + x - 4x^2$; $x = -3$

9. $4x^2 - 9$; $x = 3$

10. $25x^2 - 16$; $x = 1$

11. $(x + 4)(x - 5)$; $x = -4$

12. $(2x + 3)(x - 1)$; $x = 2$

13. $(x^2 - 4)(5x + 7)$; $x = -2$

14. $(x^2 - 9)(6x - 2)$; $x = -3$

15. $2x - 5(x + 3)$; $x = \frac{1}{2}$

16. $3x - 7(2x + 1)$; $x = \frac{1}{3}$

17. $6(x + 1) - 2(x + 3)$; $x = -4$

18. $5(2x - 3) - 3(x - 4)$; $x = \frac{1}{4}$

19. $2(x - 4) - (2x - 1)$; $x = \frac{1}{3}$

20. $3(2x + 1) - (3x + 5)$; $x = -3$

21. $x^2 + 2xy + y^2$; $x = 2, y = 5$

22. $2x^2 + 5xy + 3y^2$; $x = 3, y = 4$

23. $5(x + 2y) - 4(x - 3y)$; $x = -1, y = 3$

24. $3(2x - y) - 2(4x - y)$; $x = 3, y = -2$

25. $5 + 3(5x + y)$; $x = -2, y = -1$

26. $6 + 2(4x + 7y)$; $x = -5, y = -2$

27. $\dfrac{2x}{x + 1} + \dfrac{x - 2}{x + 4}$; $x = 5$

28. $\dfrac{4x}{x + 3} + \dfrac{3x + 3}{2x - 3}$; $x = 6$

29. $\frac{1}{2}x + \frac{2}{3}y - \frac{1}{4}z$; $x = 3, y = -2, z = 1$

30. $\frac{3}{4}x + \frac{1}{2}y - \frac{2}{3}z$; $x = -3, y = 1, z = 2$

31. $\dfrac{2x}{y} + \dfrac{3y}{z} - 2$; $x = 1, y = 3, z = 4$

32. $\dfrac{5y}{x^2} + \dfrac{2z}{y} - 3$; $x = 2, y = 3, z = -1$

33. $\frac{1}{2}(x + 3) + \frac{1}{3}(y - 4)$; $x = -2, y = -3$

34. $\frac{1}{4}(x - 5) + \frac{1}{2}(y + 3)$; $x = 2, y = 2$

35. $1.235x + 2.6(3.14x - 7.26)$; $x = 5$

36. $4.017x + 1.8(2.17x - 5.87)$; $x = 3$

37. $\dfrac{x + 2y}{\sqrt{4z}}$; $x = -7, y = 3, z = 4$

38. $\dfrac{2x - 3y}{\sqrt{5z}}$; $x = 1, y = 4, z = 20$

39. $\sqrt{43 - x^2 + 2y^2}$; $x = 5, y = -3$

40. $\sqrt{24 - 2x^2 + 3y^2}$; $x = -2, y = 4$

Simplify the following algebraic expressions by combining like terms, if possible.

41. $2x + 9x$

42. $3y + 12y$

43. $5x - 8x + 7x$

44. $2x - 4x - 6x$

45. $-8x + 4 - 2x + 1$

46. $x - 9 - 4x + 6$

47. $6x + y - 8x - 2y$

48. $-3x - 5y + 2y + 7x$

49. $x^2 - 2x + 3z + 5x^2$

50. $2a^2 + 4a - b + 7a^2$

51. $5x + 2y - 7$

52. $25x^2 + 20x - 20x - 16$

53. $9x^2 - 3x + 3x - 1$

54. $\frac{7}{8}x - 2 - \frac{1}{4}x$

55. $\frac{5}{8}x - 1 - \frac{1}{4}x$

56. $1.017x - 4.56x - 0.007x$

57. $-2.839y + 1.001y - 3.86y$

58. $2a + 3b - 12$

59. $\frac{1}{3}x^2 - \frac{1}{2}y + \frac{5}{6}x^2 + \frac{1}{4}y$

60. $\frac{2}{3}x^2 - \frac{3}{8}y - \frac{1}{6}x^2 - \frac{3}{4}y$

61. $5x + 3(2x - 6)$

62. $7x + 2(4x - 1)$

63. $8 - 3(x + 4)$

64. $9 - 2(2x + 7)$

65. $3(x + 1) + 6(x - 2)$

66. $4(2x - 1) + 2(x + 5)$

67. $5(2x - 1) - 3(3x + 4)$

68. $6(5x - 4) - 3(2x + 1)$

69. $3(2x - 6) - (x - 2)$

70. $5(3x - 2) - (x - 4)$

71. $5(x + y) - 3(x + y)$

72. $8(2x + 3y) - 4(x + y)$

73. $2(x + y) + 3(a + b)$

74. $4(x + y) + 2(a + b)$

75. $\frac{1}{3}(x + 2) - \frac{1}{2}(x - 3)$

76. $\frac{3}{4}(x - 1) - \frac{1}{3}(x + 2)$

77. $0.02(2x + 3) + 0.05(3x - 1)$

78. $0.03(4x - 1) + 0.02(2x + 3)$

79. $5x + 3[2(x + 4) - (x + 7)]$

80. $8x + 5[3(2x + 1) - (4x + 1)]$

81. The average monthly cost in dollars of keeping Y. A. the cat in good health can be represented by the algebraic expression $25x + 2y$. In this expression, x represents the number of trips to the vet and y represents the number of pounds of cat food Y. A. eats each month. Determine the monthly cost if Y. A. has 1 trip to the vet and eats 29 pounds of cat food.

82. W. D. has determined that his average monthly cost of playing tennis can be represented by the algebraic expression $3x + 42y$. In this expression, x represents the number of cans of balls and y represents the number of pairs of shoes he uses each month. Determine the monthly cost if he uses 5 cans of balls and 1 pair of shoes.

Find an algebraic expression for the perimeter of each of the following figures and simplify.

83. The medium side of a triangle is 2 in. longer than the short side, and the long side is 5 in. longer than the short side.

84. The short side of a triangle is 8 ft shorter than the long side, and the medium side is 6 ft shorter than the long side.

85. The medium side of a triangle is 5 m longer than the short side, and the long side is three times as long as the short side.

86. The medium side of a triangle is 4 yd longer than the short side, and the long side is twice as long as the medium side.

87. The length of a rectangle is 9 cm longer than its width.

88. The length of a rectangle is five times as long as its width.

89. The length of a rectangle is 3 in. longer than twice its width.

90. The length of a rectangle is 2 ft less than three times its width.

2.2 The Additive Property of Equality

An **equation** is an algebraic way of stating that two quantities are equal. Some examples of equations are

$$C = 2\pi r, \qquad x^2 + y^2 = 4, \qquad x^2 - x - 12 = 0$$

$$6x - 4 = 2, \qquad x = -8$$

In the remainder of this chapter we will examine a special group of equations called **linear equations** in one variable.

Linear Equation

Any equation that can be written in the form $ax + b = c$, where a, b, and c are real numbers with $a \neq 0$, is called a *linear equation in the variable x.*

REMARK The following are linear equations in the variable x.

1. $2x + 1 = 11$

2. $x - 2 = 3$

3. $6x - 4 = 2$

The **solution set of an equation** is the set of all numbers that yield a true statement when substituted for the variable in the equation.

EXAMPLE 2.2.1

Find the solution set of the following linear equations.

1. $2x + 1 = 11$

   By observation, $\{5\}$ is the solution set. Replacing x by 5 we obtain

   $$2(5) + 1 \overset{?}{=} 11$$
   $$10 + 1 \overset{?}{=} 11$$
   $$11 = 11 \qquad \text{A true statement}$$

NOTE ▶ 5 is often called the **solution** of $2x + 1 = 11$, because when we replace x with 5 we obtain a true statement.

2. $x - 2 = 3$

By observation, $\{5\}$ is the solution set. Replacing x by 5 we obtain

$5 - 2 \stackrel{?}{=} 3$

$\qquad 3 = 3 \qquad$ A true statement

3. $6x - 4 = 2$

By observation, $\{1\}$ is the solution set. Replacing x by 1 we obtain

$6(1) - 4 \stackrel{?}{=} 2$

$\qquad 6 - 4 \stackrel{?}{=} 2$

$\qquad\quad 2 = 2 \qquad$ A true statement

REMARK

When we want to find the solution(s) of an equation, we say we want to **solve the equation** or, if the variable is x, we want to **solve for x**.

Two or more equations are called **equivalent** if they have the same solution set. The linear equations $2x + 1 = 11$ and $x - 2 = 3$ from parts 1 and 2 of Example 2.2.1 and $x = 5$ are all equivalent, since $\{5\}$ is the solution set for all of these equations.

We can show that the equations $x - 2 = 3$ and $x = 5$ are equivalent by adding 2 to both sides of $x - 2 = 3$:

$x - 2 = 3$

$x - 2 + 2 = 3 + 2$

$x + 0 = 5$

$x = 5$

This line of reasoning illustrates an important property that can be used to find the solutions of equations.

Additive Property of Equality

Let a, b, and c denote algebraic expressions.

If $a = b$, then $a + c = b + c$.

REMARK

This property states that we can add any algebraic expression to both sides of an equation and obtain a new equation that is equivalent to the original equation. We can use this property to isolate the variable term.

EXAMPLE 2.2.2　　　　Solve the following linear equations.

1.　　$x - 4 = -16$

$x - 4 + 4 = -16 + 4$　　　Using the additive property

$x + 0 = -12$　　　　　of equality, we add 4 to both sides and thereby isolate x.

$x = -12$

Checking the solution:

$-12 - 4 \overset{?}{=} -16$

$-16 = -16$

So $\{-12\}$ is the solution set.

2.　　　$x + 3 = 12$

$x + 3 + (-3) = 12 + (-3)$　　　Since we want to isolate x,

$x + 0 = 9$　　　　we need to add -3 to both sides.

$x = 9$

NOTE ▶　Subtraction is defined as addition of the additive inverse. Because of this fact, we can also *subtract* any algebraic expression from both sides of an equation and obtain an equation that is equivalent to the original equation.

In this example,

$x + 3 = 12$　　　Now we can subtract 3 from

$x + 3 - 3 = 12 - 3$　　both sides and thereby iso-late x.

$x + 0 = 9$

$x = 9$

Using either method, we arrived at the same solution set, $\{9\}$.

NOTE ▶　For the remainder of this section, we leave it to the reader to check the solutions.

3. $5x - 7 - 4x - 2 = 11 + 3$

$5x - 4x - 7 - 2 = 11 + 3$

$x - 9 = 14$　　　Combine like terms.

$x - 9 + 9 = 14 + 9$

$x = 23$

So $\{23\}$ is the solution set.

4. $3.7x + 4x - 6.8 = 5 + 6.7x$

$\qquad 7.7x - 6.8 = 5 + 6.7x$

$\qquad 7.7x - 6.7x - 6.8 = 5 + 6.7x - 6.7x$

$\qquad x - 6.8 = 5$

$\qquad x - 6.8 + 6.8 = 5 + 6.8$

$\qquad x = 11.8$

Combine like terms. Since there are x terms on both sides of the equation, we subtract $6.7x$ from both sides to eliminate the x term from one side.

So $\{11.8\}$ is the solution set.

5. $\frac{2}{5}x - \frac{2}{3} + \frac{1}{10}x = \frac{2}{3}x + \frac{5}{6}x + \frac{1}{3} - \frac{1}{6}$

$\qquad \frac{2}{5}x + \frac{1}{10}x - \frac{2}{3} = \frac{2}{3}x + \frac{5}{6}x + \frac{1}{3} - \frac{1}{6}$

$\qquad \frac{4}{10}x + \frac{1}{10}x - \frac{2}{3} = \frac{4}{6}x + \frac{5}{6}x + \frac{2}{6} - \frac{1}{6}$

$\qquad \frac{5}{10}x - \frac{2}{3} = \frac{9}{6}x + \frac{1}{6}$

$\qquad \frac{1}{2}x - \frac{2}{3} = \frac{3}{2}x + \frac{1}{6}$

$\qquad \frac{1}{2}x - \frac{1}{2}x - \frac{2}{3} = \frac{3}{2}x - \frac{1}{2}x + \frac{1}{6}$

$\qquad -\frac{2}{3} = x + \frac{1}{6}$

$\qquad -\frac{2}{3} - \frac{1}{6} = x + \frac{1}{6} - \frac{1}{6}$

$\qquad -\frac{4}{6} - \frac{1}{6} = x$

$\qquad -\frac{5}{6} = x$

Obtain an LCD for all of the like terms on each side of the equation.

Subtract $\frac{1}{2}x$ from both sides to eliminate the x term from the left-hand side.

So $\{-\frac{5}{6}\}$ is the solution set.

6. $5x + 2(3x + 7) = 2[3(2x + 1) - (x - 7)]$

NOTE ▶ When solving equations, the order of operations still applies. So in this case we first simplify inside the brackets.

$\qquad 5x + 2(3x + 7) = 2[6x + 3 - x + 7]$ Distributive property

$\qquad 5x + 2(3x + 7) = 2[5x + 10]$

$\qquad 5x + 6x + 14 = 10x + 20$ Distributive property

$\qquad 11x + 14 = 10x + 20$

$\qquad 11x - 10x + 14 = 10x - 10x + 20$

$\qquad x + 14 = 20$

$\qquad x + 14 - 14 = 20 - 14$

$\qquad x = 6$

So $\{6\}$ is the solution set.

EXERCISES 2.2

Solve the following linear equations and check your solutions.

1. $x - 7 = 10$ **2.** $x + 3 = 11$ **3.** $4 = x - 9$ **4.** $-5 = x + 3$

5. $-8 = x + 2$ **6.** $0 = x + 7$ **7.** $x - 2 = -4$ **8.** $-6 = x - 1$

9. $x - \frac{3}{4} = \frac{1}{8}$ **10.** $x + 9 = 0$ **11.** $x + 3 = -1$ **12.** $x - 8 = -11$

13. $7.06 = x - 1.05$ **14.** $x + 5 = -2$ **15.** $3.27 = x + 7.89$ **16.** $x - 3 = -1$

17. $0 = x + 5$ **18.** $x + 2 = 2$ **19.** $x + 6 = -4$ **20.** $-7 = x - 3$

21. $x - 2 = 8$ **22.** $4 = x + 4$ **23.** $x + \frac{1}{3} = -\frac{5}{6}$ **24.** $1.42 = x - 3.65$

25. $x - 7 = -7$ **26.** $2.41 = x + 5.17$ **27.** $0 = x - 11$ **28.** $x - 6 = 12$

29. $x + 9 = 2$ **30.** $-8 = x - 1$ **31.** $-4 = x + 4$ **32.** $x - \frac{2}{3} = \frac{1}{6}$

33. $-12 = x - 15$ **34.** $x - 5 = 5$ **35.** $x + 8 = 1$ **36.** $x + \frac{1}{4} = -\frac{3}{8}$

37. $7x + 3 - 6x + 4 = 11 - 3$ **38.** $9x - 2 - 8x + 5 = 15 + 4$

39. $\frac{3}{2}x - \frac{1}{2}x - 4 - 7 = -13 - 1$ **40.** $\frac{3}{4}x + \frac{1}{4}x - 9 - 1 = -12 - 5$

41. $6x + 4 = 5x - 3$ **42.** $8x + 1 = 7x - 9$

43. $2.67x + 4.87 = 3.67x + 1.24$ **44.** $3.19x - 0.76 = 4.19x + 6.01$

45. $5x - 3x + 7 = x + 7$ **46.** $4x + 4x - 9 = 7x - 3$

47. $13x - 3x + 5 = 11x - 4$ **48.** $10x - 7 - 2x = 9x + 15$

49. $\frac{1}{8}x - \frac{2}{3} + \frac{1}{8}x = \frac{5}{4}x - \frac{5}{6}$ **50.** $\frac{5}{6}x + \frac{5}{6}x - \frac{1}{4} = \frac{2}{3}x - \frac{3}{2}$

51. $5.3x - 1.4 - 3.1x = 1.2x - 6.7$ **52.** $2.5x + 4.9 + 1.2x = 4.7x - 3.8$

53. $7x - 2 - 6 - 2x = 8x - 4x + 5 - 11$ **54.** $9x + 3x + 3 - 1 = 13x + 6 - 4 - 2x$

55. $7x + 8 - 9 - 7x = 2x + 5 - 3 - x$ **56.** $3x - 1 + 7 - 3x = 9x - 8x - 8 - 4$

57. $2x + 4x - \frac{1}{2} + \frac{3}{4} = 4x + x + \frac{5}{8} + \frac{1}{2}$ **58.** $9x + \frac{1}{3} + \frac{5}{6} - x = 2x - \frac{1}{6} + \frac{2}{3} + 5x$

59. $\frac{1}{8}x + 7 - 9 + \frac{1}{4}x = \frac{5}{8}x + \frac{3}{4}x - 1 + 8$ **60.** $\frac{1}{3}x - 4 - 5 - \frac{1}{6}x = \frac{1}{2}x - 7 - 11 + \frac{2}{3}x$

61. $\frac{1}{4}x + \frac{1}{2}x - \frac{1}{10} - \frac{2}{5} = \frac{5}{4}x + \frac{1}{2}x + \frac{9}{10} + \frac{4}{5}$ **62.** $\frac{2}{3}x + \frac{5}{6}x - \frac{1}{4} - \frac{1}{8} = \frac{1}{3}x + \frac{1}{6}x - \frac{1}{2} + \frac{3}{4}$

63. $5(2x + 3) = 11x + 17$ **64.** $4(3x + 1) = 13x + 15$

65. $2 + 3(x - 4) = 2x - 1$ **66.** $7 + 5(x + 3) = 4x + 20$

67. $6(2x + 1) + 3x = 8(2x - 3) - 1$ **68.** $5(3x - 4) + 4x = 4(5x - 2) - 12$

69. $\frac{1}{2}(8x + 3) - 2 = \frac{1}{3}(9x - 5) + 2$ **70.** $\frac{1}{4}(12x - 1) - 2 = \frac{1}{5}(10x + 3) - 1$

71. $5(2x + 7) - 3(3x + 4) = 7$ **72.** $3(5x + 1) - 2(7x - 1) = 0$

73. $4(3 - x) + 2(x + 3) = 25 - 3x$ **74.** $5(1 - x) + 3(x + 4) = 19 - 3x$

75. $3(2x + 1) + 4(x - 2) = 3(3x + 1)$

76. $5(x - 2) + 4(2x + 3) = 4(3x - 2)$

77. $8(2x - 1) - 2(x + 3) = 3(5x - 6)$

78. $4(2x + 3) - 3(x - 4) = 2(3x + 4)$

79. $8(x + 3) + 2(x - 4) = 6(x - 3) + 3(x + 2)$

80. $6(x - 4) + 5(x - 1) = 7(x - 2) + 3(x + 4)$

81. $3(2x - 1) - 4(x + 2) = 5(2x + 3) - 7(x - 2)$

82. $5(2x + 1) - 6(x - 1) = 3(3x - 2) - 4(x + 7)$

83. $2[4(2x - 1) - (5x - 3)] = 7x + 2$

84. $3[4(3x + 1) - (10x - 9)] = 7x + 9$

85. $5[3(2x + 1) - 4(x + 1)] = 3(3x + 1) - 5$

86. $4[2(5x - 3) - 9(x + 2)] = 3(x - 7) + 33$

87. $3x + 2(2x - 9) = 3[4(x - 1) - 2(x + 3)]$

88. $x + 2(x + 4) = 2[3(2x + 1) - 2(2x - 3)]$

89. $3[5(x - 3) + 2(x + 9)] = 5[2(x + 3) + 2(x - 7)]$

90. $9[3(2x - 2) - 5(x - 1)] = 2[11(x - 1) - 7(x - 1)]$

2.3 The Multiplicative Property of Equality

In the previous section we found solutions of linear equations by adding an appropriate algebraic expression to both sides of the equation. However, not all linear equations can be solved using only the additive property of equality. Consider the linear equation $2x = 14$. By observation, $x = 7$ is the solution of $2x = 14$, since $2 \cdot 7 = 14$.

Notice that if we multiply by $\frac{1}{2}$ on both sides of the equation,

$$2x = 14$$
$$\tfrac{1}{2} \cdot 2x = \tfrac{1}{2} \cdot 14$$
$$\left(\tfrac{1}{2} \cdot 2\right)x = \tfrac{1}{2} \cdot 14$$
$$1 \cdot x = 7$$
$$x = 7$$

we obtain the solution set, $\{7\}$.

This line of reasoning illustrates another important property that can be used to find the solutions of equations.

Multiplicative Property of Equality

Let a, b, and c denote algebraic expressions with $c \neq 0$.

If $a = b$, then $ac = bc$.

REMARK $\quad$ This property states that we can multiply both sides of an equation by any *nonzero* algebraic expression and obtain a new equation that is equivalent to the original equation. We can use this property to make the coefficient of the variable equal to one.

EXAMPLE 2.3.1

Solve the following linear equations.

1. $3x = 24$ The coefficient is 3, and we
         $\frac{1}{3} \cdot 3x = \frac{1}{3} \cdot 24$ multiply by its reciprocal, $\frac{1}{3}$.
         $\left(\frac{1}{3} \cdot 3\right)x = \frac{1}{3} \cdot 24$
         $1 \cdot x = 8$
         $x = 8$

Checking the solution:

$3 \cdot 8 \overset{?}{=} 24$

$24 = 24$

So $\{8\}$ is the solution set.

NOTE ▶   If we try to multiply both sides of this equation by zero, we obtain $0 = 0$, and we are unable to determine the solution.

2. $\frac{1}{4}x = -7$ The coefficient is $\frac{1}{4}$, and we
         $4 \cdot \frac{1}{4}x = 4 \cdot (-7)$ multiply by its reciprocal, 4.
         $\left(4 \cdot \frac{1}{4}\right)x = 4 \cdot (-7)$
         $1 \cdot x = -28$
         $x = -28$

So $\{-28\}$ is the solution set.

3. $\frac{2}{5}x = 10$ The coefficient is $\frac{2}{5}$, and we
         $\frac{5}{2} \cdot \frac{2}{5}x = \frac{5}{2} \cdot 10$ multiply by its reciprocal, $\frac{5}{2}$.
         $\left(\frac{5}{2} \cdot \frac{2}{5}\right)x = \frac{5}{2} \cdot \frac{10}{1}$
         $1 \cdot x = 25$
         $x = 25$

So $\{25\}$ is the solution set.

4. $\frac{x}{3} = 11$ Since division by 3 is the
         $\frac{1}{3}x = 11$ same as multiplication by $\frac{1}{3}$,
         $3 \cdot \frac{1}{3}x = 3 \cdot 11$ $x/3 = \frac{1}{3} \cdot x$. The coefficient
         $\left(3 \cdot \frac{1}{3}\right)x = 3 \cdot 11$ is $\frac{1}{3}$, and we multiply by its
         $1 \cdot x = 33$ reciprocal, 3.
         $x = 33$

So the solution set is $\{33\}$.

5.
$$-x = 5$$
$$-1 \cdot x = 5$$
$$-1 \cdot (-1) \cdot x = -1 \cdot 5$$
$$1 \cdot x = -5$$
$$x = -5$$

Since $-x = -1 \cdot x$, the coefficient is -1, and we multiply by its reciprocal, -1.

So the solution set is $\{-5\}$.

6.
$$-5x = 20$$
$$-\tfrac{1}{5} \cdot (-5x) = -\tfrac{1}{5} \cdot 20$$
$$[-\tfrac{1}{5} \cdot (-5)]x = -\tfrac{1}{5} \cdot 20$$
$$1 \cdot x = -4$$
$$x = -4$$

The coefficient is -5, and we multiply by its reciprocal, $-\tfrac{1}{5}$.

So $\{-4\}$ is the solution set.

NOTE ▶ Division is defined as multiplication by the reciprocal. Because of this fact, we also can divide both sides of an equation by any nonzero algebraic expression and obtain an equation that is equivalent to the original equation.

In the previous example,

$$-5x = 20$$
$$\frac{-5x}{-5} = \frac{20}{-5} \qquad \text{Divide both sides by } -5.$$
$$x = -4$$

So $\{-4\}$ is the solution set.

7. $7x = -42$

$$\frac{7x}{7} = \frac{-42}{7} \qquad \text{Divide both sides by 7.}$$
$$x = -6$$

So the solution set is $\{-6\}$.

In the next example we will investigate some linear equations in which we must use both the additive and multiplicative properties of equality.

EXAMPLE 2.3.2

Solve the following linear equations.

1.

$$6x - 4 = 2$$

Add 4 to both sides to isolate the variable term.

$$6x - 4 + 4 = 2 + 4$$

$$6x = 6$$

$$\frac{6x}{6} = \frac{6}{6}$$

Divide both sides by 6.

$$x = 1$$

So $\{1\}$ is the solution set.

2.

$$17 = 2 - 3x$$

Subtract 2 from both sides to isolate the variable term.

$$17 - 2 = 2 - 2 - 3x$$

$$15 = -3x$$

$$\frac{15}{-3} = \frac{-3x}{-3}$$

Divide both sides by -3.

$$-5 = x$$

So $\{-5\}$ is the solution set.

3.

$$1.4x + 2.2 = 9.2$$

$$1.4x + 2.2 - 2.2 = 9.2 - 2.2$$

$$1.4x = 7$$

$$\frac{1.4x}{1.4} = \frac{7}{1.4}$$

$$x = 5$$

So $\{5\}$ is the solution set.

4.

$$5x + 1 = 2x + 8$$

Subtract $2x$ from both sides to eliminate the x term from the right-hand side.

$$5x - 2x + 1 = 2x - 2x + 8$$

$$3x + 1 = 8$$

$$3x + 1 - 1 = 8 - 1$$

Subtract 1 from both sides to isolate the variable term.

$$3x = 7$$

$$\frac{3x}{3} = \frac{7}{3}$$

Divide both sides by 3.

$$x = \frac{7}{3}$$

So $\{\frac{7}{3}\}$ is the solution set.

NOTE ▶ Generalizing the preceding examples, we first use the additive property of equality to isolate the variable terms, combining like terms on each side of the equation. Then we use the multiplicative property of equality to make the coefficient of the variable equal to 1.

5. $\frac{1}{2}x + \frac{1}{3}x + 2 = \frac{1}{6}x + 7$

$\frac{3}{6}x + \frac{2}{6}x + 2 = \frac{1}{6}x + 7$ Obtain an LCD for the
like terms.

$\frac{5}{6}x + 2 = \frac{1}{6}x + 7$

$\frac{5}{6}x - \frac{1}{6}x + 2 = \frac{1}{6}x - \frac{1}{6}x + 7$ Subtract $\frac{1}{6}x$ from both sides
to eliminate the x term from
the right-hand side.

$\frac{4}{6}x + 2 = 7$

$\frac{2}{3}x + 2 = 7$

$\frac{2}{3}x + 2 - 2 = 7 - 2$ Subtract 2 from both sides to
isolate the variable term.

$\frac{2}{3}x = 5$

$\frac{3}{2} \cdot \frac{2}{3}x = \frac{3}{2} \cdot 5$ The coefficient is $\frac{2}{3}$, and we
multiply by its reciprocal, $\frac{3}{2}$.

$\left(\frac{3}{2} \cdot \frac{2}{3}\right) \cdot x = \frac{3}{2} \cdot \frac{5}{1}$

$1 \cdot x = \frac{15}{2}$

$x = \frac{15}{2}$

So $\left\{\frac{15}{2}\right\}$ is the solution set.

    Using the multiplicative property of equality, we can eliminate the fractions in an equation by multiplying both sides by the LCD. Let us work the preceding equation using this method.

$\frac{1}{2}x + \frac{1}{3}x + 2 = \frac{1}{6}x + 7$ The LCD is 6.

$6\left(\frac{1}{2}x + \frac{1}{3}x + 2\right) = 6\left(\frac{1}{6}x + 7\right)$ Multiply both sides by 6.

$6 \cdot \frac{1}{2}x + 6 \cdot \frac{1}{3}x + 6 \cdot 2 = 6 \cdot \frac{1}{6}x + 6 \cdot 7$ Use the distributive
property.

$\left(6 \cdot \frac{1}{2}\right)x + \left(6 \cdot \frac{1}{3}\right)x + 6 \cdot 2 = \left(6 \cdot \frac{1}{6}\right)x + 6 \cdot 7$

$3x + 2x + 12 = 1x + 42$

$5x + 12 = x + 42$

$5x - x + 12 = x - x + 42$ Subtract x from both sides
to eliminate the x term
from the right-hand side.

$4x + 12 = 42$

$4x + 12 - 12 = 42 - 12$ Subtract 12 from both sides
to isolate the variable term.

$4x = 30$

$\frac{4x}{4} = \frac{30}{4}$ Divide both sides by 4.

$x = \frac{15}{2}$

We obtain the same solution.

EXAMPLE 2.3.3

Bubba's Chinese Restaurant is on the highway from San Antonio to Houston. Bubba's is 14 miles closer to San Antonio than it is to Houston. If San Antonio and Houston are 200 miles apart, how far is Bubba's from San Antonio?

SOLUTION

We let x = distance from Bubba's to San Antonio since that is the number we are trying to find (see Figure 2.3.1).

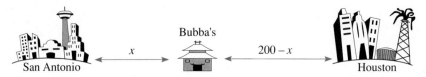

Figure 2.3.1

$$\begin{pmatrix} \text{Distance from Bubba's} \\ \text{to San Antonio} \end{pmatrix} + \begin{pmatrix} \text{Distance from Bubba's} \\ \text{to Houston} \end{pmatrix} = 200$$

$$x \qquad + \begin{pmatrix} \text{Distance from Bubba's} \\ \text{to Houston} \end{pmatrix} = 200$$

$$\text{Distance from Bubba's to Houston} = 200 - x$$

From the statement of the problem, we know that

$$\begin{pmatrix} \text{Distance from Bubba's} \\ \text{to San Antonio} \end{pmatrix} + 14 \text{ miles} = \begin{pmatrix} \text{Distance from Bubba's} \\ \text{to Houston} \end{pmatrix}$$

$$x \qquad + 14 \qquad = 200 - x$$

$$2x + 14 \qquad = 200$$

$$2x = 186$$

$$x = 93$$

Thus, it is 93 miles from Bubba's to San Antonio.

EXERCISES 2.3

Solve the following linear equations and check your solutions.

1. $4x = 32$

2. $6x = 18$

3. $\frac{3}{5}x = 9$

4. $\frac{2}{7}x = 8$

5. $-x = 3$

6. $\frac{2}{5}x = \frac{8}{15}$

7. $24 = -2x$

8. $-\frac{2}{3}x = -\frac{2}{3}$

9. $-6 = \frac{x}{3}$

10. $\frac{2}{3} = -x$

11. $7x = 0$

12. $\frac{1}{5} = -\frac{3}{4}x$

13. $3x = 1$

14. $-x = 13$

15. $\frac{3}{5}x = \frac{9}{10}$

16. $-4 = \frac{x}{9}$

17. $\frac{1}{7} = -\frac{4}{5}x$

18. $8x = \frac{4}{5}$

19. $\frac{3}{4}x = 2$

20. $0 = -\frac{6}{7}x$

21. $\frac{5}{8} = -x$

22. $27 = -3x$

23. $-\frac{3}{4} = 7x$

24. $\frac{x}{8} = -2$

25. $-x = -4$

26. $4x = 0$

27. $6x = \frac{2}{3}$

28. $\frac{5}{3}x = 2$

29. $-\frac{6}{7}x = -\frac{6}{7}$

30. $5x = 1$

31. $-6x = \frac{4}{7}$

32. $\frac{x}{7} = -5$

33. $\frac{x}{5} = -\frac{1}{9}$

34. $-\frac{7}{8} = -x$

35. $-7x = -49$

36. $-\frac{2}{3} = 5x$

37. $0 = -\frac{5}{7}x$

38. $-8x = \frac{6}{7}$

39. $8x = -5$

40. $-x = -9$

41. $-\frac{9}{10} = -x$

42. $-9x = 2$

43. $\frac{x}{8} = \frac{3}{4}$

44. $-6x = -72$

45. $1.6x = 4.8$

46. $2.9x = 14.5$

47. $5.2x = -20.8$

48. $4.8x = -33.6$

49. $3x + 5 = 14$

50. $2x + 3 = 7$

51. $7x - 8 = 13$

52. $6x - 11 = -7$

53. $3 - 4x = 11$

54. $18 - 5x = 13$

55. $-13 = 4x - 9$

56. $-14 = 3x - 2$

57. $23 = 6x + 5$

58. $-30 = 8x + 2$

59. $10 = 7 - 2x$

60. $3 = -5x + 1$

61. $-x - 2 = 1$

62. $-x - 7 = -2$

63. $\frac{2}{3}x + 1 = -3$

64. $\frac{3}{4}x + 2 = -4$

65. $\frac{1}{2}x - \frac{2}{3} = \frac{1}{6}$

66. $\frac{3}{5}x - \frac{7}{10} = \frac{7}{5}$

67. $2.6x + 1.4 = 9.2$

68. $4.3x + 6.1 = 31.9$

69. $3.7x - 1.9 = -9.3$

70. $5.3x - 6.1 = -27.3$

71. $10x = 8x - 4$

72. $7x = 4x - 18$

73. $\frac{1}{2}x + 2 = \frac{1}{3}x$

74. $\frac{1}{4}x + 3 = \frac{1}{5}x$

75. $2x + 7 = 6x - 5$

76. $4x + 1 = 9x - 19$

77. $5x - 2 = 3x - 5$

78. $7x + 4 = 5x + 1$

79. $3x - 8 = 4x - 1$

80. $5x - 9 = 6x + 2$

81. $2x + 5 = 9x - 1$

82. $5x - 8 = 2x - 3$

83. $\frac{1}{2}x + 1 = \frac{2}{3}x + 5$

84. $\frac{1}{4}x - 7 = \frac{1}{5}x - 3$

85. $\frac{3}{4}x + \frac{1}{2} = \frac{5}{2}x - \frac{1}{4}$

86. $\frac{1}{3}x - \frac{1}{6} = \frac{5}{3}x + \frac{5}{6}$

87. $4x - 3 + 2x = 8x - 9$

88. $5x - 2 + 3x = 10x - 10$

89. $7 - 2x - 4x = 7x + 72$

90. $9 - 3x - 2x = 4x + 18$

91. $3x + 7 - 9x = 2x + 4$

92. $5x + 1 - 8x = x + 6$

93. $\frac{1}{3}x + \frac{1}{6}x - \frac{13}{6} = 2x + \frac{5}{6}$

94. $\frac{1}{2}x + \frac{3}{4}x + \frac{3}{2} = 2x - \frac{9}{4}$

95. $\frac{1}{5}x + \frac{3}{10}x - \frac{1}{5} = \frac{9}{10}x - \frac{4}{5}$

96. $\frac{3}{5}x + \frac{1}{10}x - \frac{3}{5} = \frac{13}{10}x - \frac{1}{5}$

97. $\frac{5}{6}x - \frac{1}{8}x + \frac{3}{4} = \frac{2}{3}x - \frac{1}{12}$

98. $\frac{1}{6}x - \frac{3}{8}x + \frac{5}{4} = \frac{5}{3}x - \frac{5}{12}$

99. Betsy and Charlie live between Los Angeles and San Francisco. If they live twice as far from San Francisco as they do from Los Angeles, and it is 327 miles between the two cities, how far do they live from Los Angeles?

100. Marchand's Restaurant is between Baton Rouge and New Orleans. The restaurant is 35 miles closer to Baton Rouge than it is to New Orleans. If the cities are 75 miles apart, how far is Marchand's Restaurant from Baton Rouge?

101. Slick Sid's Auto City sells new cars at a price that is $\frac{7}{8}$ of the manufacturer's suggested retail price (MSRP). Sid's price on a new Doodlebug is $10,871. What is the MSRP of a Doodlebug?

102. Dora's Discount Auto World sells new cars at a price that is $\frac{8}{9}$ of the manufacturer's suggested retail price (MSRP). Dora's price on a new Junebug is $12,096. What is the MSRP of a Junebug?

2.4 Combining Properties to Solve Linear Equations

Example 2.3.2 contained linear equations in which we used both the additive and multiplicative properties of equality. In this section we will investigate equations that are further complicated with grouping symbols. Consider the following example.

EXAMPLE 2.4.1

Find the solution of the following linear equation.

$$3(4x - 1) + 5x = 5(x + 2) + 11$$

$$12x - 3 + 5x = 5x + 10 + 11 \quad \text{Distributive property}$$

$$17x - 3 = 5x + 21 \quad \text{Combine like terms.}$$

$$17x - 5x - 3 = 5x - 5x + 21 \quad \text{Subtract } 5x \text{ from both sides to eliminate the } x \text{ term on the right-hand side.}$$

$$12x - 3 = 21$$

$$12x - 3 + 3 = 21 + 3 \quad \text{Add 3 to both sides to isolate the } x \text{ term.}$$

$$12x = 24$$

$$\frac{12x}{12} = \frac{24}{12} \quad \text{Divide both sides by 12.}$$

$$x = 2$$

Checking the solution:

$$3(4 \cdot 2 - 1) + 5 \cdot 2 \stackrel{?}{=} 5(2 + 2) + 11$$

$$3(8 - 1) + 10 \stackrel{?}{=} 5(4) + 11$$

$$3(7) + 10 \stackrel{?}{=} 20 + 11$$

$$21 + 10 \stackrel{?}{=} 31$$

$$31 = 31$$

Thus, {2} is the solution set.

Generalizing the previous example, we can use the following steps to solve any linear equation.

1. Use the distributive property to eliminate any grouping symbols, and then combine all like terms on each side of the equation.

2. Eliminate the variable on one side of the equation by using the additive property of equality.

3. Eliminate the constant term that is being added to the variable term by using the additive property of equality.

> **4.** Using the multiplicative property of equality, make the coefficient of the variable equal to one.
>
> **5.** (Optional) Check the solution.

EXAMPLE 2.4.2

Solve the following linear equations.

1. $7(x + 3) + 2(2x - 4) = 46$

$$7x + 21 + 4x - 8 = 46 \qquad \text{Step 1}$$
$$11x + 13 = 46 \qquad \text{Step 1}$$
$$11x + 13 - 13 = 46 - 13 \qquad \text{Step 3}$$
$$11x = 33$$
$$\frac{11x}{11} = \frac{33}{11} \qquad \text{Step 4}$$
$$x = 3$$

Checking the solution:

$$7(3 + 3) + 2(2 \cdot 3 - 4) \stackrel{?}{=} 46$$
$$7(6) + 2(6 - 4) \stackrel{?}{=} 46$$
$$42 + 2(2) \stackrel{?}{=} 46$$
$$42 + 4 \stackrel{?}{=} 46$$
$$46 = 46$$

Thus, $\{3\}$ is the solution set.

2. $3(2x - 1) - (x + 7) = -5$

$$6x - 3 - x - 7 = -5 \qquad \begin{array}{l} \text{Step 1} \\ \text{Recall: } -(x + 7) = -1 \cdot (x + 7). \end{array}$$
$$5x - 10 = -5 \qquad \text{Step 1}$$
$$5x - 10 + 10 = -5 + 10 \qquad \text{Step 3}$$
$$5x = 5$$
$$\frac{5x}{5} = \frac{5}{5} \qquad \text{Step 4}$$
$$x = 1$$

So $\{1\}$ is the solution set.

Try to avoid this mistake:

Incorrect	Correct
$3(2x - 1) - (x + 7) = -5$ $6x - 3 - x + 7 = -5$ Here we forgot to change the sign of 7.	$3(2x - 1) - (x + 7) = -5$ $6x - 3 - x - 7 = -5$ Remember, when subtracting an expression inside parentheses you must change the signs of *all* of the terms inside the parentheses.

3. $4(x + 1) - 3 = 7(x - 1)$

$$4x + 4 - 3 = 7x - 7 \qquad \text{Step 1}$$
$$4x + 1 = 7x - 7 \qquad \text{Step 1}$$
$$4x - 4x + 1 = 7x - 4x - 7 \qquad \text{Step 2}$$
$$1 = 3x - 7$$
$$1 + 7 = 3x - 7 + 7 \qquad \text{Step 3}$$
$$8 = 3x$$
$$\frac{8}{3} = \frac{3x}{3} \qquad \text{Step 4}$$
$$\frac{8}{3} = x$$

So $\left\{\frac{8}{3}\right\}$ is the solution set.

4. $5 + 2(3 - 4x) = 23$

$$5 + 6 - 8x = 23 \qquad \text{Step 1}$$
$$11 - 8x = 23 \qquad \text{Step 1}$$
$$11 - 11 - 8x = 23 - 11 \qquad \text{Step 3}$$
$$-8x = 12$$
$$\frac{-8x}{-8} = \frac{12}{-8} \qquad \text{Step 4}$$
$$x = -\frac{3}{2}$$

So $\left\{-\frac{3}{2}\right\}$ is the solution set.

Try to avoid this mistake:

Incorrect	Correct
$5 + 2(3 - 4x) = 23$ $7(3 - 4x) = 23$	$5 + 2(3 - 4x) = 23$ $5 + 6 - 8x = 23$
Here the addition was performed before the multiplication.	Remember to perform multiplications and divisions before additions and subtractions.

5. $4[19 - 4(2x + 3)] = 5(7 - 6x)$

$\quad\quad 4[19 - 8x - 12] = 5(7 - 6x)$ Step 1

$\quad\quad\quad\quad 4[7 - 8x] = 5(7 - 6x)$ Step 1

$\quad\quad\quad\quad 28 - 32x = 35 - 30x$ Step 1

$\quad 28 - 32x + 32x = 35 - 30x + 32x$ Step 2

$\quad\quad\quad\quad\quad\quad 28 = 35 + 2x$

$\quad\quad 28 - 35 = 35 - 35 + 2x$ Step 3

$\quad\quad\quad\quad\quad -7 = 2x$

$\quad\quad\quad\quad\quad \dfrac{-7}{2} = \dfrac{2x}{2}$ Step 4

$\quad\quad\quad\quad\quad \dfrac{-7}{2} = x$

So $\left\{-\dfrac{7}{2}\right\}$ is the solution set.

6. $\quad \frac{1}{2}(3x - 2) = \frac{2}{3}(x - 9)$

$\quad\quad\quad \frac{3}{2}x - 1 = \frac{2}{3}x - 6$ Step 1

$\quad \frac{3}{2}x - \frac{2}{3}x - 1 = \frac{2}{3}x - \frac{2}{3}x - 6$ Step 2

$\quad\quad \frac{9}{6}x - \frac{4}{6}x - 1 = -6$

$\quad\quad\quad\quad \frac{5}{6}x - 1 = -6$

$\quad\quad \frac{5}{6}x - 1 + 1 = -6 + 1$ Step 3

$\quad\quad\quad\quad\quad \frac{5}{6}x = -5$

$\quad\quad \frac{6}{5} \cdot \frac{5}{6}x = \frac{6}{5} \cdot (-5)$ Step 4

$\quad\quad\quad\quad\quad\quad x = -6$

So $\{-6\}$ is the solution set.

7.

$$1.2(4.9x + 7) = 8.2x + 0.512$$

$$5.88x + 8.4 = 8.2x + 0.512 \qquad \text{Step 1}$$

$$5.88x - 5.88x + 8.4 = 8.2x - 5.88x + 0.512 \qquad \text{Step 2}$$

$$8.4 = 2.32x + 0.512$$

$$8.4 - 0.512 = 2.32x + 0.512 - 0.512 \qquad \text{Step 3}$$

$$7.888 = 2.32x$$

$$\frac{7.888}{2.32} = \frac{2.32x}{2.32} \qquad \text{Step 4}$$

$$3.4 = x$$

So {3.4} is the solution set.

There are two special types of equations that at first glance appear to be simple linear equations. The following example illustrates the solutions of these equations.

EXAMPLE 2.4.3

Solve the following equations.

1. $3(2x + 1) - 1 = 6x + 2$

$$6x + 3 - 1 = 6x + 2 \qquad \text{Step 1}$$

$$6x + 2 = 6x + 2 \qquad \text{Step 1}$$

$$6x - 6x + 2 = 6x - 6x + 2 \qquad \text{Step 2}$$

$$2 = 2 \qquad \text{A true statement}$$

REMARK   All of the variables dropped out leaving a true statement. The solution set is the set of *all real numbers* because any value for x results in a true statement.

2. $4(2x - 3) + 15 = 8x - 7$

$$8x - 12 + 15 = 8x - 7 \qquad \text{Step 1}$$

$$8x + 3 = 8x - 7 \qquad \text{Step 1}$$

$$8x - 8x + 3 = 8x - 8x - 7 \qquad \text{Step 2}$$

$$3 = -7 \qquad \text{A false statement}$$

REMARK   When all of the variables drop out leaving a false statement, the solution set is the *empty set*, $\varnothing$, because any value for x results in a false statement.

EXERCISES 2.4

Solve the following equations.

1. $5(2x + 4) = 8x - 12$
2. $7(2x + 1) = 10x - 13$
3. $4 + 2(3x - 2) = 16$

4. $5 + 3(x - 1) = -4$
5. $3x - 4(2x - 1) = 5x$
6. $5x - 2(3x - 2) = 11x$

7. $5 - (2x + 6) = -4$
8. $7 - (3x + 4) = 10$
9. $2 + 3(3x - 7) = 9x - 19$

10. $4 + 2(2x - 6) = 4x - 8$
11. $\frac{1}{4} + \frac{1}{3}(5x + 2) = -2$
12. $\frac{1}{2} + \frac{2}{3}(5x + 1) = 2$

13. $\frac{2}{3}x + \frac{1}{6}(3x - 2) = -\frac{1}{3}$
14. $\frac{5}{6}x + \frac{1}{3}(5x - 9) = -3$
15. $3(5x - 7) = 4(2x - 7)$

16. $2(3x - 1) = 7(2x + 2)$
17. $4(6 - 2x) - 2 = 9(3 - 2x) + 6$

18. $5(2 - 4x) - 1 = 7(1 - 3x) + 12$
19. $3 + 2(8x - 1) = 5 + 4(4x - 6)$

20. $2 + 4(3x - 6) = 7 + 6(2x - 1)$
21. $\frac{2}{3}(3x - 5) = \frac{1}{2}(5x - 9)$

22. $\frac{1}{3}(5x - 1) = \frac{3}{2}(5x - 8)$
23. $\frac{1}{3}(2x + 5) = \frac{3}{4}(x - 1) + \frac{5}{4}$

24. $\frac{1}{4}(5x + 7) = \frac{2}{3}(x - 5) + \frac{11}{4}$
25. $1.2(2.3x + 7.4) = 3.9(4.1x + 1.5) - 41.952$

26. $4.7(3.1x + 2.9) = 5.2(2.6x + 1.9) + 6.48$
27. $5(2x + 1) + 4(3x - 6) = -85$

28. $4(3x - 7) + 2(2x + 5) = -34$
29. $6(x - 5) - (3x + 4) = -19$

30. $7(x - 8) - (2x + 8) = -29$
31. $2(3x + 4) - 5(x + 6) = 4x - 23$

32. $4(2x + 1) - 7(x + 2) = 5x - 9$
33. $3(8x + 1) - 5(4x + 1) = 4x - 2$

34. $2(4x - 4) + 3(x + 1) = 11x + 1$
35. $\frac{1}{3}(2x - 1) + \frac{1}{6}(x - 2) = \frac{2}{3}$

36. $\frac{1}{2}(3x - 2) + \frac{4}{3}(2x - 1) = \frac{1}{6}$
37. $\frac{3}{4}(4x + 1) - \frac{1}{2}(x - 2) = \frac{5}{8}x + \frac{1}{2}$

38. $\frac{5}{8}(2x - 4) - \frac{1}{2}(x - 5) = \frac{1}{8}x + \frac{3}{4}$
39. $3.2(x - 5) + 4.2(3x + 1) = -108.18$

40. $2.7(x - 2) + 6.1(2x + 3) = -72.03$
41. $5(x - 2) - 3(x + 2) = 2(2x - 7)$

42. $6(x - 5) - 2(2x + 4) = 3(x - 10)$
43. $3(2x + 1) + 4(x - 6) = 5(2x + 3)$

44. $3(2x - 4) + 6(x - 2) = 4(3x - 8)$
45. $2(2x + 1) + 5(x - 7) = 4(2x + 1) - 23$

46. $3(x - 4) + 4(x + 1) = 3(3x - 5) + 3$
47. $5(2x - 1) - 2(4x + 3) = 2(5x - 1) + 7$

48. $6(x - 3) - 3(2x + 1) = 4(2x - 6) + 3$
49. $\frac{1}{2}(3x + 1) + \frac{2}{3}(4x - 7) = \frac{5}{6}x + \frac{5}{6}$

50. $\frac{1}{3}(4x - 1) + \frac{3}{2}(2x - 4) = \frac{1}{6}x + \frac{1}{3}$
51. $\frac{3}{4}(5x + 2) - \frac{1}{2}(3x - 7) = \frac{3}{8}x + \frac{5}{4}$

52. $\frac{3}{4}(7x - 1) - \frac{1}{4}(2x + 5) = \frac{3}{2}x + \frac{5}{4}$
53. $4.2(x - 3) + 1.7(2x + 5) = 6.4x - 6.62$

54. $3.1(x - 2) + 2.9(3x + 3) = 5.2x - 18.62$
55. $5[7 + 2(3x - 4)] = 24x - 9$

56. $4[6 + 3(2x - 1)] = 20x - 2$
57. $5[8 - 2(3x + 4)] = -60$

58. $4[10 - 5(2x + 2)] = -120$
59. $3[2(4x + 1) - 3(2x + 6)] = 5x - 52$

60. $4[3(7x - 1) - 5(4x + 2)] = 5x - 47$
61. $2[4(2x + 3) - 3(x + 3)] = 10x + 15$

62. $3[2(3x + 1) - 2(x + 4)] = 12x - 18$

63. $3[4 + 2(2x + 1)] = 5(3x + 2)$

64. $4[1 + 3(2x - 1)] = 6(3x + 1)$

65. $2[2(x + 1) + 3(x - 1)] = 4(2x - 3) + 20$

66. $3[4(x - 7) + 2(x - 1)] = 2(5x + 3)$

67. $6[3(2x + 8) - 5(x + 4)] = 3(x - 4)$

68. $5[4(3x + 1) - 11(x + 1)] = 2(x + 6) - 23$

2.5 Absolute Value Equations

In Chapter 1 we defined the absolute value of a number to be *the distance from zero to that number on the real number line.* A more formal definition of absolute value is given below.

Absolute Value

The **absolute value** of a real number x, denoted by $|x|$, is defined as follows:

$$|x| = x \quad \text{if } x \geq 0$$

$$\text{or} \quad |x| = -x \quad \text{if } x < 0$$

EXAMPLE 2.5.1

Evaluate the following.

1. $|28| = 28$ Since $28 \geq 0$

2. $\left|\frac{3}{4}\right| = \frac{3}{4}$ Since $\frac{3}{4} \geq 0$

3. $|-2| = -(-2) = 2$ Since $-2 < 0$

4. $|-6.42| = -(-6.42) = 6.42$ Since $-6.42 < 0$

5. $|0| = 0$ Since $0 \geq 0$

REMARKS

1. The absolute value of a number that is positive or zero is that number. To find the absolute value of a negative number, change its sign.

2. *The absolute value of a number is never negative.*

Let us now investigate some equations that contain absolute value expressions.

EXAMPLE 2.5.2

Find the solutions of the following equations.

1. $|x| = 2$

   There are *two* numbers that are 2 units from zero on the real number line: 2 and -2.

   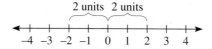

   Hence the solutions are 2 and -2.

   Checking the solutions:

   when $x = 2$ $\quad |2| \overset{?}{=} 2;$ $\quad$ when $x = -2$ $\quad |-2| \overset{?}{=} 2$

   $\qquad\qquad\qquad 2 = 2 \checkmark$ $\qquad\qquad\qquad\qquad -(-2) \overset{?}{=} 2$

   $\qquad\qquad\qquad\qquad\qquad\qquad\qquad\qquad\qquad\qquad\quad 2 = 2 \checkmark$

   Thus, $\{2, -2\}$ is the solution set.

2. $|x + 5| = 8$

   In this case, $x + 5$ represents a number that is 8 units from zero on the real number line. This number must be 8 or -8.

   This fact yields two equations that we then solve for x:

   $\qquad\quad x + 5 = 8 \qquad\qquad\text{or}\qquad\qquad x + 5 = -8$

   $\quad x + 5 - 5 = 8 - 5 \qquad\qquad\qquad x + 5 - 5 = -8 - 5$

   $\qquad\qquad\quad x = 3 \qquad\qquad\qquad\qquad\qquad\quad x = -13$

   Checking the solutions:

   when $x = 3$ $\quad |3 + 5| \overset{?}{=} 8;$ $\quad$ when $x = -13$ $\quad |-13 + 5| \overset{?}{=} 8$

   $\qquad\qquad\qquad |8| \overset{?}{=} 8$ $\qquad\qquad\qquad\qquad\qquad |-8| \overset{?}{=} 8$

   $\qquad\qquad\qquad 8 = 8 \checkmark$ $\qquad\qquad\qquad\qquad\qquad -(-8) \overset{?}{=} 8$

   $\qquad\qquad\qquad\qquad\qquad\qquad\qquad\qquad\qquad\qquad\qquad 8 = 8 \checkmark$

   Thus, $\{3, -13\}$ is the solution set.

3. $|2x - 7| = 13$

$$2x - 7 = 13 \qquad \text{or} \qquad 2x - 7 = -13$$

$$2x - 7 + 7 = 13 + 7 \qquad\qquad 2x - 7 + 7 = -13 + 7$$

$$2x = 20 \qquad\qquad\qquad 2x = -6$$

$$\frac{2x}{2} = \frac{20}{2} \qquad\qquad\qquad \frac{2x}{2} = \frac{-6}{2}$$

$$x = 10 \qquad\qquad\qquad x = -3$$

The check is left to the reader. The solution set is $\{10, -3\}$.

Try to avoid these mistakes:

Incorrect	Correct
1. $\|2x - 7\| = 13$	$\|2x - 7\| = 13$
$2x - 7 = 13$	$2x - 7 = 13 \;$ or $\; 2x - 7 = -13$
$2x = 20$	$2x = 20 \qquad\qquad 2x = -6$
$x = 10$	$x = 10 \qquad\qquad x = -3$
Here we lost one of the solutions because the second case was not considered.	
2. $\|2x - 7\| = 13$	
$2x + 7 = 13$	
Absolute value does not mean changing all minus signs to plus signs.	

The remaining examples will demonstrate how to find the solutions of more complicated absolute value equations.

EXAMPLE 2.5.3

Find the solutions of the following equations.

1. $|3x + 4| - 2 = 8$

$|3x + 4| - 2 + 2 = 8 + 2$ Isolate the absolute value

$|3x + 4| = 10$ expression.

As we isolate the variable in a linear equation, so must we isolate the absolute value expression. After this step, we follow the steps outlined in the previous example:

$3x + 4 = 10$	or	$3x + 4 = -10$
$3x + 4 - 4 = 10 - 4$		$3x + 4 - 4 = -10 - 4$
$3x = 6$		$3x = -14$
$\dfrac{3x}{3} = \dfrac{6}{3}$		$\dfrac{3x}{3} = \dfrac{-14}{3}$
$x = 2$		$x = -\dfrac{14}{3}$

Thus, $\left\{2, -\frac{14}{3}\right\}$ is the solution set.

2. $\left|\frac{2}{3}x + \frac{1}{6}\right| - \frac{1}{3} = \frac{3}{2}$

$\left|\frac{2}{3}x + \frac{1}{6}\right| - \frac{1}{3} + \frac{1}{3} = \frac{3}{2} + \frac{1}{3}$ Isolate the absolute value expression.

$\left|\frac{2}{3}x + \frac{1}{6}\right| = \frac{9}{6} + \frac{2}{6}$ Obtain an LCD.

$\left|\frac{2}{3}x + \frac{1}{6}\right| = \frac{11}{6}$

$\frac{2}{3}x + \frac{1}{6} = \frac{11}{6}$	or	$\frac{2}{3}x + \frac{1}{6} = -\frac{11}{6}$
$\frac{2}{3}x + \frac{1}{6} - \frac{1}{6} = \frac{11}{6} - \frac{1}{6}$		$\frac{2}{3}x + \frac{1}{6} - \frac{1}{6} = -\frac{11}{6} - \frac{1}{6}$
$\frac{2}{3}x = \frac{10}{6}$		$\frac{2}{3}x = -\frac{12}{6}$
$\frac{2}{3}x = \frac{5}{3}$		$\frac{2}{3}x = -2$
$\frac{3}{2} \cdot \frac{2}{3}x = \frac{3}{2} \cdot \frac{5}{3}$		$\frac{3}{2} \cdot \frac{2}{3}x = \frac{3}{2} \cdot (-2)$
$x = \frac{5}{2}$		$x = -3$

Thus, $\left\{\frac{5}{2}, -3\right\}$ is the solution set.

3. $|5x - 3| + 9 = 5$

$|5x - 3| + 9 - 9 = 5 - 9$ Isolate the absolute value

$|5x - 3| = -4$ expression.

After isolating the absolute value, we obtain a negative number on the right-hand side of the equation. However, we know that the absolute value can never be negative. Thus, the solution set is the empty set, $\varnothing$; there are no solutions to this equation.

EXERCISES 2.5

Find the solutions of the following equations.

1. $|x| = 5$ 2. $|x| = 4$ 3. $|x| = \frac{2}{3}$ 4. $|x| = 0$

5. $|x| = -4$ 6. $|x| = -8$ 7. $|-x| = -12$ 8. $|-x| = -20$

9. $|-x| = 0$ 10. $|2x| = 16$ 11. $|3x| = 15$ 12. $|4x| = 5$

13. $|-2x| = \frac{3}{5}$ 14. $|-2x| = \frac{5}{2}$ 15. $|\frac{2}{3}x| = 6$ 16. $|\frac{3}{4}x| = 9$

17. $|\frac{3}{5}x| = 4$ 18. $|\frac{3}{8}x| = 5$ 19. $|-\frac{3}{2}x| = 1$ 20. $|-\frac{4}{3}x| = 6$

21. $|\frac{5}{8}x| = -1$ 22. $|-\frac{5}{2}x| = \frac{2}{3}$ 23. $|-\frac{5}{6}x| = \frac{3}{5}$ 24. $|x + 7| = 12$

25. $|x + 3| = 9$ 26. $|x - 8| = 4$ 27. $|x - 2| = 5$ 28. $|3 - x| = 1$

29. $|8 - x| = 7$ 30. $|1 - x| = \frac{1}{2}$ 31. $|4 - x| = \frac{3}{4}$ 32. $|3 - x| = -\frac{2}{3}$

33. $|9 - x| = -\frac{1}{8}$ 34. $|2x + 1| = 13$ 35. $|2x + 5| = 9$ 36. $|3x + 4| = 6$

37. $|4x + 5| = 8$ 38. $|3x - 2| = 7$ 39. $|5x - 1| = 12$ 40. $|2 - 3x| = \frac{3}{4}$

41. $|5 - 2x| = \frac{1}{6}$ 42. $|\frac{1}{3}x + 1| = \frac{1}{4}$ 43. $|\frac{1}{2}x + 2| = \frac{1}{3}$ 44. $|\frac{1}{2}x - 3| = \frac{1}{5}$

45. $|\frac{1}{3}x - 1| = \frac{5}{6}$ 46. $|\frac{3}{4}x - \frac{2}{3}| = \frac{5}{6}$ 47. $|\frac{5}{8}x - \frac{1}{4}| = \frac{3}{8}$ 48. $|\frac{3}{10}x - \frac{2}{5}| = -\frac{4}{5}$

49. $|\frac{1}{5}x - \frac{1}{4}| = -\frac{1}{10}$ 50. $|\frac{1}{3}x + \frac{3}{5}| = 0$ 51. $|\frac{2}{3}x - \frac{1}{6}| = 0$ 52. $|2 - \frac{2}{3}x| = 4$

53. $|3 - \frac{1}{6}x| = 2$ 54. $|1 - \frac{3}{10}x| = \frac{4}{5}$ 55. $|2 - \frac{1}{5}x| = \frac{9}{10}$ 56. $|x - 1| - 2 = 9$

57. $|x - 8| - 2 = -2$ 58. $|x - 3| - 4 = -7$ 59. $|x + 3| - 4 = 1$ 60. $|8 - x| + 12 = 2$

61. $|1 - x| + 6 = -3$ 62. $|6 - x| + 1 = 8$ 63. $|x + 2| - 7 = -4$ 64. $|x - 3| + 6 = 4$

65. $|x + 6| - 10 = -3$ 66. $|7 - x| + 3 = -8$ 67. $|x - 4| + 11 = 22$ 68. $|x - 1| - 10 = -2$

69. $|x + 8| + 2 = 2$ 70. $|x + 9| - 6 = 0$ 71. $|5 - x| + 3 = 7$ 72. $|x + 5| + 3 = -2$

73. $|9 - x| + 3 = 10$ 74. $|x + 6| + 1 = -8$ 75. $|x - 8| - 2 = 3$ 76. $|5x + 3| + 1 = 8$

77. $|5x - 9| - 2 = -6$ 78. $|5 - 4x| - 1 = 7$ 79. $|3 - 2x| + 4 = 9$ 80. $|4x - 1| + 2 = 9$

81. $|6 - 4x| - 3 = 0$ 82. $|5x - 6| - 8 = -2$ 83. $|2x + 5| + 7 = 13$ 84. $|4x + 3| + 6 = -1$

85. $|9x - 3| + 8 = 2$ 86. $|3x - 1| - 2 = 9$

87. $|5x + 2| + 3 = 11$ 88. $|\frac{1}{2}x + 2| + 3 = 5$

89. $|\frac{1}{3}x + 1| + 2 = 6$ 90. $|\frac{2}{3}x - 4| - 1 = 7$

91. $|\frac{3}{4}x - 2| - 2 = 3$ 92. $|\frac{2}{3}x - \frac{1}{2}| - \frac{1}{6} = \frac{1}{3}$

93. $|\frac{3}{2}x - \frac{1}{4}| - \frac{1}{8} = \frac{3}{8}$ 94. $|\frac{1}{2}x + \frac{4}{3}| - \frac{1}{6} = \frac{1}{3}$

95. $|\frac{3}{2}x + \frac{1}{4}| - \frac{3}{8} = \frac{5}{8}$ 96. $|\frac{3}{4}x - \frac{1}{2}| + \frac{1}{2} = \frac{1}{4}$

97. $|\frac{2}{3}x - \frac{1}{6}| + \frac{5}{6} = \frac{1}{3}$

2.6 Literal Equations and Applications

To determine the area of a rectangle we use the formula $A = LW$. In this formula, A represents the area, L represents the length, and W represents the width of the rectangle. This formula gives us a way of determining the area of a rectangle when its length and width are known.

However, suppose the area and length are known and we want to determine the width. We know that

$$A = LW$$

$$\frac{A}{L} = \frac{LW}{L} \qquad \text{Using the multiplicative property of equality}$$

$$\frac{A}{L} = W$$

We now have a formula for the width. We say that we have **solved the equation for the variable** W because we have isolated W on one side of the equation.

Applying this formula, suppose that the area is 144 square feet and the length is 24 feet and we want to determine the width. We now know

$$W = \frac{A}{L}$$

$$W = \frac{144}{24}$$

$$W = 6 \text{ feet}$$

You probably have already worked with many formulas. Some common formulas are listed below.

1. $C = 2\pi r$ circumference of a circle

2. $P = 2L + 2W$ perimeter of a rectangle

3. $I = PRT$ simple interest formula

4. $A = \frac{1}{2}BH$ area of a triangle

5. $P = 4s$ perimeter of a square

6. $D = RT$ distance formula

Equations that contain more than one letter are called **literal equations**, or formulas. All of the preceding equations are examples of literal equations.

EXAMPLE 2.6.1

Solve the following literal equations for the indicated variable.

1. $D = RT$ for T

$$\frac{D}{R} = \frac{RT}{R}$$

$$\frac{D}{R} = T$$

2. $P = 2L + 2W$ for L

$$P - 2W = 2L + 2W - 2W$$

$$P - 2W = 2L$$

$$\frac{P - 2W}{2} = \frac{2L}{2}$$

$$\frac{P - 2W}{2} = L$$

3. The formula $C = \frac{5}{9}(F - 32)$ is used to convert degrees Fahrenheit to degrees Celsius. Let's solve for F:

$$C = \frac{5}{9}(F - 32)$$

$$C = \frac{5}{9}F - \frac{160}{9}$$

$$C + \frac{160}{9} = \frac{5}{9}F - \frac{160}{9} + \frac{160}{9}$$

$$C + \frac{160}{9} = \frac{5}{9}F$$

$$\frac{9}{5}(C + \frac{160}{9}) = \frac{9}{5} \cdot \frac{5}{9}F$$

$$\frac{9}{5}C + 32 = F$$

NOTE ▶ We now have a formula for converting degrees Celsius to degrees Fahrenheit.

EXAMPLE 2.6.2

Use the results of part 2 of Example 2.6.1 to find L when $P = 46$ and $W = 9$.

$$L = \frac{P - 2W}{2}$$ From part 2 of Example 2.6.1

$$= \frac{46 - 2(9)}{2}$$

$$= \frac{46 - 18}{2}$$

$$= \frac{28}{2}$$

$$= 14$$

EXAMPLE 2.6.3

The formula for approximating kilometers from miles is $K = 1.609M$, where K represents the number of kilometers and M represents the number of miles. Beth Engel is going to run in the Moss County "Trick or Treat" Halloween 50K run. Approximately how many miles will Beth run?

$$1.609M = K$$

$$\frac{1.609M}{1.609} = \frac{K}{1.609} \qquad \text{Solve for } M.$$

$$M = \frac{K}{1.609}$$

$$M = \frac{50}{1.609} \qquad \text{Substitute 50 for } K.$$

$$M \doteq 31.1$$

Beth will run approximately 31.1 miles.

EXERCISES 2.6

Solve the following literal equations for the indicated variable.

1. $A = LW$ for L
2. $D = RT$ for R
3. $F = ma$ for a
4. $W = Fs$ for F
5. $C = 2\pi r$ for r
6. $I = PRT$ for P
7. $I = PRT$ for T
8. $V = LWH$ for L
9. $V = LWH$ for H
10. $V = \pi r^2 h$ for h
11. $A = \frac{1}{2}BH$ for B
12. $A = \frac{1}{2}BH$ for H
13. $F = GmM/r^2$ for M
14. $F = GmM/r^2$ for m
15. $F = Mv^2/r$ for M
16. $PV = nRT$ for R
17. $PV = nRT$ for T
18. $y = mx + b$ for x
19. $y = mx + b$ for m
20. $P = 2L + 2W$ for W
21. $S = 2\pi r^2 + 2\pi rh$ for h
22. $A = \pi rs + \pi r^2$ for s
23. $3x + 2y = 7$ for y
24. $5x + 3y = 9$ for y
25. $2x - 7y = 12$ for y
26. $2x - 9y = 15$ for y
27. $x = (y + z)/2$ for y
28. $x = (y + z)/2$ for z
29. $a = (b - c)/3$ for b
30. $a = (b - c)/3$ for c
31. $t = (2s - 5g)/7$ for g
32. $t = (2s - 5g)/7$ for s
33. $S = \pi s(r + R)$ for R
34. $S = \pi s(r + R)$ for r
35. $A = \frac{1}{2}(b + B)h$ for b
36. $A = \frac{1}{2}(b + B)h$ for B
37. $V = r + at$ for a
38. $V^2 = r^2 + 2as$ for a
39. $V^2 = r^2 + 2as$ for s
40. $s = rt + \frac{1}{2}at^2$ for a

Solve the following literal equations for the indicated variable, and then substitute the given values to find the variable's value.

41. $D = RT$ Find R when $D = 135$ and $T = 3$.   **42.** $D = RT$ Find T when $D = 385$ and $R = 55$.

43. $W = Fs$ Find s when $W = 20$ and $F = 15$.   **44.** $W = Fs$ Find F when $W = 64$ and $s = 80$.

45. $A = \frac{1}{2}BH$ Find B when $A = 84$ and $H = 7$.   **46.** $A = \frac{1}{2}BH$ Find H when $A = 80$ and $B = 32$.

47. $V = LWH$ Find L when $V = 60$, $W = 3$, and $H = 4$.

48. $V = LWH$ Find H when $V = 84$, $L = 7$, and $W = 2$.

49. $P = 2L + 2W$ Find L when $P = 36$ and $W = 5$.

50. $P = 2L + 2W$ Find W when $P = 42$ and $L = 15$.

51. $y = mx + b$ Find x when $y = -10$, $m = 3$, and $b = 2$.

52. $y = mx + b$ Find m when $y = -13$, $x = -3$, and $b = 2$.

53. The boiling point of water is 100° Celsius. What is the boiling point of water in degrees Fahrenheit? (*Hint:* Use the results of part 3 of Example 2.6.1.)

54. The freezing point of water is 0° Celsius. What is the freezing point of water in degrees Fahrenheit? (*Hint:* Use the results of part 3 of Example 2.6.1.)

55. Hooke's law states that $F = KX$. F represents the force on a spring, X represents the distance the spring is stretched, and K is some undetermined number. Find K when $F = 84$ and $X = 36$.

56. Hooke's law states that $F = KX$. (See problem 55.) Find K when $F = 112$ and $X = 128$.

57. The formula for the area of a trapezoid is $A = \frac{1}{2}(b + B)h$. Find b when $B = 14$, $h = 9$, and $A = 99$.

58. The formula for the area of a trapezoid is $A = \frac{1}{2}(b + B)h$. Find B when $b = 12$, $h = 7$, and $A = 98$.

59. The formula for the surface area of a rectangular parallelepiped (that is, box) is $S = 2WL + 2LH + 2WH$. Find W when $S = 228$, $L = 6$, and $H = 9$.

60. The formula for the surface area of a rectangular parallelepiped (that is, box) is $S = 2WL + 2LH + 2WH$. Find L when $S = 208$, $W = 8$, and $H = 4$.

2.7 Applications of Linear Equations

Whenever the topic of word or application problems is introduced in an algebra class, it is most unusual to hear the class say, "Hooray, we are finally at our favorite topic—word problems!" Unfortunately, word problems usually intimidate and puzzle many students. In this section we will show how to convert words and phrases into algebraic expressions. In addition, a sequence of steps will be given that will enable us to solve most word problems.

Some commonly used words and phrases are

- sum—indicates addition

- difference—indicates subtraction

- product—indicates multiplication
- quotient—indicates division
- is—indicates equality
- is the same as—indicates equality

Listed below are examples of some phrases you might encounter. To the right of the phrase is its algebraic equivalent.

Phrase	*Algebraic Equivalent*
An unknown number	x
4 more than a number	$x + 4$
A number increased by 7	$x + 7$
The sum of a number and 2	$x + 2$
A number plus 11	$x + 11$
A number minus 9	$x - 9$
8 minus a number	$8 - x$
The difference of a number and 12	$x - 12$
A number decreased by 4	$x - 4$
2 less than a number	$x - 2$
3 times a number	$3x$
Twice a number	$2x$
The product of a number and 5	$5x$
$\frac{2}{3}$ of a number ("of" indicates multiplication)	$\frac{2}{3}x$
A number divided by 7	$\frac{x}{7}$
10 divided by a number	$\frac{10}{x}$
The quotient of a number and 3	$\frac{x}{3}$
3 times a number plus 4	$3x + 4$
Twice a number minus 7	$2x - 7$
9 minus 5 times a number	$9 - 5x$
Twice the difference of a number and 3	$2(x - 3)$
4 times the sum of a number and 1	$4(x + 1)$
The sum of a number and 2 all divided by 9	$(x + 2)/9$
The difference of a number and 6 all divided by 8	$(x - 6)/8$

The following steps can be used to find the solutions of most word problems.

---

1. Determine what you are trying to find, and let a variable represent that quantity. If the problem asks for more than one quantity, there must be something within the problem that tells how to represent all the different quantities in terms of one variable.

2. Draw a picture if appropriate.

3. Write an equation involving the variable, using the information given in the problem. (This is usually the most difficult step.)

4. Solve the equation found in step 3.

5. Check your solution. Go back to the statement of the problem and make sure your answer is reasonable.

In this section we will investigate three different types of word problems. The first group we will examine is number problems.

Number Problems

EXAMPLE 2.7.1

Twice a number minus 7 is 5. What is the number?

SOLUTION

Let x = the unknown number.   Step 1

Then $2x - 7$ represents twice the unknown number minus 7. So

$$2x - 7 = 5 \qquad \text{Step 3}$$
$$2x - 7 + 7 = 5 + 7 \qquad \text{Step 4}$$
$$2x = 12$$
$$\frac{2x}{2} = \frac{12}{2}$$
$$x = 6$$

Thus, the number that satisfies the original problem is 6. To check, reread the problem and observe that

twice 6 minus 7 is 5, that is,   Step 5

$$2 \cdot 6 - 7 = 5$$
$$12 - 7 = 5$$
$$5 = 5 \checkmark$$

EXAMPLE 2.7.2

Find three consecutive integers whose sum is 27.

SOLUTION

Consecutive integers are integers that succeed one another in counting order. Examples of consecutive integers are 32, 33, 34; 19, 20, 21; and -13, -12, -11. We let

x = first integer

$x + 1$ = second integer

$x + 2$ = third integer

Then

$$x + (x + 1) + (x + 2) = 27$$
$$x + x + 1 + x + 2 = 27$$
$$3x + 3 = 27$$
$$3x = 24$$
$$x = 8$$

The three consecutive integers are 8, 9, and 10. Note that

$$8 + 9 + 10 \stackrel{?}{=} 27$$
$$27 = 27 \checkmark$$

Geometric Problems

EXAMPLE 2.7.3

A 65-in. board is cut into three pieces. The second piece is twice as long as the first piece. The third piece is 9 in. longer than the first piece. How long is each piece?

SOLUTION

In geometric problems it is a good idea to draw a picture to represent the problem (see Figure 2.7.1).

Since the lengths of the second and third pieces are described in terms of the first piece, we let

x = length of first piece

$2x$ = length of second piece

$x + 9$ = length of third piece

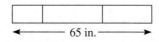

65 in.

Figure 2.7.1

Now from Figure 2.7.1 we can observe that

(Length of   (Length of   (Length of
first piece) + second piece) + third piece) = 65

$$(x) \quad + \quad (2x) \quad + \quad (x + 9) \quad = 65$$

$$x + 2x + x + 9 = 65$$

$$4x + 9 = 65$$

$$4x = 56$$

$$x = 14$$

So $2x = 2 \cdot 14 = 28$

and $x + 9 = 14 + 9 = 23$

Thus, the lengths of the three pieces are 14 in., 28 in., and 23 in. The check is left to the reader.

EXAMPLE 2.7.4

The perimeter of a rectangle is 46 ft. The length of the rectangle is 5 ft less than three times the width. What are the dimensions of the rectangle?

SOLUTION

Again it is a good idea to draw a picture to represent the problem (see Figure 2.7.2).

Since the length is described in terms of the width, we let

x = width of rectangle

$3x - 5$ = length of rectangle

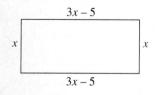

Figure 2.7.2

Now we can form an equation using the fact that the perimeter is the sum of the measures of the sides:

$$(x) + (3x - 5) + (x) + (3x - 5) = 46$$

$$x + 3x - 5 + x + 3x - 5 = 46$$

$$8x - 10 = 46$$

$$8x = 56$$

$$x = 7$$

The width is 7 ft. Then

$$\text{length} = 3x - 5$$

$$= 3(7) - 5$$

$$= 21 - 5$$

$$= 16$$

Thus, we have a 7 ft by 16 ft rectangle.

Money Problems

EXAMPLE 2.7.5

Syed has $181 in ones, fives, and tens. He has two more $10 bills than $5 bills. The number of $1 bills is one less than three times the number of $5 bills. How many of each type of bill does he have?

SOLUTION

Since the number of $1 bills and $10 bills is described in terms of the number of $5 bills, we let

$$x = \text{number of \$5 bills}$$
$$x + 2 = \text{number of \$10 bills}$$
$$3x - 1 = \text{number of \$1 bills}$$

NOTE ▶

If you had eight $5 bills, you would have $40. The way we obtained $40 was by multiplying 8 by 5 (the number of bills times the value of each bill). Repeating this process for each type of bill, we obtain the information in the following chart.

	$5	$10	$1	Total
Number of bills	x	$x + 2$	$3x - 1$	
Value (in dollars)	$5 \cdot x$	$10 \cdot (x + 2)$	$1 \cdot (3x - 1)$	181

When filling in the box, don't confuse the *number* of bills with the *value* of the bills. The equation is generated by the bottom row:

$$\begin{pmatrix} \text{Value in} \\ \text{\$5 bills} \end{pmatrix} + \begin{pmatrix} \text{Value in} \\ \text{\$10 bills} \end{pmatrix} + \begin{pmatrix} \text{Value in} \\ \text{\$1 bills} \end{pmatrix} = \text{Total value}$$

$$5x + 10(x + 2) + 1(3x - 1) = 181$$
$$5x + 10x + 20 + 3x - 1 = 181$$
$$18x + 19 = 181$$
$$18x = 162$$
$$x = 9$$

From earlier, $x + 2 = 9 + 2 = 11$ and $3x - 1 = 3 \cdot 9 - 1 = 27 - 1 = 26$.

Checking: 9 $5 bills = 9 · $5 = $ 45
11 $10 bills = 11 · $10 = $110
26 $1 bills = 26 · $1 = $ 26
$181

Thus, Syed has 9 $5 bills, 11 $10 bills, and 26 $1 bills.

EXAMPLE 2.7.6

Mabeline has $3.35 in quarters and dimes. She has a total of 23 coins. How many quarters and how many dimes does she have?

NOTE ▶

If she had 10 quarters, she would have 13 dimes. The way we obtained 13 dimes was by subtracting 10 from 23 (the total number of coins minus the number of quarters).

SOLUTION

Since neither coin is described in terms of the other, we can let x be either the number of quarters or the number of dimes. Let

$$x = \text{number of quarters}$$
Then $23 - x = \text{number of dimes}$

because $\dfrac{(\text{Number}}{\text{of quarters)}} + \dfrac{(\text{Number}}{\text{of dimes)}} = \dfrac{(\text{Total number}}{\text{of coins)}}$

$$x + (\text{Number of dimes}) = 23$$
$$\text{Number of dimes} = 23 - x$$

We now generate a chart similar to the one in the previous example.

	25¢	10¢	Total
Number of coins	x	$23 - x$	23
Value (in cents)	$25 \cdot x$	$10 \cdot (23 - x)$	335

Again the equation is generated by the bottom row:

$$25x + 10(23 - x) = 335$$
$$25x + 230 - 10x = 335$$
$$15x + 230 = 335$$
$$15x = 105$$
$$x = 7$$

From earlier, $23 - x = 23 - 7 = 16$. Thus, Mabeline has 7 quarters and 16 dimes. The check is left to the reader.

EXERCISES 2.7

In problems 1–20, convert each phrase to an algebraic expression. Let x be the variable.

1. A number increased by 2

2. 9 more than a number

3. 4 times a number

4. $\frac{3}{4}$ of a number

5. The difference of a number and 13

6. A number divided by 16

7. The quotient of a number and 4

8. A number plus 6

9. The sum of a number and 5

10. A number decreased by 1

11. $\frac{1}{2}$ of a number

12. The product of a number and 7

13. 12 divided by a number

14. A number minus 4

15. 2 more than 3 times a number

16. 6 times a number minus 8

17. 4 minus 7 times a number

18. 8 times a number plus 1

19. 5 times the sum of a number and 9

20. 4 times the difference of a number and 2

Use equations to find the solutions of the following problems.

21. Three times a number plus 1 is 19. What is the number?

22. Four times a number minus 2 is 34. What is the number?

23. Eight times a number minus 1 is 3. What is the number?

24. Nine times a number plus 2 is 5. What is the number?

25. Four minus 3 times a number is 10. What is the number?

26. Five minus 2 times a number is 13. What is the number?

27. Find three consecutive integers whose sum is 39.

28. Find three consecutive integers whose sum is 54.

29. Find three consecutive integers whose sum is -63.

30. Find three consecutive integers whose sum is -75.

31. Find three consecutive odd integers whose sum is 93. (*Hint:* If x is an odd integer, the next odd integer is $x + 2$.)

32. Find three consecutive odd integers whose sum is 81.

33. Find three consecutive even integers whose sum is -30. (*Hint:* If x is an even integer, the next even integer is $x + 2$.)

34. Find three consecutive even integers whose sum is -36.

35. A 64-in. board is cut into three pieces. The second piece is three times as long as the first piece. The third piece is 4 in. longer than the first piece. How long is each piece?

36. A 51-in. board is cut into three pieces. The second piece is four times as long as the first piece. The third piece is 3 in. longer than the first piece. How long is each piece?

37. A 37-in. board is cut into three pieces. The second piece is 3 in. longer than the first piece. The third piece is 4 in. longer than the second piece. How long is each piece?

38. A 53-in. board is cut into three pieces. The second piece is 2 in. longer than the first piece. The third piece is 7 in. longer than the second piece. How long is each piece?

39. A 98-cm board is cut into three pieces. The first piece is twice as long as the third piece. The second piece is one-half as long as the third piece. How long is each piece?

40. A 40-cm board is cut into three pieces. The first piece is twice as long as the third piece. The second piece is one-third as long as the third piece. How long is each piece?

41. The perimeter of a rectangle is 74 ft. The length of the rectangle is 7 ft more than the width. What are the dimensions of the rectangle?

42. The perimeter of a rectangle is 80 ft. The length of the rectangle is 6 ft more than the width. What are the dimensions of the rectangle?

43. The perimeter of a rectangle is 80 ft. The length is 4 ft more than 3 times the width. What are the dimensions of the rectangle?

44. The perimeter of a rectangle is 58 ft. The length is 5 ft more than 2 times the width. What are the dimensions of the rectangle?

45. The perimeter of a rectangle is 64 ft. The length is 7 ft less than 2 times the width. What are the dimensions of the rectangle?

46. The perimeter of a rectangle is 100 ft. The length is 6 ft less than 3 times the width. What are the dimensions of the rectangle?

47. Eric has $120 in fives and tens. The number of fives is 1 less than 3 times the number of tens. How many of each type of bill does he have?

48. Erin has $80 in ones and fives. The number of ones is 3 more than twice the number of fives. How many of each type of bill does she have?

49. Nathan has $4.55 in nickels and quarters. He has twice as many nickels as quarters. How many of each type of coin does he have?

50. Katy has $5.70 in pennies and nickels. The number of nickels is 2 more than 3 times the number of pennies. How many of each type of coin does she have?

51. Roena has $166 in ones, fives, and tens. She has 4 more $1 bills than $10 bills. The number of $5 bills is 3 more than twice the number of $10 bills. How many of each type of bill does she have?

52. Margie has $224 in ones, fives, and tens. She has 1 more $10 bill than $1 bills. The number of $5 bills is 4 less than 3 times the number of $1 bills. How many of each type of bill does she have?

53. W. D. has $5.50 in nickels, dimes, and quarters. The number of dimes is 3 less than the number of nickels. The number of quarters is 7 greater than the number of dimes. How many of each type of coin does he have?

54. Theresa has $3.65 in pennies, nickels, and dimes. The number of nickels is 2 less than the number of dimes. The number of pennies is 9 greater than the number of nickels. How many of each type of coin does she have?

55. B. K. has $200 in fives, tens, and twenties. The number of $20 bills is one-third the number of $5 bills. The number of $10 bills is 1 greater than twice the number of $5 bills. How many of each type of bill does she have?

56. Carlos has $520 in fives, tens, and twenties. The number of $5 bills is one-third the number of $20 bills. The number of $10 bills is 2 greater than twice the number of $20 bills. How many of each type of bill does he have?

57. Karla has $4.60 in nickels and quarters. She has a total of 36 coins. How many nickels and how many quarters does she have?

58. Lisa has $3.90 in nickels and dimes. She has a total of 42 coins. How many nickels and how many dimes does she have?

59. Kim has $210 in $5 bills and $20 bills. She has a total of 24 bills. How many $5 bills and how many $20 bills does she have?

60. Marcus has $300 in $5 bills and $10 bills. He has a total of 36 bills. How many $5 bills and how many $10 bills does he have?

61. Huong has $215 in ones, fives, and tens. The number of $1 bills is 1 more than 3 times the number of $5 bills. He has a total of 48 bills. How many of each type of bill does he have?

62. Ramon has $225 in ones, fives, and tens. The number of $1 bills is 2 less than 3 times the number of $10 bills. He has a total of 56 bills. How many of each type of bill does he have?

2.8 More Applications of Linear Equations

In this section we will consider three more types of application problems. The first group is called motion problems. Motion problems use the formula $D = RT$, where D represents the distance an object travels, R is its rate or speed, and T is the amount of time it is traveling. For example, if you drive your car at a rate of 60 mph for 4 hours, then the distance you travel is

$$D = RT$$
$$= 60 \cdot 4$$
$$= 240 \text{ miles}$$

Motion Problems

EXAMPLE 2.8.1

Two bank robbers leave the First National Bank of Moss County in a car traveling north at a rate of 65 mph. At the same time the sheriff of Moss County, "Wrong-Way" Pearson, chases after them in a car traveling south at a rate of 55 mph. In how many hours will the sheriff and the robbers be 180 miles apart?

SOLUTION

Let x = number of hours each car travels. We will organize our information with a chart similar to the charts we developed in Section 2.7.

	Rate	·	Time	=	Distance
Robbers	65		x		$65x$
Sheriff	55		x		$55x$

NOTE ▶

We first fill in the rate and time columns and then determine the distance by using the formula $D = RT$.

$$\begin{array}{c}\text{(Distance the} \\ \text{robbers travel)}\end{array} + \begin{array}{c}\text{(Distance the} \\ \text{sheriff travels)}\end{array} = \begin{array}{c}\text{Total} \\ \text{distance}\end{array} \quad \text{(See Figure 2.8.1.)}$$

$$65x + 55x = 180$$

$$120x = 180$$

$$\frac{120x}{120} = \frac{180}{120}$$

$$x = 1.5$$

Figure 2.8.1

Thus, 1.5 hours after the robbery the sheriff and the robbers are 180 miles apart.

EXAMPLE 2.8.2

Bubba leaves the Moss County Rodeo traveling west at a rate of 30 mph. Two hours later his brother Hubba leaves the rodeo traveling in the same direction at a rate of 55 mph. How many hours will it take Hubba to catch his brother Bubba?

SOLUTION

Let x = number of hours Hubba travels

then $x + 2$ = number of hours Bubba travels
(since Bubba had a 2-hour head start)

	Rate	· Time	= Distance
Bubba	30	$x + 2$	$30(x + 2)$
Hubba	55	x	$55x$

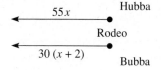

Figure 2.8.2

When Hubba catches Bubba, the distance Hubba traveled must equal the distance Bubba traveled.

(Distance Bubba traveled) = (Distance Hubba traveled)   (See Figure 2.8.2.)

$$30(x + 2) = 55x$$
$$30x + 60 = 55x$$
$$30x - 30x + 60 = 55x - 30x$$
$$60 = 25x$$
$$\frac{60}{25} = \frac{25x}{25}$$
$$2.4 = x$$

Thus, it takes Hubba 2.4 hours to catch Bubba.

The second group of problems we will investigate is called interest problems. Interest problems use the formula $I = PRT$. In this formula, I represents the amount of simple interest earned, P represents the principal (that is, the amount of money invested), R represents the rate of interest, and T stands for the time measured in years. For example, if you invest $1500 in a bond paying a rate of 8% for 3 years, then the amount of simple interest you earn is

$$I = P \cdot R \cdot T$$
$$= (\$1500)(8\%)(3)$$
$$= (1500)(0.08)(3)$$
$$= \$360$$

Interest Problems

EXAMPLE 2.8.3

After coming back from Las Vegas, "Aces" Perez decided to place his winnings in the Moss County Savings and Loan. He invested part of the winnings in a Christmas Club account paying 8% simple interest. He invested $3000 less than the amount he invested in the Christmas Club account in a money market account paying 6% simple interest. After 1 year his interest income was $520. How much did Aces invest in each account?

SOLUTION

Let x = amount invested in Christmas Club account

then $x - 3000$ = amount invested in money market account

Again we organize our information with a chart.

	Principal ·	Rate ·	Time =	Interest
Christmas Club	x	8% = 0.08	1	$0.08x$
Money market	$x - 3000$	6% = 0.06	1	$0.06(x - 3000)$

NOTE ▶ We first fill in the principal, rate, and time columns and then determine the interest by using the formula $I = PRT$.

$$
\begin{pmatrix} \text{Interest in} \\ \text{Christmas Club} \\ \text{account} \end{pmatrix} + \begin{pmatrix} \text{Interest in} \\ \text{money market} \\ \text{account} \end{pmatrix} = \begin{matrix} \text{Total} \\ \text{interest} \end{matrix}
$$

$$0.08x + 0.06(x - 3000) = 520$$
$$0.08x + 0.06x - 180 = 520$$
$$0.14x - 180 = 520$$
$$0.14x = 700$$
$$\frac{0.14x}{0.14} = \frac{700}{0.14}$$
$$x = 5000$$

So $x - 3000 = 5000 - 3000 = 2000$. Thus, Aces invested $5000 in the Christmas Club account and $2000 in the money market account.

EXAMPLE 2.8.4
───────────

Selma invested $8000 in the United Bank and Trust of Moss County. She invested part of her money in a lifeline account paying 9% simple interest and invested the rest of her money in a checking account paying 4% simple interest. After 1 year her interest income was $620. How much did Selma invest in each account?

SOLUTION

Let x = amount invested in lifeline account

then $8000 - x$ = amount invested in checking account

	Principal ·	Rate ·	Time =	Interest
Lifeline	x	9% = 0.09	1	$0.09x$
Checking	$8000 - x$	4% = 0.04	1	$0.04(8000 - x)$

$$\begin{array}{c}\text{(Interest in} \\ \text{lifeline account)}\end{array} + \begin{array}{c}\text{(Interest in} \\ \text{checking account)}\end{array} = \text{Total interest}$$

$$0.09x + 0.04(8000 - x) = 620$$
$$0.09x + 320 - 0.04x = 620$$
$$0.05x + 320 = 620$$
$$0.05x = 300$$
$$\frac{0.05x}{0.05} = \frac{300}{0.05}$$
$$x = 6000$$

So $8000 - x = 8000 - 6000 = 2000$. Thus, Selma invested $6000 in the lifeline account and $2000 in the checking account.

The final group of problems we will investigate is called mixture problems.

Mixture Problems

EXAMPLE 2.8.5

"Mr. Macho" aftershave contains 84% alcohol. "Try a Little Tenderness" aftershave contains 45% alcohol. How many ounces of each aftershave must be used to make 65 oz of aftershave that is 60% alcohol?

SOLUTION

Let x = amount of Mr. Macho

then $65 - x$ = amount of Try a Little Tenderness   (See Figure 2.8.3.)

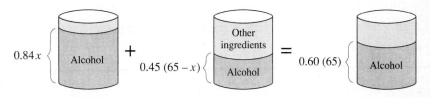

Figure 2.8.3

Again we organize our information with a chart.

	Amount ·	Percent alcohol =	Amount of alcohol
Mr. Macho	x	$84\% = 0.84$	$0.84x$
Try a Little Tenderness	$65 - x$	$45\% = 0.45$	$0.45(65 - x)$

NOTE ▶ We first fill in the amount and percent alcohol columns. The amount of alcohol is formed by multiplying the amount by the percent column.

$$\begin{matrix} \text{(Amount of alcohol} \\ \text{in Mr. Macho)} \end{matrix} + \begin{matrix} \text{(Amount of alcohol in} \\ \text{Try a Little Tenderness)} \end{matrix} = \begin{matrix} \text{Total amount} \\ \text{of alcohol} \end{matrix}$$

$$0.84x + 0.45(65 - x) = 0.60(65)$$
$$0.84x + 29.25 - 0.45x = 39$$
$$0.39x + 29.25 = 39$$
$$0.39x = 9.75$$
$$\frac{0.39x}{0.39} = \frac{9.75}{0.39}$$
$$x = 25$$

So $65 - x = 65 - 25 = 40$. Thus, it takes 25 oz of Mr. Macho and 40 oz of Try a Little Tenderness.

EXAMPLE 2.8.6

Aunt Nancy's potato salad is selling for $1.75 a pint, and Uncle John's potato salad is selling for $1.20 a pint. How many pints of each type of potato salad must be purchased to create 110 pints of potato salad that will sell for $1.60 a pint?

SOLUTION

Let x = amount of Aunt Nancy's potato salad
then $110 - x$ = amount of Uncle John's potato salad

	Amount	Cost per pint	Cost
Aunt Nancy's	x	$1.75	$1.75x$
Uncle John's	$110 - x$	$1.20	$1.20(110 - x)$

$$\begin{pmatrix}\text{Cost of Aunt Nancy's}\\\text{potato salad}\end{pmatrix} + \begin{pmatrix}\text{Cost of Uncle John's}\\\text{potato salad}\end{pmatrix} = \text{Total cost}$$

$$1.75x + 1.20(110 - x) = 1.60(110)$$

$$1.75x + 132 - 1.20x = 176$$

$$0.55x + 132 = 176$$

$$0.55x = 44$$

$$\frac{0.55x}{0.55} = \frac{44}{0.55}$$

$$x = 80$$

So $110 - x = 110 - 80 = 30$. Thus, it takes 80 pints of Aunt Nancy's potato salad and 30 pints of Uncle John's potato salad.

EXERCISES 2.8

Use equations to find the solutions of the following problems.

1. Debbie leaves school in a car traveling north at a rate of 45 mph. At the same time, Jim leaves school in a car traveling south at a rate of 55 mph. In how many hours will Debbie and Jim be 150 miles apart?

2. Ivy leaves school on a bicycle traveling west at a rate of 18 mph. At the same time, Beth leaves school on a bicycle traveling east at a rate of 22 mph. In how many hours will Ivy and Beth be 100 miles apart?

3. A train leaves San Francisco at noon bound for Los Angeles at a rate of 30 mph. At the same time, another train leaves Los Angeles bound for San Francisco (on the same route) at a rate of 35 mph. The distance between San Francisco and Los Angeles is 325 miles. At what time will the two trains meet?

4. A train leaves Atlanta at noon bound for Chicago at a rate of 45 mph. At the same time, another train leaves Chicago bound for Atlanta (on the same route) at a rate of 55 mph. The distance between Atlanta and Chicago is 600 miles. At what time will the two trains meet?

5. Roscoe leaves Jim Wells County traveling southwest at a rate of 35 mph. Three hours later Junior leaves Jim Wells County traveling in the same direction at a rate of 60 mph. How many hours will it take Junior to catch up to Roscoe?

6. Rich leaves The Mall of Moss County traveling northeast at a rate of 25 mph. Three hours later Vicki leaves The Mall of Moss County traveling in the same direction at a rate of 45 mph. How many hours will it take Vicki to catch up to Rich?

7. Bob leaves the bowling alley on a bicycle traveling west at a rate of 14 mph. Four hours later Judy leaves the bowling alley in a car traveling in the same direction at a rate of 54 mph. How many hours will it take Judy to catch up to Bob?

8. Lee leaves the amusement park on a bicycle traveling south at a rate of 12 mph. Four hours later Dora leaves the amusement park in a car traveling in the same direction at a rate of 52 mph. How many hours will it take Dora to catch up to Lee?

9. Mike leaves the bookstore at noon traveling north at a rate of 42 mph. Two hours later Sam leaves the bookstore traveling south at a rate of 54 mph. At what time will Mike and Sam be 276 miles apart?

10. Rose Mary leaves the Moss County International Karate Tournament at noon traveling north at a rate of 38 mph. Two hours later Melissa leaves the tournament traveling south at a rate of 56 mph. At what time will Rose Mary and Melissa be 170 miles apart?

11. Charles leaves the Moss County Chili Contest at noon traveling west at a rate of 32 mph. One hour later Bernadette leaves the contest traveling east at a rate of 50 mph. At what time will Charles and Bernadette be 401 miles apart?

12. Elmer leaves the Moss County Tractor Show at noon traveling south at a rate of 44 mph. One hour later Jolene leaves the show traveling north at a rate of 62 mph. At what time will Elmer and Jolene be 203 miles apart?

13. Siria invested part of her savings in a Moss County public utility bond paying 11% simple interest. She invested $2000 less than the amount she invested in the bond in a passbook savings account paying 6% simple interest. After 1 year her interest income was $730. How much did Siria invest in the bond and in the savings account?

14. Hamid invested part of his savings in a Moss County school bond paying 9% simple interest. He invested $5000 less than the amount he invested in the bond in a checking account paying 4% simple interest. After 1 year his interest income was $710. How much did Hamid invest in the bond and in the checking account?

15. Joe invested part of his savings in a retirement account paying 5% simple interest. He invested four times the amount he invested in the retirement account in a bond paying 12% simple interest. After 1 year his interest income was $1060. How much did Joe invest in the retirement account and in the bond?

16. Julie invested part of her savings in a checking account paying 3% simple interest. She invested three times the amount she invested in the checking account in a bond paying 10% simple interest. After 1 year her interest income was $1320. How much did Julie invest in the checking account and in the bond?

17. Juan invested part of his savings in a certificate of deposit paying 7% simple interest. He invested $1000 less than the amount he invested in the certificate of deposit in a money market account paying 4% simple interest. After 2 years his interest income was $1240. How much did Juan invest in the certificate of deposit and in the money market account?

18. Louisa invested part of her savings in a certificate of deposit paying 10% simple interest. She invested $3000 less than the amount she invested in the certificate of deposit in a checking account paying 5% simple interest. After 2 years her interest income was $2400. How much did Louisa invest in the certificate of deposit and in the checking account?

19. Maggie invested $7000 in the Citizens Bank of Moss County. She invested part of her money in a checking account paying 5% simple interest and invested the rest of her money in a savings account paying 8% simple interest. After 1 year her interest income was $500. How much did Maggie invest in each account?

20. Terry invested $10,000 in the Moss County Bank of Commerce. He invested part of his money in a money market account paying 9% simple interest and invested the rest of his money in a checking account paying 4% simple interest. After 1 year his interest income was $700. How much did Terry invest in each account?

21. Hubert invested $12,000 in Moss County Savings and Loan. He invested part of his money in a retirement account paying 9% simple interest and invested the rest of his money in a lifeline account paying 5% simple interest. After 1 year his interest income was $960. How much did Hubert invest in each account?

22. Leona invested $13,000 in Moss County Savings and Loan. She invested part of her money in a money market account paying 7% simple interest and invested the rest of her money in a checking account paying 4% simple interest. After 1 year her interest income was $820. How much did Leona invest in each account?

23. Felicia invested $8000 in the Allied Bank of Moss County. She invested part of her money in a passbook account paying 4% simple interest and invested the rest of her money in a retirement account paying 9% simple interest. After 2 years her interest income was $840. How much did Felicia invest in each account?

24. Marvin invested $11,000 in the Allied Bank of Moss County. He invested part of his money in a money market account paying 7% simple interest and invested the rest of his money in a lifeline account paying 5% simple interest. After 2 years his interest income was $1220. How much did Marvin invest in each account?

25. Big E toothpaste contains 45% fluoride. Big M toothpaste contains 36% fluoride. How many ounces of each toothpaste must be used to make 18 oz of toothpaste that contains 42% fluoride?

26. Big C toothpaste contains 63% fluoride. Big K toothpaste contains 27% fluoride. How many ounces of each toothpaste must be used to make 12 oz of toothpaste that contains 57% fluoride?

27. "Bald Away" shampoo contains 46% alcohol. "Old Skin Head" shampoo contains 69% alcohol. How many ounces of each shampoo must be used to make 23 oz of shampoo that contains 55% alcohol?

28. "Soft and Smooth" shampoo contains 28% alcohol. "Rough and Tough" shampoo contains 42% alcohol. How many ounces of each shampoo must be used to make 14 oz of shampoo that contains 39% alcohol?

29. "Sweetie" low-calorie sweetener contains 32% dextrose. "This Might Be Nectar" low-calorie sweetener contains 48% dextrose. How many ounces of each sweetener must be used to make 16 oz of sweetener that is 42% dextrose?

30. "Sweet-Tooth" low-calorie sweetener contains 24% dextrose. "Honey-Dew" low-calorie sweetener contains 60% dextrose. How many ounces of each sweetener must be used to make 36 oz of sweetener that is 46% dextrose?

31. Bobo's strawberry preserves are selling for $1.20 a pint. Red's strawberry preserves are selling for $1.80 a pint. How many pints of each type must be purchased to create 24 pints of strawberry preserves that will sell for $1.55 a pint?

32. Bobo's orange marmalade is selling for $2.40 a pint. Red's orange marmalade is selling for $1.60 a pint. How many pints of each type must be purchased to create 32 pints of orange marmalade that will sell for $1.95 a pint?

33. Texas peanuts are selling for $2.80 a pound, and Georgia peanuts are selling for $2.10 a pound on the New York Peanut Exchange. How many pounds of each type of peanut must be purchased to create 14 pounds of peanuts that will sell for $2.50 a pound?

34. California peanuts are selling for $2.40 a pound, and Moss County peanuts are selling for $0.80 a pound on the New York Peanut Exchange. How many pounds of each type of peanut must be purchased to create 16 pounds of peanuts that will sell for $1.90 a pound?

35. West Texas heating oil is selling for $0.72 a gallon, and Moss County heating oil is selling for $0.54 a gallon. How many gallons of each type of heating oil must be purchased to create 18 gallons of heating oil that will sell for $0.67 a gallon?

36. Pennsylvania heating oil is selling for $0.90 a gallon, and East Asian heating oil is selling for $0.60 a gallon. How many gallons of each type of heating oil must be purchased to create 30 gallons of heating oil that will sell for $0.81 a gallon?

2.9 Linear Inequalities

An **inequality** is an algebraic way of stating that one quantity is less than another. Some examples of inequalities are

$$x^2 + y^2 < 9, \quad 3x - 5y > 2, \quad x \leq 0, \quad 4x + 7 \geq 3$$

In this section we will examine only linear inequalities in one variable.

Linear Inequality

An inequality that can be written in the form

$$ax + b < c \quad \text{or} \quad ax + b \leq c$$

where a, b, and c are real numbers with $a \neq 0$ is called a **linear inequality** in the variable x.

REMARK   The following are linear inequalities in the variable x:

$$x + 2 < 7 \qquad \text{read } x + 2 \text{ is less than } 7$$
$$3x \geq 4 \qquad \text{read } 3x \text{ is greater than or equal to } 4$$

The **solution set of an inequality** is the set of all numbers that yield a true statement when substituted for the variable in the inequality. For example, the inequality $x \geq 5$ has solutions of 5, $6\frac{1}{2}$, 9.8, and 130. In fact, there are an infinite number of solutions to this inequality. We usually represent them by a shaded region on a number line. For example, $x \geq 5$ is represented by

which is called the **graph** of the solution set. This graph says that x can be any number to the right of, and including, 5. The solid circle at 5 indicates that 5 is a solution of the inequality.

REMARK   The point at 5 is called a *boundary point*. The boundary point is determined by replacing the inequality symbol with an equals sign.

The inequality $x < 2$ is represented by

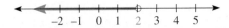

This graph says that x can be any number to the left of 2. The open circle at 2 indicates that 2 is not a solution of the inequality.

Sometimes two inequalities are used to define a solution set. For example, the inequality $-1 < x < 3$ says that x is any number that is both greater than

−1 and less than 3. This is the set of all numbers *between* −1 and 3. The solution set of the inequality −1 < x < 3 is illustrated by

This graph is a "bounded segment" with endpoints at −1 and 3. It says that x can be any number between −1 and 3.

Try to avoid this mistake:

Incorrect	Correct
3 < x < −1	−1 < x < 3
This inequality says that x is both greater than 3 and less than −1, which is impossible.	The only other way this inequality can be written is 3 > x > −1 Both of these inequalities indicate that x is between −1 and 3. In both cases the inequality symbols point to the smaller number.

In finding the solutions of linear inequalities, we employ tools similar to those used in solving linear equations. The additive and multiplicative properties of inequalities are used to simplify an inequality to an **equivalent inequality**, that is, one with the same solution set. Consider the following inequalities.

We know

$$-4 < 7 \qquad \text{Add 5 to both sides.}$$
$$-4 + 5 \overset{?}{<} 7 + 5$$
$$1 < 12 \qquad \text{A true statement}$$

In a similar fashion,

$$-4 < 7 \qquad \text{Subtract 2 from both sides.}$$
$$-4 - 2 \overset{?}{<} 7 - 2$$
$$-6 < 5 \qquad \text{A true statement}$$

These two facts illustrate the following property.

Additive Property of Inequality

Let a, b, and c denote algebraic expressions.

If $a < b$ then $a + c < b + c$.

REMARK   This property states that we can add or subtract the same number on both sides of an inequality and obtain a new inequality that is equivalent to the original one.

EXAMPLE 2.9.1   Find the solutions of the following inequalities and graph the solution sets on a number line.

1. $x - 5 > -2$

$x - 5 + 5 > -2 + 5$ Use the additive property of inequality to isolate x.

$x > 3$

The solution set consists of all numbers x such that $x > 3$.

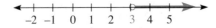

2. $x + \frac{3}{4} \leq \frac{5}{2}$

$x + \frac{3}{4} - \frac{3}{4} \leq \frac{5}{2} - \frac{3}{4}$ Use the additive property of inequality to isolate x.

$x \leq \frac{7}{4}$

The solution set consists of all numbers x such that $x \leq \frac{7}{4}$.

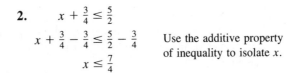

3. $-5 < x - 1 < -1$

Here we are simply solving two inequalities at the same time. The goal is to isolate x in the middle.

$-5 + 1 < x - 1 + 1 < -1 + 1$ Use the additive property of inequality to isolate x in the middle.

$-4 < x < 0$

The solution set consists of all numbers x such that $-4 < x < 0$.

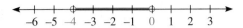

In solving linear equations, we often have to multiply both sides of the equation by some number. What would happen if we multiplied both sides of an inequality by some number? Again consider the following.

We know

$$-4 < 7 \qquad \text{Multiply both sides by 3.}$$
$$3(-4) \overset{?}{<} 3(7)$$
$$-12 < 21 \qquad \text{A true statement}$$

However,

$$-4 < 7 \qquad \text{Multiply both sides by } -3.$$
$$-3(-4) \overset{?}{<} -3(7)$$
$$12 > -21 \qquad \text{The inequality sign must}$$
$$\text{be reversed to make a true}$$
$$\text{statement.}$$

These two facts illustrate the following property.

Multiplicative Property of Inequality

Let a, b, and c denote algebraic expressions, with $c \neq 0$.

1. If $a < b$ and $c > 0$, then $ac < bc$.

2. If $a < b$ and $c < 0$, then $ac > bc$.

REMARK   Part 1 of this property says that we can multiply (or divide) both sides of an inequality by any positive number and obtain a new inequality that is equivalent to the original one. Part 2 says that if we *multiply (or divide) both sides by a negative number,* we must *reverse the inequality sign* to obtain an equivalent inequality.

EXAMPLE 2.9.2

Find the solutions of the following inequalities and graph the solution sets on a number line.

1. $3x < 12$ As with linear equations, we

$\frac{1}{3} \cdot 3x < \frac{1}{3} \cdot 12$ multiply both sides by the reciprocal of the coefficient.

$1 \cdot x < 4$

$x < 4$

The solution set consists of all numbers x such that $x < 4$.

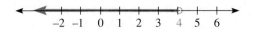

2. $-\frac{5}{4}x \le 20$ Here we must multiply both

$-\frac{4}{5} \cdot \left(-\frac{5}{4}x\right) \ge -\frac{4}{5} \cdot 20$ sides by $-\frac{4}{5}$. *Remember to reverse the inequality*

$x \ge -16$ *symbol.*

The solution set consists of all numbers x such that $x \ge -16$.

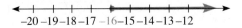

3. $2 \le -4x \le 12$ Here we must multiply

$-\frac{1}{4} \cdot 2 \ge -\frac{1}{4} \cdot (-4x) \ge -\frac{1}{4} \cdot 12$ all three sides by $-\frac{1}{4}$. *Remember to reverse both*

$-\frac{1}{2} \ge x \ge -3$ *inequality symbols.*

or $-3 \le x \le -\frac{1}{2}$ This inequality is equivalent to the previous one but is written with the numbers in the same order as they appear on the number line.

The solution set consists of all numbers x such that $-3 \le x \le -\frac{1}{2}$.

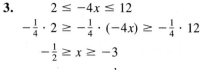

4. $10 \ge 7x - 4$ To solve this inequality we must use both the additive and multiplicative properties of inequality.

$10 + 4 \ge 7x - 4 + 4$ Add 4 to both sides to isolate the x term.

$$14 \geq 7x$$

$$\tfrac{1}{7} \cdot 14 \geq \tfrac{1}{7} \cdot 7x$$

Multiply both sides by $\tfrac{1}{7}$ to make the coefficient of x equal to 1.

$$2 \geq x$$

or $\quad x \leq 2$

The solution set consists of all numbers x such that $x \leq 2$.

Suppose we had a more complicated inequality, such as $6x - 5 > 4x - 1$. What process is used to find the solution set?

EXAMPLE 2.9.3

Find the solutions of the following inequalities and graph the solution sets on a number line.

1. $\qquad 6x - 5 > 4x - 1$

$$6x - 4x - 5 > 4x - 4x - 1$$

Notice that we follow exactly the same steps outlined in Section 2.4 to solve linear equations.

$$2x - 5 > -1$$

$$2x - 5 + 5 > -1 + 5$$

$$2x > 4$$

$$\tfrac{1}{2} \cdot 2x > \tfrac{1}{2} \cdot 4$$

$$x > 2$$

The solution set consists of all numbers x such that $x > 2$.

2. $5(2x + 1) + 4 < 8x + 6$

$$10x + 5 + 4 < 8x + 6$$

$$10x + 9 < 8x + 6$$

$$10x - 8x + 9 < 8x - 8x + 6$$

$$2x + 9 < 6$$

$$2x + 9 - 9 < 6 - 9$$

$$2x < -3$$

$$\tfrac{1}{2} \cdot 2x < \tfrac{1}{2} \cdot (-3)$$

$$x < -\tfrac{3}{2}$$

The solution set consists of all numbers x such that $x < -\frac{3}{2}$.

$$-\frac{3}{2}$$

3. $\frac{1}{4}\left(6 - \frac{2}{3}x\right) \le 3\left(\frac{1}{2}x - \frac{7}{6}\right)$

$\frac{3}{2} - \frac{1}{6}x \le \frac{3}{2}x - \frac{7}{2}$

$\frac{3}{2} - \frac{1}{6}x - \frac{3}{2}x \le \frac{3}{2}x - \frac{3}{2}x - \frac{7}{2}$

$\frac{3}{2} - \frac{5}{3}x \le -\frac{7}{2}$

$\frac{3}{2} - \frac{3}{2} - \frac{5}{3}x \le -\frac{7}{2} - \frac{3}{2}$

$-\frac{5}{3}x \le -5$

$-\frac{3}{5} \cdot \left(-\frac{5}{3}x\right) \ge -\frac{3}{5} \cdot (-5)$ Don't forget to reverse
                                            the inequality symbol.

$x \ge 3$

The solution set consists of all numbers x such that $x \ge 3$.

4. $5 < 3x + 11 < 14$

$5 - 11 < 3x + 11 - 11 < 14 - 11$

$-6 < 3x < 3$

$\frac{1}{3} \cdot (-6) < \frac{1}{3} \cdot 3x < \frac{1}{3} \cdot 3$

$-2 < x < 1$

The solution set consists of all numbers x such that $-2 < x < 1$.

5. $\frac{5}{3} \le 4 - \frac{1}{2}x \le 3$

$\frac{5}{3} - 4 \le 4 - 4 - \frac{1}{2}x \le 3 - 4$

$-\frac{7}{3} \le -\frac{1}{2}x \le -1$

$-2 \cdot \left(-\frac{7}{3}\right) \ge -2 \cdot \left(-\frac{1}{2}x\right) \ge -2 \cdot (-1)$ Reverse both inequality
                                                                      symbols.

$\frac{14}{3} \ge x \ge 2$

or $2 \le x \le \frac{14}{3}$

The solution set consists of all numbers x such that $2 \le x \le \frac{14}{3}$.

Remember when solving linear inequalities to follow the same steps used to solve linear equations. The only difference is in *reversing the inequality symbols when multiplying (or dividing) by a negative number.*

EXERCISES 2.9

Find the solutions of the following inequalities and graph the solution sets on a number line.

1. $x + 4 < 6$

2. $x + 7 \le 4$

3. $x - 5 \ge 2$

4. $x - 6 > -1$

5. $x + \frac{3}{5} \le \frac{2}{3}$

6. $x - \frac{4}{3} \ge \frac{1}{2}$

7. $3 < x - 7$

8. $-2 > x - 11$

9. $1 < x + 3 < 5$

10. $-6 < x + 1 < 3$

11. $-4 \le x - 8 \le 2$

12. $-9 \le x - 5 \le -1$

13. $-2 \le x + \frac{3}{8} \le -\frac{3}{2}$

14. $1 \le x + \frac{3}{4} \le \frac{7}{2}$

15. $2x > 6$

16. $5x \ge -10$

17. $-6x \ge 4$

18. $-3x > 8$

19. $\frac{2}{3}x > \frac{4}{15}$

20. $\frac{4}{5}x \le -2$

21. $3 \le 5x$

22. $6 < 15x$

23. $-4 < 3x < 6$

24. $-8 \le 4x \le 3$

25. $-2 \le 4x \le 5$

26. $-3 \le -6x \le 2$

27. $-6 < \frac{3}{4}x < -3$

28. $-8 < \frac{2}{7}x < -2$

29. $2x + 5 \ge 11$

30. $4x + 3 \le 9$

31. $3x - 7 < 5$

32. $5x - 4 > 3$

33. $1 - 6x \le 3$

34. $2 - 3x > 11$

35. $\frac{4}{5}x + 2 > \frac{2}{5}$

36. $\frac{5}{3}x - \frac{1}{2} \ge 2$

37. $10 < 2x + 5$

38. $14 \le 3x + 2$

39. $1 \ge 8x - 3$

40. $1 \ge 12x - 5$

41. $6 \ge 1 - 3x$

42. $-3 < 7 - 4x$

43. $1.8x - 3.14 < 4.06$

44. $4.2x + 1.7 > 8$

45. $4x - 3 \le x + 7$

46. $5x - 2 \ge 3x + 8$

47. $x + 9 > 5x - 7$

48. $2x + 13 \le 7x - 2$

49. $6 - x \ge 7x - 12$

50. $5 - 3x < 3x - 3$

51. $4x + 11 + 6x < 6 + 3x - 2$

52. $9x - 4 - 2x \ge 7 + x + 1$

53. $x - 8 + 7x > 5 + 3x - 13$

54. $6x + 5 - x < 8 + 8x - 3$

55. $\frac{1}{2}x - 2 + \frac{2}{3}x \ge \frac{5}{4}x - \frac{5}{2}$

56. $\frac{5}{3}x + 1 - \frac{2}{9}x \le \frac{3}{2}x + \frac{5}{6}$

57. $3(4x - 1) \le 7x + 12$

58. $2(3x - 7) > 2x - 4$

59. $7(x + 3) - 8 > 3x + 11$

60. $4(x - 5) + 6 \ge x + 1$

61. $5(4x + 3) - 2x < 6x - 9$

62. $3(2x + 9) - x > 10x + 7$

63. $2(1 - 3x) \le 5(x - 4)$

64. $7(6 - x) < 3(3x - 2)$

65. $6 - 3(5x - 4) \ge 9 - 3x$

66. $5 - 2(3x - 7) \ge 4 - x$

67. $\frac{1}{3}(2x - 5) + 2 > \frac{3}{2}(x + 3)$

68. $\frac{1}{4}(5x + 5) + 1 \le \frac{1}{2}(x + 9)$

69. $2\left(\frac{3}{4}x + \frac{3}{5}\right) - \frac{1}{2}x < \frac{5}{4}x + 1$

70. $3\left(\frac{1}{2}x - \frac{5}{6}\right) - \frac{4}{3}x > \frac{1}{4}x - 2$

71. $-1 < 3x - 7 < 2$

72. $-11 < 4x - 3 < 3$

73. $6 \le 5x + 1 \le 13$

74. $-5 \le 7x + 2 \le 16$

75. $-7 \le 5 - 2x \le 3$

76. $-17 < 1 - 6x < 4$

77. $4 < 4 - \frac{1}{3}x < 5$

78. $0 \le 2 - \frac{1}{4}x \le 2$

79. $-4 < \frac{5}{2}x - \frac{1}{4} < 1$

80. $-\frac{1}{3} < \frac{5}{6}x - 2 < \frac{1}{2}$

Chapter 2 Review

Properties

Let a, b, and c denote algebraic expressions.

Additive property of equality:

$$\text{If } a = b \text{ then } a + c = b + c$$

Multiplicative property of equality:

$$\text{If } a = b \text{ then } ac = bc, \text{ given } c \ne 0$$

Additive property of inequality:

$$\text{If } a < b \text{ then } a + c < b + c$$

Multiplicative property of inequality:

1. If $a < b$ and $c > 0$ then $ac < bc$

2. If $a < b$ and $c < 0$ then $ac > bc$

Review Exercises

2.1 Evaluate the following algebraic expressions for the given values of the variables.

1. $3 - 5x; \quad x = -4$

2. $x^2 + 7x - 4; \quad x = 3$

3. $(3x + 1)(x - 9); \quad x = \frac{3}{2}$

4. $2x + 6(x - 4); \quad x = -2$

5. $3(x - 5) - (2x + 4); \quad x = -\frac{1}{3}$

6. $5x^2 + 7xy - y^2; \quad x = 5, y = -1$

7. $\dfrac{2x}{x - 4} + \dfrac{9x}{x + 2}; \quad x = 3$

8. $\frac{2}{3}(x + 5) - \frac{1}{4}(y - 8); \quad x = 4, y = -4$

9. $\dfrac{3x + 7y}{\sqrt{8z}}; \quad x = 1, y = -5, z = 2$

10. $\sqrt{28 - 2x^2 + 5y^2}; \quad x = 2, y = 4$

Simplify the following algebraic expressions by combining like terms, if possible.

11. $5x - 7 + 3x - 4$

12. $2x^2 + 7x + 9x + 10$

13. $\frac{1}{2}x^2 + \frac{2}{3}y^2 - \frac{3}{8}x^2 - \frac{8}{9}y^2$

14. $2x^2 + 3x - 4y + 12$

15. $5(x + 3) + 2(x - 4)$

16. $7(2x - 4) - (3x - 17)$

17. $3(2x + 9y) + 9(3x - 4y)$

18. $0.3(2x - 9) + 0.4(5x - 1)$

19. $\frac{2}{3}(x + 4) - \frac{1}{4}(x + 5)$

20. $7x + 7[4(2x + 1) - (x + 3)]$

2.2 Solve the following linear equations and check your solutions.

21. $x + 7 = 2$

22. $\frac{5}{9} = x - \frac{1}{3}$

23. $8x + 4 - 7x + 2 = 5 - 17$

24. $\frac{3}{4}x - \frac{7}{8} + \frac{1}{4}x + \frac{1}{4} = 9$

25. $4x + 9 = 3x + 11$

26. $4.65x - 1.92 = 3.65x - 13.75$

27. $5x + 2x + 4 = 6x + 4$

28. $3x + 7 - 12x = 4 - 8x - 2$

29. $3x - \frac{1}{2} - x = x + \frac{3}{4}$

30. $\frac{2}{3}x + \frac{5}{6}x - \frac{1}{8} = \frac{1}{2}x - \frac{3}{4}$

31. $5x - 3 + 4x + 17 = 2x + 5 + 8x + 1$

32. $\frac{3}{4}x + \frac{1}{3} + x - 2 = \frac{1}{2}x + \frac{5}{6} + \frac{1}{4}x + 1$

33. $4(5x - 1) = 21x - 4$

34. $5(2x + 3) + 3x = 6(2x - 1) + 9$

35. $\frac{1}{3}(12x + 9) - 4 = \frac{1}{4}(12x - 8) + 5$

36. $4(x + 2) - 3(x - 7) = 24$

37. $3(2x + 5) + 2(x - 8) = 7(x - 1)$

38. $5(2x - 4) + 3(x + 1) = 4(2x - 1) + 2(2x + 6)$

39. $2[4(x + 4) - 2(x + 3)] = 3(x + 5)$

40. $4[3(2x + 4) - 2(x - 1)] = 5[2(2x + 1) - (x - 2)]$

2.3 Solve the following linear equations and check your solutions.

41. $4x = -28$

42. $3x = 8$

43. $\frac{2}{3}x = 4$

44. $-5x = 0$

45. $\frac{5}{4} = \frac{1}{10}x$

46. $-\dfrac{1}{3} = \dfrac{x}{4}$

47. $-x = -12$

48. $\frac{3}{4} = 7x$

49. $4x + 1 = 7$

50. $13 = 5x - 2$

51. $\frac{5}{6}x + 2 = 4$

52. $\frac{1}{4}x - 3 = -4$

53. $5x + 4 = 3x$

54. $7x - 2 = 4x$

55. $3x + 14 = 8x - 1$

56. $9x + 2 = 2x + 11$

57. $\frac{3}{4}x + \frac{2}{3} = \frac{1}{2}x - \frac{5}{6}$

58. $11 - 3x - 4x = x + 5$

59. $\frac{4}{3}x + \frac{3}{8} - 2x = \frac{1}{9}x + 3$

60. $\frac{3}{10}x + \frac{5}{2} + \frac{4}{5}x = \frac{1}{2}x + \frac{5}{6}$

2.4 Solve the following equations.

61. $3(4x - 1) = 5x + 6$

62. $5 + 2(3x - 2) = 10x + 2$

63. $8 - (3x + 4) = 6x$

64. $3 + 2(2x + 5) = 4x - 9$

65. $\frac{1}{3} + \frac{3}{4}(5x + 2) = \frac{28}{3}$

66. $\frac{2}{7} - \frac{1}{4}(2x - 3) = \frac{11}{14}$

67. $\frac{2}{3}(6x - 4) = \frac{1}{2}(7x - 5)$

68. $5(2x + 1) - 3(x + 4) = -21$

69. $5(4 - 7x) = 2(3x - 2) - x$

70. $\frac{3}{4}(6x - 1) - \frac{3}{2}(x - 2) = \frac{3}{4}x + \frac{3}{4}$

71. $2.6(x + 5) + 4.1(2x - 4) = 3.6x + 0.2$

72. $3(4x + 5) - 5(2x - 7) = 2(6x + 1)$

73. $5(x - 1) - 2(3x + 1) = 4(2x + 1) - 11$

74. $2(3x + 1) + 4(2x - 2) = 2(7x - 3)$

75. $2[5x - (4x - 7)] = 3(5 - x) + 6$

76. $5[2(3x - 1) - 6(x + 2)] = 3x - 7$

2.5 Find the solutions of the following equations.

77. $|x| = 7$

78. $|-3x| = \frac{5}{2}$

79. $|\frac{3}{4}x| = 9$

80. $|x - 3| = 6$

81. $|2 - x| = \frac{3}{4}$

82. $|5 - 3x| = 4$

83. $|\frac{1}{2}x + 1| = \frac{5}{6}$

84. $|2x - 5| = 7$

85. $|3x + 4| = -8$

86. $|\frac{1}{3}x - \frac{1}{6}| = 0$

87. $|x - 5| + 2 = 13$

88. $|2x - 3| - 1 = 7$

89. $|7 - x| + 3 = 10$

90. $|5 - x| + 9 = 2$

91. $|4x + 2| - 1 = 4$

92. $|2x - 5| + 3 = 3$

93. $|2x - 4| - 9 = 0$

94. $|\frac{1}{2}x + 1| + 3 = 8$

95. $|\frac{3}{2}x - \frac{1}{8}| + \frac{1}{4} = \frac{3}{4}$

96. $|\frac{1}{2}x - \frac{1}{3}| + \frac{1}{6} = \frac{7}{6}$

2.6 Solve the following literal equations for the indicated variable.

97. $I = PRT$, for R

98. $V = LWH$, for W

99. $m = \dfrac{(3n - 4p)}{5}$, for p

100. $5x - 4y = 3$, for y

101. $A = 2x^2 + 4xy$, for y

102. $c = 3a + 5b$, for b

Solve the following literal equations for the indicated variable, and substitute the given values to find the variable's value.

103. $D = RT$. Find R when $D = 252$ and $T = 6$.

104. $A = \frac{1}{2}BH$. Find H when $A = 6$ and $B = 16$.

105. $P = 2L + 2W$. Find L when $P = 32$ and $W = 4$.

106. $y = mx + b$. Find x when $y = 13$, $m = 4$, and $b = -7$.

107. Hooke's law states that $F = kx$. Find k when $F = 120$ and $x = 192$.

108. The formula for the area of a trapezoid is $A = \frac{1}{2}(b + B)h$. Find B when $b = 5$, $h = 6$, and $A = 51$.

2.7 Use algebraic expressions to find the solutions of the following problems.

109. Three times a number plus 14 is 8. What is the number?

110. Five minus 4 times a number is 3. What is the number?

111. Find three consecutive integers whose sum is 66.

112. Find three consecutive odd integers whose sum is −21.

113. A 34-in. board is cut into three pieces. The second piece is 1 in. longer than twice the length of the first piece. The third piece is 3 in. shorter than the first piece. How long is each piece?

114. A 43-in. board is cut into three pieces. The second piece is one-half as long as the first piece. The third piece is 7 in. longer than the second piece. How long is each piece?

115. The perimeter of a rectangle is 44 m. The length of the rectangle is 8 m more than the width. What are the dimensions of the rectangle?

116. The perimeter of a rectangle is 36 ft. The length of the rectangle is 2 ft more than 3 times the width. What are the dimensions of the rectangle?

117. Jeffrey has $11.75 in nickels and quarters. The number of quarters is 5 more than 4 times the number of nickels. How many of each type of coin does he have?

118. Melissa has $220 in $1 bills, $5 bills, and $10 bills. The number of $1 bills is 3 times the number of $5 bills. The number of $10 bills is 4 more than the number of $5 bills. How many of each type of bill does she have?

119. Michael has $7.95 in dimes and quarters. He has a total of 39 coins. How many dimes and how many quarters does he have?

120. Deborah has $55 in $1 bills and $5 bills. She has a total of 23 bills. How many $1 bills and how many $5 bills does she have?

2.8

121. Joey leaves school on a bicycle traveling east at a rate of 12 mph. At the same time, Amy leaves school on a bicycle traveling west at a rate of 10 mph. In how many hours will Joey and Amy be $16\frac{1}{2}$ miles apart?

122. Juan leaves church in a car traveling south at a rate of 70 mph. At the same time, Gilbert leaves church in a car traveling north at a rate of 30 mph. In how many hours will Juan and Gilbert be 120 miles apart?

123. Susie leaves the bank at 3:00 P.M. traveling at a rate of 42 mph. An hour later Earl leaves the bank traveling in the same direction at 56 mph. At what time will Earl overtake Susie?

124. Jack leaves the college campus at 4:30 P.M. running at a rate of 7 mph. Ten minutes later Jerry leaves the campus running the same route at a rate of 8 mph. At what time will Jerry overtake Jack?

125. Louise invested part of her income tax return in a checking account paying 5% simple interest. She invested $500 less than the amount she invested in the checking account in a savings account paying 6% simple interest. After 1 year her interest income was $102. How much did Louise invest in the checking account and in the savings account?

126. Luann invested part of her bingo winnings in a Christmas account paying 7% simple interest. She invested 3 times the amount she invested in the Christmas account in a certificate of deposit paying 9% simple interest. After 1 year her interest income was $142.80. How much did Luann invest in the Christmas account and in the certificate of deposit?

127. Lulu invested $5000 in two certificates of deposit. She invested part of her money in one certificate paying 8% simple interest and invested the rest of her money in a certificate paying 9% simple interest. After 1 year her interest income was $436.50. How much did Lulu invest in each certificate?

128. Lupe invested $9000 in two certificates of deposit. She invested part of her money in one certificate paying 10% simple interest and invested the rest of her money in a certificate paying 8%. After 2 years her interest income was $1752. How much did Lupe invest in each certificate?

129. "Real Stuff" soft drinks contain 18% fruit juice. "Sugar Water" soft drinks contain 3% fruit juice. How many ounces of each drink must be used to make 12 oz of soft drink that is 10% fruit juice?

130. Big G's chili mix contains 12% red pepper. Little D's chili mix contains 2% red pepper. How many ounces of each mix must be used to make 16 oz of chili mix that is 9% red pepper?

131. East Connecticut honey sells for $4.00 a quart. Moss County honey sells for $1.50 a quart. How many quarts of each type must be purchased to create 25 quarts of honey that will sell for $2.60 a quart?

132. Beth's peanut butter sells for $2.60 a pound. Dennis's peanut butter sells for $1.40 a pound. How many pounds of each type must be purchased to create 86 pounds of peanut butter that will sell for $2.15 a pound?

2.9   Find the solutions of the following inequalities, and graph the solution sets on a number line.

133. $x - 7 \le -4$

134. $5 < x + 4$

135. $-1 < x + 2 < 3$

136. $1 \le x - \frac{2}{5} \le \frac{5}{2}$

137. $-5x > 20$

138. $\frac{3}{4}x \ge -6$

139. $-6 \le 2x \le \frac{2}{3}$

140. $-9 < -3x < -\frac{3}{2}$

141. $5x - 7 > 3$

142. $4x + 9 \le 3$

143. $1 - 9x \ge 4$

144. $-2 < 3x - 14$

145. $x - 8 < 5x + 2$

146. $2x + 15 > 7x - 1$

147. $2x - 7 - 8x > 3 - 3x - 1$

148. $7 - 4x + 2 \ge 6x - 3 - x$

149. $\frac{2}{3}(5x - 1) \ge 4x - 2$

150. $4(5x - 7) < 2 + 5(4 - 3x)$

151. $-1 < 3x + 8 < 5$

152. $-4 \le 4 - x \le 5$

Chapter 2 Test   (You should be able to complete this test in 60 minutes.)

I. Evaluate the following algebraic expressions for the given values of the variables.

1. $\frac{4}{5}(x + 2) - (3x - 5);$ $x = -7$

2. $\dfrac{5x - 2y}{\sqrt{3x}};$ $x = 12, y = -2$

II. Simplify the following algebraic expressions by combining like terms.

3. $5x^2 - 7x + 3x^2 + 2x$

4. $\frac{3}{4}(3x - 5) - 2\left(x + \frac{7}{2}\right)$

III. Solve the following literal equations for the indicated variable.

5. $I = pm + V$, for m

6. $V = \frac{1}{3}\pi r^2 h$, for h

IV. Solve the following equations.

7. $5x - 2 = 11$

8. $3x + 7 - 5x = 2x - 5$

9. $\frac{4}{5}x + \frac{2}{3} = \frac{5}{6}x + 1$

10. $2(3x - 9) + x = 4(3 - 2x)$

11. $|4x - 3| = 5$

12. $\left|\frac{1}{2}x - 5\right| - 2 = 1$

V. Find the solutions of the following inequalities, and graph the solution sets on a number line.

13. $2x + 17 \leq 9$

14. $5 - 2(3x - 7) > 2x + 11$

15. $-\frac{5}{2} < 3x - \frac{1}{4} < 2$

VI. Use algebraic expressions to find the solutions of the following problems.

16. The perimeter of a rectangle is 23 ft. The length of the rectangle is 1 ft more than twice the width. What are the dimensions of the rectangle?

17. Jennifer has $230 in $5 bills and $20 bills. She has a total of 25 bills. How many $5 bills and how many $20 bills does she have?

18. Trey leaves home on a bicycle traveling south at a rate of 8 mph. At the same time, Dusty leaves home on a bicycle traveling north at a rate of 6 mph. In how many hours will Trey and Dusty be $3\frac{1}{2}$ miles apart?

Test Your Memory

These problems review Chapters 1 and 2.

I. Graph each set on the real number line.

1. $\{-\frac{7}{4}, 1.3, \frac{5}{2}\}$

2. $\{-\frac{13}{3}, -\frac{2}{5}, 3.8\}$

II. Reduce each fraction to lowest terms.

3. $\frac{70}{42}$

4. $\frac{19}{95}$

III. Perform the indicated operations. When necessary, use the rules governing the order of operations.

5. $-\frac{5}{4} + \frac{7}{12}$

6. $6\left(\frac{3}{5} - \frac{7}{2}\right)$

7. $5 - |-2 - 7|$

8. $\dfrac{9 + \sqrt{36}}{3}$

9. $2.8 - (-7.5)(0.44)$

10. $8.82 \div (-3.6)$

11. $5(7) + 3(9)$

12. $5(7 + 3)9$

13. $-36 \div 4 \cdot 3$

14. $2 \cdot 5^2 - 7 \cdot 5$

15. $\left(-\frac{3}{10}\right)\left(\frac{35}{6}\right)$

16. $-\frac{3}{10} + \frac{35}{6}$

17. $\dfrac{1 - \sqrt{(-1)^2 - 4(-6)}}{2}$

18. $\dfrac{5 \cdot 8 - 2 \cdot 7}{3^2 + 2^2}$

19. $(-3)(2)(-5)(-4)$

20. $-48 \div (-3)$

21. $\frac{7}{4} - \left(\frac{5}{6} + \frac{2}{15}\right)$

22. $\left(5\frac{4}{9}\right)\left(2\frac{1}{7}\right)$

23. $8 - 2[6 + 4(3 - 11)]$

24. $[3(-2) - 4]^3$

IV. Evaluate the following algebraic expressions for the given values of the variables.

25. $\frac{2}{3}(5x + 2) - \left(7x - \frac{1}{3}\right); \quad x = 3$

26. $\dfrac{\sqrt{4x + 5}}{3x + 8y}; \quad x = -1, y = 2$

V. Simplify the following algebraic expressions by combining like terms.

27. $8x^2 - 11x + x - 5x^2$

28. $\frac{5}{2}(3x - 1) - 4\left(x + \frac{3}{4}\right)$

VI. Solve the following literal equations for the indicated variable.

29. $3x - 4y = 8$, for y

30. $y = \frac{1}{2}x + 3$, for x

VII. Find the solutions of the following equations.

31. $3x - 4 = 7$

32. $5x + 2 - 3x = 7x + 6$

33. $2(4x - 3) = 9x + 7$

34. $2 + 3(x - 4) = x - 9$

35. $7x - 3(2x - 1) = 8$

36. $x - (5x + 4) = 2(2x + 3) - 9$

37. $\frac{1}{4}x - \frac{5}{6} = \frac{1}{6}x - \frac{1}{2}$

38. $\frac{2}{3}\left(x + \frac{1}{2}\right) = \frac{3}{4}\left(2x - \frac{1}{3}\right)$

VIII. Find the solutions of the following inequalities, and graph the solution sets on a number line.

39. $3x + 10 \leq 1$

40. $11 - 2x > 3$

41. $\frac{1}{2}x - \frac{1}{3} > 2$

42. $2(x - 5) + 3 \geq 7x - 1$

43. $-9 < 2x - 1 < 9$

44. $\frac{1}{2}(3x - 4) \leq \frac{2}{3}(2x - 1)$

IX. Use algebraic expressions to find the solutions of the following problems.

45. Find the perimeter of a square whose sides each have a length of $4\frac{2}{3}$ ft.

46. Anne drove her car for $1\frac{1}{2}$ hours, covering $79\frac{1}{5}$ miles. What was her average speed, that is, how many miles per hour did she average?

47. The perimeter of a rectangle is 54 yd. The length is 1 yd less than 3 times the width. What are the dimensions of the rectangle?

48. José has $1500 in $50 bills and $100 bills. He has three more $50 bills than $100 bills. How many $50 bills and how many $100 bills does José have?

49. Ike has $4.40 in dimes and quarters. He has a total of 32 coins. How many quarters and how many dimes does he have?

50. Butch leaves the "Hole-in-the-Wall" hideout traveling east at a rate of 10 mph. At the same time, Sundance leaves the "Hole-in-the-Wall" hideout traveling west at a rate of 12 mph. In how many hours will Butch and Sundance be 55 miles apart?

Exponents and Polynomials

3.1 Natural Number Exponents

In Chapter 1 we introduced the concept of exponential notation. Recall that the expression 3^4, read three to the fourth power, was defined to be

$$3^4 = 3 \cdot 3 \cdot 3 \cdot 3 = 81$$

In the expression 3^4, 3 is the **base** and 4 is the **exponent**, or **power**. The exponent specifies the number of factors of the base.

In general we define

$$a^n = \underbrace{a \cdot a \cdot a \cdot \ldots \cdot a}_{n \text{ factors}} \qquad \text{where } a \text{ is any number} \\ \text{and } n \text{ is a natural number}$$

NOTE ▶ Remember, when the base is negative, we must place parentheses around the base and write the exponent outside of the parentheses. Consider the following examples. In $(-6)^2 = (-6)(-6) = 36$, the base is -6. However, in $-6^2 = -(6 \cdot 6) = -36$, the exponent 2 applies only to 6, the base, and not to the negative sign in front. We write the negative sign in front of the answer after we have multiplied the product 6^2.

In this section we will apply exponential notation to products and quotients of algebraic expressions. Suppose that you were asked to simplify the expression $x^3 \cdot x^4$. We know that $x^3 = x \cdot x \cdot x$ and $x^4 = x \cdot x \cdot x \cdot x$, so

$$x^3 \cdot x^4 = (x \cdot x \cdot x) \cdot (x \cdot x \cdot x \cdot x)$$
$$= x^7$$

Thus, $x^3 \cdot x^4 = x^{3+4} = x^7$. This example leads to the following law.

Product Law of Exponents

Let m and n be natural numbers; then

$$x^m \cdot x^n = x^{m+n}$$

EXAMPLE 3.1.1

Use the product law of exponents to simplify the following expressions.

1. $x^6 \cdot x^{10} = x^{6+10} = x^{16}$

2. $x^{23} \cdot x^{45} = x^{23+45} = x^{68}$

3. $x^4 \cdot x = x^4 \cdot x^1 = x^{4+1} = x^5$

When there is no number in the exponent position, the exponent is understood to be 1, because there is just one factor of the base.

4. $2^5 \cdot 2^9 = 2^{5+9} = 2^{14}$

Try to avoid this mistake:

Incorrect	Correct
$2^5 \cdot 2^9 = 4^{5+9}$ $\quad = 4^{14}$ The exponents were added, and the bases were multiplied.	$2^5 \cdot 2^9 = 2^{5+9}$ $\quad = 2^{14}$ When we multiply, we add the exponents, but the bases do not change.

5. $x^9 \cdot y^2 \qquad$ cannot be simplified

If we want to use the product law of exponents, the *bases must be exactly the same*, which is not the case in part 5.

Suppose that we wanted to simplify the algebraic expression $(x^8)^3$. We know that

$$(x^8)^3 = x^8 \cdot x^8 \cdot x^8$$
$$= x^{8+8+8}$$
$$= x^{24}$$

Thus, $(x^8)^3 = x^{8\cdot3} = x^{24}$. This example leads to the next law of exponents.

Power to a Power Law of Exponents

Let m and n be natural numbers; then

$$(x^m)^n = x^{m \cdot n}$$

EXAMPLE 3.1.2

Use the power to a power law of exponents to simplify the following expressions.

1. $(2^6)^8 = 2^{6\cdot8} = 2^{48}$

2. $[(-3)^4]^{11} = (-3)^{4\cdot11} = (-3)^{44}$

3. $-(5^3)^6 = -5^{3\cdot6} = -5^{18}$

4. $(x^7)^9 = x^{7\cdot9} = x^{63}$

Suppose that we wanted to simplify the algebraic expression $(xy)^5$. We know that

$$(xy)^5 = (xy) \cdot (xy) \cdot (xy) \cdot (xy) \cdot (xy)$$
$$= x \cdot x \cdot x \cdot x \cdot x \cdot y \cdot y \cdot y \cdot y \cdot y$$
$$= x^5y^5$$

Thus, $(xy)^5 = x^5y^5$. This example suggests another law of exponents.

Product to a Power Law of Exponents

Let n be a natural number; then

$$(xy)^n = x^ny^n$$

EXAMPLE 3.1.3

Use the product to a power law of exponents to simplify the following expressions.

1. $(xy)^9 = x^9 y^9$

2. $(2x^3 y)^5 = 2^5 (x^3)^5 y^5$ By the product to a power law

           $= 32x^{15} y^5$ By the power to a power law

3. $(x^7 y^2)^8 = (x^7)^8 (y^2)^8$

           $= x^{56} y^{16}$

4. $(-2x^5 y^9)^3 = (-2)^3 (x^5)^3 (y^9)^3$

               $= (-2)^3 x^{15} y^{27}$

               $= -8x^{15} y^{27}$

In the first part of this section, we investigated the laws of exponents used in multiplying algebraic expressions. In the remainder of this section, we will examine the laws of exponents used in dividing algebraic expressions. First we consider the following problem from arithmetic.

$$\frac{3^6}{3^4} = \frac{3 \cdot 3 \cdot 3 \cdot 3 \cdot 3 \cdot 3}{3 \cdot 3 \cdot 3 \cdot 3}$$

$$= 3^2 \text{ or } 9$$

Suppose that we wanted to simplify the algebraic expression x^6/x^4. We know that

$$\frac{x^6}{x^4} = \frac{x \cdot x \cdot x \cdot x \cdot x \cdot x}{x \cdot x \cdot x \cdot x}$$

$$= \frac{\not{x} \cdot \not{x} \cdot \not{x} \cdot \not{x} \cdot x \cdot x}{\not{x} \cdot \not{x} \cdot \not{x} \cdot \not{x}} \qquad \text{Generalizing the preceding argument}$$

$$= x^2$$

Thus, $x^6/x^4 = x^{6-4} = x^2$. This example leads to another law of exponents.

Quotient Law of Exponents

Let m and n be natural numbers with $m > n$ and $x \neq 0$; then

$$\frac{x^m}{x^n} = x^{m-n}$$

REMARK   In Section 3.2 we will expand the quotient law of exponents to expressions x^m/x^n where $m \leq n$.

EXAMPLE 3.1.4

Use the quotient law of exponents to simplify the following expressions.

1. $\dfrac{2^{10}}{2^4} = 2^{10-4}$ **2.** $\dfrac{x^{18}}{x^2} = x^{18-2} = x^{16}$

       $= 2^6$ or 64

Try to avoid these mistakes:

Incorrect	Correct
1. $\dfrac{2^{10}}{2^4} = \dfrac{2^{10}}{2^4}$ $= \dfrac{1^{10}}{1^4}$ $= \dfrac{1}{1} = 1$ The bases were divided	$\dfrac{2^{10}}{2^4} = 2^{10-4}$ $= 2^6$ $= 64$ When we divide, we subtract the exponents, but the bases do not change.
2. $\dfrac{x^{18}}{x^2} = x^{18/2}$ $= x^9$ The exponents were divided instead of subtracted.	$\dfrac{x^{18}}{x^2} = x^{18-2}$ $= x^{16}$

3. $\dfrac{y^{23}}{y} = y^{23-1} = y^{22}$

Our last law of exponents is suggested in the following problem from arithmetic.

$$\left(\frac{5}{7}\right)^2 = \frac{5}{7} \cdot \frac{5}{7}$$

$$= \frac{5 \cdot 5}{7 \cdot 7}$$

$$= \frac{5^2}{7^2} \text{ or } \frac{25}{49}$$

Suppose that we wanted to simplify the algebraic expression $(x/y)^2$. We know that

$$\left(\frac{x}{y}\right)^2 = \frac{x}{y} \cdot \frac{x}{y}$$

$$= \frac{x \cdot x}{y \cdot y} \qquad \text{Generalizing the preceding argument}$$

$$= \frac{x^2}{y^2}$$

Thus, $(x/y)^2 = x^2/y^2$. This example suggests the last law of exponents.

Quotient to a Power Law of Exponents

Let n be a natural number and $y \neq 0$; then

$$\left(\frac{x}{y}\right)^n = \frac{x^n}{y^n}$$

EXAMPLE 3.1.5

Use the quotient to a power law of exponents to simplify the following expressions.

1. $\left(\dfrac{-3}{5}\right)^4 = \dfrac{(-3)^4}{5^4}$ or $\dfrac{81}{625}$

2. $\left(\dfrac{x}{y}\right)^{13} = \dfrac{x^{13}}{y^{13}}$

3. $\left(\dfrac{x^3}{y^8}\right)^6 = \dfrac{(x^3)^6}{(y^8)^6}$ By the quotient to a power law

$\qquad\qquad = \dfrac{x^{18}}{y^{48}}$ By the power to a power law

---

In the last example of this section, we will examine algebraic expressions that require the use of several laws of exponents.

EXAMPLE 3.1.6

Using the laws of exponents, simplify the following expressions.

1. $(2x^4)^3(5x)^2 = 2^3 \cdot (x^4)^3 \cdot 5^2 \cdot x^2$

$\qquad = 2^3 \cdot x^{12} \cdot 5^2 \cdot x^2$

$\qquad = 8 \cdot x^{12} \cdot 25 \cdot x^2$

$\qquad = 8 \cdot 25 \cdot x^{12} \cdot x^2$

$\qquad = 200x^{14}$

2. $(xy^8)^4(2x^3y^5)^2 = x^4 \cdot (y^8)^4 \cdot 2^2 \cdot (x^3)^2 \cdot (y^5)^2$

$\qquad = x^4 \cdot y^{32} \cdot 2^2 \cdot x^6 \cdot y^{10}$

$\qquad = 2^2 \cdot x^4 \cdot x^6 \cdot y^{32} \cdot y^{10}$

$\qquad = 4x^{10}y^{42}$

3. $\dfrac{15x^8y^4}{3x^2y} = \dfrac{15}{3} \cdot \dfrac{x^8}{x^2} \cdot \dfrac{y^4}{y}$

$\qquad = 5 \cdot x^{8-2} \cdot y^{4-1}$

$\qquad = 5x^6y^3$

4. $\left(\dfrac{6x^3}{7y}\right)^2 = \dfrac{(6x^3)^2}{(7y)^2}$

$\qquad = \dfrac{6^2(x^3)^2}{7^2y^2}$

$\qquad = \dfrac{36x^6}{49y^2}$

5. $\left(\dfrac{8x^3y^{10}}{2x^2y^5}\right)^3 = \left(\dfrac{8}{2} \cdot \dfrac{x^3}{x^2} \cdot \dfrac{y^{10}}{y^5}\right)^3$

$\qquad = (4 \cdot x^{3-2} \cdot y^{10-5})^3$

$\qquad = (4xy^5)^3$

$\qquad = 4^3x^3(y^5)^3$

$\qquad = 64x^3y^{15}$

6. $\dfrac{(4xy^2)(3x^3y^5)^2}{6x^4y^2} = \dfrac{4xy^2 \cdot 3^2(x^3)^2(y^5)^2}{6x^4y^2}$

$\qquad = \dfrac{4 \cdot x \cdot y^2 \cdot 9 \cdot x^6 \cdot y^{10}}{6x^4y^2}$

$\qquad = \dfrac{4 \cdot 9 \cdot x \cdot x^6 \cdot y^2 \cdot y^{10}}{6x^4y^2}$

$\qquad = \dfrac{36x^7y^{12}}{6x^4y^2}$

$\qquad = \dfrac{36}{6} \cdot \dfrac{x^7}{x^4} \cdot \dfrac{y^{12}}{y^2}$

$\qquad = 6x^{7-4}y^{12-2}$

$\qquad = 6x^3y^{10}$

As an aid, the laws of exponents of this section are summarized here.

Let m and n be natural numbers:

$x^m \cdot x^n = x^{m+n}$ Product law

$(x^m)^n = x^{m \cdot n}$ Power to a power law

$(xy)^n = x^n \cdot y^n$ Product to a power law

$\dfrac{x^m}{x^n} = x^{m-n};\ m > n,\ x \neq 0$ Quotient law

$\left(\dfrac{x}{y}\right)^n = \dfrac{x^n}{y^n};\ y \neq 0$ Quotient to a power law

EXERCISES 3.1

Name the base and the exponent in each of the following expressions.

1. 8^3 **2.** 6^4 **3.** 2 **4.** 9 **5.** x^4 **6.** y^7

7. -7^5 **8.** $(-3)^6$ **9.** $(-4)^{10}$ **10.** -5^6 **11.** $3x^8$ **12.** $5y^2$

Rewrite each of the following expressions using exponents.

13. $3 \cdot 3 \cdot 3 \cdot 3 \cdot 3$ **14.** $2 \cdot 2 \cdot 2 \cdot 2 \cdot 2 \cdot 2 \cdot 2$ **15.** $(-5)(-5)(-5)(-5)$

16. $(-8)(-8)(-8)(-8)$ **17.** $-6 \cdot 6 \cdot 6 \cdot 6 \cdot 6 \cdot 6$ **18.** $-4 \cdot 4 \cdot 4 \cdot 4$

19. $x \cdot x \cdot x \cdot x \cdot x$ **20.** $y \cdot y \cdot y$ **21.** $(2z)(2z)(2z)$

22. $(7g)(7g)(7g)(7g)$

Evaluate each of the following expressions.

23. 8^2 **24.** 3^3 **25.** $(-3)^4$ **26.** $(-2)^6$ **27.** -3^4 **28.** -2^6

29. -5^4 **30.** -4^4 **31.** $5^2 + 5^3$ **32.** $3^3 + 3^4$ **33.** $2^5 - 2^2$ **34.** $2^6 - 2^3$

35. $2 \cdot 2^4$ **36.** $3 \cdot 3^4$ **37.** $3^3 \cdot 3^2$ **38.** $2^2 \cdot 2^3$ **39.** $(2^4)^2$ **40.** $(2^2)^3$

41. $\dfrac{5^6}{5^3}$ **42.** $\dfrac{2^7}{2^2}$ **43.** $\dfrac{7^3}{7}$ **44.** $\dfrac{3^4}{3}$ **45.** $\left(\dfrac{2}{3}\right)^3$ **46.** $\left(\dfrac{3}{5}\right)^3$

47. $2^3 \cdot 3^2$ **48.** $5^2 \cdot 2^2$

Using the laws of exponents, simplify the following expressions, if possible.

49. $x^8 \cdot x^{13}$ **50.** $y^9 \cdot y^{24}$ **51.** $x^{42} \cdot x^{29}$ **52.** $z^{15} \cdot z$

53. $x^5 \cdot y^9$ **54.** $m^{12} \cdot n^2$ **55.** $x^5 \cdot x^{12} \cdot x^4$ **56.** $2x^7 \cdot 5x^9$

57. $(-7x^3)(2x^6)$ **58.** $(x^3)^5$ **59.** $(x^4)^6$ **60.** $(x^{11})^{13}$

61. $[(x^3)^6]^9$ **62.** $[(x^4)^3]^8$ **63.** $[(x^2)^5]^7$ **64.** $(x^8 \cdot x^3)^7$

65. $(x \cdot x^4 \cdot x^7)^{10}$ **66.** $(xy)^7$ **67.** $(6xy)^2$ **68.** $(4xy)^3$

69. $(pt^7)^{12}$ **70.** $(x^2y^4)^7$ **71.** $(xy^3z^7)^2$ **72.** $(x^2yz^6)^6$

73. $(5x^7y^4)^2$ **74.** $[(x^3y^8z)^7]^3$ **75.** $\dfrac{x^{12}}{x^3}$ **76.** $\dfrac{x^{23}}{x^{16}}$

77. $\dfrac{x^{31}}{x^8}$ **78.** $\dfrac{y^6}{y}$ **79.** $\dfrac{x^8}{y^2}$ **80.** $\left(\dfrac{x^5}{x^2}\right)^4$

81. $\left(\dfrac{x^9}{x^4}\right)^5$ **82.** $\left(\dfrac{x^{15}}{x^3}\right)^9$ **83.** $\dfrac{(x^4)^6}{(x^3)^5}$ **84.** $\left(\dfrac{x}{y}\right)^{23}$

85. $\left(\dfrac{x^9}{y}\right)^5$ **86.** $\left(\dfrac{x^6}{y^4}\right)^5$ **87.** $\left(\dfrac{x^8}{y^2}\right)^4$ **88.** $\left(\dfrac{3x^2}{y^4}\right)^3$

89. $\left(\dfrac{6x^8}{y}\right)^2$ **90.** $(5x^4)(12x^7)$ **91.** $(3x^7)^2(2x^6)^3$ **92.** $(4x^8y)^2(x^6y^2)^3$

93. $[(x^3y^5)^3]^4[(x^3y)^4]^3$ **94.** $\dfrac{15x^8y^7}{3x^2y^3}$ **95.** $\dfrac{-12x^7y^9}{3x^3y}$ **96.** $\dfrac{15x^4y^7}{40y^3}$

97. $\left(\dfrac{3x^4}{7y^8}\right)^2$ **98.** $\left(\dfrac{2x^8}{3y}\right)^3$ **99.** $\left(\dfrac{6x^9y^{12}}{3x^5y^4}\right)^2$ **100.** $\left(\dfrac{15x^{10}y^5}{5x^2y}\right)^3$

101. $\left(\dfrac{16xy^7}{24y^3}\right)^3$ **102.** $\dfrac{(6x^4y^8)^2}{(x^2y)^3}$ **103.** $\dfrac{(2x^3y^6)^5}{(4xy^{15})^2}$ **104.** $\left(\dfrac{12x^6y^4}{4xy^3}\right)^2(4x^8y^2)^2$

105. $\dfrac{(5xy^3)(8x^4y^9)^2}{10x^3y^7}$ **106.** $\dfrac{(4x^8y^3)(6xy^7)^2}{12x^5y^6}$ **107.** $\dfrac{(9xy^3)(2x^2y^6)^3}{(6x^2y)^2}$ **108.** $\dfrac{(6x^3y^5)(3x^4y)^3}{(9x^3y^2)^2}$

109. $\dfrac{(2x^2y^3)^4(3xy^4)^2}{(4x^3y)^2}$ **110.** $\dfrac{(6x^6y^5)^2(2x^3y^2)^2}{(3x^4y^3)^2}$

3.2 Integer Exponents

In Section 3.1 we learned how to work with natural number exponents. In this section we extend our work to integer exponents. Consider the following problem from arithmetic.

$$\frac{3^4}{3^4} = \frac{\cancel{3} \cdot \cancel{3} \cdot \cancel{3} \cdot \cancel{3}}{\cancel{3} \cdot \cancel{3} \cdot \cancel{3} \cdot \cancel{3}}$$

$$= \frac{1}{1}$$

$$= 1$$

Suppose that we wanted to simplify the algebraic expression x^4/x^4. We know that

$$\frac{x^4}{x^4} = \frac{x \cdot x \cdot x \cdot x}{x \cdot x \cdot x \cdot x}$$

$$= \frac{\cancel{x} \cdot \cancel{x} \cdot \cancel{x} \cdot \cancel{x}}{\cancel{x} \cdot \cancel{x} \cdot \cancel{x} \cdot \cancel{x}} \qquad \text{Generalizing the preceding argument}$$

$$= \frac{1}{1}$$

$$= 1$$

However,

$$\frac{x^4}{x^4} = x^{4-4} \qquad \text{By the quotient law of exponents}$$

$$= x^0$$

Thus, for the quotient law of exponents to be consistent, x^0 must be 1. This fact, plus the fact that division by zero is undefined, motivates the following definition.

Zero Exponent
If x is any real number other than zero, then $x^0 = 1$.

REMARK   This definition says that any nonzero real number raised to the zero power is 1.

EXAMPLE 3.2.1   Evaluate each of the following.

1. $5^0 = 1$

Try to avoid this mistake:

Incorrect	Correct
$5^0 = 0$	$5^0 = 1$
	The definition states that any non-zero number raised to the zero power is 1, not 0.

2. $(-8)^0 = 1$

3. $-6^0 = -1 \cdot (6^0)$
$$= -1 \cdot 1$$
$$= -1$$

4. $\left(\dfrac{3}{7}\right)^0 = 1$

5. $(9xy^3)^0 = 1; x \neq 0, y \neq 0$

6. $2^0 + 3^0 = 1 + 1 = 2$, whereas $(2 + 3)^0 = 5^0 = 1$

7. 0^0 is undefined.

Consider the following arithmetic problem.

$$\frac{5^3}{5^5} = \frac{\overset{1}{\cancel{5}} \cdot \cancel{5} \cdot \cancel{5}}{\cancel{5} \cdot \cancel{5} \cdot \cancel{5} \cdot 5 \cdot 5}$$

$$= \frac{1}{5 \cdot 5}$$

$$= \frac{1}{5^2} \text{ or } \frac{1}{25}$$

Suppose we want to simplify the algebraic expression x^3/x^5. We know that

$$\frac{x^3}{x^5} = \frac{x \cdot x \cdot x}{x \cdot x \cdot x \cdot x \cdot x}$$

$$= \frac{\overset{1}{\cancel{x}} \cdot \cancel{x} \cdot \cancel{x}}{\cancel{x} \cdot \cancel{x} \cdot \cancel{x} \cdot x \cdot x} \qquad \text{Generalizing the preceding argument}$$

$$= \frac{1}{x^2}$$

However,

$$\frac{x^3}{x^5} = x^{3-5} \qquad \text{If we could apply the quotient law of exponents}$$

$$= x^{-2}$$

Thus, for the quotient law of exponents to be consistent, $x^{-2} = 1/x^2$. This observation leads to the following definition.

Negative Exponent
If x is any real number other than zero, then $$x^{-n} = \frac{1}{x^n}$$

EXAMPLE 3.2.2

Evaluate each of the following.

1. $6^{-2} = \dfrac{1}{6^2}$

$\phantom{6^{-2}} = \dfrac{1}{36}$

Try to avoid this mistake:

Incorrect	**Correct**
$6^{-2} = -36$ A negative exponent on a positive base *never* makes the answer negative.	$6^{-2} = \dfrac{1}{6^2} = \dfrac{1}{36}$

2. $(-2)^{-3} = \dfrac{1}{(-2)^3}$

$\phantom{(-2)^{-3}} = \dfrac{1}{-8}$

$\phantom{(-2)^{-3}} = -\dfrac{1}{8}$

3. $-7^{-2} = -(7)^{-2}$ -2, the exponent, applies only to 7, the base.

$\phantom{-7^{-2}} = -\dfrac{1}{7^2}$

$\phantom{-7^{-2}} = -\dfrac{1}{49}$

4. $2^{-1} + 2^{-2} = \dfrac{1}{2^1} + \dfrac{1}{2^2}$

$$= \dfrac{1}{2} + \dfrac{1}{4}$$

$$= \dfrac{3}{4}$$

NOTE ▶ Using the definitions of this section, we can show that all of the laws of exponents hold for all integer exponents. In addition, we now can state the quotient law of exponents in a more general form.

Quotient Law of Exponents

Let m and n be integers and $x \neq 0$; then

$$\frac{x^m}{x^n} = x^{m-n}$$

EXAMPLE 3.2.3

Simplify each of the following. (Unless otherwise indicated, we *always* write answers with only positive exponents. Also, we assume that none of the variables is zero.)

1. $x^8 \cdot x^{-2} = x^{8+(-2)}$ By the product law

$\qquad = x^6$

2. $(x^4)^{-3} = x^{4(-3)}$ By the power to a power law

$\qquad = x^{-12}$

$\qquad = \dfrac{1}{x^{12}}$

3. $(x^{-2}y)^{-6} = (x^{-2})^{-6}y^{-6}$ By the product to a power law

$\qquad = x^{12} \cdot y^{-6}$ By the power to a power law

$\qquad = \dfrac{x^{12}}{1} \cdot \dfrac{1}{y^6}$

$\qquad = \dfrac{x^{12}}{y^6}$

4. $\dfrac{x^{-5}}{x^{-3}} = x^{-5-(-3)}$ By the quotient law

$\qquad = x^{-5+3}$

$\qquad = x^{-2}$

$\qquad = \dfrac{1}{x^2}$

5. $\left(\dfrac{3}{7}\right)^{-2} = \dfrac{3^{-2}}{7^{-2}}$ By the quotient to a power law

$\qquad = \dfrac{\dfrac{1}{3^2}}{\dfrac{1}{7^2}}$

$\qquad = \dfrac{1}{3^2} \cdot \dfrac{7^2}{1}$

$\qquad = \dfrac{7^2}{3^2} \text{ or } \dfrac{49}{9}$

NOTE ▶ This problem illustrates the following property. If a negative exponent is on a *factor* in the denominator of a fraction, the exponent can be made positive by taking that factor and making it a factor of the numerator. We have a similar property if the negative exponent is on a *factor* in the numerator.

6. $\dfrac{x^{-7}}{y^{-9}} = \dfrac{x^{-7}}{y^{-9}}$

$\qquad = \dfrac{y^9}{x^7}$

Here we had negative exponents on factors and used the preceding rule.

NOTE ▶ $$\dfrac{1}{x^{-n}} = \dfrac{1}{\dfrac{1}{x^n}} = 1 \div \dfrac{1}{x^n} = 1 \cdot \dfrac{x^n}{1} = x^n$$

Try to avoid this mistake:

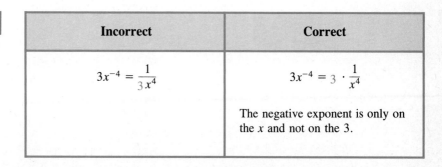

Incorrect	Correct
$3x^{-4} = \dfrac{1}{3x^4}$	$3x^{-4} = 3 \cdot \dfrac{1}{x^4}$
	The negative exponent is only on the x and not on the 3.

7. $\dfrac{5x^{-2}}{y^{-10}} = \dfrac{5\,x^{-2}}{y^{-10}}$

$\quad\quad = \dfrac{5y^{10}}{x^2}$

8. $\dfrac{3x^6}{20x^{10}} = \dfrac{3}{20x^{10-6}}$

$\quad\quad = \dfrac{3}{20x^4}$

NOTE ▶ When a variable has a greater exponent in the denominator, we subtract the exponent in the numerator from the exponent in the denominator. In this way we avoid introducing negative exponents.

9. $\dfrac{10x^5y^{-7}}{15x^2y^3} = \dfrac{10x^5\,y^{-7}}{15x^2y^3}$

$\quad\quad = \dfrac{10x^5}{15x^2y^3y^7}$

$\quad\quad = \dfrac{2 \cdot 5 \cdot x^5}{3 \cdot 5 \cdot x^2y^{10}}$

$\quad\quad = \dfrac{2x^5}{3x^2y^{10}}$

$\quad\quad = \dfrac{2x^3}{3y^{10}}$

10. $(-3x^{-2}y^3)^{-2}(2xy^{-4})^3 = (-3)^{-2}(x^{-2})^{-2}(y^3)^{-2}2^3x^3(y^{-4})^3$

$\quad\quad\quad\quad = (-3)^{-2} \cdot x^4 \cdot y^{-6} \cdot 2^3 \cdot x^3 \cdot y^{-12}$

$\quad\quad\quad\quad = (-3)^{-2} \cdot 2^3 \cdot x^4 \cdot x^3 \cdot y^{-6} \cdot y^{-12}$

$\quad\quad\quad\quad = (-3)^{-2} \cdot 2^3 \cdot x^7 \cdot y^{-18}$

$\quad\quad\quad\quad = \dfrac{1}{(-3)^2} \cdot \dfrac{2^3}{1} \cdot \dfrac{x^7}{1} \cdot \dfrac{1}{y^{18}}$

$\quad\quad\quad\quad = \dfrac{8x^7}{9y^{18}}$

11. $\left(\dfrac{7x^{-3}y^4}{8x^{-8}y^6}\right)^2 = \left(\dfrac{7x^8y^4}{8x^3y^6}\right)^2$

$\quad\quad\quad = \left(\dfrac{7x^5}{8y^2}\right)^2$

$\quad\quad\quad = \dfrac{(7x^5)^2}{(8y^2)^2}$

$\quad\quad\quad = \dfrac{7^2(x^5)^2}{8^2(y^2)^2}$

$\quad\quad\quad = \dfrac{49x^{10}}{64y^4}$

12. $\dfrac{(5x^2y^{-3})^{-2}}{(2xy^{-4})^3} = \dfrac{5^{-2}(x^2)^{-2}(y^{-3})^{-2}}{2^3x^3(y^{-4})^3}$

$\qquad\qquad = \dfrac{5^{-2}x^{-4}y^6}{2^3x^3y^{-12}}$

$\qquad\qquad = \dfrac{y^{12}y^6}{2^35^2x^3x^4}$

$\qquad\qquad = \dfrac{y^{18}}{200x^7}$

13. $\dfrac{(3xy^{-2})(2x^{-2}y)^{-3}}{(5x^{-1}y^{-3})^2} = \dfrac{3xy^{-2}2^{-3}(x^{-2})^{-3}y^{-3}}{5^2(x^{-1})^2(y^{-3})^2}$

$\qquad\qquad = \dfrac{3\cdot x\cdot y^{-2}\cdot 2^{-3}\cdot x^6\cdot y^{-3}}{5^2\cdot x^{-2}\cdot y^{-6}}$

$\qquad\qquad = \dfrac{3\cdot 2^{-3}\cdot x\cdot x^6\cdot y^{-2}\cdot y^{-3}}{5^2\cdot x^{-2}\cdot y^{-6}}$

$\qquad\qquad = \dfrac{3\cdot 2^{-3}\cdot x^7y^{-5}}{5^2\cdot x^{-2}\cdot y^{-6}}$

$\qquad\qquad = \dfrac{3\cdot x^7\cdot x^2\cdot y^6}{5^2\cdot 2^3\cdot y^5}$

$\qquad\qquad = \dfrac{3x^9y}{200}$

Scientific Notation

In science and mathematics we sometimes come across problems with very large or very small numbers. The laws of exponents and exponential notation can be used to simplify the work in these types of problems. For example, it is known that the average distance from the earth to the moon is approximately 384,000,000 m. Note that

$$384{,}000{,}000 = 3.84 \times 100{,}000{,}000$$
$$= 3.84 \times 10^8$$

We now have the 384,000,000 expressed in a concise form. This example illustrates the following definition.

Scientific Notation

A number written in the form $n \times 10^p$ where $1 \le n < 10$ and p is an integer is said to be written in **scientific notation**.

One centimeter is approximately 0.0000062 mile. Let's write 0.0000062 in scientific notation:

$$0.0000062 = \frac{62}{10,000,000}$$

$$= \frac{6.2}{1,000,000}$$

$$= 6.2 \times \frac{1}{1,000,000}$$

$$= 6.2 \times \frac{1}{10^6}$$

$$= 6.2 \times 10^{-6}$$

To write a number in scientific notation we use the following steps.

1. Move the decimal point to the right of the first nonzero digit in the number.

2. Count the number of places the decimal point has been moved. This number is used to determine the exponent on 10.

3. If the decimal point is moved to the right, the exponent is the negative of the integer found in step 2. If the decimal point is moved to the left, the exponent is the positive integer found in step 2.

EXAMPLE 3.2.4

Write the following numbers in scientific notation.

1. $54,890,000 = 5.4890000 \times 10^7$ Move the decimal point seven
   $\qquad\qquad\quad = 5.489 \times 10^7$ places to the left, which
   makes the exponent positive.

2. $0.000071 = 000007.1 \times 10^{-5}$ Move the decimal point five
   $\qquad\quad = 7.1 \times 10^{-5}$ places to the right, which
   makes the exponent negative.

EXAMPLE 3.2.5

Use scientific notation to evaluate the following expression.

$$\frac{(142,000,000)(3,600,000)}{7,200,000,000} = \frac{(1.42 \times 10^8)(3.6 \times 10^6)}{7.2 \times 10^9}$$

$$= \frac{(1.42)(3.6) \times (10^8)(10^6)}{7.2 \times 10^9}$$

$$= \frac{(1.42)(3.6) \times 10^{14}}{7.2 \times 10^9}$$

$$= \frac{(1.42)(3.6) \times 10^5}{\underset{2}{7.2}}$$

$$= \frac{1.42}{2} \times 10^5$$

$$= 0.71 \times 10^5$$

$$= 71{,}000 \text{ or } 7.1 \times 10^4$$

EXAMPLE 3.2.6

Farmer Finney grows "The Pride of Moss County" watermelons. His watermelons have approximately 12,400 seeds in each melon. This summer, Farmer Finney expects a crop of 15,000 watermelons. Use scientific notation to find the number of seeds in Farmer Finney's crop.

SOLUTION

We first express each number in scientific notation:

$$12{,}400 = 1.24 \times 10^4$$

$$15{,}000 = 1.5 \times 10^4$$

(Number of seeds per melon) · (Number of melons) = (Total number of seeds)

$$1.24 \times 10^4 \qquad \cdot \; 1.5 \times 10^4 = (1.24)(1.5) \times (10^4)(10^4)$$

$$= 1.86 \times 10^8$$

$$\text{or } 186{,}000{,}000 \text{ seeds}$$

EXERCISES 3.2

Evaluate each of the following.

1. 3^0

2. $(-2)^0$

3. -8^0

4. $\left(-\dfrac{2}{3}\right)^0$

5. 7^{-1}

6. $(-3)^{-1}$

7. 4^{-2}

8. $(-3)^{-4}$

9. -3^{-4}

10. 2^{-3}

11. $(-5)^{-3}$

12. -5^{-2}

13. -7^{-2}

14. $\left(\dfrac{2}{3}\right)^{-1}$

15. $\left(\dfrac{6}{5}\right)^{-2}$

16. $\left(\dfrac{-3}{4}\right)^{-2}$

17. $-\left(\dfrac{3}{2}\right)^{-2}$

18. $2^0 + 2^{-1}$

19. $3^{-1} - 3^{-2}$

20. $4^{-1} + 2^{-2}$

21. $3^0 + 3 + 3^{-1}$

22. $\left(\dfrac{1}{2}\right)^{-1} - \left(\dfrac{1}{2}\right)^{-2}$

23. $5^8 \cdot 5^{-6}$

24. $3^{-1} \cdot 3^{-2}$

25. $(-4)^{-1} \cdot (-4)^{-2}$

26. $(2^{-2})^{-3}$

27. $\left[\left(\dfrac{2}{3}\right)^{-1}\right]^{-2}$

28. $\left[\left(\dfrac{5}{2}\right)^{-1}\right]^{2}$

29. $(3^{-2})^2$

30. $[(-2)^{-2}]^{-2}$

31. $\dfrac{7^{-4}}{7^{-2}}$

32. $\dfrac{2^{-1}}{2^3}$

33. $\dfrac{4}{4^{-2}}$

34. $\dfrac{3^{-2}}{2^{-3}}$

35. $\dfrac{4^2}{3^{-1}}$

36. $\dfrac{3^{-2}}{2^2}$

Simplify each of the following. (Remember to write your answers with only positive exponents.)

37. $x^7 \cdot x^{-4}$

38. $x^{-3} \cdot x^{-5}$

39. $x^{-4} \cdot x$

40. $x^3 \cdot x^{-6}$

41. $x^{-2} \cdot x^{-3} \cdot x^4$

42. $x^{-3} \cdot x^{-5} \cdot x^8$

43. $(x^3)^{-2}$

44. $(x^{-5})^3$

45. $(x^{-3})^{-5}$

46. $[(x^2)^3]^{-4}$

47. $[(x^{-2})^{-3}]^{-2}$

48. $(x^{-3}y^4)^{-2}$

49. $(x^{-2}y^{-1})^{-2}$

50. $(x^2y^3)^{-5}$

51. $(x^{-2}y^{-4})^5$

52. $\dfrac{x^{-5}}{x^{-7}}$

53. $\dfrac{x^5}{x^{-2}}$

54. $\dfrac{x^{-2}}{x^8}$

55. $\dfrac{x^{-2}}{x^{-2}}$

56. $\left(\dfrac{x^{-2}}{x^{-4}}\right)^{-3}$

57. $\left(\dfrac{x^{-3}}{x^{-1}}\right)^4$

58. $\left(\dfrac{x^3}{x^5}\right)^{-2}$

59. $\left(\dfrac{x^{-3}}{x^4}\right)^{-2}$

60. $\dfrac{(x^{-2})^3}{(x^4)^{-5}}$

61. $\dfrac{x^{-8}}{y^{-3}}$

62. $\dfrac{7x^{-4}}{y^{-6}}$

63. $\dfrac{6x^{-2}}{y^3}$

64. $\dfrac{8x^{10}}{9x^{15}}$

65. $\dfrac{12x}{16x^9}$

66. $\dfrac{21x^7}{14x}$

67. $\dfrac{20x^4}{5x^2}$

68. $\dfrac{9x^2y^{-8}}{12x^4y^{-4}}$

69. $\dfrac{24x^{-3}y^{-5}}{20x^{-1}y^{-11}}$

70. $\dfrac{28x^{-4}y^{-3}}{8x^3y^5}$

71. $(-5xy^{-3})^{-2}(3x^{-4}y^{-1})^2$

72. $(3xy^4)^{-2}(-2x^2y^3)^{-3}$

73. $(6^2x^{-3}y^8)^0(5x^0y^4)^3$

74. $\left(\dfrac{5x^{-4}y^3}{8x^{-7}y^6}\right)^2$

75. $\left(\dfrac{2x^{-2}y^4}{3x^5y^{-5}}\right)^3$

76. $\left(\dfrac{4x^{-2}y^4}{5x^{-3}y^{-4}}\right)^{-2}$

77. $\dfrac{(4x^{-2}y^{-3})^2}{(3x^4y^{-1})^3}$

78. $\dfrac{(5xy^4)^{-2}}{(2x^{-2}y^3)^{-3}}$

79. $\dfrac{(2x^3y^{-4})^3}{(4x^{-2}y^{-5})^{-2}}$

80. $\dfrac{(5x^3y^{-3})(2xy^{-4})^{-2}}{(3x^2y^3)^2}$

81. $\dfrac{(2x^{-2}y^{-3})^2(3x^4y^{-2})^{-2}}{(5x^4y^{-2})^{-1}}$

82. $\dfrac{(8x^3y^2)(x^{-3}y^{-4})^2}{(2x^{-4}y^3)^2(x^5y^{-8})^{-3}}$

Write the following numbers in scientific notation.

83. 5,376,000

84. 8,492,000

85. 75,100,000

86. 69,400,000

87. 1023

88. 4108

89. 101,000,000

90. 230,000,000

91. 0.00079

92. 0.00084

93. 0.000000605

94. 0.000000701

95. 0.024

96. 0.058

97. 0.004009

98. 0.007007

Use scientific notation to evaluate the following.

99. $\dfrac{(42,000,000)(1,500,000)}{45,000,000,000}$

100. $\dfrac{(2,400,000)(160,000,000)}{6,400,000,000}$

101. $\dfrac{(0.00000054)(0.00041)}{0.000000018}$

102. $\dfrac{(0.0000039)(0.00067)}{0.00000013}$

103. $\dfrac{(32,000)(5,100)}{(160,000)(1,700,000)}$

104. $\dfrac{(4400)(52,000)}{(11,000)(1,300,000)}$

105. $\dfrac{(0.000000072)(0.000084)}{(0.0000000096)(0.0000042)}$

106. $\dfrac{(0.00000081)(0.00036)}{(0.000027)(0.000000048)}$

107. $\dfrac{(124,000,000)(0.0000054)}{(0.0000000009)(3,100,000)}$

108. $\dfrac{(30,800,000)(0.000051)}{(0.000000017)(770,000)}$

109. $\dfrac{(0.00000532)(605,000,000)}{(9,680,000)(0.000133)}$

110. $\dfrac{(0.000728)(12,500,000)}{(5,000,000,000)(0.000091)}$

111. Chemists use Avogadro's number

$$602,300,000,000,000,000,000,000$$

which is the number of molecules in a mole of any substance. How many molecules are in 850 moles?

112. A light-year is the distance that light travels in 1 year. It is approximately equal to 5,870,000,000,000 miles. If a parsec is 3.26 light-years, how many miles are in 1 parsec?

113. Proxima Centauri, the nearest star to the sun, is about 4.3 light-years from the sun (see problem 112). What is the distance in miles from the sun to Proxima Centauri?

114. Squinty McPherson, the Moss County agricultural agent, has determined that the average mature Angus bull has approximately 15,800,000 bull hairs. There are 28 mature Angus bulls in Moss County. Approximately how many bull hairs do they have? Squinty wants your answer in scientific notation.

115. The average hare family in Moss County has 12 members. There are approximately 289,000 hare families in Moss county. Approximately how many hares are in Moss County? Express your answer in scientific notation.

3.3 Adding and Subtracting Polynomials

In Chapter 2 we introduced the concept of algebraic expressions. The following are examples of algebraic expressions.

$$5x^2 + 2x + 13, \qquad 8x^3 - \frac{7}{x^4}, \qquad \sqrt{2x^2 + 3y^2}, \qquad 25x^2 - 9y^2, \qquad \tfrac{7}{8}x^5$$

In the remainder of this chapter, we will study a special type of algebraic expression called a polynomial.

A **polynomial** is a finite sum of terms. Each term is a number, or the product of a number and one or more variable factors, raised to a whole number power.

REMARKS

1. In general, if an expression has variables:

a. under square root signs, $\sqrt{2x^2 + 3y^2}$

b. in denominators, $8x^3 - \dfrac{7}{x^4}$

c. with negative exponents, $9x^{-2} - 5x^{-1} + 12$

then the expression is *not* a polynomial.

2. The following expressions are polynomials.

$$5x^2 + 2x + 13, \qquad 25x^2 - 9y^2 = 25x^2 + (-9y^2), \qquad \tfrac{7}{8}x^5, \qquad 4$$

If a polynomial contains only one term, it is called a **monomial**; if it contains two terms, it is a **binomial**; and if it contains three terms it is a **trinomial**. If a polynomial has more than three terms, normally we use no special name. From the previous remark, $\tfrac{7}{8}x^5$ and 4 are monomials, $25x^2 - 9y^2$ is a binomial, and $5x^2 + 2x + 13$ is a trinomial. The monomial 4 is called a **constant**, because it has no variable factors.

Recall from Chapter 2 that the numerical factor of each term in an algebraic expression is called the numerical coefficient, or just **coefficient,** of that term. The monomial $2x$ has a coefficient of 2 and a variable factor of x. The monomial $\tfrac{5}{6}x^2y$ has a coefficient of $\tfrac{5}{6}$ and two variable factors of x and one variable factor of y. In the polynomial

$$x^4 - 7x^3 + \frac{2x^2}{5} - x + 37$$

the coefficients are $1, -7, \dfrac{2}{5}, -1$, and 37 since

$$x^4 - 7x^3 + \frac{2x^2}{5} - x + 37 = 1x^4 - 7x^3 + \frac{2}{5}x^2 + (-1)x + 37$$

Another important characteristic of a monomial is its number of variable factors. The **degree of a monomial** is the sum of the exponents of the variables.

EXAMPLE 3.3.1

Find the degree of each of the following monomials.

1. $\tfrac{7}{8}x^5$ The degree is 5.

2. $\tfrac{5}{6}x^2y$ Since $y = y^1$, the degree is $2 + 1 = 3$.

3. $7x^8y^2z^4$ The degree is $8 + 2 + 4 = 14$.

4. $15 = 15 \cdot 1 = 15 \cdot x^0$ The degree is 0.

NOTE ▶   The degree of a constant, other than zero, is said to be zero.

5. $4^2x^3y^4$ The degree is $3 + 4 = 7$.

NOTE ▶   The exponent of the coefficient does not count toward the degree.

Because a polynomial is a sum of terms or monomials, we define the **degree of a polynomial** to be the same as that of its term of highest degree.

EXAMPLE 3.3.2

Find the degree of each of the following polynomials.

1. $2xy^3 + 8x^3 - 9x^5y^6$

The degrees of the three terms are 4, 3, and 11, so the degree of the polynomial is 11.

2. $5x^3 - 7x^2 + 3x - 4$

The degree is 3, the same as that of the first term.

When a polynomial contains only one variable, it is common practice to arrange the terms so that the degrees are in descending order, as in $5x^3 - 7x^2 + 3x - 4$. We call this the **standard form for a polynomial in one variable**. The coefficient of the highest degree term is called the **leading coefficient**, because it is written first when the polynomial is in standard form. The leading coefficient of $5x^3 - 7x^2 + 3x - 4$ is 5.

EXAMPLE 3.3.3

Write each of the following polynomials in standard form, and determine the leading coefficient.

1. $x^2 - 7x + 4$

This polynomial is in standard form; the leading coefficient is 1.

2. $-7x^2 + 8x^5 - 4x + 12$

Standard form is $8x^5 - 7x^2 - 4x + 12$; the leading coefficient is 8.

In Chapter 2 we used the distributive property to combine like terms. We now use the distributive property to simplify some polynomials by combining like terms.

EXAMPLE 3.3.4

Simplify each of the following polynomials by combining like terms, if possible. If the polynomial contains only one variable, write your answer in standard form.

1. $5x^2 + 7x + 3x - 2x^2 + 29$

$= 5x^2 - 2x^2 + 7x + 3x + 29$ Commutative property of addition

$= (5 - 2)x^2 + (7 + 3)x + 29$ Distributive property

$= 3x^2 + 10x + 29$

2. $7m - 2n - 12m + 2n$

$= 7m - 12m - 2n + 2n$ Commutative property of addition

$= (7 - 12)m + (-2 + 2)n$ Distributive property

$= -5m + 0n$

$= -5m$ Since $0n = 0$

3. $-5x^2y + 9xy^2 + 8x + 2x - 15$

$= -5x^2y + 9xy^2 + (8 + 2)x - 15$

$= -5x^2y + 9xy^2 + 10x - 15$

NOTE ▶ $-5x^2y$ and $9xy^2$ cannot be combined, because they are *not* like terms. Remember, for terms to be like terms, the exponents on the variable factors must be *exactly* the same.

Let's use the distributive property to find multiples of polynomials.

EXAMPLE 3.3.5

Use the distributive property to perform the following multiplications.

1. $7(2x + 5y) = 7(2x) + 7(5y)$

$= 14x + 35y$

2. $4(3x^2 - 2x + 12) = 4(3x^2) - 4(2x) + 4(12)$

$= 12x^2 - 8x + 48$

3. $-5(2x^2 - 3xy + 9y^2) = -5(2x^2) - (-5)(3xy) + (-5)(9y^2)$

$= -10x^2 - (-15xy) + (-45y^2)$

$= -10x^2 + 15xy - 45y^2$

We now are ready to add or subtract polynomials or multiples of polynomials.

EXAMPLE 3.3.6

Perform the indicated operations, and simplify each answer by combining like terms.

1. $(3x^2 + 7x - 8) + (-5x^2 + 12x - 9)$

$= 3x^2 + 7x - 8 - 5x^2 + 12x - 9$

$= 3x^2 - 5x^2 + 7x + 12x - 8 - 9$

$= -2x^2 + 19x - 17$

2. $\left(\frac{1}{2}x^2 - \frac{2}{3}x + 1\right) + \left(\frac{3}{4}x^2 - \frac{1}{6}x - \frac{5}{7}\right)$

$= \frac{1}{2}x^2 - \frac{2}{3}x + 1 + \frac{3}{4}x^2 - \frac{1}{6}x - \frac{5}{7}$

$= \frac{1}{2}x^2 + \frac{3}{4}x^2 - \frac{2}{3}x - \frac{1}{6}x + 1 - \frac{5}{7}$ Obtain an LCD for all

$= \frac{2}{4}x^2 + \frac{3}{4}x^2 - \frac{4}{6}x - \frac{1}{6}x + \frac{7}{7} - \frac{5}{7}$ the like terms.

$= \frac{5}{4}x^2 - \frac{5}{6}x + \frac{2}{7}$

3. $(5x^2 - 2xy) - (7x^2 - 3xy)$

This expression is the same as $(5x^2 - 2xy) + (-1)(7x^2 - 3xy)$. *Now we simply distribute a coefficient of -1 across the second polynomial:*

$(5x^2 - 2xy) + (-1)(7x^2 - 3xy) = 5x^2 - 2xy - 7x^2 + 3xy$

$= 5x^2 - 7x^2 - 2xy + 3xy$

$= -2x^2 + xy$

NOTE ▶   *When subtracting a polynomial, we change the signs in the second polynomial and then add.*

Try to avoid this mistake:

Incorrect	Correct
$(5x^2 - 2xy) - (7x^2 - 3xy)$ $= 5x^2 - 2xy - 7x^2 - 3xy$	$(5x^2 - 2xy) - (7x^2 - 3xy)$ $= 5x^2 - 2xy - 7x^2 + 3xy$
The signs of both $7x^2$ and $-3xy$ should have been changed.	When you subtract, you must change all of the signs in the second polynomial.

4. $(x^2 + 3x - 4) - (5x^2 + 2x - 9) = x^2 + 3x - 4 - 5x^2 - 2x + 9$

$= x^2 - 5x^2 + 3x - 2x - 4 + 9$

$= -4x^2 + x + 5$

5. $3(2x - 7) + 4(9x + 8) = 6x - 21 + 36x + 32$

$$= 6x + 36x - 21 + 32$$

$$= 42x + 11$$

6. $5(4x - 9) - 6(3x + 5) = 20x - 45 - 18x - 30$

$$= 20x - 18x - 45 - 30$$

$$= 2x - 75$$

Try to avoid this mistake:

Incorrect	Correct
$5(4x - 9) - 6(3x + 5)$ $= 20x - 45 - 18x \overset{+}{} 30$ Remember that we are distributing -6 across the entire binomial $3x + 5$.	$5(4x - 9) - 6(3x + 5)$ $= 20x - 45 - 18x \overset{-}{} 30$

When adding or subtracting polynomials, we can sometimes make the addition or subtraction less complicated by arranging like terms vertically.

EXAMPLE 3.3.7

Perform the indicated operations.

1.
$$\begin{array}{r} 5x^2 - 9x + 2 \\ + \underline{3x^2 + x - 9} \\ 8x^2 - 8x - 7 \end{array}$$
Note that the like terms are in the same column.

2.
$$\begin{array}{r} 3x^3 - 2x^2 + 4x - 1 \\ + \underline{\quad - 6x^2 - 4x + 7} \\ 3x^3 - 8x^2 \qquad + 6 \end{array}$$
Note that $4x + (-4x) = 0x$ $= 0$.

3.
$$\begin{array}{r} 9x^2 - 10x - 3 \\ - \underline{4x^2 + 3x - 6} \end{array}$$
An easy way to do this subtraction is to change the signs of *all* of the terms of the bottom polynomial, and add.
$$\begin{array}{r} 9x^2 - 10x - 3 \\ + \underline{-4x^2 - 3x + 6} \\ 5x^2 - 13x + 3 \end{array}$$

4. $-2x^2 + 5xy - 8y^2$
    $\underline{\ 6x^2 - 2xy - \ y^2}$

    $\ -2x^2 + 5xy - 8y^2$
    $\underline{+\ -6x^2 + 2xy + \ y^2}$
    $\ -8x^2 + 7xy - 7y^2$ Change the signs and add.

EXERCISES 3.3

Simplify each of the following polynomials by combining like terms, if possible. After simplifying, state the degree and specify which are monomials, binomials, and trinomials.

1. $5x - 7x^2 + 11$

2. $-3x - 5x^2 + 2$

3. $6x - 8 + 7x$

4. $9x - 12 + 4$

5. $7xy + 4x^2 - 3xy + y^3 - 7$

6. $3x^2 + 12xy - 9x^2 + 2y^4 - 1$

7. 9

8. 14

9. $5xy + x^4 - 5xy - 4y^2$

10. $9y^2 - 8xy - x^6 + 8xy$

11. $\frac{2}{3}x^6y^2z$

12. $\frac{4}{7}xy^4z^8$

13. $5x^2 + 7x - 4 - 9x^2 - 2x - 8$

14. $9x^2 + 3x - 7 + 4x^2 - 11x - 4$

15. $\frac{2}{3}x^2 - \frac{1}{9}x + 2 + \frac{1}{2}x^2 + \frac{1}{3}x - \frac{4}{5}$

16. $\frac{3}{4}x^2 + \frac{1}{5}x + 3 + \frac{1}{3}x^2 - \frac{1}{2}x - \frac{8}{9}$

17. $-1.21x^3 + 3.24z^2 - 6.48x^3 - 0.002z^2$

18. $-4.68x^3 + 1.19z^2 - 2.07x^3 - 0.05z^2$

19. $2x - 3y + 4z - 8 + 7x - 8y + 11$

20. $-5x + 2y - 11z - 14 + 2y - 9z + 6$

21. $\frac{1}{8}x^3y - \frac{2}{3}x^4y^2 - \frac{3}{4}x^3y + \frac{1}{2}x^4y^2$

22. $\frac{3}{8}x^3y - \frac{1}{3}x^4y^2 - \frac{1}{4}x^3y + \frac{3}{2}x^4y^2$

23. $\frac{5}{8}xz^2 - \frac{2}{3}x^2y^2 + \frac{1}{4}xz^2 + \frac{4}{6}x^2y^2$

24. $\frac{1}{2}xz^2 - \frac{3}{4}x^3y^3 + \frac{1}{3}xz^2 + \frac{6}{8}x^3y^3$

Simplify each of the following polynomials by combining like terms, if possible. After simplifying, write the polynomial in standard form and determine the leading coefficient.

25. $-7 - 2x + 5x^2$

26. $-9 - x + 4x^2$

27. $3x^2 - 4x^4 - x^9 - 2$

28. $5x^2 - 3x^3 - x^8 - 6$

29. $112x^5 + 19x^{47}$

30. $206x^2 + 23x^{35}$

31. $\dfrac{x^7}{5} + \dfrac{3x^9}{4} + \dfrac{3x^3}{7} + 1$

32. $\dfrac{x^8}{6} + \dfrac{2x^{10}}{9} + \dfrac{3x^2}{4} + 8$

33. $4x - 9 - 3 - 6x$

34. $7x - 12 - 5 - 10x$

35. $3x^2 - 2x - 9 + 5x^2 + 2x - 14$

36. $9x^2 + 8x - 1 + 2x^2 - 8x - 6$

37. $2x - 5 + 3x^2 + x^3 + 4x - 9$

38. $5x - 12 + 6x + x^3 + 3x^2 - 11$

39. $\frac{1}{5}x + \frac{1}{9} - \frac{2}{3}x + \frac{1}{6}$

40. $\frac{1}{4}x + \frac{1}{6} - \frac{3}{5}x - \frac{2}{9}$

41. $\frac{1}{2}x^2 - \frac{1}{3}x - \frac{3}{4} + \frac{3}{10}x^2 - \frac{5}{6}x - \frac{1}{2}$

42. $\frac{1}{3}x^2 - \frac{3}{8}x - \frac{5}{9} + \frac{5}{6}x^2 - \frac{1}{4}x - \frac{1}{3}$

43. $\frac{1}{9} + \frac{2}{3}x^5 - \frac{1}{4}x - \frac{3}{8}x + \frac{5}{6}$

44. $\frac{3}{10} + \frac{4}{9}x^7 - \frac{2}{3}x^2 + \frac{1}{5} + \frac{1}{6}x^2$

45. $1.24 - 1.21x^2 - 2.36x^2 - 3.06x^2 + 9.56x^4$

46. $3.76 - 2.06x - 5.12x - 1.76x + 7.41x^3$

Perform the indicated operations, and simplify your answer by combining like terms.

47. $(x^2 + 7x - 4) + (5x^2 - 11x - 12)$

48. $(-2x^2 - 9) + (-4x - 16)$

49. $(x^4 + 5x^2 - 3x + 4) + (9x^3 - 6x^2 - 19)$

50. $(2x^2 - 3xy + 4y^2) + (-6x^2 + 9xy - 9y^2)$

51. $(6x^2 - 7xy - y^2) + (9x^2 + 7xy + y^2)$

52. $(x^2 + 6x - 1) - (x^2 - 3x + 9)$

53. $(2x^3 + 11x^2 + x - 4) - (5x^3 - 4x - 12)$

54. $(-x^2 - 3x - 4) - (-5x^2 - x - 2)$

55. $(8x^2 + 2xy - 4y^2) - (5x^2 + 2xy + 9y^2)$

56. $(x^2 - 2xy - y^2) - (5x^2 + xy - y^2)$

57. $\left(\frac{2}{3}x^2 - \frac{1}{2}x + \frac{3}{5}\right) + \left(\frac{3}{4}x^2 - \frac{1}{8}x + \frac{7}{10}\right)$

58. $\left(\frac{5}{9}x^2 + \frac{1}{4}x - \frac{1}{6}\right) + \left(-\frac{2}{3}x^2 - \frac{1}{12}x + 1\right)$

59. $\left(\frac{1}{2}x^2 - \frac{1}{3}x - \frac{1}{4}\right) - \left(\frac{2}{3}x^2 - \frac{3}{4}x - \frac{5}{6}\right)$

60. $\left(-\frac{5}{4}x^2 + \frac{1}{3}x - \frac{3}{8}\right) - \left(-\frac{5}{3}x^2 + \frac{1}{3}x - \frac{1}{4}\right)$

61. $(1.12x^2 - 3.04x - 4.79) + (6.19x^2 + 1.42x - 6.34)$

62. $(-4.06x^2 + 1.16x - 3.41) - (2.12x^2 - 3.47x - 6.77)$

63.
$$\begin{array}{r} 5x^2 + 3x - 4 \\ + \underline{-2x^2 - 7x + 12} \end{array}$$

64.
$$\begin{array}{r} x^3 + 3x^2 - 2x - 9 \\ + \underline{- 2x^2 + 5x + 1} \end{array}$$

65.
$$\begin{array}{r} -8x^2 + 7xy - 4y^2 \\ + \underline{-2x^2 - 9xy + 16y^2} \end{array}$$

66.
$$\begin{array}{r} 8x^2 - 4x + 2 \\ - \underline{5x^2 + 3x - 7} \end{array}$$

67.
$$\begin{array}{r} 4x^3 - 5x^2 - 2x - 3 \\ - \underline{- 3x^2 - 4x + 1} \end{array}$$

68.
$$\begin{array}{r} -2x^2 + 3xy - 4y^2 \\ - \underline{-5x^2 - 3xy - y^2} \end{array}$$

69.
$$\begin{array}{r} \frac{4}{3}x^2 + \frac{1}{2}x - \frac{1}{4} \\ + \underline{\frac{1}{6}x^2 - \frac{3}{4}x - \frac{5}{8}} \end{array}$$

70.
$$\begin{array}{r} -\frac{3}{2}x^2 - \frac{1}{4}x - \frac{5}{6} \\ + \underline{-\frac{1}{3}x^2 + \frac{1}{5}x + \frac{2}{3}} \end{array}$$

71.
$$\begin{array}{r} \frac{3}{4}x^2 + \frac{7}{8}x + \frac{1}{3} \\ - \underline{\frac{1}{3}x^2 - \frac{1}{2}x + \frac{2}{5}} \end{array}$$

72.
$$\begin{array}{r} -\frac{3}{4}x^2 - \frac{2}{5}x - \frac{5}{6} \\ - \underline{-\frac{5}{8}x^2 + \frac{3}{10}x - \frac{1}{9}} \end{array}$$

73.
$$\begin{array}{r} 1.24x^2 - 3.96x + 4.09 \\ + \underline{5.61x^2 + 1.45x - 6.34} \end{array}$$

74.
$$\begin{array}{r} -3.14x^2 - 6.24x + 6.89 \\ - \underline{-2.47x^2 + 5.01x + 3.02} \end{array}$$

75. $5(3x + 4) + 6(2x - 7)$

76. $6(x - 8) - 2(3x - 5)$

77. $-7(3x + 4) - 2(x - 3)$

78. $5(x^2 - 3x + 4) + 2(2x^2 + 7x + 6)$

79. $2(-4x^2 - 3x + 7) - 5(2x^2 + x - 2)$

80. $-5(8x^2 - y^2) - 2(4x^2 - 9y^2)$

81. $4(x^3 - 3x^2 + 7) - 5(-2x^2 + 3x - 8)$

82. $3(4x^2 - 7x + 4) - (x^2 - 9x + 2)$

83. $-2(4x^2 + 3xy - 9y^2) - (2x^2 + 7xy + y^2)$

84. $\frac{2}{3}(5x + 7) + \frac{1}{4}(2x - 1)$

85. $\frac{1}{2}(7x - 3) - \frac{1}{3}(4x + 2)$

86. $\frac{1}{3}\left(\frac{2}{3}x - \frac{1}{2}\right) + \frac{1}{2}\left(\frac{5}{6}x + \frac{2}{3}\right)$

87. $-\frac{1}{2}(4x^2 - 7x + 2) - \frac{2}{3}(3x^2 + 5x - 6)$

88. $6.3(5x - 7) - 4.2(2x - 8)$

89. $3.6(1.9x - 4.3) + 2.7(5.5x - 9.3)$

3.4 Multiplying Polynomials

At the beginning of this chapter the laws of exponents were introduced. In the previous section we learned how to add and subtract polynomials. In this section we use both of these topics when we multiply polynomials. Let's first investigate multiplication problems in which all of the factors are monomials.

EXAMPLE 3.4.1

Perform the indicated multiplications.

1. $3x^4 \cdot 7x^8 = 3 \cdot 7 \cdot x^4 \cdot x^8$ Commutative property

$\qquad = (3 \cdot 7) \cdot (x^4 \cdot x^8)$ Associative property

$\qquad = 21x^{12}$ Product law of exponents

Try to avoid these mistakes:

Incorrect	Correct
1. $3x^4 \cdot 7x^8 = 21x^{32}$	$3x^4 \cdot 7x^8 = 21x^{12}$
	Remember to add the exponents of like bases when multiplying.
2. $5x + 4x = 9x^2$	$5x + 4x = 9x$
	When we add or subtract like terms, only the coefficient changes, not the variable factors.

2. $(-9x^5y^3)(2x^4y^2)$

$\quad = -9 \cdot 2 \cdot x^5 \cdot x^4 \cdot y^3 \cdot y^2$ Commutative property

$\quad = -18x^{5+4}y^{3+2}$ Product law of exponents

$\quad = -18x^9y^5$

3. $(-4x^3y^4z^2)(5x^2y^8z^2)(-2x^9yz^6)$

$\quad = (-4)(5)(-2) \cdot x^3 \cdot x^2 \cdot x^9 \cdot y^4 \cdot y^8 \cdot y \cdot z^2 \cdot z^2 \cdot z^6$

$\quad = 40x^{3+2+9}y^{4+8+1}z^{2+2+6}$

$\quad = 40x^{14}y^{13}z^{10}$

4. $(x^3y^4)^2 = (x^3)^2(y^4)^2$ Product to a power law

$\qquad = x^{3\cdot2}y^{4\cdot2}$ Power to a power law

$\qquad = x^6y^8$

5. $(5xy^3)^2(-3x^4y^7) = 5^2x^2(y^3)^2 \cdot (-3)x^4y^7$

$$= 25x^2y^6 \cdot (-3)x^4y^7$$

$$= 25 \cdot (-3) \cdot x^2 \cdot x^4 \cdot y^6 \cdot y^7$$

$$= -75x^6y^{13}$$

In the previous example we multiplied only monomials. In the next example we see how to multiply a monomial by a polynomial.

EXAMPLE 3.4.2

Perform the indicated multiplications.

1. $4x(3x + 7)$

To do this multiplication, we use both the product law of exponents and the distributive property:

$$4x(3x + 7) = 4x \cdot 3x + 4x \cdot 7 \qquad \text{Distributive property}$$

$$= 4 \cdot 3 \cdot x \cdot x + 4 \cdot 7 \cdot x \qquad \text{Commutative property}$$

$$= 12x^2 + 28x \qquad \text{Product law of exponents}$$

2. $7x^2(3x^2 - 4x - 2) = 7x^2 \cdot 3x^2 - 7x^2 \cdot 4x - 7x^2 \cdot 2$

$$= 7 \cdot 3 \cdot x^2 \cdot x^2 - 7 \cdot 4 \cdot x^2 \cdot x - 7 \cdot 2 \cdot x^2$$

$$= 21x^4 - 28x^3 - 14x^2$$

3. $\quad -8x^3y^2(2x + 3y^3 - 4x^2y^5)$

$$= (-8x^3y^2) \cdot 2x + (-8x^3y^2) \cdot 3y^3 - (-8x^3y^2) \cdot 4x^2y^5$$

$$= -8 \cdot 2 \cdot x^3 \cdot x \cdot y^2 + (-8) \cdot 3 \cdot x^3 \cdot y^2 \cdot y^3$$
$$\quad -(-8) \cdot 4 \cdot x^3 \cdot x^2 \cdot y^2 \cdot y^5$$

$$= -16x^4y^2 + (-24x^3y^5) - (-32x^5y^7)$$

$$= -16x^4y^2 - 24x^3y^5 + 32x^5y^7$$

We have considered only multiplication problems involving at least one monomial factor. In the remainder of this section we investigate multiplication problems in which none of the factors are monomials.

EXAMPLE 3.4.3

Perform the indicated multiplications.

1. $(2x + 9)(3x + 7)$

This multiplication problem is somewhat similar to part 1 of Example 3.4.2. However, in this problem the first factor is a binomial. Therefore, we distribute the binomial $2x + 9$ across $3x + 7$:

$$(2x + 9)(3x + 7)$$

$= (2x + 9) \cdot 3x + (2x + 9) \cdot 7$	Distributive property
$= 2x \cdot 3x + 9 \cdot 3x + 2x \cdot 7 + 9 \cdot 7$	Distributive property
$= 6x^2 + 27x + 14x + 63$	
$= 6x^2 + 41x + 63$	Combine like terms.

NOTE ▶ In this problem we had to apply the distributive property two times.

2. $(x - 4)(x - 5) = (x - 4)x - (x - 4)5$

To avoid making a sign error, let's switch the order of the factors on the right-hand side:

$= x(x - 4) - 5(x - 4)$	Commutative property of multiplication
$= x \cdot x - x \cdot 4 - 5 \cdot x - 5(-4)$	
$= x^2 - 4x - 5x + 20$	
$= x^2 - 9x + 20$	

3. $(2x - 3y)(5x + 8y) = (2x - 3y)5x + (2x - 3y)8y$
$$= 5x(2x - 3y) + 8y(2x - 3y)$$
$$= 10x^2 - 15xy + 16xy - 24y^2$$
$$= 10x^2 + xy - 24y^2$$

4. $(4x + 7y)(4x - 7y) = (4x + 7y)4x - (4x + 7y)7y$
$$= 4x(4x + 7y) - 7y(4x + 7y)$$
$$= 16x^2 + 28xy - 28xy - 49y^2$$
$$= 16x^2 - 49y^2$$

5. $\quad (2x - 3)(5x^2 - 4x + 9)$
$$= (2x - 3)5x^2 - (2x - 3)4x + (2x - 3)9$$
$$= 5x^2(2x - 3) - 4x(2x - 3) + 9(2x - 3)$$
$$= 10x^3 - 15x^2 - 8x^2 + 12x + 18x - 27$$
$$= 10x^3 - 23x^2 + 30x - 27$$

6. $\quad (3x - 4y)(2x^2 - 5xy - 8y^2)$
$$= (3x - 4y)2x^2 - (3x - 4y)5xy - (3x - 4y)8y^2$$
$$= 2x^2(3x - 4y) - 5xy(3x - 4y) - 8y^2(3x - 4y)$$
$$= 6x^3 - 8x^2y - 15x^2y + 20xy^2 - 24xy^2 + 32y^3$$
$$= 6x^3 - 23x^2y - 4xy^2 + 32y^3$$

When multiplying polynomials, we can sometimes make the multiplication less complicated by arranging the polynomials vertically, as we did in the previous section. Consider the following example.

EXAMPLE 3.4.4

Multiply the following polynomials.

1.

$$\begin{array}{r} 2x + 9 \\ 3x + 7 \\ \hline 14x + 63 \\ 6x^2 + 27x \\ \hline 6x^2 + 41x + 63 \end{array}$$

We first multiply $2x + 9$ by 7 and then multiply $2x + 9$ by $3x$, placing like terms underneath one another. Then we add the rows together and obtain the same result as in part 1 of Example 3.4.3.

2.

$$\begin{array}{r} 4x - 7 \\ x - 6 \\ \hline -24x + 42 \\ 4x^2 - 7x \\ \hline 4x^2 - 31x + 42 \end{array}$$

Multiply by -6.

Multiply by x.

Add the rows.

3.

$$\begin{array}{r} 2x + 5y \\ -3x - 7y \\ \hline -14xy - 35y^2 \\ -6x^2 - 15xy \\ \hline -6x^2 - 29xy - 35y^2 \end{array}$$

4.

$$\begin{array}{r} 2x^2 - 5x + 4 \\ 3x - 6 \\ \hline -12x^2 + 30x - 24 \\ 6x^3 - 15x^2 + 12x \\ \hline 6x^3 - 27x^2 + 42x - 24 \end{array}$$

5.

$$\begin{array}{r} x^2 + 3xy + 9y^2 \\ x - 3y \\ \hline -3x^2y - 9xy^2 - 27y^3 \\ x^3 + 3x^2y + 9xy^2 \\ \hline x^3 + 0x^2y + 0xy^2 - 27y^3 = x^3 - 27y^3 \end{array}$$

6.

$$\begin{array}{r} 3x^2 - 7x + 4 \\ x^2 + 2x - 6 \\ \hline -18x^2 + 42x - 24 \\ 6x^3 - 14x^2 + 8x \\ 3x^4 - 7x^3 + 4x^2 \\ \hline 3x^4 - x^3 - 28x^2 + 50x - 24 \end{array}$$

Multiply by -6.

Multiply by $2x$.

Multiply by x^2.

Add the rows.

EXERCISES 3.4

Multiply the following polynomials.

1. $5x^2 \cdot 8x^6$

2. $(-5x^3)(4x)$

3. $(-9x^3)(7x^2)$

4. $(4x^3y^2)(9x^3y^8)$

5. $(-8xy^3)(6x^3y^6)$

6. $(-x^8y)(-10xy)$

7. $(5x^2y^4)(2x^3y^6)(4x^3y^5)$

8. $(-8xy^4)(2x^4y^3)(-6x^5y^5)$

9. $(-5xy^3z)(3x^4y^2z^2)(3x^2y^3z^6)$

10. $(x^2y^3)^2(2x^4y^5)$

11. $(6xy^2)^2(3xy^4)$

12. $(x^3y^2z^2)^2(4x^2y^3z)^2$

13. $5x(2x + 4)$

14. $2x^2(3x - 7)$

15. $-6x^3(4x - 8)$

16. $-5x^2(2x^2 + 4x)$

17. $8x^2y(2x + 3y)$

18. $8x(2x^2 - 5x - 1)$

19. $-2x^2(x^2 + 4x - 4)$

20. $2x^2(3x - 4y + 7)$

21. $-4xy^2(1 + 2x - 5xy)$

22. $2x^3y^4(3x - 4y^4 - 6x^3y^7)$

23. $(3x + 4)(2x + 6)$

24. $(4x + 7)(2x + 9)$

25. $(x + 5)(x - 3)$

26. $(x + 7)(x - 9)$

27. $(x - 8)(x - 1)$

28. $(x - 6)(x - 1)$

29. $(2x + 3)(x - 5)$

30. $(3x + 7)(x - 6)$

31. $(2x - 3)(2x - 3)$

32. $(3x - 7)(3x - 7)$

33. $(4x + 7)(4x + 7)$

34. $(2x + 5)(2x + 5)$

35. $(2x - 3)(2x - 4)$

36. $(3x - 5)(3x - 6)$

37. $(6x + 7)(6x - 7)$

38. $(5x + 4)(5x - 4)$

39. $(x + y)(x + y)$

40. $(x - y)(x - y)$

41. $(2x - y)(x - 3y)$

42. $(3x - y)(x - 5y)$

43. $(3x + 2y)(5x + 4y)$

44. $(7x + 3y)(2x + 9y)$

45. $(5x - 2y)(2x + 4y)$

46. $(6x - 5y)(5x + 2y)$

47. $(x - 2y)(3x - 4y)$

48. $(x - 3y)(4x - 7y)$

49. $(-x + 4y)(3x - 2y)$

50. $(-x + 6y)(5x - 3y)$

51. $6x - 2y$
   $x - 4y$

52. $5x - 3y$
   $x - 6y$

53. $4x + 5y$
   $4x + 5y$

54. $3x + 7y$
   $3x + 7y$

55. $(6x^2 - 7)(3x^2 - 1)$

56. $(4x^2 - 9)(2x^2 - 1)$

57. $(5x^2 + 3)(2x - 4)$

58. $(3x^2 + 7)(5x - 8)$

59. $(8x^2 + 4x)(2x + 9)$

60. $(7x^2 + 3x)(5x + 2)$

61. $(2x^2 + 5)(2x^2 - 5)$

62. $(3x^2 + 8)(3x^2 - 8)$

63. $(6x^2 + 1)(6x^2 + 1)$

64. $(7x^2 + 1)(7x^2 + 1)$

65. $(3x + 2)(5x^2 - 7x + 4)$

66. $(5x + 3)(4x^2 - 9x + 6)$

67. $(2x - 1)(x^2 - 2x - 7)$

68. $(4x - 1)(x^2 - 5x - 8)$

69. $(x - 2)(x^2 + 2x + 4)$

70. $(x - 5)(x^2 + 5x + 25)$

71. $(3x + 4)(-2x^2 - 6x + 7)$

72. $(6x + 4)(-3x^2 - 2x + 6)$

73. $(x^2 + 1)(2x^2 + 4x - 9)$

74. $(x^2 - 3)(5x^2 + 2x - 7)$

75. $(4x^2 - 3x)(x^2 - 2x + 3)$

76. $(2x^2 + 7x)(x^2 - 9x + 4)$

77. $(5x + 2)(2x^3 - 7x + 3)$

78. $(2x + 6)(3x^3 - 8x + 6)$

79. $(4x - 1)(4x^3 - 3x^2 + 2)$

80. $(3x - 2)(2x^3 - 6x^2 - 5)$

81. $(6x - 5y)(2x^2 - 8xy + y^2)$

82. $(5x - 2y)(3x^2 - xy + 7y^2)$

83. $(x + 4y)(x^2 - 4xy + 16y^2)$

84. $(x + 2y)(x^2 - 2xy + 4y^2)$

85. $x^2 + 6xy + 3y^2$
   $4x + 7y$

86. $x^2 + 8xy + 2y^2$
   $9x + 5y$

87. $3x^2 - 4xy - 4y^2$
   $x - 2y$

88. $5x^2 - xy - 9y^2$
   $4x - y$

89. $(2x + 4)(5x^3 - 7x^2 + 3x - 2)$

90. $(3x + 2)(4x^3 - 9x^2 + 2x - 6)$

91. $(x - 3)(2x^3 + 6x^2 - 7x - 1)$

92. $(x - 2)(5x^3 + 7x^2 - x - 4)$

93. $(3x - 2)(x^3 - 6x^2 - 6x + 5)$

94. $(2x - 3)(x^3 - 5x^2 - 5x + 4)$

95. $(4x + 2)(2x^4 + 3x^2 - 2x - 5)$

96. $(5x + 2)(3x^4 + 4x^2 - 2x + 3)$

97. $(x^2 - 7x + 4)(x^2 + 3x + 5)$

98. $(x^2 - 6x + 8)(x^2 + 2x + 9)$

99. $(2x^2 + 4x - 5)(3x^2 + 5x - 1)$

100. $(4x^2 + 2x - 7)(3x^2 + 7x - 2)$

101. $(-3x^2 - 6x + 4)(x^2 + x + 3)$

102. $(-2x^2 - 5x + 9)(x^2 + 6x + 1)$

3.5 The FOIL Method and Special Products of Binomials

As we have seen, multiplication of polynomials is sometimes a lengthy task. However, when the multiplication involves just two binomials, we can use a more convenient method of multiplying, called the **FOIL method**. The FOIL method is illustrated below.

$$
\begin{aligned}
(2x + 9)(3x + 7) &= \overset{F}{2x \cdot 3x} + \overset{O}{2x \cdot 7} + \overset{I}{9 \cdot 3x} + \overset{L}{9 \cdot 7} \\
&= 6x^2 + 14x + 27x + 63 \\
&= 6x^2 + 41x + 63
\end{aligned}
$$

This technique is called the FOIL method because we multiply the **f**irst terms of each binomial together, then the **o**uter terms together, next the **i**nner terms, and finally the **l**ast terms.

EXAMPLE 3.5.1

Perform the indicated multiplications using the FOIL method.

1.
$$
\begin{aligned}
(5x + 2)(x + 7) &= \overset{F}{5x \cdot x} + \overset{O}{5x \cdot 7} + \overset{I}{2 \cdot x} + \overset{L}{2 \cdot 7} \\
&= 5x^2 + 35x + 2x + 14 \\
&= 5x^2 + 37x + 14
\end{aligned}
$$

NOTE ▶ We draw in the arrows to help us perform the multiplications correctly.

2.
$$
\begin{aligned}
(2x + 3)(4x - 9) &= \overset{F}{2x \cdot 4x} + \overset{O}{2x(-9)} + \overset{I}{3 \cdot 4x} + \overset{L}{3(-9)} \\
&= 8x^2 + (-18x) + 12x + (-27) \\
&= 8x^2 - 6x - 27
\end{aligned}
$$

NOTE ▶ Remember, we can use the FOIL method only when we are multiplying two binomials.

$$\overset{\text{F}\quad\ \ \text{O}\quad\ \text{I}\quad\ \ \text{L}}{\textbf{3. } (4x - 6)(x - 3) = 4x^2 - 12x - 6x + 18}$$
$$= 4x^2 - 18x + 18$$

NOTE ▶ Here, some of the multiplication was done mentally to save steps.

$$\overset{\text{F}\quad\quad\ \ \text{O}\quad\ \ \ \text{I}\quad\quad\ \ \text{L}}{\textbf{4. } (5x - 7y)(x + 3y) = 5x^2 + 15xy - 7xy - 21y^2}$$
$$= 5x^2 + 8xy + 21y^2$$

In the remainder of this section we investigate special products of binomials. These special products follow a simple pattern and are relatively easy to memorize.

EXAMPLE 3.5.2

Perform the indicated multiplications.

$$\textbf{1. } (x - 6)(x + 6) = x^2 + 6x - 6x - 36$$
$$= x^2 - 36$$

$$\textbf{2. } (3x + 5y)(3x - 5y) = 9x^2 - 15xy + 15xy - 25y^2$$
$$= 9x^2 - 25y^2$$

Both parts of Example 3.5.2 illustrate the first special product that we want to consider: the **difference of two squares**.

Difference of Two Squares
$(a + b)(a - b) = a^2 - b^2$

REMARKS **1.** This formula is easily verified using the FOIL method:

$$(a + b)(a - b) = a^2 - ab + ab - b^2$$
$$= a^2 - b^2$$

2. Two binomials that have the same first terms and the same last terms but opposite signs in the middle are called *conjugates*.

3. To find the product of two conjugates, *subtract the square of the last term from the square of the first term.*

EXAMPLE 3.5.3

Perform the indicated multiplications.

1. $(x + 7)(x - 7) = x^2 - 7^2$ Using the difference of two
$$= x^2 - 49$$ squares formula

2. $(2x - 9)(2x + 9) = (2x)^2 - 9^2$
$$= 4x^2 - 81$$

3. $(6x + 1)(6x - 1) = (6x)^2 - 1^2$
$$= 36x^2 - 1$$

4. $(8x - 5y)(8x + 5y) = (8x)^2 - (5y)^2$
$$= 64x^2 - 25y^2$$

5. $\left(\frac{1}{2}a + \frac{1}{3}b\right)\left(\frac{1}{2}a - \frac{1}{3}b\right) = \left(\frac{1}{2}a\right)^2 - \left(\frac{1}{3}b\right)^2$
$$= \frac{1}{4}a^2 - \frac{1}{9}b^2$$

The last type of special product we want to investigate is illustrated by the following problems.

1. $(2x + 5)^2 = (2x + 5)(2x + 5)$
$$= 4x^2 + 10x + 10x + 25$$
$$= 4x^2 + 20x + 25$$

2. $(4x - 7)^2 = (4x - 7)(4x - 7)$
$$= 16x^2 - 28x - 28x + 49$$
$$= 16x^2 - 56x + 49$$

In each of these problems we start with a binomial squared. This type of special product is called the **square of a binomial**.

Square of a Binomial
1. $(a + b)^2 = a^2 + 2ab + b^2$ **2.** $(a - b)^2 = a^2 - 2ab + b^2$

REMARKS

1. These formulas are easily verified using the FOIL method. Part 1 is verified below; the verification of part 2 is left to the reader.

$$(a + b)^2 = (a + b)(a + b)$$
$$= a^2 + ab + ab + b^2$$
$$= a^2 + 2ab + b^2$$

2. To find the square of a binomial that is the *sum* of two terms, *add the square of the first term plus twice the product of the two terms plus the square of the last term*. The rule is similar for finding the square of a binomial that is the difference of two terms.

EXAMPLE 3.5.4

Perform the indicated multiplications.

1. $(x + 7)^2 = \quad x^2 \quad + \quad 2(x)(7) \quad + \quad 7^2$

                        ↑   ↑   ↑

              The first   Twice the product   The last

            term squared   of the two terms   term squared

           $= x^2 + 14x + 49$

2. $(5x + 6)^2 = (5x)^2 + 2(5x)(6) + 6^2$
               $= 25x^2 + 60x + 36$

Try to avoid this mistake:

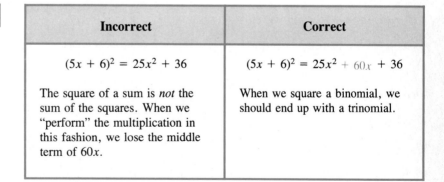

Incorrect	Correct
$(5x + 6)^2 = 25x^2 + 36$	$(5x + 6)^2 = 25x^2 + 60x + 36$
The square of a sum is *not* the sum of the squares. When we "perform" the multiplication in this fashion, we lose the middle term of $60x$.	When we square a binomial, we should end up with a trinomial.

3. $(2x + y)^2 = (2x)^2 + 2(2x)(y) + y^2$
           $= 4x^2 + 4xy + y^2$

4. $(x - 9)^2 = x^2 - 2(x)(9) + 9^2$
          $= x^2 - 18x + 81$

5. $(3x - 7y)^2 = (3x)^2 - 2(3x)(7y) + (7y)^2$
             $= 9x^2 - 42xy + 49y^2$

$$6. \left(\tfrac{2}{3}x - \tfrac{1}{5}y\right)^2 = \left(\tfrac{2}{3}x\right)^2 - 2\left(\tfrac{2}{3}x\right)\left(\tfrac{1}{5}y\right) + \left(\tfrac{1}{5}y\right)^2$$
$$= \tfrac{4}{9}x^2 - \tfrac{4}{15}xy + \tfrac{1}{25}y^2$$

EXERCISES 3.5

Perform the indicated multiplications.

1. $(x + 5)(x + 2)$ **2.** $(x + 7)(x + 4)$ **3.** $(x - 9)(x + 1)$ **4.** $(x - 6)(x + 3)$

5. $(5x + 6)(x - 2)$ **6.** $(7x + 3)(x - 8)$ **7.** $(8x + 4)(3x + 1)$ **8.** $(9x + 2)(2x + 4)$

9. $(3x + 5)(2x - 3)$ **10.** $(2x + 6)(5x - 7)$ **11.** $(5x - 4)(3x - 1)$ **12.** $(6x - 7)(2x - 1)$

13. $(x - 2y)(x - 9y)$ **14.** $(x - 3y)(x + 11y)$ **15.** $(5x + 6y)(2x + y)$ **16.** $(3x + 4y)(x + 2y)$

17. $(6x - 5y)(3x + y)$ **18.** $(4x - 7y)(2x + y)$ **19.** $(2x - 7y)(4x - 3y)$ **20.** $(3x - 8y)(4x - 2y)$

21. $(x^2 + 5)(x^2 - 3)$ **22.** $(x^2 + 7)(x^2 - 9)$ **23.** $(x^2 - 3)(x + 6)$ **24.** $(x^2 - 5)(x + 7)$

25. $(2x^2 + 3x)(x - 4)$ **26.** $(3x^2 + x)(x - 9)$ **27.** $\left(\tfrac{1}{3}x + 4\right)\left(\tfrac{2}{3}x + 5\right)$ **28.** $\left(\tfrac{1}{2}x + 7\right)\left(\tfrac{3}{2}x + 3\right)$

29. $\left(\tfrac{1}{4}x - 5\right)\left(\tfrac{3}{4}x - 9\right)$ **30.** $\left(\tfrac{2}{3}x - 4\right)\left(\tfrac{1}{3}x - 5\right)$ **31.** $(2.3x - 1.4)(3.5x + 4.7)$

32. $(5.1x - 6.7)(2.9x + 3.5)$ **33.** $(1.4x - 2.2)(9.8x - 6.3)$ **34.** $(3.7x - 7.2)(5.1x - 4.4)$

35. $(x + 5)(x - 5)$ **36.** $(x + 3)(x - 3)$ **37.** $(8 - x)(8 + x)$ **38.** $(11 - x)(11 + x)$

39. $(2x - 6)(2x + 6)$ **40.** $(4x - 2)(4x + 2)$ **41.** $(3x + 5)(3x - 5)$ **42.** $(7x + 6)(7x - 6)$

43. $(9x + 1)(9x - 1)$ **44.** $(10x + 1)(10x - 1)$ **45.** $(x - 3y)(x + 3y)$ **46.** $(x - 8y)(x + 8y)$

47. $(x + y)(x - y)$ **48.** $(x + 2y)(x - 2y)$ **49.** $(2x + y)(2x - y)$ **50.** $(5x + y)(5x - y)$

51. $(4x - 3y)(4x + 3y)$ **52.** $(6x - 7y)(6x + 7y)$ **53.** $(9x - 2y)(9x + 2y)$

54. $(10x - 3y)(10x + 3y)$ **55.** $\left(\tfrac{1}{2}x + 5\right)\left(\tfrac{1}{2}x - 5\right)$ **56.** $\left(\tfrac{1}{3}x + 7\right)\left(\tfrac{1}{3}x - 7\right)$

57. $\left(1 - \tfrac{5}{6}x\right)\left(1 + \tfrac{5}{6}x\right)$ **58.** $\left(1 - \tfrac{6}{7}x\right)\left(1 + \tfrac{6}{7}x\right)$ **59.** $\left(\tfrac{3}{4}x - \tfrac{1}{5}\right)\left(\tfrac{3}{4}x + \tfrac{1}{5}\right)$

60. $\left(\tfrac{3}{5}x - \tfrac{1}{3}\right)\left(\tfrac{3}{5}x + \tfrac{1}{3}\right)$ **61.** $\left(2x - \tfrac{1}{5}\right)\left(2x + \tfrac{1}{5}\right)$ **62.** $\left(5x - \tfrac{1}{4}\right)\left(5x + \tfrac{1}{4}\right)$

63. $\left(x - \tfrac{3}{5}y\right)\left(x + \tfrac{3}{5}y\right)$ **64.** $\left(x - \tfrac{3}{4}y\right)\left(x + \tfrac{3}{4}y\right)$ **65.** $\left(\tfrac{1}{2}x + 3y\right)\left(\tfrac{1}{2}x - 3y\right)$

66. $\left(\tfrac{1}{4}x + 5y\right)\left(\tfrac{1}{4}x - 5y\right)$ **67.** $\left(\tfrac{2}{3}x - \tfrac{3}{4}y\right)\left(\tfrac{2}{3}x + \tfrac{3}{4}y\right)$ **68.** $\left(\tfrac{3}{5}x - \tfrac{2}{3}y\right)\left(\tfrac{3}{5}x + \tfrac{2}{3}y\right)$

69. $(1.2x + 7)(1.2x - 7)$ **70.** $(3.4x + 8)(3.4x - 8)$ **71.** $(6.5x - 2.1y)(6.5x + 2.1y)$

72. $(4.1x - 7.8y)(4.1x + 7.8y)$ **73.** $(x + 8)^2$ **74.** $(x + 9)^2$

75. $(x - 1)^2$ **76.** $(x - 4)^2$ **77.** $(2x - 5)^2$ **78.** $(2x - 7)^2$

79. $(3x + 7)^2$ **80.** $(5x + 1)^2$ **81.** $(2x + 10)^2$ **82.** $(3x + 9)^2$

83. $(x - 2y)^2$ **84.** $(x - y)^2$ **85.** $(x + 7y)^2$ **86.** $(x + 9y)^2$

87. $(4x - y)^2$ **88.** $(5x - y)^2$ **89.** $(3x + 2y)^2$ **90.** $(3x + 5y)^2$

91. $\left(\frac{1}{5}x - \frac{3}{4}y\right)^2$ **92.** $\left(\frac{1}{2}x - \frac{3}{5}y\right)^2$ **93.** $\left(\frac{1}{2}x + \frac{2}{3}y\right)^2$ **94.** $\left(\frac{1}{3}x + \frac{2}{5}y\right)^2$

95. $(1.2x - 3)^2$ **96.** $(1.5x - 2)^2$ **97.** $(7.6x + 3.4y)^2$ **98.** $(6.8x + 5.1y)^2$

3.6 Dividing Polynomials

We have seen that when polynomials are added, subtracted, or multiplied, the result is another polynomial. This is not the case when polynomials are divided. As is the case with integers, sometimes polynomials divide evenly,

$$\frac{84}{7} = 12$$

and sometimes there is a remainder,

$$\frac{72}{5} = 14\frac{2}{5}$$

Let us first investigate division problems where the divisor is a monomial. In Chapters 1 and 2 we noted that the process of dividing by a number could also be accomplished by multiplying by the number's reciprocal. For example,

$$(12 + 9) \div 3 = (12 + 9) \cdot \frac{1}{3}$$
$$= 12 \cdot \frac{1}{3} + 9 \cdot \frac{1}{3}$$
$$= 4 + 3 = 7$$

We use the distributive property in this example because we are dividing into an expression with two terms. This problem could also be worked as follows:

$$\frac{12 + 9}{3} = \frac{12}{3} + \frac{9}{3}$$
$$= 4 + 3 = 7$$

We use the second format when dividing a polynomial by a monomial.

EXAMPLE 3.6.1

Perform the indicated divisions.

1. $\dfrac{8y^2 - 12y + 20}{4} = \dfrac{8y^2}{4} - \dfrac{12y}{4} + \dfrac{20}{4}$

$\qquad\qquad = 2y^2 - 3y + 5 \qquad$ Divide the coefficients.

2. $\dfrac{5ab^3 + 7a^4b^2}{b} = \dfrac{5ab^3}{b^1} + \dfrac{7a^4b^2}{b^1}$

$\qquad\qquad = 5ab^2 + 7a^4b \qquad$ Subtract the exponents on b.

3. $\dfrac{10x^9 - 14x^6 + 6x^5}{2x^3} = \dfrac{10x^9}{2x^3} - \dfrac{14x^6}{2x^3} + \dfrac{6x^5}{2x^3}$

$\qquad\qquad = 5x^6 - 7x^3 + 3x^2$ $\qquad$ Divide the coefficients and subtract the exponents.

4. $\dfrac{25x^2 + 15xy - 2y^2}{-5y^2} = \dfrac{25x^2}{-5y^2} + \dfrac{15xy}{-5y^2} - \dfrac{2y^2}{-5y^2}$

$\qquad\qquad = -\dfrac{5x^2}{y^2} - \dfrac{3x}{y} + \dfrac{2}{5}$

We leave the answer as a sum or difference of terms each of which is a ratio of monomials. Therefore, when a variable has a greater exponent in the denominator, we subtract the exponent in the numerator from the exponent in the denominator. In this way we avoid introducing negative exponents.

Try to avoid this mistake:

Incorrect	Correct
$\dfrac{3x^2 - y^2}{2x^2} = \dfrac{3x^2 - y^2}{2x^2}$	$\dfrac{3x^2 - y^2}{2x^2} = \dfrac{3x^2}{2x^2} - \dfrac{y^2}{2x^2}$
$\qquad = \dfrac{3 - y^2}{2}$	$\qquad = \dfrac{3}{2} - \dfrac{y^2}{2x^2}$
The distributive property was not used.	The monomial in the denominator must be divided into *each* term of the numerator.

5. $\dfrac{21n^3p^2 + 9n^2p^4 - 4np^5}{6np^3} = \dfrac{21n^3p^2}{6np^3} + \dfrac{9n^2p^4}{6np^3} - \dfrac{4np^5}{6np^3}$

$\qquad\qquad = \dfrac{7n^2}{2p} + \dfrac{3np}{2} - \dfrac{2p^2}{3}$

Division of a polynomial by a polynomial is more involved than division by a monomial. However, because division by a polynomial resembles long division from arithmetic, it is not difficult. Let's refresh our memory on long division by considering the problem $8495 \div 27$.

$$27\overline{)8495}$$

First, determine how many integral times 27 will divide into 84; it will go 3 times.

$$\begin{array}{r} 3 \\ 27\overline{)8495} \\ 81 \\ \hline 39 \end{array}$$

Place the 3 in the quotient and multiply 3 by 27, obtaining 81. Subtract 81 from 84, obtaining 3. Bring down the next digit, 9, and divide 27 into 39, obtaining 1.

$$\begin{array}{r} 31 \\ 27\overline{)8495} \\ 81 \\ \hline 39 \\ 27 \\ \hline 125 \end{array}$$

Place the 1 in the quotient and multiply 1 by 27, obtaining 27. Subtract 27 from 39, obtaining 12. Bring down the last digit, 5, and divide 27 into 125, obtaining 4.

$$\begin{array}{r} 314 \\ 27\overline{)8495} \\ 81 \\ \hline 39 \\ 27 \\ \hline 125 \\ 108 \\ \hline 17 \end{array}$$

Place the 4 in the quotient and multiply 4 by 27, obtaining 108. Subtract 108 from 125 to determine the remainder, 17. The answer is $314\frac{17}{27}$.

Recall that $314\frac{17}{27} = 314 + \frac{17}{27}$; this form is used in algebraic problems. We will use the preceding process from arithmetic to divide polynomials by binomials and trinomials.

EXAMPLE 3.6.2

Perform the indicated divisions.

1. $\dfrac{3x^2 + 8x + 5}{x + 2}$

Rewrite in long division format.

$$x + 2\overline{)3x^2 + 8x + 5}$$

Divide x into $3x^2$, obtaining $3x$.

$$\begin{array}{r} 3x \\ x + 2\overline{)3x^2 + 8x + 5} \\ 3x^2 + 6x \end{array}$$

Place $3x$ in the quotient, and multiply $3x$ by $x + 2$, obtaining $3x^2 + 6x$.

$$\begin{array}{r} 3x \\ x + 2\overline{)3x^2 + 8x + 5} \\ 3x^2 + 6x \\ \hline 2x + 5 \end{array}$$

Subtract $3x^2 + 6x$ from $3x^2 + 8x$, obtaining $2x$. Note that the leading terms always combine to zero. Bring down the next term, 5, and divide x into $2x$, obtaining 2.

$$
\begin{array}{r}
3x + 2 \\
x + 2\overline{)\,3x^2 + 8x + 5} \\
3x^2 + 6x \\
\hline
2x + 5 \\
2x + 4 \\
\hline
1
\end{array}
$$

Write $+\,2$ in the quotient to the right of $3x$. Multiply 2 by $x + 2$, obtaining $2x + 4$. Subtract $2x + 4$ from $2x + 5$, yielding the remainder of 1.

Thus, the answer to this division problem is

$$
3x + 2 + \frac{1}{x + 2}
$$

2. $\dfrac{2x^2 + 3x - 27}{{}^{\prime}x - 3}$

$$
x - 3\overline{)\,2x^2 + 3x - 27}
$$

Rewrite in long division format.

$$
\begin{array}{r}
2x \\
x - 3\overline{)\,2x^2 + 3x - 27} \\
2x^2 - 6x \\
\end{array}
$$

Be careful here; remember that you are *subtracting* $2x^2 - 6x$ from $2x^2 + 3x$. An easy way to do the subtraction correctly is to change the signs of all the terms of the bottom polynomial and add. We indicate the new signs by circling them. Add $2x^2 + 3x$ and $-2x^2 + 6x$, obtaining $9x$.

$$
\begin{array}{r}
2x \\
x - 3\overline{)\,2x^2 + 3x - 27} \\
\ominus \qquad \oplus \\
2x^2 - 6x \\
\hline
9x - 27
\end{array}
$$

$$
\begin{array}{r}
2x + 9 \\
x - 3\overline{)\,2x^2 + 3x - 27} \\
\ominus \qquad \oplus \\
2x^2 - 6x \\
\hline
9x - 27 \\
\ominus \quad \oplus \\
9x - 27 \\
\hline
0
\end{array}
$$

Change the signs and add.

Because the remainder is 0, $x - 3$ divides exactly into $2x^2 + 3x - 27$, yielding a quotient of $2x + 9$.

REMARK Both the divisor polynomial, $x - 3$, and the dividend polynomial, $2x^2 + 3x - 27$, should be arranged in descending powers of the variable. Also, if a power of the variable is missing, insert that power into the polynomial with a coefficient of zero. This zero term acts as a placeholder, just as zero does in arithmetic problems.

3. $\dfrac{3x^3 + 14x^2 - 38}{3x - 4}$

$$
3x - 4\overline{)\,3x^3 + 14x^2 + \ 0x - 38}
$$

Insert $0x$ as a placeholder.

$$\begin{array}{r} x^2 + 6x + 8 \\ 3x - 4\overline{)3x^3 + 14x^2 + 0x - 38} \end{array}$$

$$3x^3 - 4x^2 \qquad \text{Change the signs and add.}$$

$$18x^2 + 0x$$

$$18x^2 - 24x \qquad \text{Change the signs and add.}$$

$$24x - 38$$

$$24x - 32 \qquad \text{Change the signs and add.}$$

$$- 6$$

Thus, the solution is

$$x^2 + 6x + 8 + \frac{-6}{3x - 4} \qquad \text{or} \qquad x^2 + 6x + 8 - \frac{6}{3x - 4}$$

4. $\dfrac{15x^3 + 6x^2 - 35x - 4}{5x + 2}$

$$\begin{array}{r} 3x^2 \qquad\qquad - 7 \\ 5x + 2\overline{)15x^3 + 6x^2 - 35x - 4} \end{array}$$

$$15x^3 + 6x^2$$

$$- 35x - 4 \qquad \text{Because the subtraction yields 0, bring down the next } \textit{two} \text{ terms.}$$

$$- 35x - 14 \qquad \text{Change the signs and add.}$$

$$10$$

Hence, the solution is

$$3x^2 - 7 + \frac{10}{5x + 2}$$

5. $\dfrac{6x^4 + 5x^3 + 9x^2 + 20x + 5}{2x^2 - x + 5}$

$$\begin{array}{r} 3x^2 + 4x - 1 \\ 2x^2 - x + 5\overline{)6x^4 + 5x^3 + 9x^2 + 20x + 5} \end{array}$$

$$6x^4 - 3x^3 + 15x^2 \qquad \text{Change the signs and add.}$$

$$8x^3 - 6x^2 + 20x$$

$$8x^3 - 4x^2 + 20x \qquad \text{Change the signs and add.}$$

$$-2x^2 \qquad\qquad + 5$$

$$-2x^2 + \qquad x - 5 \qquad \text{Change the signs and add.}$$

$$-x + 10$$

Thus, the solution is

$$3x^2 + 4x - 1 + \frac{-x + 10}{2x^2 - x + 5}$$

Try to avoid this mistake:

Incorrect	Correct
$$\begin{array}{r} 3x \\ 2x - 1\overline{)6x^2 + 5x + 7} \\ 6x^2 - 3x \\ \hline 2x + 7 \end{array}$$	$$\begin{array}{r} 3x + 4 \\ 2x - 1\overline{)6x^2 + 5x + 7} \\ \ominus \quad \oplus \\ 6x^2 - 3x \\ \hline 8x + 7 \\ \ominus \quad \oplus \\ 8x - 4 \\ \hline 11 \end{array}$$
Remember that we are *subtracting* the bottom polynomial from the top polynomial.	Change the signs and add.

EXERCISES 3.6

Perform the indicated divisions.

1. $\dfrac{7x + 35}{7}$

2. $\dfrac{8y - 24}{4}$

3. $\dfrac{5a^2 + 20ab - 40}{5}$

4. $\dfrac{6n - 14 + 10w^3}{2}$

5. $\dfrac{9p^3 - 18p - 27}{-9}$

6. $\dfrac{-15x + 3x^5 - 30}{-3}$

7. $\dfrac{4m^4 - 11m^2 + m}{m}$

8. $\dfrac{2k^3 + k^2 - 8k}{k}$

9. $\dfrac{x^5 + 9x^4 - 2x^3}{x^2}$

10. $\dfrac{4q^7 - q^5 + 12q^4}{q^4}$

11. $\dfrac{16n^3 - 4n^3p + n^2p^2}{n^2}$

12. $\dfrac{7x^4 + x^3y^2 + 17x^5y}{x^3}$

13. $\dfrac{6r^4 - 8r^3 + 2r^2}{2r^2}$

14. $\dfrac{12a^6 + 30a^5 - 24a^4}{6a^3}$

15. $\dfrac{-20y^5 + 28y^3 - 4y^2}{4y}$

16. $\dfrac{-21x^7 - 14x^5 + 56x^4}{7x^4}$

17. $\dfrac{15k^8 + 3k^7 - 30k^5 + 36k^4}{3k^3}$

18. $\dfrac{20n^7 - 50n^6 - 30n^5 - 100n^3}{10n^2}$

19. $\dfrac{8q^5t^2 - 32q^4t^3 - 40q^3t^4}{8qt^2}$

20. $\dfrac{10a^7b^3 + 18a^6b^4 - 22a^5b^5}{2a^2b}$

21. $\dfrac{22x^3 - 55y^2}{11x^3y^2}$

22. $\dfrac{28m^4 + 42n^5}{14m^4n^5}$

23. $\dfrac{6p^3 - 15p^2q + 8q^4}{18p^2q}$

24. $\dfrac{4w^5z^2 + 9w^4z^3 + 48wz^5}{12wz^3}$

25. $\dfrac{5b^7 - 45b^6 - 30b^5 - 12b^4}{15b^4}$

26. $\dfrac{8v^5 + 48v^4 - 12v^3 - 80v}{16v^3}$

27. $\dfrac{42r^5 + 36r^4t^2 + 8r^3t^3}{6r^2}$

28. $\dfrac{4a^3y^4 - 40a^2y^5 + 25ay^6}{10y^5}$

29. $\dfrac{24x^3m + 27m^4}{12xm}$

30. $\dfrac{12xz^3 - 40x^5z}{8x^2z}$

31. $\dfrac{26y^8p^3 - 39y^6p^5 - 52y^4p^7}{-13y^2p^3}$

32. $\dfrac{-34b^4m^4 + 17b^3m^5 - 51b^2m^8}{-17bm^4}$

33. $\dfrac{-45x^2 + 30xy - y^4}{-5xy^3}$

34. $\dfrac{35a^3p^2 - 7ap^4 + p^6}{-7a^2p^5}$

35. $\dfrac{9a^2b^5c^4 + 54a^3b^4c^2 - 6a^4b^3c}{18ab^3c^2}$

36. $\dfrac{6x^4yz^5 - x^3y^2z^3 - 15x^2y^3z}{3x^2yz^4}$

37. $\dfrac{4mn^8p^4 + 10m^2n^6p^3 + m^3n^4p^2}{2m^2n^4p^3}$

38. $\dfrac{11t^3w^6 - 22tv^2w^5 - v^4w^4}{11tv^3w^5}$

39. $\dfrac{12k^5r^2 - 2r^4q^3 + 64k^2q^8}{-8k^3rq^6}$

40. $\dfrac{-16a^4y^3 - 6x^2y + 72ax^8}{-24ax^4y^3}$

41. $\dfrac{3x^2 + 13x - 6}{x + 5}$

42. $\dfrac{4x^2 + 9x + 8}{x + 2}$

43. $\dfrac{5x^2 - 18x - 9}{x - 4}$

44. $\dfrac{2x^2 - 13x + 13}{x - 3}$

45. $\dfrac{5x^2 + 9x - 18}{x + 3}$

46. $\dfrac{4x^2 - 21x - 18}{x - 6}$

47. $\dfrac{2x^2 - 117}{x - 8}$

48. $\dfrac{3x^2 - 39}{x + 4}$

49. $\dfrac{10x^2 + 13x - 1}{2x + 3}$

50. $\dfrac{6x^2 + x - 10}{3x + 5}$

51. $\dfrac{20x^2 - 3x - 9}{5x - 2}$

52. $\dfrac{18x^2 + 27x - 8}{6x - 1}$

53. $\dfrac{8x^2 + 22x + 5}{4x + 1}$

54. $\dfrac{18x^2 - 67x + 14}{2x - 7}$

55. $\dfrac{18x^2 + 6}{3x - 1}$

56. $\dfrac{32x^2 - 20}{4x - 3}$

57. $\dfrac{x^3 + 2x^2 - 13x + 8}{x - 2}$

58. $\dfrac{x^3 - 6x^2 + x + 14}{x - 5}$

59. $\dfrac{2x^3 + 13x^2 + 21x + 7}{x + 4}$

60. $\dfrac{3x^3 + 22x^2 + 5x - 10}{x + 7}$

61. $\dfrac{4x^3 - 33x + 4}{x + 3}$

62. $\dfrac{2x^3 - 77x - 12}{x + 6}$

63. $\dfrac{x^3 - 8x^2 + 8x - 7}{x - 7}$

64. $\dfrac{4x^3 - 9x^2 - 27}{x - 3}$

65. $\dfrac{12x^3 + 5x^2 - 34x + 10}{4x - 1}$

66. $\dfrac{18x^3 - 9x^2 - 8x + 6}{3x + 1}$

67. $\dfrac{14x^3 + 37x^2 + 9x + 6}{2x + 5}$

68. $\dfrac{16x^3 - 74x^2 + 7x + 6}{2x - 9}$

69. $\dfrac{5x^3 + 21x^2 - 16}{5x - 4}$

70. $\dfrac{4x^3 - 13x^2 + 9}{4x + 3}$

71. $\dfrac{27x^3 + 2}{3x - 2}$

72. $\dfrac{8x^3 + 6}{2x + 1}$

73. $\dfrac{2x^3 + 7x^2 - 11x + 10}{x^2 + 5x + 2}$

74. $\dfrac{3x^3 - 10x^2 - 13x + 20}{x^2 - 2x - 7}$

75. $\dfrac{4x^3 - 15x^2 - 27x - 3}{x^2 - 4x - 6}$

76. $\dfrac{6x^3 + 49x^2 + 22x + 7}{x^2 + 8x + 2}$

77. $\dfrac{6x^3 + 7x^2 + 10x + 25}{2x^2 - x + 5}$

78. $\dfrac{12x^3 + 17x^2 - 26x - 24}{3x^2 + 2x - 8}$

79. $\dfrac{8x^3 - 16x - 8}{4x^2 + 6x + 1}$

80. $\dfrac{10x^3 + 21x^2 - 5}{2x^2 + 5x + 2}$

81. $\dfrac{3x^4 + 19x^3 - 4x^2 - 24x + 13}{x^2 + 6x - 2}$

82. $\dfrac{2x^4 - 6x^3 - 3x^2 + 24x - 17}{x^2 - x - 5}$

83. $\dfrac{2x^4 - 19x^3 + 17x^2 + 4x - 10}{2x^2 - 3x + 1}$

84. $\dfrac{4x^4 - 3x^3 + 21x^2 + 3x + 13}{4x^2 + x + 2}$

85. $\dfrac{6x^4 + 2x^3 - 29x^2 - 2}{3x^2 - 5x - 3}$

86. $\dfrac{8x^4 + 24x^3 + 19x + 8}{2x^2 + 7x + 1}$

87. $\dfrac{5x^4 + 19x^3 + 18x + 25}{x^2 + 5x + 4}$

88. $\dfrac{6x^4 - 23x^2 - 28x - 12}{2x^2 - 4x - 3}$

89. $\dfrac{5x^3 - 17x^2 + 26x - 3}{2 - 5x}$

90. $\dfrac{6x^3 - 17x^2 + 22x - 24}{7 - 3x}$

91. $\dfrac{8x^3 - 6x^2y - 31xy^2 + 9y^3}{4x - 9y}$

92. $\dfrac{8x^3 - 30x^2y + 11xy^2 - 14y^3}{2x - 7y}$

93. $\dfrac{19x - 14x^3 + 3x^4 - 2x^2}{2 - 5x + x^2}$

94. $\dfrac{-x^3 - 30x^2 + 2x^4 + 28x}{3x + x^2 - 4}$

95. $\dfrac{-3x^4 - 8x^3 + 3x^2 + 10x - 4}{4 - 2x - 3x^2}$

96. $\dfrac{3x^4 + 13x^3 - 11x^2 - 34x + 19}{7 - 3x - x^2}$

97. $\dfrac{2x^4 - 8x^3 + 13x^2 - 36x + 3}{x^2 + 5}$

98. $\dfrac{3x^4 + 4x^3 - 4x^2 - 10x - 15}{x^2 - 3}$

99. $\dfrac{12x^4 - 21x^3 - 26x^2 + 35x + 10}{3x^2 - 5}$

100. $\dfrac{4x^4 - 2x^3 + 8x^2 - 7x - 21}{2x^2 + 7}$

Chapter 3
Review

**Definitions and
Laws of Exponents**

Let m and n be integers:

$x^m \cdot x^n = x^{m+n}$ Product law

$(x^m)^n = x^{m \cdot n}$ Power to a power law

$(xy)^n = x^n y^n$ Product to a power law

$\dfrac{x^m}{x^n} = x^{m-n},\ x \neq 0$ Quotient law

$$\left(\frac{x}{y}\right)^n = \frac{x^n}{y^n}, y \neq 0 \qquad \text{Quotient to a power law}$$

$$x^0 = 1, x \neq 0 \qquad \text{Zero exponent}$$

$$x^{-n} = \frac{1}{x^n}, x \neq 0 \qquad \text{Negative exponent}$$

Formulas

$$(a + b)(a - b) = a^2 - b^2 \qquad \text{Difference of two squares}$$

$$\left.\begin{array}{l}(a + b)^2 = a^2 + 2ab + b^2 \\ (a - b)^2 = a^2 - 2ab + b^2\end{array}\right\} \quad \text{Square of a binomial}$$

Review Exercises

3.1 Evaluate each of the following expressions.

1. -7^2 　　　 **2.** $4^3 + 4^2$ 　　　 **3.** $4^3 \cdot 4^2$ 　　　 **4.** $(3^2)^3$ 　　　 **5.** $\dfrac{3^6}{3^2}$ 　　　 **6.** $\left(\dfrac{5}{4}\right)^3$

Using the laws of exponents, simplify the following expressions, if possible.

7. $k^5 \cdot k^8$ 　　　 **8.** $(-2x^3)(5x^2)(3x^7)$ 　　　 **9.** $(y^4)^6$ 　　　 **10.** $(x^2 \cdot x^7)^3$

11. $(2x^3y^5)^4$ 　　　 **12.** $\dfrac{m^6}{m^3}$ 　　　 **13.** $\dfrac{(x^2)^6}{(x^3)^3}$ 　　　 **14.** $\left(\dfrac{3x^4}{y^8}\right)^3$

15. $(6xy^4)^2[(x^5y^2)^3]^2$ 　　　 **16.** $\dfrac{39x^5y^{12}}{3x^4y^6}$ 　　　 **17.** $\left(\dfrac{8x^5y^6}{2x^2y^2}\right)^3$ 　　　 **18.** $\dfrac{(2x^7y^3)^4}{(6x^4y^6)^2}$

19. $\left(\dfrac{15x^6y^9}{3xy^3}\right)^3$ 　　　 **20.** $\dfrac{(3x^2y^4)^2(4xy^5)^3}{(6x^5y^2)^2}$

3.2 Evaluate each of the following.

21. -5^0 　　　 **22.** $(-3)^{-2}$ 　　　 **23.** $\left(\dfrac{5}{2}\right)^{-3}$ 　　　 **24.** $4^0 + 4^{-1}$

25. $2^{-4} \cdot 2^{-1}$ 　　　 **26.** $[(-2)^{-3}]^{-2}$ 　　　 **27.** $\dfrac{3^{-1}}{3^{-5}}$

Simplify each of the following. (Remember to write your answers with only positive exponents.)

28. $x^4 \cdot x^{-7}$ 　　　 **29.** $(x^{-9})^2$ 　　　 **30.** $(x^{-2}y^5)^{-3}$ 　　　 **31.** $\dfrac{x^{-5}}{x^2}$

32. $\left(\dfrac{x^4}{x^{-1}}\right)^{-3}$ 　　　 **33.** $\dfrac{16x^3}{12x^5}$ 　　　 **34.** $\dfrac{14x^{-2}y^6}{21x^{-10}y^{-3}}$

35. $(4x^{-2}y^{-7})^{-3}(6x^3y^{-2})^{-2}$ 　　　 **36.** $\left(\dfrac{4x^{-4}y^{-3}}{7x^2y^{-6}}\right)^{-2}$ 　　　 **37.** $\dfrac{(3x^{-8}y^{-2})^{-3}(2x^{-2}y^{-1})}{(8x^{-2}y)^{-2}}$

Write the following numbers in scientific notation.

38. $483,000,000$

39. 0.0000206

Use scientific notation to evaluate the following expressions.

40. $\dfrac{(0.0049)(0.000081)}{(0.0000063)}$

41. $\dfrac{(0.00042)(11,200)}{(0.00000035)(160,000)}$

3.3 Simplify each of the following polynomials by combining like terms, if possible. After simplifying, state the degree and specify which are monomials, binomials, and trinomials.

42. $5x - 2x^2 - 7x + 5 - 6x^2$

43. $14x - 6y + 3x + 6y$

44. $\frac{1}{2}x^2 - \frac{2}{3}xy - \frac{1}{6}x^2 + \frac{1}{3}y - \frac{1}{12}xy$

45. $5.04xz^3 - 3.6x^2y^4 - 2.8xz^3$

Simplify each of the following polynomials by combining like terms, if possible. After simplifying, write the polynomial in standard form and determine the leading coefficient.

46. $-3x + x^3 - 7x^2 - 6x + 5 + x^2$

47. $2.6x - x + 3.05x^2 + 5 - 5x^2$

48. $\frac{1}{10} - \frac{3}{4}x + x^2 - \frac{4}{5}x^2 + 2x - \frac{2}{5}$

49. $\frac{7}{8}x - \frac{11}{12} - \frac{3}{2}x - \frac{1}{4}$

Perform the indicated operations and simplify your answers by combining like terms.

50. $(x^2 - 5x - 12) + (7x^2 - 3x + 4)$

51. $(5x^2 - x + 6) - (x^2 - 8x - 2)$

52. $(2x^3 + 11x^2 - 6x - 3) + (x^3 - 4x^2 - 9x + 10)$

53. $(x^3 - 15x^2 + 3x - 7) + (4x^3 - 8x^2 + 5x + 1)$

54. $\left(\frac{5}{2}x^2 - \frac{2}{3}xy - \frac{7}{4}y^2\right) - \left(\frac{1}{4}x^2 - \frac{1}{2}xy + \frac{5}{6}y^2\right)$

55. $\left(\frac{4}{3}x^2 + xy - \frac{7}{15}y^2\right) - \left(\frac{2}{15}x^2 - \frac{4}{5}xy + \frac{1}{3}y^2\right)$

56.
$\begin{array}{r} 2x^3 - 13x^2 - 5x + 11 \\ + \quad \underline{ 4x^2 - 14x - 6} \end{array}$

57.
$\begin{array}{r} 5x^2 + 17x - 8 \\ - \quad \underline{x^3 - x^2 + 2x - 9} \end{array}$

58.
$\begin{array}{r} \frac{3}{4}x^2 - x - \frac{2}{7} \\ - \quad \underline{\frac{7}{2}x^2 - \frac{1}{3}x + \frac{5}{14}} \end{array}$

59.
$\begin{array}{r} \frac{2}{3}x^2 + \frac{1}{5}x - \frac{5}{4} \\ + \quad \underline{\frac{1}{2}x^2 - \frac{7}{20}x - \frac{9}{10}} \end{array}$

60. $7(4x - 2) + 2(3x^2 - 5x - 8)$

61. $5(2x^3 - 6x + 3) - 4(4x^2 + 3x - 7)$

62. $\frac{2}{3}(9x^2 + 5x - 6) - \frac{3}{8}(\frac{2}{9}x - \frac{28}{15})$

63. $2.4(4x + 7) + 3.5(1.6x - 5.2)$

3.4 Multiply the following polynomials.

64. $(-5x^3)(8x^7)$

65. $(2x^4)(7x^2)(-3x^8)$

66. $(3x^5y^4)^2(6x^7y^{10})$

67. $-8x^2(3x^2 - 5)$

68. $3x^2y^5(4x + 9y^2)$

69. $4x(-2x^2 + 8x - 1)$

70. $(x - 6)(x - 9)$

71. $(3x + 8)(6x - 5)$

72. $(4x - 1)(4x - 1)$

73. $(5x - 2y)(8x + 7y)$

74.
$\begin{array}{r} 5x + 1 \\ \underline{3x - 5} \end{array}$

75.
$\begin{array}{r} x + 6y \\ \underline{3x + 10y} \end{array}$

76. $(3x^2 - 4)(5x^2 + 4)$

77. $(2x + 5)(3x^2 - x + 4)$

78. $(x^2 + 4)(-4x^2 + 7x - 2)$

79. $(5x + 6)(x^3 + 2x^2 - 2)$

80. $5x^2 - 8x + 1$
     $\quad\quad 3x - 4$

81. $4x^2 - 5xy - y^2$
     $\quad\quad 6x - 7y$

82. $(4x - 7)(x^3 + 5x^2 - 2x - 4)$

83. $(3x^2 + 6x - 5)(2x^2 - x + 5)$

3.5 Perform the indicated multiplications.

84. $(x + 7)(x - 9)$

85. $(x - 10)(x - 15)$

86. $(3x - 10)(2x + 7)$

87. $(5x + 1)(4x + 9)$

88. $(x - 8y)(x - 12y)$

89. $(6x - 5y)(2x + 15y)$

90. $(2x^2 - 7)(4x^2 + 5)$

91. $\left(\frac{3}{5}x - 2\right)\left(\frac{7}{6}x + 5\right)$

92. $\left(\frac{5}{2}x - \frac{3}{4}\right)\left(\frac{2}{3}x - \frac{4}{5}\right)$

93. $(3.2x + 1.5)(6.5x - 4.8)$

94. $(2x - 9)(2x + 9)$

95. $(x - 11y)(x + 11y)$

96. $(5x + 8y)(5x - 8y)$

97. $\left(2 + \frac{3}{8}x\right)\left(2 - \frac{3}{8}x\right)$

98. $\left(\frac{4}{3}x - \frac{1}{5}y\right)\left(\frac{4}{3}x + \frac{1}{5}y\right)$

99. $(1.3x - 5)(1.3x + 5)$

100. $(x + 12)^2$

101. $(4x - 7)^2$

102. $(x - 6y)^2$

103. $\left(x + \frac{1}{3}\right)^2$

104. $\left(\frac{2}{5}x + \frac{5}{4}y\right)^2$

105. $(3.5x - 4)^2$

3.6 Perform the indicated divisions.

106. $\dfrac{12x^2 - 40x + 28}{4}$

107. $\dfrac{5p^6 + 12p^4 - 3p^2}{p^2}$

108. $\dfrac{15y^4 - 21y^3 + 6y^2}{3y^2}$

109. $\dfrac{7x^5 + 35x^4 - 14x^3 + 42x^2}{7x}$

110. $\dfrac{18a^4x - 15a^3x^2 - 20a^2x^3}{6ax}$

111. $\dfrac{18m^2y^2 - 6m^4y}{9my^2}$

112. $\dfrac{14ab^6 - 20a^2b^4 - 4a^3b^2}{-4a^2b^3}$

113. $\dfrac{24x^5yz + 6x^4y^2z^3 - 4x^3y^3z^5}{3x^4yz^2}$

114. $\dfrac{3x^2 - 23x + 18}{x - 6}$

115. $\dfrac{4x^2 + 35x + 24}{x + 8}$

116. $\dfrac{6x^2 + 22x - 5}{2x + 8}$

117. $\dfrac{3x^3 - 7x^2 - 25x + 11}{x - 4}$

118. $\dfrac{4x^3 + 7x^2 - 11x - 5}{4x - 5}$

119. $\dfrac{6x^3 + x^2 - 20x - 8}{3x + 2}$

120. $\dfrac{4x^3 + 11x^2 + 16x - 27}{x^2 + 4x + 9}$

121. $\dfrac{6x^3 - 23x^2 - 9x + 56}{3x^2 - x - 8}$

122. $\dfrac{2x^4 + 8x^3 + 7x^2 - 7}{2x^2 + 2x - 5}$

123. $\dfrac{2x^4 + 9x^3y - 3x^2y^2 + 21xy^3 - 4y^4}{x^2 + 5xy - y^2}$

124. $\dfrac{23x^3 + 33x - 4x^4 + 13x^2 + 9}{6x + 3 - x^2}$

125. $\dfrac{10x^4 + 4x^3 + 11x^2 + 6x}{2x^2 + 3}$

Chapter 3 Test (You should be able to complete this test in 60 minutes.)

I. Evaluate each of the following.

1. -10^2

2. $2^3 + 2^{-2}$

3. $\left(\dfrac{4}{5}\right)^{-2}$

4. $5^6 \cdot 5^{-4}$

5. $\dfrac{4^{-3}}{4}$

6. -4^0

II. Simplify each of the following. (Remember to write your answers with only positive exponents.)

7. $\dfrac{4x^{-3}}{x^5}$

8. $(5x^{-2})^3$

9. $\dfrac{6x^{-2}y^4}{4x^6y^{-2}}$

10. $\left(\dfrac{x^4}{x^{-3}}\right)^{-2}$

11. $(4x^3y^{-2})^{-2}(x^5y^3)$

12. $\left(\dfrac{2x^4y^8}{3x^0y^{-2}}\right)^{-3}$

III. Use scientific notation to evaluate the following expression.

13. $\dfrac{(8000)(450,000)}{24,000}$

IV. Simplify each of the following polynomials by combining like terms, if possible. After simplifying, state the degree and specify which are monomials, binomials, and trinomials.

14. $-5xy - y^2 + 2x^2 - 3y^2 + xy$

15. $\frac{3}{2}x - \frac{4}{5}x^3 + \frac{5}{6}x - x^2 + 2x^3$

V. Perform the indicated operations.

16. $3x^4(5x^2 - 2x + 7)$

17. $\dfrac{6x^3 + 31x^2 + 23x - 23}{2x + 7}$

18. $6(x^2 - 3x - 5) - 2(4x^2 + x - 7)$

19. $(3x + 5)(x - 4)$

20. $(7x - 2)^2$

21. $\left(\frac{5}{8}x^2 - \frac{1}{2}xy - \frac{4}{9}y^2\right) + \left(\frac{3}{2}x^2 + xy - \frac{1}{6}y^2\right)$

22. $\begin{array}{r} 3x^2 - 8x - 7 \\ \times\ \underline{\qquad 4x + 2\qquad} \end{array}$

23. $\begin{array}{r} 5x^3 - 7x^2 + 3x - 3 \\ -\ \underline{\qquad 2x^2 - 6x + 9\qquad} \end{array}$

24. $(6x - 11y)(6x + 11y)$

25. $\dfrac{18x^4 + 27x^2y - 15y^2}{3x^2y}$

26. $(-4x^3)(5x^2)(2x^6)$

27. $\left(\frac{3}{4}x + \frac{6}{5}\right)^2$

Test Your Memory

These problems review Chapters 1–3.

I. Perform the indicated operations. When necessary, use the rules governing the order of operations.

1. $\frac{11}{8}\left(\frac{3}{7} + \frac{5}{21}\right)$

2. $\dfrac{-12}{\frac{4}{9}}$

3. $\sqrt{(-1 - 5)^2 + (10 - 2)^2}$

4. $(7 - 2 \cdot 4)^4$

5. $60 \div 10 \div (-2)$

6. $14 - 20 \div (-5) + 5$

7. $-4.08 + 3.9 - 0.625$

8. $3[4^2 - (3 - 11)]$

9. $\frac{7}{6} - 3 - \frac{4}{15}$

10. $\dfrac{5 - 4 \cdot 11}{5 - 8}$

II. Find the solutions of the following equations.

11. $5 + 3(x - 2) = 7 - 2(2x - 4)$

12. $6 - (3x - 5) = 8(x + 1) - 2$

13. $3(x - 4) = 2(x - 7) + 2$

14. $2(3x + 4) - 3(x - 4) = 11$

15. $\frac{1}{3}x + \frac{3}{2} = \frac{2}{3}x - \frac{1}{4}$

16. $\frac{1}{4}(3x - 2) = \frac{2}{3}\left(2x + \frac{1}{4}\right)$

III. Find the solutions of the following inequalities, and graph the solution sets on a number line.

17. $-4x + 2 \le 26$

18. $7 - (2x + 3) \le 5 - 3x$

19. $2 + 2(x + 3) < 5x + 4$

20. $\frac{1}{2}x + 1 > \frac{2}{3}x + \frac{3}{4}$

21. $-7 \le 2x + 3 \le 13$

22. $\frac{1}{4}(x + 3) > \frac{2}{9}(x + 4)$

IV. Evaluate each of the following.

23. 6^{-2} **24.** 8^0 **25.** $\left(\frac{3}{4}\right)^{-3}$ **26.** $4^2 \cdot 4^{-3}$ **27.** $(2^3)^{-2}$ **28.** $3^2 + 3^{-2}$

V. Simplify each of the following. (Remember to write your answers with only positive exponents.)

29. $\dfrac{x^4}{8x^{-2}}$

30. $(4x^{-5})^2$

31. $\dfrac{10x^3y^{-5}}{18x^{-4}y^{10}}$

32. $(x^8 \cdot x^{-2})^{-1}$

33. $(3x^4y^{-2})^{-2}(x^3y^{-3})$

34. $\dfrac{(2x^{-2})^{-2}}{(2^0x^{-4})^3}$

VI. Perform the indicated operations.

35. $3x^2y(5x^2 - 6xy - y^2)$

36. $4(2x^2 - x + 7) - (7x^2 + 3x - 6)$

37. $(4x^2 + 5x - 1)(3x - 2)$

38. $(5x - 2)^2$

39. $\dfrac{6x^3 + 7x^2 - 17x + 20}{3x + 5}$

40.
$$\begin{array}{r} 2x^3 - 5x^2 + 7x - 4 \\ -\ \ x^3 - 2x^2 - 6x + 3 \\ \hline \end{array}$$

41. $(4x - 3y)(5x + 2y)$

42. $\dfrac{12x^3 + 8x^2y - 10xy^4}{4xy^2}$

43. $(-5x^3)(4x^2)(6x^4)$

44. $\left(\tfrac{1}{2}x + 3\right)\left(\tfrac{1}{2}x - 3\right)$

VII. Use algebraic expressions to find the solutions of the following problems.

45. Find the perimeter of an equilateral triangle whose sides have a length of $5\tfrac{3}{4}$ feet. (*Hint:* In an equilateral triangle all sides have the same length.)

46. Kent the kayaker paddled his kayak for $\tfrac{3}{4}$ hour covering $7\tfrac{7}{20}$ miles. What was his average speed, that is, how many miles per hour did he average?

47. The perimeter of a rectangle is 96 kilometers. The length is 3 kilometers more than four times the width. What are the dimensions of the rectangle?

48. Cliff has $5.81 in 22¢ stamps and 39¢ stamps. He has twice as many 22¢ stamps as 39¢ stamps. How many 22¢ stamps and how many 39¢ stamps does Cliff have?

49. Norma buys $4.10 in 22¢ stamps and 39¢ stamps from Cliff. She buys a total of 14 stamps. How many 22¢ stamps and how many 39¢ stamps does Norma buy?

50. Ron and Juan plan to go fishing at the coast, after school. Ron leaves first, traveling at 50 mph. Juan leaves $\tfrac{1}{2}$ hour later, traveling at 65 mph and following the same route. How many hours will it take Juan to catch up to Ron?

CHAPTER 4

Factoring Polynomials

4.1 The Greatest Common Factor

A large portion of Chapter 3 was devoted to the topic of multiplication of polynomials. In this chapter we will learn how to reverse the multiplication of polynomials with a process called factoring polynomials.

First we must introduce some terminology. Recall from Chapter 1 that 3 and 5 were called *factors* of 15 since $3 \cdot 5 = 15$. We also saw that 1 and 15 were factors of 15 since $1 \cdot 15 = 15$.

In general, if a, b, and c are natural numbers and $a \cdot b = c$, then a and b are called **factors** of c.

Since $a \cdot b = c$ implies that $c \div a = b$ and $c \div b = a$, we say that c is **divisible** by a, or c is divisible by b.

REMARKS

1. We will also consider negative factors. For example, 15 also can be factored as $(-3)(-5)$. But we will not consider factors that are fractions, because the factorization can go on forever: $15 = 30 \cdot \frac{1}{2} = 60 \cdot \frac{1}{4}, \ldots$.

2. Recall the difference between terms and factors:

$$\text{In } 3 + 12 = 15, 3 \text{ and } 12 \text{ are terms.}$$
$$\text{In } 3 \cdot 5 = 15, 3 \text{ and } 5 \text{ are factors.}$$

In Chapter 1 we defined a **prime number** to be any natural number other than 1 that has only itself and 1 as factors. A **composite number** is any natural number other than 1 that is not prime.

Any composite number can be written as a product of prime numbers. For example,

$$140 = 14 \cdot 10$$
$$= 2 \cdot 7 \cdot 2 \cdot 5$$

$$= 2 \cdot 2 \cdot 5 \cdot 7$$
$$= 2^2 \cdot 5 \cdot 7 \qquad \text{Using exponential notation}$$

This last line is called the **prime factored form**. It can be shown that there is *one and only one* prime factored form for each natural number (not considering the order of the factors).

EXAMPLE 4.1.1

Find the prime factored form for each of the following numbers.

1. $36 = 3 \cdot 12$
$\qquad = 3 \cdot 4 \cdot 3$
$\qquad = 3 \cdot 2 \cdot 2 \cdot 3$
$\qquad = 2 \cdot 2 \cdot 3 \cdot 3$
$\qquad = 2^2 \cdot 3^2$

2. $54 = 6 \cdot 9$
$\qquad = 2 \cdot 3 \cdot 3 \cdot 3$
$\qquad = 2 \cdot 3^3$

3. $180 = 4 \cdot 45$
$\qquad = 2 \cdot 2 \cdot 9 \cdot 5$
$\qquad = 2 \cdot 2 \cdot 3 \cdot 3 \cdot 5$
$\qquad = 2^2 \cdot 3^2 \cdot 5$

We can use several tests from arithmetic to help us find the prime factored form. These tests are listed below.

A number has a factor of	if it satisfies the condition that	Examples
2	the last digit is even	$3014 = 2 \cdot 1507$
3	the sum of its digits is a multiple of 3	$225 = 3 \cdot 75$ $(2 + 2 + 5 = 9)$
4	the last two digits form a number that is divisible by 4	$5312 = 4 \cdot 1328$ (12 is divisible by 4)
5	the last digit is 0 or 5	$435 = 5 \cdot 87$

After a number is placed in prime factored form, it is relatively easy to determine its factors.

Since $36 = 2^2 \cdot 3^2$, the factors of 36 are:

1	2	2^2	3	3^2	$2 \cdot 3$	$2^2 \cdot 3$	$2 \cdot 3^2$	$2^2 \cdot 3^2$
1	2	4	3	9	6	12	18	36

Sometimes we are given a group of numbers and must find the largest number that is a factor of all of the given numbers. This number is called the **greatest common factor**, which is abbreviated **GCF**.

Finding the GCF of a Group of Numbers

1. Find the prime factored form of each of the numbers.

2. Find the prime numbers that occur in *all* of the prime factored forms. Raise each of these prime numbers to the *smallest* power that occurs in any prime factored form.

3. Form the GCF by multiplying the numbers found in step 2.

EXAMPLE 4.1.2

Find the greatest common factor for 36, 54, and 180.

$$36 = 2^2 \cdot 3^2 \qquad \text{From Example 4.1.1}$$
$$54 = 2 \cdot 3^3$$
$$180 = 2^2 \cdot 3^2 \cdot 5$$

To find the GCF we can use only the numbers that occur in *all* of the prime factored forms. In this case we can use only 2 and 3. In addition, the exponent of each of these numbers must be the *smallest* power that occurs in any prime factored form. Thus,

$$\text{GCF} = 2^1 \cdot 3^2$$
$$= 2 \cdot 9$$
$$= 18$$

This result states that 18 is the largest natural number that is a factor of 36, 54, and 180. Note that $36 = 18 \cdot 2$, $54 = 18 \cdot 3$, and $180 = 18 \cdot 10$.

In a similar fashion we can find the GCF for a group of monomials. However, when we are dealing with monomials we have to include variable as well as prime number factors in the GCF. Consider the following example.

EXAMPLE 4.1.3

Find the greatest common factor for the following groups of monomials.

1. $10x^4$, $6x^3$, $14x^7$

The GCF of the coefficients is 2. The variable x occurs in all three monomials, and its smallest power is 3. Thus,

$$GCF = 2 \cdot x^3$$
$$= 2x^3$$

Note that $10x^4 = 2x^3 \cdot 5x$, $6x^3 = 2x^3 \cdot 3$, and $14x^7 = 2x^3 \cdot 7x^4$.

2. $20x^2y^4$, $8x^3y$, $12x^4y^2$

The GCF of the coefficients is 4. The variables x and y occur in all three monomials. The smallest exponent on x is 2; the smallest exponent on y is 1. Thus,

$$GCF = 4 \cdot x^2 \cdot y^1$$
$$= 4x^2y$$

3. $20x^3y^4$, $5x^4y^2$, $15x^2z$

The GCF of the coefficients is 5. Only x occurs in all three monomials, and its smallest power is 2. Thus,

$$GCF = 5 \cdot x^2$$
$$= 5x^2$$

We use the GCF and the distributive property to express a polynomial as a product. For example, in the polynomial $15x + 9y$ the GCF of the two terms is 3. We now rewrite each term as a product with 3, the GCF, being one of the factors:

$$15x + 9y = 3 \cdot 5x + 3 \cdot 3y$$
$$= 3(5x + 3y) \qquad \text{By the distributive property}$$

The polynomial $15x + 9y$ has been expressed as a product. The process of writing a polynomial as a product is called **factoring a polynomial**.

We have seen that all natural numbers can be written uniquely as the product of prime numbers. Similarly, all polynomials can be written uniquely as the product of prime polynomials, where a **prime polynomial** is a polynomial with *integer* coefficients that has only itself and 1 as factors.

We can factor the polynomial $7x^2 + 14x$ in four ways:

$$7x^2 + 14x = 1 \cdot (7x^2 + 14x)$$
$$= 7(x^2 + 2x)$$
$$= x(7x + 14)$$
$$= 7x(x + 2)$$

Only the last factorization is a product of prime polynomials. This is called the **complete factorization**. A polynomial is **completely factored** when all of its factors are prime polynomials.

EXAMPLE 4.1.4

Completely factor the following polynomials.

1. $18x^3 - 24x^2$

The GCF of the coefficients is 6. The variable x occurs in both terms, and its smallest power is 2. Thus, the GCF is $6x^2$.

$$18x^3 - 24x^2 = 6x^2 \cdot 3x - 6x^2 \cdot 4$$
$$= 6x^2(3x - 4)$$

NOTE ▶ To check this problem, multiply the factors of $6x^2$ and $3x - 4$. After performing the multiplication, we obtain $18x^3 - 24x^2$, our original polynomial.

2. $9x^3 - 27x^2 + 36x$

The GCF is $9x$.

$$9x^3 - 27x^2 + 36x = 9x \cdot x^2 - 9x \cdot 3x + 9x \cdot 4$$
$$= 9x(x^2 - 3x + 4)$$

3. $20x^2y^4 - 8x^3y + 12x^4y^2$

The GCF is $4x^2y$.

$$20x^2y^4 - 8x^3y + 12x^4y^2 = 4x^2y \cdot 5y^3 - 4x^2y \cdot 2x + 4x^2y \cdot 3x^2y$$
$$= 4x^2y(5y^3 - 2x + 3x^2y)$$

4. $15x^2 - 20xy + 5x$

The GCF is $5x$.

$$15x^2 - 20xy + 5x = 5x \cdot 3x - 5x \cdot 4y + 5x \cdot 1$$
$$= 5x(3x - 4y + 1)$$

Try to avoid this mistake:

Incorrect	Correct
$15x^2 - 20xy + 5x$ $= 5x(3x - 4y)$	$15x^2 - 20xy + 5x$ $= 5x \cdot 3x - 5x \cdot 4y + 5x \cdot 1$ $= 5x(3x - 4y + 1)$
If you check your answer by multiplying, $5x(3x - 4y) = 15x^2 - 20xy$ $\neq 15x^2 - 20xy + 5x$	Remember that $5x = 5x \cdot 1$. Check: $5x(3x - 4y + 1)$ $= 15x^2 - 20xy + 5x$
the term of $5x$ is missing.	

5. $7x^2 + 10y^3z$

   In this case the GCF is 1. The polynomial $7x^2 + 10y^3z$ is a prime polynomial.

We have seen how to factor out a common monomial factor. In the next example we will see how to handle a common binomial factor.

EXAMPLE 4.1.5

Completely factor the following polynomials.

1. $3a(x + 7) + 2b(x + 7)$

   Do not perform the indicated multiplication. Here we have two terms. The first term is $3a(x + 7)$, and the second term is $2b(x + 7)$. There is a common factor of $x + 7$ in each term. We apply the distributive property to factor out $x + 7$:

   $$3a(x + 7) + 2b(x + 7) = 3a(x + 7) + 2b(x + 7)$$
   $$= (x + 7)(3a \underline{\quad\quad} + 2b \underline{\quad\quad})$$
   $$= (x + 7)(3a + 2b)$$

2. $9x(y - 2) - 5(y - 2)$

   Again we have a common binomial factor; this time it is $y - 2$. Applying the distributive property, we obtain

$$9x(y - 2) - 5(y - 2) = 9x(y - 2) - 5(y - 2)$$
$$= (y - 2)(9x - 5)$$

3. $3x^2(y + 5) - 6x(y + 5)$

First, remove the common binomial factor of $y + 5$.

$$3x^2(y + 5) - 6x(y + 5) = 3x^2(y + 5) - 6x(y + 5)$$
$$= (y + 5)(3x^2 - 6x)$$

Don't stop here! The second binomial, $3x^2 - 6x$, is not a prime polynomial. It has a GCF of $3x$.

$$= (y + 5)(3x \cdot x - 3x \cdot 2)$$
$$= (y + 5)3x(x - 2)$$
or $$= 3x(y + 5)(x - 2)$$

Remember to always find the complete factorization.

EXERCISES 4.1

Find the prime factored form for each of the following numbers.

1. 18	**2.** 45	**3.** 175	**4.** 98
5. 100	**6.** 1225	**7.** 441	**8.** 196
9. 105	**10.** 70	**11.** 90	**12.** 60
13. 525	**14.** 315	**15.** 252	**16.** 700

Find the greatest common factor for each of the following groups of numbers.

17. 18, 90, 252	**18.** 45, 105, 315	**19.** 98, 441, 70	**20.** 45, 100, 70
21. 1225, 441, 525	**22.** 98, 196, 70	**23.** 90, 60, 315	**24.** 18, 60, 252

Find the greatest common factor for each of the following groups of monomials.

25. $8x^4$, $12x^3$ **26.** $12x^5$, $21x^4$ **27.** $18x^2y^2$, $27xy^3$

28. $16x^2y^2$, $40x^3y$ **29.** $12xy$, $18yz$, $24xz$ **30.** $45xy$, $63yz$, $63xz$

31. x^4, $3x^3$, $15x^2$ **32.** x^4, $4x^3$, $16x^2$ **33.** $15x^3y^2$, $6x^3y^3$, $3x^2y$

34. $6x^2y^3$, $8x^3y^3$, $2xy^2$ **35.** $21x^3$, $28x^3y$, $35x^2z$, $21x^2yz^2$ **36.** $10xy^2$, $25y^3z$, $15y^3$, $20x^2y^2z$

37. $24x^3y^4$, $16x^4y^4$, $40x^3y^3$, $8x^2y^4$ **38.** $18x^4y^4$, $27x^4y^5$, $9x^3y^4$, $36x^4y^3$

Completely factor the following polynomials.

39. $12x - 28y$ **40.** $18x - 45y$ **41.** $15x^2 + 20x$

42. $18x^2 + 21x$ **43.** $14x^3 - 7x^2$ **44.** $15x^3 - 5x^2$

45. $16x^4y + 24x^2y^3$ **46.** $24x^3y^2 + 32xy^4$ **47.** $10xy^3 - 11ab^2$

48. $6xy^2 - 13ab^2$ **49.** $36abcd + 8abc$ **50.** $21abcd + 15abc$

51. $12x^2 + 42x + 12$ **52.** $16x^2 + 72x + 16$ **53.** $4x^2 - 24x - 4$

54. $3x^2 - 12x - 3$ **55.** $5x^3 + 2x^2y - 3x^2z$ **56.** $3xy^2 + 7y^3 - 9y^2z$

57. $10x^3y - 6x^2y^2 - 8xy$ **58.** $12x^3y - 6x^2y^2 + 21xy$ **59.** $8x^2y^2 + 9yz^2 - 4xz$

60. $9x^2y + 10y^2z - 5x^2z^2$ **61.** $-12x^2 + 18y^2 - 24z^2$ **62.** $-16x^2 + 32y^2 - 24z^2$

63. $15x^3y - 9x^2y^2 + 21x^2y$ **64.** $8xy^3 - 4x^2y^2 + 18xy^2$ **65.** $12x^2y^3 - 4x^3y^3 - 8x^3y^2$

66. $6x^2y^3 - 18x^3y^3 + 24x^3y^2$ **67.** $10x^3y^2 + 15x^3y^3 + 5x^2y$ **68.** $28x^2y^3 + 14x^2y^4 + 7xy^2$

69. $6x^3y^2 - 8y^3 + 15x^2$ **70.** $9x^2y^3 - 11x^3 + 12y$ **71.** $8x^4y^2 + 8x^3y^3 - 40x^2y^4$

72. $9x^4y^2 - 9x^3y^3 + 81x^2y^4$ **73.** $-6xy^3z - 30x^2y^2z + 18xyz^3$ **74.** $-8x^3yz - 12xy^2z^2 + 4xyz^3$

75. $18x - 81y + 27z + 63$ **76.** $32x - 16y + 72z + 40$

77. $6x^5 - 8x^4 + 14x^3 - 2x^2$ **78.** $6x^5 - 24x^4 + 27x^3 - 3x^2$

79. $8x^3y - 10x^2y^2 - 12xy^3 - 8x^3y^3$ **80.** $12x^3y - 24x^2y^2 - 8xy^3 - 20x^3y^3$

81. $15x^3y^2 + 6x^3y - 12x^2y^2 + 3x^2y^3$ **82.** $12x^3y^2 + 6x^2y^2 - 4x^3y + 2x^4y$

83. $5a(x + 2) + 3b(x + 2)$ **84.** $7a(x + 7) + 2b(x + 7)$

85. $8x(x - 3) + 5(x - 3)$ **86.** $4x(x - 8) + 9(x - 8)$

87. $2a(x - 1) - 7b(x - 1)$ **88.** $5a(x - 4) - 8b(x - 4)$

89. $3x(x + 2) - 7(x + 2)$ **90.** $2x(x + 5) - 5(x + 5)$

91. $5y(3y + 4) + (3y + 4)$ **92.** $3y(6y + 5) + (6y + 5)$

93. $4x^2(x - 6) + 8x(x - 6)$ **94.** $3x^2(x - 3) + 6x(x - 3)$

95. $9x^2(2x - 1) - 3x(2x - 1)$ **96.** $10x^2(3x - 4) - 5x(3x - 4)$

4.2 Factoring the Difference of Two Squares and Perfect Square Trinomials

In the previous section we learned that writing a polynomial as a product of prime polynomials is called *completely factoring the polynomial*. In this section we will use the results from Section 3.5 to completely factor some special polynomials as products of binomials.

Recall the **difference of two squares** formula from page 151:

$$(a + b)(a - b) = a^2 - b^2$$

We used this formula to show that

$$(x + 7)(x - 7) = x^2 - 7^2$$
$$= x^2 - 49$$

Now suppose that you were asked to completely factor the binomial $x^2 - 49$:

$$x^2 - 49 = x^2 - 7^2$$
$$= (x + 7)(x - 7)$$

The product on the right-hand side does yield $x^2 - 49$, which follows from the difference of two squares formula.

We can express the difference of two squares formula as follows:

Difference of Two Squares
$a^2 - b^2 = (a + b)(a - b)$

EXAMPLE 4.2.1

Completely factor the following polynomials.

1. $x^2 - 81 = x^2 - 9^2$
$$= (x + 9)(x - 9)$$

NOTE ▶ We recognized that we had a *difference* of two terms, each of which was the *square* of a quantity. Then we applied the difference of two squares formula.

2. $36y^2 - 1 = (6y)^2 - 1^2$
$$= (6y + 1)(6y - 1)$$

3. $49x^2 - 100y^2 = (7x)^2 - (10y)^2$
$$= (7x + 10y)(7x - 10y)$$

4. $6x^2 - 24 = 6(x^2 - 4)$ Remove the common factor
$$= 6(x^2 - 2^2)$$ of 6.
$$= 6(x + 2)(x - 2)$$

NOTE ▶ *Whenever we factor a polynomial, we always begin by removing a common factor, if possible.*

5. $3x^4y - 27x^2y^3 = 3x^2y(x^2 - 9y^2)$ Remove the common
$$= 3x^2y(x^2 - (3y)^2)$$ factor of $3x^2y$.
$$= 3x^2y(x + 3y)(x - 3y)$$

6. $16x^2 + 25$

NOTE ▶ This binomial is a prime polynomial because it is a *sum* of two squares that have no common factor except 1.

Try to avoid this mistake:

Incorrect	Correct
$16x^2 + 25 = (4x + 5)(4x + 5)$ Checking your "answer": $(4x + 5)(4x + 5) = 16x^2 + 20x$ $\qquad\qquad\qquad + 20x + 25$ $\qquad\qquad = 16x^2 + 40x$ $\qquad\qquad\quad + 25$ $\qquad\qquad \neq 16x^2 + 25$	$16x^2 + 25$ cannot be factored. Remember: A sum of squares that have no common factor except 1 is prime.

7. $81x^4 - 16y^4 = (9x^2)^2 - (4y^2)^2$
$$= (9x^2 + 4y^2)(9x^2 - 4y^2)$$

Don't stop here! The second binomial, $9x^2 - 4y^2$, is not a prime polynomial—it is another difference of two squares:

$$= (9x^2 + 4y^2)[(3x)^2 - (2y)^2]$$
$$= (9x^2 + 4y^2)(3x + 2y)(3x - 2y)$$

The other type of special polynomial we will investigate is called a *perfect square trinomial*. A **perfect square trinomial** is any trinomial that can be expressed as the square of a binomial. Recall the square of a binomial formulas from page 152:

$(a + b)^2 = a^2 + 2ab + b^2$
$(a - b)^2 = a^2 - 2ab + b^2$

In Section 3.5 we used the first formula to show that

$(x + 7)^2 = x^2 + 2(x)(7) + 7^2$
$\qquad\qquad = x^2 + 14x + 49$

Now suppose that you were asked to completely factor the trinomial $x^2 + 14x + 49$:

$$x^2 + 14x + 49 = x^2 + 2(x)(7) + 7^2$$
$$= (x + 7)^2$$

The product on the right-hand side does yield $x^2 + 14x + 49$. Since $x^2 + 14x + 49$ can be expressed as the square of a binomial, $x^2 + 14x + 49$ is a perfect square trinomial.

We now restate the square of a binomial formulas from Section 3.5:

Perfect Square Trinomials
1. $a^2 + 2ab + b^2 = (a + b)^2$
2. $a^2 - 2ab + b^2 = (a - b)^2$

REMARKS

1. In a perfect square trinomial, the first and last terms are the squares of two quantities and the middle term is twice the product of the quantities being squared in the first and last terms.

2. The factorization of a perfect square trinomial that is a *sum* of three terms is the square of the *sum* of the two quantities that are being squared. We have a similar rule for finding the factorization when the middle term is being subtracted.

EXAMPLE 4.2.2

Completely factor the following polynomials.

1. $x^2 + 10x + 25 = x^2 + 2(x)(5) + 5^2$
$$= (x + 5)^2$$

NOTE ▶

We recognized that we had a perfect square trinomial. Then we applied part 1 of the perfect square trinomial formula.

2. $x^2 - 6x + 9 = x^2 - 2(x)(3) + 3^2$
$$= (x - 3)^2$$

3. $4x^2 + 28x + 49 = (2x)^2 + 2(2x)(7) + 7^2$
$$= (2x + 7)^2$$

4. $25x^2 - 30xy + 9y^2 = (5x)^2 - 2(5x)(3y) + (3y)^2$
$$= (5x - 3y)^2$$

5. $2x^2 + 24x + 72 = 2(x^2 + 12x + 36)$ Remove the common
$$= 2(x^2 + 2(x)(6) + 6^2)$$ factor of 2.
$$= 2(x + 6)^2$$

6. $9x^4y + 6x^3y^2 + x^2y^3$

$= x^2y(9x^2 + 6xy + y^2)$ Remove the common

$= x^2y((3x)^2 + 2(3x)(y) + y^2)$ factor of x^2y.

$= x^2y(3x + y)^2$

7. $4x^2 + 6xy + 9y^2 = (2x)^2 + 1(2x)(3y) + (3y)^2$

Here the middle term is *not* twice the product of the quantities being squared. The trinomial $4x^2 + 6xy + 9y^2$ is a prime polynomial.

EXERCISES 4.2

Completely factor the following polynomials.

1. $x^2 - 100$ **2.** $x^2 - 64$ **3.** $y^2 - 36$ **4.** $y^2 - 9$

5. $t^2 - 1$ **6.** $t^2 - 4$ **7.** $m^2 - 6$ **8.** $m^2 - 12$

9. $x^4 - 81$ **10.** $x^4 - 16$ **11.** $36 - x^2$ **12.** $25 - x^2$

13. $121 - y^2$ **14.** $49 - y^2$ **15.** $4 - t^2$ **16.** $1 - t^2$

17. $x^2 + 49$ **18.** $x^2 + 9$ **19.** $x^2 - y^2$ **20.** $4x^2 - y^2$

21. $9x^2 - y^2$ **22.** $x^2 - 25y^2$ **23.** $x^2 - 49y^2$ **24.** $64x^2 - 9y^2$

25. $100x^2 - 9y^2$ **26.** $16x^2 + 49y^2$ **27.** $4x^2 + 81y^2$ **28.** $121x^2 - 100y^2$

29. $49m^2 - 121n^2$ **30.** $49m^2 - 25n^2$ **31.** $x^4 - 81y^4$ **32.** $x^4 - 16y^4$

33. $16x^4 - 625y^4$ **34.** $81x^4 - 256y^4$ **35.** $5x^2 - 45$ **36.** $7x^2 - 28$

37. $4x^2 + 100$ **38.** $9x^2 + 36$ **39.** $6x^2 - 600y^2$ **40.** $7x^2 - 700y^2$

41. $32x^2 - 50y^2$ **42.** $12x^2 - 27y^2$ **43.** $147x^2 - 12y^2$ **44.** $162x^2 - 8y^2$

45. $2x^3 - 50x$ **46.** $5x^3 - 180x$ **47.** $9x^3 + 81x$ **48.** $4x^3 + 16x$

49. $4x^3y - 9xy^3$ **50.** $16x^3y - 9xy^3$ **51.** $75x^3y^3 - 3xy^5$ **52.** $24x^5y - 6x^3y^3$

53. $18x^4y^3 - 50x^2y^5$ **54.** $48x^5y^2 - 75x^3y^4$ **55.** $32x^4y - 162y^5$ **56.** $48x^5 - 243xy^4$

57. $3x^6y - 48x^2y^5$ **58.** $2x^5y^2 - 162xy^6$ **59.** $x^2 + 6x + 9$ **60.** $x^2 + 8x + 16$

61. $x^2 + 20x + 25$ **62.** $x^2 + 2x + 1$ **63.** $y^2 - 14y + 49$ **64.** $y^2 - 18y + 81$

65. $y^2 - 2y + 1$ **66.** $y^2 - 6y + 9$ **67.** $4x^2 + 20x + 25$ **68.** $4x^2 + 12x + 9$

69. $9x^2 + 6x + 1$ **70.** $16x^2 + 8x + 1$ **71.** $16x^2 - 24x + 9$ **72.** $9x^2 - 30x + 25$

73. $4x^2 + 10x + 25$ **74.** $9x^2 + 6x + 4$ **75.** $25x^2 - 20x + 4$ **76.** $25x^2 - 40x + 16$

77. $x^2 + 4xy + 4y^2$ **78.** $x^2 + 10xy + 25y^2$ **79.** $x^2 + 14xy + 49y^2$ **80.** $x^2 + 6xy + 9y^2$

81. $x^2 - 12xy + 36y^2$ **82.** $x^2 - 2xy + y^2$ **83.** $x^2 - 20xy + 100y^2$ **84.** $x^2 - 16xy + 64y^2$

85. $9x^2 + 6xy + y^2$ **86.** $4x^2 + 4xy + y^2$ **87.** $16x^2 + 40xy + 25y^2$ **88.** $16x^2 + 24xy + 9y^2$

89. $25x^2 - 15xy + 9y^2$ **90.** $25x^2 - 35xy + 49y^2$ **91.** $25x^2 - 20xy + 4y^2$

92. $9x^2 - 42xy + 49y^2$ **93.** $16x^2 - 40xy + 25y^2$ **94.** $49x^2 - 28xy + 4y^2$

95. $3x^2 + 12x + 12$ **96.** $3x^2 + 18x + 27$ **97.** $4x^2 - 48x + 144$

98. $4x^2 - 8x + 4$ **99.** $75x^2 + 30xy + 3y^2$ **100.** $18x^2 + 12xy + 2y^2$

101. $16x^2 - 48xy + 36y^2$ **102.** $16x^2 - 80xy + 100y^2$ **103.** $2x^3 - 16x^2 + 32x$

104. $3x^3 - 6x^2 + 3x$ **105.** $6x^4y + 24x^3y^2 + 24x^2y^3$ **106.** $6x^3y^2 + 36x^2y^3 + 54xy^4$

107. $16x^2y + 24xy^2 + 36y^3$ **108.** $8x^3 + 20x^2y + 50xy^2$ **109.** $12x^3y - 36x^2y^2 + 27xy^3$

110. $18x^3y - 24x^2y^2 + 8xy^3$ **111.** $64x^4y^2 + 32x^3y^3 + 4x^2y^4$ **112.** $27x^4y^2 + 18x^3y^3 + 3x^2y^4$

4.3 Factoring Trinomials with a Leading Coefficient of 1

The ability to factor polynomials is one of the most important skills to be learned in a beginning algebra course. In this section we will learn how to factor a special type of polynomial—a trinomial with a leading coefficient of 1.

Let us first review the FOIL method of multiplying binomials:

$$(x + 4)(x + 3) = x^2 + 3x + 4x + 12$$
$$= x^2 + 7x + 12$$

Now suppose that you were asked to completely factor the trinomial $x^2 + 7x + 12$:

$$x^2 + 7x + 12 = (x + 4)(x + 3)$$

From above, the product of the prime polynomials on the right-hand side does yield $x^2 + 7x + 12$; so $x^2 + 7x + 12$ has been completely factored. In this problem we knew the factorization, since we had just performed the multiplication of $(x + 3)(x + 4)$. However, suppose that we were asked to completely factor another trinomial, say, $x^2 + 8x + 15$. To solve this problem, observe what happens when we multiply two binomials of the form $(x + m)(x + n)$, where m and n are integers:

$$(x + m)(x + n) = x^2 + nx + mx + mn$$
$$= x^2 + (n + m)x + mn$$

Note that the product is a trinomial with a leading coefficient of 1. To factor a trinomial of this type, we need to find two integers m and n whose product is the constant term and whose sum is the coefficient of x.

In the example of $x^2 + 8x + 15$, we are looking for two integers whose product is 15 and whose sum is 8. We start the search for m and n by looking at the integer factors of 15.

Factors of 15	Sum of the Factors
1 · 15	1 + 15 = 16
(−1) · (−15)	(−1) + (−15) = −16
3 · 5	3 + 5 = 8 √
(−3) · (−5)	(−3) + (−5) = −8

By examining this list we see that 3 and 5 satisfy our conditions. Thus, the factorization is

$$x^2 + 8x + 15 = (x + 3)(x + 5)$$

NOTE ▶ We can check this factorization by multiplying the factors of $x + 3$ and $x + 5$. After performing the multiplication, we obtain $x^2 + 8x + 15$, our original polynomial.

EXAMPLE 4.3.1 Completely factor the following trinomials.

1. $x^2 + 9x + 14 = (x + m)(x + n)$

We are looking for two integers whose product is 14 and whose sum is 9.

Factors of 14	Sum of the Factors
1 · 14	1 + 14 = 15
(−1) · (−14)	(−1) + (−14) = −15
2 · 7	2 + 7 = 9 √
(−2) · (−7)	(−2) + (−7) = −9

The third row meets our conditions. Thus,

$$x^2 + 9x + 14 = (x + 2)(x + 7)$$

2. $x^2 - 10x + 9 = (x + m)(x + n)$

We are looking for two integers whose product is 9 and whose sum is −10.

Factors of 9	Sum of the Factors
$1 \cdot 9$	$1 + 9 = 10$
$(-1) \cdot (-9)$	$(-1) + (-9) = -10 \checkmark$
$3 \cdot 3$	$3 + 3 = 6$
$(-3) \cdot (-3)$	$(-3) + (-3) = -6$

The second row meets our conditions. In fact, once we have found the answer there is no need to complete the above table. Thus,

$$x^2 - 10x + 9 = (x - 1)(x - 9)$$

3. $x^2 + 4x - 12 = (x + m)(x + n)$

We are looking for two integers whose product is -12 and whose sum is 4.

Factors of -12	Sum of the Factors
$1 \cdot (-12)$	$1 + (-12) = -11$
$(-1) \cdot (12)$	$(-1) + 12 = 11$
$2 \cdot (-6)$	$2 + (-6) = -4$
$(-2) \cdot 6$	$-2 + 6 = 4$ Stop here.

Thus, $x^2 + 4x - 12 = (x - 2)(x + 6)$.

4. $x^2 - x - 6 = (x + m)(x + n)$

We are looking for two integers whose product is -6 and whose sum is -1. (Remember that $-x = -1 \cdot x$.)

Factors of -6	Sum of the Factors
$1 \cdot (-6)$	$1 + (-6) = -5$
$(-1) \cdot 6$	$(-1) + 6 = 5$
$2 \cdot (-3)$	$2 + (-3) = -1 \checkmark$

Thus, $x^2 - x - 6 = (x + 2)(x - 3)$.

5. $x^2 + 7x + 4 = (x + m)(x + n)$

We are looking for two integers whose product is 4 and whose sum is 7.

Factors of 4	Sum of the Factors
$1 \cdot 4$	$1 + 4 = 5$
$(-1) \cdot (-4)$	$(-1) + (-4) = -5$
$2 \cdot 2$	$2 + 2 = 4$
$(-2) \cdot (-2)$	$(-2) + (-2) = -4$

There are no two integers whose product is 4 and whose sum is 7. The trinomial $x^2 + 7x + 4$ is a *prime polynomial.*

6. $2x^2 + 2x - 40 = 2(x^2 + x - 20)$

NOTE ▶ *Recall from Section 4.2 that whenever we factor a polynomial, we always begin by removing a common factor, if possible:*

$$= 2(x + m)(x + n)$$

To factor the trinomial inside the parentheses, we are looking for two integers whose product is -20 and whose sum is 1.

Factors of -20	Sum of the Factors
$1 \cdot (-20)$	$1 + (-20) = -19$
$(-1) \cdot 20$	$(-1) + 20 = 19$
$2 \cdot (-10)$	$2 + (-10) = -8$
$(-2) \cdot 10$	$(-2) + 10 = 8$
$4 \cdot (-5)$	$4 + (-5) = -1$
$(-4) \cdot 5$	$(-4) + 5 = 1 \checkmark$

Thus, $2x^2 + 2x - 40 = 2(x - 4)(x + 5)$.

7. $-x^2 + 7x + 8$

NOTE ▶ The trinomial $-x^2 + 7x + 8$ does not have a leading coefficient of 1. However, we can obtain a leading coefficient of 1 by factoring out -1.

$$-x^2 + 7x + 8 = -(x^2 - 7x - 8)$$
$$= -(x + m)(x + n)$$

To factor the trinomial inside the parentheses, we are looking for two integers whose product is -8 and whose sum is -7.

Factors of -8	Sum of the Factors
$1 \cdot (-8)$	$1 + (-8) = -7 \checkmark$

Thus, $-x^2 + 7x + 8 = -(x + 1)(x - 8)$.

8. $x^2 + 8xy + 12y^2$

Since the middle term has a factor of y and the last term has a factor of y^2, the binomial factors of $x^2 + 8xy + 12y^2$ will be of the form $x + my$ and $x + ny$:

$$x^2 + 8xy + 12y^2 = (x + my)(x + ny)$$

Now we are looking for two integers whose product is 12 and whose sum is 8.

Factors of 12	Sum of the Factors
$1 \cdot 12$	$1 + 12 = 13$
$(-1) \cdot (-12)$	$(-1) + (-12) = -13$
$2 \cdot 6$	$2 + 6 = 8 \checkmark$

Thus, $x^2 + 8xy + 12y^2 = (x + 2y)(x + 6y)$.
Checking the factorization:

$$(x + 2y)(x + 6y) = x^2 + 6xy + 2xy + 12y^2$$
$$= x^2 + 8xy + 12y^2$$

9. $4x^2y^2 - 16xy^3 - 84y^4 = 4y^2(x^2 - 4xy - 21y^2)$ Remove the common factor of $4y^2$.
$$= 4y^2(x + my)(x + ny)$$

We are looking for two integers whose product is -21 and whose sum is -4.

Factors of -21	Sum of the Factors
$1 \cdot (-21)$	$1 + (-21) = -20$
$(-1) \cdot 21$	$(-1) + 21 = 20$
$3 \cdot (-7)$	$3 + (-7) = -4 \checkmark$

Thus, $4x^2y^2 - 16xy^3 - 84y^4 = 4y^2(x + 3y)(x - 7y)$.

EXERCISES 4.3

Completely factor the following trinomials.

1. $x^2 + 9x + 18$ **2.** $x^2 + 9x + 20$ **3.** $x^2 - 11x + 18$ **4.** $x^2 + x - 30$

5. $x^2 - 2x + 1$ **6.** $x^2 - 10x + 21$ **7.** $x^2 - 5x - 6$ **8.** $x^2 - 4x + 4$

9. $x^2 + 8x + 6$ **10.** $x^2 - x - 2$ **11.** $x^2 - 5x - 36$ **12.** $x^2 + 11x + 24$

13. $x^2 + 11x - 12$ **14.** $x^2 - 8x - 9$ **15.** $x^2 + x - 72$ **16.** $x^2 + 9x + 7$

17. $x^2 - 3x - 54$ **18.** $x^2 + 7x - 8$ **19.** $x^2 + 12x + 11$ **20.** $x^2 + 3x - 18$

21. $x^2 - 6x + 9$ **22.** $x^2 - 8x + 12$ **23.** $x^2 + x - 2$ **24.** $x^2 - 6x - 16$

25. $x^2 - x - 12$ **26.** $x^2 - x + 10$ **27.** $3x^2 + 9x - 30$ **28.** $2x^2 + 4x - 48$

29. $2x^2 - 14x + 24$ **30.** $3x^2 - 15x + 18$ **31.** $4x^2 - 36x + 32$ **32.** $4x^2 - 28x + 24$

33. $3x^2 - 3x - 90$ **34.** $3x^2 - 3x - 36$ **35.** $x^4 - 3x^3 - 18x^2$ **36.** $x^4 - 2x^3 - 8x^2$

37. $4x^3 + 36x^2 + 32x$ **38.** $3x^3 + 30x^2 + 27x$ **39.** $2x^3y + 8x^2y - 64xy$

40. $2x^3y + 14x^2y - 36xy$ **41.** $3x^3y^2 - 12x^2y^2 + 12xy^2$ **42.** $3x^4y - 18x^3y + 27x^2y$

43. $-x^2 + 4x + 21$ **44.** $-x^2 - 6x - 8$ **45.** $-x^2 + 8x - 12$

46. $-x^2 + 9x + 10$ **47.** $-x^2 - 8x - 15$ **48.** $-x^2 + 6x + 16$

49. $-x^2 - 11x + 12$ **50.** $-x^2 + 7x - 12$ **51.** $-2x^2 - 16x - 30$

52. $-2x^2 + 14x - 24$ **53.** $-3x^2 + 27x - 60$ **54.** $-3x^2 - 18x - 24$

55. $x^2 - 5xy - 24y^2$ **56.** $x^2 + 4xy + 4y^2$ **57.** $x^2 + 9xy + 8y^2$

58. $x^2 + 2xy - 15y^2$ **59.** $x^2 + 14xy + 45y^2$ **60.** $x^2 + 3xy - 28y^2$

61. $x^2 + 3xy - 40y^2$ **62.** $x^2 + 5xy + y^2$ **63.** $x^2 - 2xy + y^2$

64. $x^2 - 6xy + 9y^2$ **65.** $x^2 + 3xy + 2y^2$ **66.** $x^2 + 2xy + y^2$

67. $x^2 + 7xy - 8y^2$ **68.** $x^2 + 9xy + y^2$ **69.** $x^2 + 3xy - 18y^2$

70. $x^2 - 10xy + 25y^2$ **71.** $x^2 - 7xy + 10y^2$ **72.** $x^2 + 10xy + 24y^2$

73. $x^2 - 6xy - 3y^2$ **74.** $x^2 + 8xy + 16y^2$ **75.** $x^2 - 7xy - 18y^2$

76. $x^2 - 8xy - 4y^2$ **77.** $x^2 + 3xy - 4y^2$ **78.** $x^2 - 10xy + 21y^2$

79. $2x^2 + 6xy - 36y^2$ **80.** $3x^2 + 6xy - 24y^2$ **81.** $3x^2 - 27xy + 60y^2$

82. $2x^2 - 26xy + 84y^2$ **83.** $4x^3 + 24x^2y + 32xy^2$ **84.** $4x^3 + 36x^2y + 72xy^2$

85. $2x^3y + 16x^2y^2 - 18xy^3$ **86.** $2x^3y + 18x^2y^2 - 20xy^3$ **87.** $-3x^2 + 12xy + 36y^2$

88. $-3x^2 + 15xy + 72y^2$ **89.** $-2x^3 + 8x^2y - 8xy^2$ **90.** $-2x^3 + 12x^2y - 18xy^2$

4.4 Factoring General Trinomials

In the previous section we learned how to factor trinomials with a leading coefficient of 1. In this section we will extend our factoring techniques to include trinomials with a leading coefficient not equal to 1. Let's again review the FOIL method of multiplying binomials:

$$(2x + 3)(x + 1) = 2x^2 + 2x + 3x + 3$$
$$= 2x^2 + 5x + 3$$

Now suppose that you were asked to completely factor the trinomial $2x^2 + 5x + 3$:

$$2x^2 + 5x + 3 = (2x + 3)(x + 1)$$

The product of the prime polynomials on the right-hand side does yield $2x^2 + 5x + 3$; so $2x^2 + 5x + 3$ has been completely factored.

Suppose that you were asked to completely factor $3x^2 + 13x + 14$. Since the first term is $3x^2$, the first terms of the binomial factors must be $3x$ and x. Thus,

$$3x^2 + 13x + 14 = (3x + h)(x + k)$$

where h and k are integers yet to be determined. However, by the FOIL method the product of the last two terms, $h \cdot k$, must be 14. Therefore, h and k must be factors of 14. The following table lists some of the possible factorizations of $3x^2 + 13x + 14$.

Factors of 14 $h \cdot k$	$(3x + h)(x + k)$	Expanded Trinomials
$1 \cdot 14$	$(3x + 1)(x + 14)$	$3x^2 + 43x + 14$
$14 \cdot 1$	$(3x + 14)(x + 1)$	$3x^2 + 17x + 14$
$(-1) \cdot (-14)$	$(3x - 1)(x - 14)$	$3x^2 - 43x + 14$
$(-14) \cdot (-1)$	$(3x - 14)(x - 1)$	$3x^2 - 17x + 14$
$2 \cdot 7$	$(3x + 2)(x + 7)$	$3x^2 + 23x + 14$
$7 \cdot 2$	$(3x + 7)(x + 2)$	$3x^2 + 13x + 14 \checkmark$

Since the sixth row meets our conditions, there is no need to complete the table. Thus,

$$3x^2 + 13x + 14 = (3x + 7)(x + 2)$$

NOTE ▶ Since all of the terms of the polynomial had positive coefficients, h and k had to be positive integers. Therefore, we did not even need to try the possible factorizations in the third and fourth rows.

EXAMPLE 4.4.1

Completely factor the following trinomials.

1. $2x^2 - 13x + 15 = (2x + h)(x + k)$

Since the last term is positive 15, h and k must have the same sign. The coefficient of the middle term is negative, so both h and k must be negative integers.

Factors of 15 $h \cdot k$	$(2x + h)(x + k)$	Expanded Trinomials
$(-1) \cdot (-15)$	$(2x - 1)(x - 15)$	$2x^2 - 31x + 15$
$(-15) \cdot (-1)$	$(2x - 15)(x - 1)$	$2x^2 - 17x + 15$
$(-3) \cdot (-5)$	$(2x - 3)(x - 5)$	$2x^2 - 13x + 15 \checkmark$

Thus, $2x^2 - 13x + 15 = (2x - 3)(x - 5)$.

2. $3x^2 - 11x - 4 = (3x + h)(x + k)$

Since the last term is negative, h and k must have opposite signs.

Factors of -4 $h \cdot k$	$(3x + h)(x + k)$	Expanded Trinomial
$1 \cdot (-4)$	$(3x + 1)(x - 4)$	$3x^2 - 11x - 4 \checkmark$

Sometimes we get lucky and find the factorization with the first try. Thus,

$$3x^2 - 11x - 4 = (3x + 1)(x - 4)$$

3. $6x^2 + 7x + 2$

Here we are not even sure how to start. We could use

$$6x^2 + 7x + 2 = (6x + h)(x + k) \qquad \text{or}$$
$$6x^2 + 7x + 2 = (3x + h)(2x + k)$$

We will try the $(6x + h)(x + k)$ form first. We do know that h and k must be positive integers, since all of the terms of the polynomial have positive coefficients.

Factors of 2 $h \cdot k$	$(6x + h)(x + k)$	Expanded Trinomials
$1 \cdot 2$	$(6x + 1)(x + 2)$	$6x^2 + 13x + 2$
$2 \cdot 1$	$(6x + 2)(x + 1)$	$6x^2 + 8x + 2$

Now we must try $(3x + h)(2x + k)$.

$1 \cdot 2$	$(3x + 1)(2x + 2)$	$6x^2 + 8x + 2$
$2 \cdot 1$	$(3x + 2)(2x + 1)$	$6x^2 + 7x + 2 \checkmark$

Thus, $6x^2 + 7x + 2 = (3x + 2)(2x + 1)$.

NOTE ▶ In the product $(6x + 2)(x + 1)$, the first binomial factor has a common factor of 2. Our original polynomial does not have a common factor of 2. Thus, it is impossible for the product $(6x + 2)(x + 1)$ to yield $6x^2 + 7x + 2$. We have a similar argument in the case of the product $(3x + 1)(2x + 2)$.

4. $5x^2 + 10x - 3 = (5x + h)(x + k)$

Since the last term is negative, h and k must have opposite signs.

Factors of -3 $h \cdot k$	$(5x + h)(x + k)$	Expanded Trinomials
$1 \cdot (-3)$	$(5x + 1)(x - 3)$	$5x^2 - 14x - 3$
$-3 \cdot 1$	$(5x - 3)(x + 1)$	$5x^2 + 2x - 3$
$(-1) \cdot 3$	$(5x - 1)(x + 3)$	$5x^2 + 14x - 3$
$3 \cdot (-1)$	$(5x + 3)(x - 1)$	$5x^2 - 2x - 3$

After trying every possibility, we determine that none of the potential factorizations works. The trinomial $5x^2 + 10x - 3$ is a *prime polynomial.*

5. $4x^2 - 22x + 30 = 2(2x^2 - 11x + 15)$

NOTE ▶ *Remember to always begin a factorization problem by removing a common factor, if possible:*

$$= 2(2x + h)(x + k)$$

Inside the parentheses, the coefficient of the middle term is negative and the last term is positive. Thus, both h and k must be negative integers.

Factors of 15 $h \cdot k$	$(2x + h)(x + k)$	Expanded Trinomials
$(-1) \cdot (-15)$	$(2x - 1)(x - 15)$	$2x^2 - 31x + 15$
$(-15) \cdot (-1)$	$(2x - 15)(x - 1)$	$2x^2 - 17x + 15$
$(-3) \cdot (-5)$	$(2x - 3)(x - 5)$	$2x^2 - 13x + 15$
$(-5) \cdot (-3)$	$(2x - 5)(x - 3)$	$2x^2 - 11x + 15$ ✓

Thus, $4x^2 - 22x + 30 = 2(2x - 5)(x - 3)$.

NOTE ▶ Remember to include the common factor of 2.

6. $-4x^2 + 19x + 5$

NOTE ▶ The trinomial $-4x^2 + 19x + 5$ does not have a positive leading coefficient.

However, we can obtain a positive leading coefficient by factoring out -1:

$$-4x^2 + 19x + 5 = -(4x^2 - 19x - 5)$$
$$-4x^2 + 19x + 5 = -(4x + h)(x + k) \qquad \text{or}$$
$$-4x^2 + 19x + 5 = -(2x + h)(2x + k)$$

We will try the $-(4x + h)(x + k)$ form first. Inside the parentheses the last term is negative. Thus, h and k must have opposite signs.

Factors of -5 $h \cdot k$	$(4x + h)(x + k)$	Expanded Trinomial
$1 \cdot (-5)$	$(4x + 1)(x - 5)$	$4x^2 - 19x - 5 \checkmark$

Thus, $-4x^2 + 19x + 5 = -(4x + 1)(x - 5)$.

7. $6x^2 + 23x - 4 = (6x + h)(x + k) \qquad \text{or}$

$6x^2 + 23x - 4 = (3x + h)(2x + k)$

We will try the $(6x + h)(x + k)$ form first. Note that h and k must have opposite signs.

Factors of -4 $h \cdot k$	$(6x + h)(x + k)$	Expanded Trinomials
$1 \cdot (-4)$	$(6x + 1)(x - 4)$	$6x^2 - 23x - 4$
$(-4) \cdot 1$	$(6x - 4)(x + 1)$	$6x^2 + 2x - 4$
$(-1) \cdot 4$	$(6x - 1)(x + 4)$	$6x^2 + 23x - 4 \checkmark$

Thus, $6x^2 + 23x - 4 = (6x - 1)(x + 4)$.

REMARK   In the first trial, the middle term $-23x$ is the opposite of what we want, $+23x$. When we get the opposite of what we are looking for, all we need to do is change the signs of h and k to obtain the correct factorization.

8. $6x^2 + xy - 2y^2$

Since the middle term has a factor of y and the last term has a factor of y^2, the last terms of the binomial factors must have factors of y:

$$6x^2 + xy - 2y^2 = (6x + hy)(x + ky) \qquad \text{or}$$
$$6x^2 + xy - 2y^2 = (3x + hy)(2x + ky)$$

Again h and k must have opposite signs.

Factors of -2 $h \cdot k$	$(6x + hy)(x + ky)$	Expanded Trinomials
$1 \cdot (-2)$	$(6x + y)(x - 2y)$	$6x^2 - 11xy - 2y^2$
$(-2) \cdot 1$	$(6x - 2y)(x + y)$	Do not even try this. Why?
$(-1) \cdot 2$	$(6x - y)(x + 2y)$	$6x^2 + 11xy - 2y^2$
$2 \cdot (-1)$	$(6x + 2y)(x - y)$	This won't work either.

Now we must try $(3x + hy)(2x + ky)$.

$1 \cdot (-2)$	$(3x + y)(2x - 2y)$	No chance.
$(-2) \cdot 1$	$(3x - 2y)(2x + y)$	$6x^2 - xy - 2y^2$

Since the middle term is $-xy$, switch the signs of h and k.

$2 \cdot (-1)$	$(3x + 2y)(2x - y)$	$6x^2 + xy - 2y^2$ ✓

Thus, $6x^2 + xy - 2y^2 = (3x + 2y)(2x - y)$.

9. $15x^3y - 33x^2y^2 + 6xy^3 = 3xy(5x^2 - 11xy + 2y^2)$ Remove the
$$= 3xy(5x + hy)(x + ky)$$ common factor of $3xy$.

Note that both h and k must be negative integers.

Factors of 2 $h \cdot k$	$(5x + hy)(x + ky)$	Expanded Trinomial
$(-1) \cdot (-2)$	$(5x - y)(x - 2y)$	$5x^2 - 11xy + 2y^2$ ✓

Thus, $15x^3y - 33x^2y^2 + 6xy^3 = 3xy(5x - y)(x - 2y)$.

The technique used to factor the trinomials in this section is sometimes called the "trial and error" method. However, if you use the hints contained in the examples and make intelligent guesses, you can eliminate many of the trials.

EXERCISES 4.4

Completely factor the following trinomials.

1. $2x^2 + 15x + 7$
2. $3x^2 + 10x + 7$
3. $5x^2 + 16x + 11$
4. $7x^2 + 36x + 5$

5. $7x^2 + 20x - 3$
6. $5x^2 + 8x - 13$
7. $3x^2 + 4x - 7$
8. $11x^2 + 21x - 2$

9. $7x^2 + 3x - 5$
10. $2x^2 - 15x + 7$
11. $5x^2 - 26x + 5$
12. $7x^2 - 10x + 3$

13. $3x^2 - 8x + 5$
14. $5x^2 - 3x - 2$
15. $7x^2 - 2x - 5$
16. $2x^2 - 5x - 7$

17. $5x^2 - 8x - 13$
18. $5x^2 - 3x - 7$
19. $5x^2 - 22x + 8$
20. $3x^2 - 11x + 6$

21. $7x^2 - 71x + 10$
22. $2x^2 - 25x + 12$
23. $6x^2 + x - 12$
24. $6x^2 - x - 15$

25. $4x^2 - 7x + 6$
26. $9x^2 - 16x - 4$
27. $8x^2 - 26x + 15$
28. $6x^2 - 5x + 10$

29. $25x^2 - 20x + 4$
30. $6x^2 + 19x + 14$
31. $14x^2 - 19x + 6$
32. $4x^2 + 16x - 9$

33. $9x^2 + 9x - 10$
34. $9x^2 - 55x + 6$
35. $6x^2 + 17x + 10$
36. $6x^2 - 13x + 6$

37. $8x^2 + 9x - 14$
38. $14x^2 + 25x + 9$
39. $12x^2 + 34x + 10$
40. $6x^2 + 22x - 8$

41. $24x^2 - 44x - 40$
42. $10x^2 - 25x + 30$
43. $12x^2 - 48x + 45$
44. $18x^2 - 21x - 30$

45. $10x^2 + 25x - 35$
46. $60x^2 - 124x + 40$
47. $12x^2 - 32x + 4$
48. $16x^2 + 28x + 6$

49. $-3x^2 + 8x - 5$
50. $-4x^2 - 17x + 15$
51. $-6x^2 - 11x + 10$
52. $-7x^2 + 10x + 8$

53. $-5x^2 - 14x - 8$
54. $-2x^2 + 5x - 3$
55. $-9x^2 + 3x + 20$
56. $-7x^2 - 17x - 6$

57. $-24x^2 - 34x - 12$
58. $-24x^2 + 52x + 60$
59. $-70x^2 - 95x + 15$
60. $-21x^2 + 90x - 24$

61. $8x^4 - 14x^3 + 5x^2$
62. $5x^4 + 12x^3 + 5x^2$
63. $10x^3y - 7x^2y - 12xy$

64. $6x^3y - 11x^2y - 10xy$
65. $5x^2 + 8xy + 3y^2$
66. $6x^2 + 7xy - 3y^2$

67. $9x^2 + 62xy - 7y^2$
68. $5x^2 - 22xy + 8y^2$
69. $7x^2 - 17xy + 6y^2$

70. $10x^2 + 49xy - 5y^2$
71. $6x^2 + 7xy - 5y^2$
72. $7x^2 + 9xy + 2y^2$

73. $9x^2 - 12xy + 4y^2$
74. $8x^2 - 47xy - 6y^2$
75. $8x^2 + 26xy + 15y^2$

76. $6x^2 - 5xy - 21y^2$
77. $9x^2 - 35xy - 4y^2$
78. $6x^2 - 19xy + 10y^2$

79. $10x^2 - 31xy - 14y^2$
80. $4x^2 + 12xy + 9y^2$
81. $-6x^2 + 11xy - 4y^2$

82. $-7x^2 + 18xy + 9y^2$
83. $-12x^2 + 17xy + 7y^2$
84. $-5x^2 - 7xy - 2y^2$

85. $16x^2 + 12xy - 10y^2$
86. $16x^2 - 44xy + 18y^2$
87. $21x^2 + 36xy + 15y^2$

88. $15x^2 - 18xy - 24y^2$
89. $6x^2 + 21xy - 18y^2$
90. $12x^2 + 20xy - 24y^2$

91. $8x^4 - 26x^3y + 15x^2y^2$

92. $7x^4 - 10x^3y - 8x^2y^2$

93. $4x^3y - 2x^2y^2 - 42xy^3$

94. $9x^3y - 3x^2y^2 - 42xy^3$

95. $15x^4y + 24x^3y^2 + 9x^2y^3$

96. $24x^3y^2 - 8x^2y^3 - 10xy^4$

97. $-60x^3 + 76x^2y - 24xy^2$

98. $-20x^3 + 108x^2y + 72xy^2$

99. $-16x^3y^2 + 20x^2y^3 + 14xy^4$

100. $-14x^4y - 18x^3y^2 - 4x^2y^3$

4.5 Factoring by Grouping

The last two sections illustrated some techniques for factoring trinomials. In this section we will investigate a factoring technique commonly used to factor four-term polynomials. This technique is called **factoring by grouping**. At the end of this section we will see how to use factoring by grouping on trinomials whose leading coefficient is not 1.

Again let's review a multiplication problem:

$$(x + 2)(y + 3) = xy + 3x + 2y + 6$$

Now suppose that you were asked to completely factor the polynomial $xy + 3x + 2y + 6$:

$$xy + 3x + 2y + 6 = (x + 2)(y + 3)$$

The product of the prime polynomials on the right-hand side does yield $xy + 3x + 2y + 6$; so $xy + 3x + 2y + 6$ has been completely factored. In this problem we knew the factorization, since we had just performed the multiplication of $(x + 2)(y + 3)$. However, suppose that we were asked to completely factor $xy + 9x + 5y + 45$. Given a four-term polynomial, it is a good idea to try factoring by grouping, which uses the following steps.

1. Group the first two terms together and remove their common factor. In addition, group the last two terms together and remove their common factor.

2. Determine whether or not a common binomial factor has been generated. If it has, remove that common binomial factor as we did in Example 4.1.5 of Section 4.1.

3. If a common binomial factor has not been generated, interchange the second and third terms (reorder the polynomial) and repeat steps 1 and 2.

REMARK In some polynomials it is useful to group three terms together. However, in this book we will focus on grouping terms in pairs.

Returning to the problem of factoring $xy + 9x + 5y + 45$:

$$xy + 9x + 5y + 45 = (xy + 9x) + (5y + 45) \qquad \text{Step 1}$$

$$= x(y + 9) + 5(y + 9) \qquad \text{Step 1}$$

$$= (y + 9)(x + 5) \qquad \text{Step 2}$$

NOTE ▶ We can check this factorization by multiplying the factors of $y + 9$ and $x + 5$.

EXAMPLE 4.5.1

Completely factor the following polynomials.

1. $xy - 8x + 2y - 16 = (xy - 8x) + (2y - 16)$ Group in pairs.

$\qquad\qquad\qquad\qquad = x(y - 8) + 2(y - 8)$ Remove common factors.

$\qquad\qquad\qquad\qquad = (y - 8)(x + 2)$ Factor the polynomial.

2. $8xy - 20x + 2y - 5 = (8xy - 20x) + (2y - 5)$

$\qquad\qquad\qquad\qquad = 4x(2y - 5) + 1(2y - 5)$

$\qquad\qquad\qquad\qquad = (2y - 5)(4x + 1)$

NOTE ▶ The second binomial does not have a common factor other than 1. To avoid making a careless error, it is a good idea to write the factor of 1 outside the parentheses.

3. $5ay - 5by - 3a + 3b = (5ay - 5by) + (-3a + 3b)$

$\qquad\qquad\qquad\qquad = 5y(a - b) + 3(-a + b)$

The binomials $a - b$ and $-a + b$ are opposites. Since we need them to be the same, we should factor -3 instead of $+3$ out of the last two terms:

$5ay - 5by - 3a + 3b = (5ay - 5by) + (-3a + 3b)$

$\qquad\qquad\qquad\qquad = 5y(a - b) - 3(a - b)$

$\qquad\qquad\qquad\qquad = (a - b)(5y - 3)$

Try to avoid this mistake:

Incorrect	Correct
$5ay - 5by - 3a + 3b$	$5ay - 5by - 3a + 3b$
$= (5ay - 5by) - (3a + 3b)$	$= (5ay - 5by) + (-3a + 3b)$
$= 5y(a - b) - 3(a + b)$	$= 5y(a - b) - 3(a - b)$
	$= (a - b)(5y - 3)$
-3 was factored out of the third term, but $+3$ was factored out of the fourth term.	When a minus sign is placed in front of parentheses, we must change the signs of *all* of the terms inside the parentheses.

4. $2x^2 + 18y + 3x + 12xy = (2x^2 + 18y) + (3x + 12xy)$

$$= 2(x^2 + 9y) + 3x(1 + 4y)$$

We have different binomial factors, so we must reorder the polynomial as indicated in step 3:

$2x^2 + 18y + 3x + 12xy = 2x^2 + 3x + 18y + 12xy$

$$= (2x^2 + 3x) + (18y + 12xy)$$

$$= x(2x + 3) + 6y(3 + 2x)$$

$$= (2x + 3)(x + 6y)$$

NOTE ▶ The binomials $2x + 3$ and $3 + 2x$ are the same. Why?

5. $x^3 + 6x^2 - 4x - 24 = (x^3 + 6x^2) + (-4x - 24)$

$$= x^2(x + 6) - 4(x + 6)$$

We should factor a -4, instead of a $+4$, out of the second binomial:

$$= (x + 6)(x^2 - 4)$$

Don't stop here! The second binomial, $x^2 - 4$, is not a prime polynomial— it is a difference of two squares:

$$= (x + 6)(x + 2)(x - 2)$$

The following example will show us how to use factoring by grouping on trinomials with a leading coefficient not equal to 1.

Note: The remainder of Section 4.5 is optional.

Suppose that we were asked to factor the trinomial $6x^2 + 19x + 8$. Factoring by grouping makes use of the following steps.

1. Multiply the leading coefficient by the constant term. In this case the product is $6 \cdot 8 = 48$.

2. Find the two factors of this product whose sum is the coefficient of x. In this case the coefficient of x is 19.

Factors of 48	Sums of the Factors
$1 \cdot 48$	$1 + 48 = 49$
$2 \cdot 24$	$2 + 24 = 26$
$3 \cdot 16$	$3 + 16 = 19$ √

(*Note:* We need to consider only positive factors of 48.) Thus, the two factors are 3 and 16.

3. Split the x term into two terms whose coefficients are the two numbers found in step 2. In this case

$$6x^2 + 19x + 8 = 6x^2 + 3x + 16x + 8$$

4. Factor the resulting polynomial by grouping:

$$\begin{aligned}
6x^2 + 19x + 8 &= 6x^2 + 3x + 16x + 8 \\
&= (6x^2 + 3x) + (16x + 8) \\
&= 3x(2x + 1) + 8(2x + 1) \\
&= (2x + 1)(3x + 8)
\end{aligned}$$

As we noted in the previous section, the trial and error method for factoring trinomials is sometimes lengthy. Frequently, factoring by grouping yields the factorization more quickly.

EXAMPLE 4.5.2

Completely factor the following trinomials by grouping.

1. $8x^2 + 10x + 3$

   Multiply: $8 \cdot 3 = 24$. Now we are looking for two positive integers whose product is 24 and whose sum is 10.

Factors of 24	Sums of the Factors
$1 \cdot 24$	$1 + 24 = 25$
$2 \cdot 12$	$2 + 12 = 14$
$3 \cdot 8$	$3 + 8 = 11$
$4 \cdot 6$	$4 + 6 = 10 \checkmark$

$$\begin{aligned}
8x^2 + 10x + 3 &= 8x^2 + 4x + 6x + 3 \\
&= (8x^2 + 4x) + (6x + 3) \\
&= 4x(2x + 1) + 3(2x + 1) \\
&= (2x + 1)(4x + 3)
\end{aligned}$$

2. $6x^2 - 23x + 15$

   Multiply: $6 \cdot 15 = 90$. Now we are looking for two negative integers whose product is 90 and whose sum is -23.

Factors of 90	Sums of the Factors
$(-1) \cdot (-90)$	$-1 + (-90) = -91$
$(-2) \cdot (-45)$	$-2 + (-45) = -47$
$(-3) \cdot (-30)$	$-3 + (-30) = -33$
$(-5) \cdot (-18)$	$-5 + (-18) = -23$ $\checkmark$

$$6x^2 - 23x + 15 = 6x^2 - 5x - 18x + 15$$
$$= (6x^2 - 5x) + (-18x + 15)$$
$$= x(6x - 5) - 3(6x - 5)$$
$$= (6x - 5)(x - 3)$$

3. $10x^2 - 11x - 6$

Multiply: $10 \cdot (-6) = -60$. Now we are looking for two integers whose product is -60 and whose sum is -11.

Factors of -60	Sums of the Factors
$1 \cdot (-60)$	$1 + (-60) = -59$
$(-1) \cdot 60$	$-1 + 60 = 59$
$2 \cdot (-30)$	$2 + (-30) = -28$
$(-2) \cdot 30$	$-2 + 30 = 28$
$3 \cdot (-20)$	$3 + (-20) = -17$
$(-3) \cdot 20$	$-3 + 20 = 17$
$4 \cdot (-15)$	$4 + (-15) = -11$ $\checkmark$ Hooray!

$$10x^2 - 11x - 6 = 10x^2 + 4x - 15x - 6$$
$$= (10x^2 + 4x) + (-15x - 6)$$
$$= 2x(5x + 2) - 3(5x + 2)$$
$$= (5x + 2)(2x - 3)$$

EXERCISES 4.5

Completely factor the following polynomials.

1. $xy + 2x + 5y + 10$ 2. $xy + 6x + 4y + 24$ 3. $2xy + 8x + 7y + 28$

4. $2xy + 10x + 3y + 15$ 5. $6xy + 4x + 9y + 6$ 6. $6xy + 8x + 9y + 12$

7. $12xy + 42x + 10y + 35$ 8. $24xy + 20x + 18y + 15$ 9. $xy - 6x + 3y - 18$

10. $xy - 8x + 4y - 32$ 11. $3xy - 9x + 4y - 12$ 12. $2xy - 8x + 3y - 12$

13. $xy - 6y + 2x - 12$ 14. $xy - 9y + 6x - 54$ 15. $4xy - 14y + 6x - 21$

16. $6xy - 10y + 27x - 45$ 17. $2xy + 3y + 2x + 3$ 18. $3xy + 4x + 3y + 4$

19. $10xy - 35x + 2y - 7$ 20. $12xy - 20x + 3y - 5$ 21. $xy + 5y - x - 5$

22. $xy + 4y - x - 4$ 23. $6xy - 8y - 3x + 4$ 24. $15xy - 9y - 5x + 3$

25. $2xy + 3y - 10x - 15$ 26. $3xy + 4y - 6x - 8$ 27. $10xy - 15x - 4y + 6$

28. $8xy - 20x - 14y + 35$ 29. $3ax + 4ay - 3bx - 4by$ 30. $2ax + 3ay - 2bx - 3by$

31. $6ax - 4bx - 21ay + 14by$ 32. $8ax - 20bx - 10ay + 25by$ 33. $xy + 32 + 8x + 4y$

34. $xy + 12 + 2x + 6y$ 35. $15xy + 8 + 20x + 6y$ 36. $14xy + 15 + 10x + 21y$

37. $2x^2 - 12y - 8xy + 3x$ 38. $3x^2 - 15y - 9xy + 5x$ 39. $6x^2 + 14y - 4xy - 21x$

40. $21x^2 + 30y - 18xy - 35x$ 41. $12x^2 + 2y - 8xy - 3x$ 42. $12x^2 + 9y - 27xy - 4x$

43. $x^3 + 7x^2 - 9x - 63$ 44. $x^3 + 2x^2 - 25x - 50$ 45. $3x^3 + 7x^2 - 3x - 7$

46. $7x^3 + 2x^2 - 7x - 2$ 47. $20x^3 + 32x^2 - 45x - 72$ 48. $18x^3 + 63x^2 - 8x - 28$

49. $75x^3 - 50x^2 - 3x + 2$ 50. $48x^3 - 64x^2 - 3x + 4$

Completely factor the following trinomials by grouping. (*Note: The remaining exercises are optional*.)

51. $10x^2 + 19x + 6$ 52. $12x^2 - 25x + 7$ 53. $5x^2 + 6x - 8$ 54. $9x^2 + 27x + 8$

55. $9x^2 - 29x + 6$ 56. $9x^2 - 5x - 4$ 57. $4x^2 + 4x - 35$ 58. $9x^2 + 18x + 8$

59. $10x^2 - 27x + 5$ 60. $9x^2 - 24x + 16$ 61. $6x^2 + 23x + 10$ 62. $3x^2 - 10x - 8$

63. $12x^2 - 17x + 6$ 64. $8x^2 - 35x + 12$ 65. $12x^2 + 20x + 7$ 66. $8x^2 - 26x + 15$

67. $4x^2 + 20x + 25$ 68. $4x^2 + 3x - 7$ 69. $6x^2 + x - 7$ 70. $8x^2 + 26x + 11$

71. $8x^2 - 10x - 3$ 72. $6x^2 - 19x + 15$ 73. $4x^2 + 5x - 9$ 74. $4x^2 + 8x - 21$

75. $5x^2 + 16x + 12$ 76. $9x^2 + 24x + 16$ 77. $9x^2 - 18x + 8$ 78. $8x^2 + 18x - 5$

79. $9x^2 - 6x - 8$ 80. $7x^2 + 25x + 12$ 81. $8x^2 + 35x + 12$ 82. $7x^2 - 30x + 8$

83. $4x^2 - 12x + 9$ 84. $4x^2 + 4x - 15$ 85. $5x^2 - 13x + 6$ 86. $6x^2 + 17x + 10$

4.6 Factoring the Sum and Difference of Two Cubes

This chapter has focused on learning how to factor polynomials. In this section we will learn how to factor two special types of binomials. Let's first consider a multiplication problem:

$$
\begin{array}{r}
x^2 + 3x + 9 \\
\times \quad\quad x - 3 \\
\hline
-3x^2 - 9x - 27 \\
x^3 + 3x^2 + 9x \\
\hline
x^3 + 0x^2 + 0x - 27 = x^3 - 27
\end{array}
$$

The binomial $x^3 - 27$ is called a **difference of two cubes** since $x^3 - 27 = x^3 - 3^3$.

Now suppose that you were asked to completely factor the binomial $x^3 - 27$:

$$x^3 - 27 = (x - 3)(x^2 + 3x + 9)$$

The product on the right-hand side does yield $x^3 - 27$. This example illustrates one of the two formulas we will investigate in this section.

Recall that in Section 4.2 a formula was found for factoring the difference of two squares. We also have formulas that enable us to factor *both* the sum and the difference of two cubes.

Sum and Difference of Two Cubes
1. $x^3 + y^3 = (x + y)(x^2 - xy + y^2)$
2. $x^3 - y^3 = (x - y)(x^2 + xy + y^2)$

Let's verify the **sum of two cubes** formula:

$$
\begin{array}{r}
x^2 - xy + y^2 \\
\times \quad\quad x + y \\
\hline
x^2y - xy^2 + y^3 \\
x^3 - x^2y + xy^2 \\
\hline
x^3 + 0x^2y + 0xy^2 + y^3 = x^3 + y^3
\end{array}
$$

Thus, $x^3 + y^3 = (x + y)(x^2 - xy + y^2)$.

The verification of the difference of two cubes formula is left to the reader.

EXAMPLE 4.6.1

Completely factor the following binomials.

1. $x^3 + 8$

This binomial is equivalent to $x^3 + 2^3$, so we can apply formula 1. In formula 1 the role of y is played by 2:

$$x^3 + 8 = x^3 + 2^3$$
$$= (x + 2)(x^2 - x \cdot 2 + 2^2)$$
$$= (x + 2)(x^2 - 2x + 4)$$

Try to avoid this mistake:

Incorrect	Correct
$x^3 + 8 = x^3 + 2^3$ $= (x + 2)^3$ Note that $(x + 2)^3 = (x + 2)(x + 2)(x + 2)$ $= (x + 2)(x^2 + 4x + 4)$ $= x^3 + 6x^2 + 12x + 8$ $\neq x^3 + 8$ An exponent cannot distribute over a sum or difference: $(4 + 2)^3 \overset{?}{=} 4^3 + 2^3$ $(6)^3 \overset{?}{=} 64 + 8$ $216 \neq 72$	$x^3 + 8 = x^3 + 2^3$ $= (x + 2)(x^2 - x \cdot 2 + 2^2)$ $= (x + 2)(x^2 - 2x + 4)$

2. $r^3 - 64$

This binomial is the same as $r^3 - 4^3$, so we can apply formula 2. In formula 2 the role of x is played by r and the role of y by 4:

$$r^3 - 64 = r^3 - 4^3$$
$$= (r - 4)(r^2 + r \cdot 4 + 4^2)$$
$$= (r - 4)(r^2 + 4r + 16)$$

NOTE ▶ The trinomial $r^2 + 4r + 16$ is a prime polynomial. After we apply the sum or difference of two cubes formula, *usually* the trinomial factor is a prime polynomial.

3. $64m^3 + 27n^3$

This binomial is the same as $(4m)^3 + (3n)^3$, so again we can use formula 1. In formula 1 the role of x is played by $4m$ and the role of y by $3n$:

$$64m^3 + 27n^3 = (4m)^3 + (3n)^3$$
$$= (4m + 3n)[(4m)^2 - (4m) \cdot (3n) + (3n)^2]$$
$$= (4m + 3n)(16m^2 - 12mn + 9n^2)$$

4. $32x^4y - 500xy^4$

First, remove the common factor of $4xy$:

$$32x^4y - 500xy^4 = 4xy(8x^3 - 125y^3)$$

Note that the binomial inside the parentheses is a difference of two cubes and we can use formula 2:

$$32x^4y - 500xy^4 = 4xy(8x^3 - 125y^3)$$
$$= 4xy[(2x)^3 - (5y)^3]$$
$$= 4xy(2x - 5y)[(2x)^2 + (2x) \cdot (5y) + (5y)^2]$$
$$= 4xy(2x - 5y)(4x^2 + 10xy + 25y^2)$$

5. $27p^6 - q^3$

This binomial is the same as $(3p^2)^3 - q^3$, so again we can use formula 2. In formula 2 the role of x is played by $3p^2$ and the role of y by q:

$$27p^6 - q^3 = (3p^2)^3 - q^3$$
$$= (3p^2 - q)[(3p^2)^2 + (3p^2) \cdot q + q^2]$$
$$= (3p^2 - q)(9p^4 + 3p^2q + q^2)$$

NOTE ▶ To work this problem you have to mind your p's and q's.

EXERCISES 4.6

Completely factor the following binomials.

1. $x^3 + 27$ **2.** $x^3 - 1$ **3.** $125 - t^3$ **4.** $x^3 + 64$

5. $r^3 - 216$ **6.** $r^3 - 125$ **7.** $x^3 - 8$ **8.** $27 - t^3$

9. $x^3 + 1$ **10.** $x^3 + 125$ **11.** $8x^3 + 125$ **12.** $27y^3 + 1$

13. $64y^3 - 1$ **14.** $125x^3 - 64$ **15.** $27 - 64m^3$ **16.** $8 - 27m^3$

17. $x^3 + 8y^3$ **18.** $x^3 - 27y^3$ **19.** $125x^3 - 8y^3$ **20.** $125m^3 + 27n^3$

21. $125p^3 + q^3$ **22.** $x^3 + 27y^3$ **23.** $x^3 - 64y^3$ **24.** $8p^3 + q^3$

25. $64m^3 + 125n^3$ **26.** $8x^3 - 125y^3$ **27.** $2x^3 + 250$ **28.** $2r^3 - 128$

29. $3r^3 - 81$ **30.** $3x^3 + 648$ **31.** $32m^3 - 108n^3$ **32.** $625m^3 - 135n^3$

33. $432p^3 + 2q^3$ **34.** $256p^3 + 4q^3$ **35.** $128x^4y - 2xy^4$ **36.** $81x^4y - 3xy^4$

37. $81x^4y^2 - 375xy^5$ **38.** $192x^5y - 81x^2y^4$ **39.** $8x6 - y^3$ **40.** $64x^6 + y^3$

41. $27x^3 + 8y^6$ **42.** $125x^3 - 64y^6$ **43.** $8x^9 + y^3$ **44.** $27x^9 - y^3$

45. $x^9 - y^6$ **46.** $x^9 + y^6$ **47.** $81x^6 + 192y^3$ **48.** $250x^6 + 16y^3$

49. $128x^7y - 250xy^7$ **50.** $54x^7y - 128xy^7$

4.7 Factoring Summary

In this section we combine all of the methods of factoring that we have covered, and we give a strategy for factoring an arbitrary polynomial.

Given a polynomial to factor:

1. Examine the terms to determine if there is a common factor. If so, factor out the greatest common factor.

2. Consider the number of terms.

Two terms	Difference of squares? Difference of cubes? Sum of cubes?
Three terms	Perfect square trinomial? Leading coefficient of 1? Leading coefficient not equal to 1?
Four or more terms	Factor by grouping

3. Examine all of the factors to determine if any factor can be factored further.

EXERCISES 4.7

Completely factor the following polynomials.

1. $x^2 - 3x + 2$ **2.** $x^2 - 49$ **3.** $x^2 - 2x - 3$ **4.** $y^2 + 2y - 8$

5. $2x^3 + 7x + 3$ **6.** $x^2 - 3x + 4$ **7.** $9x^2 - 1$ **8.** $16x^2 + 72x + 81$

9. $x^2 + 11x + 24$ **10.** $3y^2 - 8y + 4$ **11.** $x^2 + 10x + 25$ **12.** $4x^2 + 16$

13. $y^2 - y - 12$

14. $x^2 + 6x + 9$

15. $x^2 - 36$

16. $2x^2 - x - 3$

17. $x^2 - 7x + 8$

18. $y^2 + 3y - 10$

19. $2x^2 - 4x$

20. $x^2 + 19x + 48$

21. $9x^2 - 42x + 49$

22. $25y^2 - 16$

23. $x^2 + 9x - 22$

24. $4x^2 + 9$

25. $y^2 - 11y + 28$

26. $4x^2 + 4x - 3$

27. $2x^3 + 5x^2 + 6x + 15$

28. $5x^2 - 3x - 2$

29. $21 + 4x - x^2$

30. $12x^4 + 21x^3 - 9x^3y - 15x^2y^3$

31. $64x^2 - 81$

32. $y^2 - 11y - 26$

33. $4x^2 + 4x - 15$

34. $3 - 8x - 3x^2$

35. $24x^2 - 37x - 5$

36. $4 - 81y^2$

37. $5x^5 - 80xy^4$

38. $7x^2 - 4x - 3$

39. $2x^2 + 4x + 1$

40. $x^2 - 19x + 70$

41. $3x^2(2x - 7) + 6x(2x - 7)$

42. $x^8 - 81$

43. $8y^2 - 6y - 9$

44. $x^2 + 16xy + 60y^2$

45. $18a^3b^3 - 12ab^5 - 6ab^3$

46. $2x^3 - 20x^2 + 18x$

47. $6x^2 + x - 12$

48. $72x^2 + 42x - 9$

49. $x^2 + 20xy + 75y^2$

50. $3x^3 - 27x$

51. $3x^3 + x^2y - 3x - y$

52. $x^3 - 12x^2 + 32x$

53. $6x^3 + 27x^2 + 12x$

54. $12x^3 - 48x^2 - 30x + 120$

55. $12x^4 + 11x^2 - 15$

56. $4x^4 + 35x^2 - 9$

57. $5x^3 - 125x$

58. $9x^4 + 30x^2 + 25$

59. $2x^3 + 5x^2 - 2x - 5$

60. $24xy^2 + 36xy + 8y^2 + 12y$

61. $x^4 + 19x^2 - 20$

62. $18x^3y^3 - 6x^2y^4 - 4xy^5$

63. $x^4 + 3x^3y - 40x^2y^2$

64. $3x^4y - x^3y^2 + 5x^2y^3$

65. $8x^3 + 27y^3$

66. $64x^3 - 1$

67. $2x^5 - 16x^2$

68. $2x^4 - x^3 + 2x - 1$

69. $x^6 + 7x^3 - 8$

70. $y^6 - 64$

4.8 Factorable Quadratic Equations

In Chapter 2 we learned how to find the solutions of linear equations. In this section we will use the tools developed in Chapter 2 and investigate another group of equations called *quadratic equations* in one variable.

Quadratic Equation
Any equation that can be written in the form $$ax^2 + bx + c = 0$$ where a, b, and c are real numbers with $a \neq 0$, is called a **quadratic equation** *in the variable x.*

REMARKS

1. If $a = 0$, $ax^2 + bx + c = 0$ becomes $bx + c = 0$ (a linear equation). Thus, we have the restriction $a \neq 0$.

2. The form $ax^2 + bx + c = 0$ is called **standard form**. It is usually easier to find the solutions of a quadratic equation if we first place the quadratic equation in standard form.

3. The following expressions are examples of quadratic equations in the variable x.

    a. $2x^2 + 5x - 3 = 0$

    b. $x^2 + 7x - 6 = 5x + 2$, since this equation is equivalent to

$$x^2 + 2x - 8 = 0$$

    c. $4x^2 = 25$, since this equation is equivalent to $4x^2 - 25 = 0$.

    d. $(x - 5)(x + 3) = 0$, since this equation is equivalent to

$$x^2 - 2x - 15 = 0$$

When one side of the equation is zero, it is more useful to have the polynomial in factored form. Then, to find the solutions, we use the following theorem.

Zero Product Theorem

Let a and b be real numbers. If $a \cdot b = 0$, then $a = 0$ or $b = 0$, or both.

REMARK

This theorem states that if the product of two numbers is zero, then one or both of the two numbers must be zero.

Suppose that we wanted to find the solutions of the quadratic equation $x^2 - 2x - 15 = 0$:

$$x^2 - 2x - 15 = 0$$
$(x - 5)(x + 3) = 0$ Factor the left-hand side.

Now, $x - 5 = 0$ or $x + 3 = 0$ by the zero product theorem so $x = 5$ or $x = -3$.

    Checking our solutions:

when $x = 5$

$$5^2 - 2 \cdot 5 - 15 \overset{?}{=} 0$$
$$25 - 10 - 15 \overset{?}{=} 0$$
$$0 = 0 \qquad x = 5 \text{ is a solution.}$$

when $x = -3$

$(-3)^2 - 2 \cdot (-3) - 15 \stackrel{?}{=} 0$

$9 + 6 - 15 \stackrel{?}{=} 0$

$0 = 0 \quad x = -3$ is a solution.

Thus, the solution set is $\{5, -3\}$.

NOTE ▶ After applying the zero product theorem, we can find the solutions of the quadratic equation by finding the solutions of two linear equations.

Generalizing the preceding example, we can use the following steps when trying to find the solutions of factorable quadratic equations.

1. Place the quadratic equation in standard form.

2. Factor the resulting polynomial.

3. Use the zero product theorem, and set each factor equal to zero.

4. Solve the resulting equations.

5. Check your solutions.

EXAMPLE 4.8.1

Find the solutions of the following equations; also, check your solutions.

1. $(x - 7)(x + 1) = 0$ We are already at step 2.

Thus, $x - 7 = 0$ or $x + 1 = 0$ Step 3

$x = 7$ or $x = -1$ Step 4

Checking our solutions:

when $x = 7$

$(7 - 7)(7 + 1) \stackrel{?}{=} 0$

$0 \cdot 8 \stackrel{?}{=} 0$

$0 = 0 \quad x = 7$ is a solution.

when $x = -1$

$(-1 - 7)(-1 + 1) \stackrel{?}{=} 0$

$(-8) \cdot 0 \stackrel{?}{=} 0$

$0 = 0 \quad x = -1$ is a solution.

Thus, the solution set is $\{7, -1\}$.

2. $2x^2 + 5x - 3 = 0$ Step 1

$(2x - 1)(x + 3) = 0$ Step 2

$2x - 1 = 0$ or $x + 3 = 0$ Step 3

$$2x = 1 \qquad\qquad x = -3 \qquad \text{Step 4}$$
$$x = \tfrac{1}{2}$$

Checking our solutions:

when $x = \tfrac{1}{2}$

$$2\left(\tfrac{1}{2}\right)^2 + 5\left(\tfrac{1}{2}\right) - 3 \overset{?}{=} 0$$
$$2\left(\tfrac{1}{4}\right) + \tfrac{5}{2} - 3 \overset{?}{=} 0$$
$$\tfrac{1}{2} + \tfrac{5}{2} - 3 \overset{?}{=} 0$$
$$0 = 0 \qquad x = \tfrac{1}{2} \text{ is a solution.}$$

when $x = -3$

$$2 \cdot (-3)^2 + 5(-3) - 3 \overset{?}{=} 0$$
$$2 \cdot 9 - 15 - 3 \overset{?}{=} 0$$
$$18 - 15 - 3 \overset{?}{=} 0$$
$$0 = 0 \qquad x = -3 \text{ is a solution.}$$

Thus, the solution set is $\left\{\tfrac{1}{2}, -3\right\}$.

3. $6x^2 - 15x = 0$ $\qquad\qquad$ Step 1

 $3x(2x - 5) = 0$ $\qquad\qquad$ Step 2

 $3x = 0 \quad$ or $\quad 2x - 5\ = 0$ $\qquad$ Step 3

 $x = 0 \qquad\qquad\quad 2x\ \ = 5$ $\qquad$ Step 4

$ \qquad\qquad\qquad\qquad\quad x = \tfrac{5}{2}$

By checking these solutions, the reader can verify that the solution set is $\left\{0, \tfrac{5}{2}\right\}$.

4. $\qquad\qquad\qquad 4x^2 = 25$ $\qquad\qquad$ First, place the equation in standard form.

$\qquad\qquad\qquad 4x^2 - 25 = 0$ $\qquad\qquad$ Step 1

$\qquad (2x + 5)(2x - 5) = 0$ $\qquad\qquad$ Step 2

$\qquad 2x + 5 = 0 \quad$ or $\quad 2x - 5 = 0$ $\qquad$ Step 3

$\qquad\quad 2x = -5 \qquad\qquad 2x = 5$ $\qquad$ Step 4

$\qquad\qquad x = -\tfrac{5}{2} \qquad\qquad x = \tfrac{5}{2}$

By checking these solutions, the reader can verify that the solution set is $\left\{-\tfrac{5}{2}, \tfrac{5}{2}\right\}$.

5. $\quad 4x^2 + 16x + 1 = 4(x - 2)$ $\qquad\qquad$ Perform the indicated multiplication on the right-hand side.

$\qquad 4x^2 + 16x + 1 = 4x - 8$ $\qquad\qquad$ Place the equation in standard form.

$\quad 4x^2 + 16x - 4x + 1 + 8 = 4x - 4x - 8 + 8$

$\qquad 4x^2 + 12x + 9 = 0$ $\qquad\qquad$ Step 1

$$(2x + 3)(2x + 3) = 0 \qquad \text{Step 2}$$

$$2x + 3 = 0 \quad \text{or} \quad 2x + 3 = 0 \qquad \text{Step 3}$$

$$2x = -3$$

Since these equations are identical, we obtain only one solution.

$$x = -\tfrac{3}{2}$$

To check a solution, we always use the original equation:

$$4\left(-\tfrac{3}{2}\right)^2 + 16\left(-\tfrac{3}{2}\right) + 1 \stackrel{?}{=} 4\left(-\tfrac{3}{2} - 2\right)$$

$$4\left(\tfrac{9}{4}\right) + (-24) + 1 \stackrel{?}{=} 4\left(-\tfrac{7}{2}\right)$$

$$9 + (-24) + 1 \stackrel{?}{=} -14$$

$$-14 = -14$$

Thus, the solution set is $\left\{-\tfrac{3}{2}\right\}$.

6. $\qquad (3x - 5)(x + 2) = 20$

Perform the indicated multiplication on the left-hand side.

$$3x^2 + x - 10 = 20$$

Place the equation in standard form.

$$3x^2 + x - 30 = 0 \qquad \text{Step 1}$$

$$(3x + 10)(x - 3) = 0 \qquad \text{Step 2}$$

$$3x + 10 = 0 \quad \text{or} \quad x - 3 = 0 \qquad \text{Step 3}$$

$$3x = -10 \qquad\qquad x = 3 \qquad \text{Step 4}$$

$$x = -\tfrac{10}{3}$$

By checking these solutions, the reader can verify that the solution set is $\left\{-\tfrac{10}{3}, 3\right\}$.

Try to avoid this mistake:

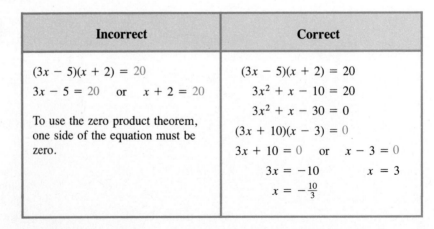

Incorrect	Correct
$(3x - 5)(x + 2) = 20$	$(3x - 5)(x + 2) = 20$
$3x - 5 = 20 \quad \text{or} \quad x + 2 = 20$	$3x^2 + x - 10 = 20$
	$3x^2 + x - 30 = 0$
To use the zero product theorem, one side of the equation must be zero.	$(3x + 10)(x - 3) = 0$
	$3x + 10 = 0 \quad \text{or} \quad x - 3 = 0$
	$3x = -10 \qquad\qquad x = 3$
	$x = -\tfrac{10}{3}$

7. $x^2 - 4x + 1 = 0$

This quadratic equation is already in standard form. However, the polynomial $x^2 - 4x + 1$ cannot be factored (that is, it is prime), so we cannot use the zero product theorem. This quadratic equation does have solutions, and we will learn how to find them in Chapter 9.

8.

$$4x^2 - 2x - 30 = 0 \qquad \text{Step 1}$$
$$2(2x^2 - x - 15) = 0 \qquad \text{Step 2}$$
$$2(2x + 5)(x - 3) = 0 \qquad \text{Step 2}$$
$$2x + 5 = 0 \quad \text{or} \quad x - 3 = 0 \qquad \text{Step 3}$$
$$2x = -5 \qquad\qquad x = 3$$
$$x = -\tfrac{5}{2}$$

By checking these solutions, the reader can verify that the solution set is $\left\{-\tfrac{5}{2}, 3\right\}$.

NOTE ▶ Do not use the zero product theorem to state that $2 = 0$ or $2x + 5 = 0$ or $x - 3 = 0$. *Clearly, $2 \neq 0$!* Only a factor containing a variable can equal zero.

Try to avoid these mistakes:

Incorrect	Correct
1. Factor the polynomial $$x^2 - x - 12:$$ $$x^2 - x - 12 = (x - 4)(x + 3)$$ $$x - 4 = 0 \quad \text{or} \quad x + 3 = 0$$ $$x = 4 \qquad\qquad x = -3$$ No equation is given. The value of x cannot be determined.	$$x^2 - x - 12 = (x - 4)(x + 3)$$
2. Solve the following equation. $$2x^2 - x - 3 = 0$$ $$(2x - 3)(x + 1)$$ Don't stop here! There is an equation to solve.	$$2x^2 - x - 3 = 0$$ $$(2x - 3)(x + 1) = 0$$ $$2x - 3 = 0 \quad \text{or} \quad x + 1 = 0$$ $$2x = 3 \qquad\qquad x = -1$$ $$x = \tfrac{3}{2}$$

The following problem illustrates how we can extend the zero product theorem to include polynomials in standard form that can be factored into more than two polynomial factors. (This is *not* a quadratic equation.)

9. $4x^3 + 26x^2 + 30x = 0$ Step 1

$2x(2x^2 + 13x + 15) = 0$

$2x(2x + 3)(x + 5) = 0$ Step 2

$2x = 0$ or $2x + 3 = 0$ or $x + 5 = 0$ Step 3

$x = 0$ $2x = -3$ $x = -5$ Step 4

$x = -\frac{3}{2}$

By checking these solutions, the reader can verify that the solution set is $\{0, -\frac{3}{2}, -5\}$.

EXERCISES 4.8

Find the solutions of the following equations; also, check your solutions.

1. $(x + 3)(x + 7) = 0$ **2.** $(x + 2)(x + 4) = 0$ **3.** $(2x - 5)(x + 12) = 0$

4. $(3x - 4)(x + 8) = 0$ **5.** $(4x - 1)(4x - 1) = 0$ **6.** $(2x - 9)(2x - 9) = 0$

7. $(3x - 4)(2x + 7) = 0$ **8.** $(5x - 1)(3x + 1) = 0$ **9.** $5x(x - 2)(x + 3) = 0$

10. $4x(x - 1)(x + 6) = 0$ **11.** $(2x + 7)(x - 3)(3x + 1) = 0$ **12.** $(2x + 9)(x - 4)(3x + 2) = 0$

13. $(x + 1)(x - 7)(x + 8)(x - 1) = 0$ **14.** $(x + 3)(x - 5)(x + 6)(x - 2) = 0$

15. $x^2 - 9x + 18 = 0$ **16.** $x^2 - 10x + 9 = 0$

17. $2x^2 + x - 15 = 0$ **18.** $4x^2 + 8x = 0$

19. $5x^2 + 15x = 0$ **20.** $9x^2 - 1 = 0$

21. $x^2 - 4x - 12 = 0$ **22.** $x^2 - 4x - 21 = 0$

23. $6x^2 - 3x = 0$ **24.** $3x^2 + 2x - 8 = 0$

25. $10x^2 - 19x + 7 = 0$ **26.** $10x^2 - 11x + 3 = 0$

27. $9x^2 + 12x + 4 = 0$ **28.** $6x^2 - 11x - 10 = 0$

29. $8x^2 - 22x - 21 = 0$ **30.** $6x^2 - 2x = 0$

31. $6x^2 + 13x + 5 = 0$ **32.** $9x^2 + 30x + 25 = 0$

33. $16x^2 - 1 = 0$ **34.** $12x^2 + 7x + 1 = 0$

35. $x^2 + 11x + 38 = 8$ **36.** $x^2 + 7x + 18 = 6$

37. $8x^2 + x - 10 = 2x - 3$

38. $7x^2 + 22x - 10 = 3x - 4$

39. $13x^2 - 13x + 9 = x^2 + 6$

40. $7x^2 - x + 3 = x^2 + 2$

41. $7x^2 + 4x + 1 = 2x^2 - 4x + 5$

42. $14x^2 - 22x + 8 = 2x^2 - 6x + 3$

43. $x^2 + 12x + 2 = 3(x - 4)$

44. $x^2 + 7x - 6 = 2(x - 5)$

45. $5x^2 - x - 50 = 2(2x^2 + 3)$

46. $7x^2 - x - 27 = 3(2x^2 + 1)$

47. $3x^2 - 18x + 21 = x(x - 5)$

48. $4x^2 - 17x + 10 = x(x - 6)$

49. $13x^2 + 17x + 2 = 4x(x + 2)$

50. $18x^2 + 26x + 3 = 2x(x + 5)$

51. $10x^2 - 2x - 64 = (x + 3)(x - 5)$

52. $26x^2 - 3x - 44 = (x + 5)(x - 8)$

53. $9x^2 - 71x + 5 = (2x + 1)(x - 4)$

54. $7x^2 - 19x - 8 = (2x + 3)(x - 5)$

55. $(2x + 3)(x + 1) = 1$

56. $(3x + 2)(x + 2) = -1$

57. $(2x - 7)(x - 5) = 2$

58. $(3x - 8)(x - 4) = 7$

59. $(3x + 1)(2x - 5) = 39$

60. $(7x - 4)(2x - 1) = 33$

61. $(2x + 5)(4x - 5) = -13$

62. $(6x - 5)(3x - 4) = -1$

63. $(x - 9)(2x + 3) = (x - 9)(x + 4)$

64. $(x - 8)(3x + 4) = (x - 8)(x + 1)$

65. $(2x - 3)(3x + 4) = (3x + 4)(x - 5)$

66. $(5x - 1)(3x + 1) = (3x + 1)(x - 3)$

67. $(2x - 1)(x + 3) = (x + 2)(x - 5)$

68. $(2x + 1)(x + 4) = (x - 2)(x + 2)$

69. $(3x + 4)(x - 2) = (x - 1)(x + 8)$

70. $(2x - 3)(x + 2) = (x + 1)(x - 6)$

71. $2x^3 + 4x^2 - 6x = 0$

72. $2x^3 + 6x^2 - 20x = 0$

73. $27x^3 + 18x^2 + 3x = 0$

74. $48x^3 + 24x^2 + 3x = 0$

75. $30x^3 - 65x^2 + 30x = 0$

76. $48x^3 - 100x^2 + 48x = 0$

77. $(x + 2)(12x^2 - 32x + 5) = 0$

78. $(x + 1)(12x^2 - 44x + 7) = 0$

79. $(2x - 5)(3x^2 - 29x + 18) = 0$

80. $(3x - 4)(2x^2 - 13x - 7) = 0$

81. $(x^2 - 4)(4x^2 - 4x - 15) = 0$

82. $(x^2 - 9)(6x^2 - 7x - 5) = 0$

4.9 Applications of Quadratic Equations

In Chapter 2 we studied applications of linear equations. In this section we will study word problems that lead to quadratic equations. We will follow the same format for working word problems developed in Section 2.7.

Number Problems

EXAMPLE 4.9.1   Find two consecutive odd integers whose product is 23 more than their sum.

SOLUTION   Note that consecutive odd integers such as 3 and 5 or 11 and 13 are 2 apart. Therefore, if

$$x = \text{first odd integer}$$

then $x + 2 = $ next consecutive odd integer

Since   product   is 23 more than   sum,

$$\text{then}\quad x \cdot (x + 2) = \quad 23 + \quad x + (x + 2)$$

$$x^2 + 2x = 25 + 2x$$

$$x^2 - 25 = 0 \qquad\qquad\quad \text{Subtract } 2x \text{ and } 25 \text{ from both sides.}$$

$$(x + 5)(x - 5) = 0 \qquad\qquad \text{Factor.}$$

$$x + 5 = 0 \quad \text{or} \quad x - 5 = 20 \qquad \text{Set each factor equal}$$

$$x = -5 \qquad\qquad\quad x = 5 \qquad \text{to zero and solve.}$$

If $x = -5$, then $x + 2 = -3$. If $x = 5$, then $x + 2 = 7$. Since $(-5)(-3) = 15$ is 23 more than $(-5) + (-3) = -8$, and $5 \cdot 7 = 35$ is 23 more than $5 + 7 = 12$, two sets of numbers are solutions to the word problem: -5 and -3 or 5 and 7.

EXAMPLE 4.9.2   The sum of two numbers is 4 and the sum of their squares is $\frac{25}{2}$. What are the two numbers?

SOLUTION x can represent only one number. How can we represent the other number?

Since $x + $ (other number) $= 4$

then   (other number) $= 4 - x$

Now, sum of squares is $\frac{25}{2}$ so

$$x^2 + (4 - x)^2 = \tfrac{25}{2}$$

$$x^2 + 16 - 8x + x^2 = \tfrac{25}{2} \qquad \text{Use the formula for squaring a binomial.}$$

$$2x^2 - 8x + 16 = \tfrac{25}{2}$$

$$4x^2 - 16x + 32 = 25 \qquad \text{Multiply both sides by 2.}$$

$$4x^2 - 16x + 7 = 0$$

$$(2x - 1)(2x - 7) = 0$$

$$2x - 1 = 0 \quad \text{or} \quad 2x - 7 = 0$$

$$2x = 1 \qquad\qquad 2x = 7$$

$$x = \tfrac{1}{2} \qquad\qquad x = \tfrac{7}{2}$$

If $x = \frac{1}{2}$, then $4 - x = 4 - \frac{1}{2} = \frac{7}{2}$. Therefore, the two solutions of the equation give us the solution of the word problem.

$$\text{Check:} \quad \left(\tfrac{1}{2}\right)^2 + \left(\tfrac{7}{2}\right)^2 = \tfrac{1}{4} + \tfrac{49}{4} = \tfrac{50}{4} = \tfrac{25}{2}$$

Geometric Problems

EXAMPLE 4.9.3

The length of a rectangle is 1 foot more than twice the width. If the area of the rectangle is 105 square feet (sq ft), what are the dimensions of the rectangle?

SOLUTION

Since the length is described in terms of the width, we let x = width of the rectangle and $2x + 1$ = length of the rectangle. (See Figure 4.9.1.)

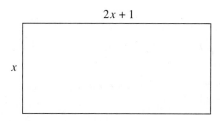

$2x + 1$

x

Figure 4.9.1

Recall that the area of a rectangle is given by the formula

$$\text{Area} = \text{width} \cdot \text{length}$$

$$\text{Thus, } 105 = x \cdot (2x + 1)$$

$$105 = 2x^2 + x$$

$$0 = 2x^2 + x - 105$$

$$0 = (2x + 15)(x - 7)$$

$$2x + 15 = 0 \quad \text{or} \quad x - 7 = 0$$

$$2x = -15 \qquad\qquad x = 7$$

$$x = -\tfrac{15}{2}$$

We reject $-\tfrac{15}{2}$, since the width of a rectangle cannot be negative. Therefore, the dimensions are

$$\text{width} = 7 \text{ ft}$$

$$\text{length} = 2 \cdot 7 + 1 = 15 \text{ ft}$$

Check: Area = 7 ft · 15 ft = 105 sq ft

EXAMPLE 4.9.4

The height of a triangle is 3 inches shorter than the base. Find the base and the height if the area of the triangle is 20 square inches (sq in.).

SOLUTION

If we let x = base of the triangle, then $x - 3$ = height of the triangle. (See Figure 4.9.2.)

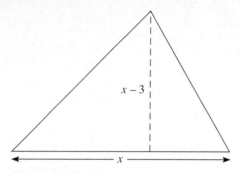

Figure 4.9.2

For a triangle,

$$\text{Area} = \tfrac{1}{2} \cdot \text{base} \cdot \text{height}$$

$$20 = \tfrac{1}{2} \cdot x \cdot (x - 3)$$

$$20 = \tfrac{1}{2}x^2 - \tfrac{3}{2}x$$

$$40 = x^2 - 3x \qquad \text{Multiply both sides by 2.}$$

$$0 = x^2 - 3x - 40$$

$$0 = (x - 8)(x + 5)$$

$$x - 8 = 0 \qquad \text{or} \qquad x + 5 = 0$$

$$x = 8 \qquad\qquad x = -5$$

Reject -5, since the base cannot be negative. Therefore, if base = 8 inches, then height = $8 - 3 = 5$ inches.

$$\text{Check:} \quad \text{Area} = \tfrac{1}{2} \cdot 8 \text{ in.} \cdot 5 \text{ in.} = 4 \text{ in.} \cdot 5 \text{ in.} = 20 \text{ sq in.}$$

EXAMPLE 4.9.5

The longer leg of a right triangle is 1 m less than twice the shorter leg. The hypotenuse is 1 m more than twice the shorter leg. Find the lengths of the three sides of the right triangle.

SOLUTION

Since the longer leg and the hypotenuse are both described in terms of the shorter leg,

$$\text{let } x = \text{length of shorter leg}$$

$$\text{then } 2x - 1 = \text{length of longer leg}$$

$$\text{and } 2x + 1 = \text{length of hypotenuse}$$

(See Figure 4.9.3.)

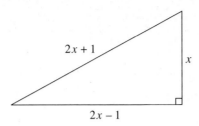

Figure 4.9.3

To set up the equation for this problem, we need the Pythagorean theorem. This theorem is covered in Appendix II. Here, we will merely state the theorem:

$$(\text{Hypotenuse})^2 = (\text{Leg 1})^2 + (\text{Leg 2})^2$$
$$(2x + 1)^2 = x^2 + (2x - 1)^2$$
$$4x^2 + 4x + 1 = x^2 + 4x^2 - 4x + 1 \qquad \text{Square the binomials.}$$
$$4x^2 + 4x + 1 = 5x^2 - 4x + 1$$
$$0 = x^2 - 8x$$
$$0 = x(x - 8)$$
$$x = 0 \qquad \text{or} \qquad x - 8 = 0$$
$$x = 8$$

Reject 0, since each side must be positive. Therefore, if

$$\text{short leg} = 8 \text{ m}$$
$$\text{then long leg} = 2 \cdot 8 - 1 = 15 \text{ m}$$
$$\text{and hypotenuse} = 2 \cdot 8 + 1 = 17 \text{ m}$$
$$\text{Check:} \quad 17^2 \overset{?}{=} 8^2 + 15^2$$
$$289 \overset{?}{=} 64 + 225$$
$$289 = 289$$

Physics Problem

EXAMPLE 4.9.6

An object is projected vertically upward, from ground level, with an initial speed of 120 feet per second. The equation that gives the object's height above ground level is $h = -16t^2 + 120t$, where h is the height (measured in feet) and t is the time (measured in seconds). When will the object be 200 feet above the ground?

SOLUTION   To answer this question, we substitute 200 for h and solve the resulting equation:

$$200 = -16t^2 + 120t$$
$$16t^2 - 120t + 200 = 0 \quad \text{Divide by 8.}$$
$$2t^2 - 15t + 25 = 0$$
$$(2t - 5)(t - 5) = 0$$

$$2t - 5 = 0 \qquad \text{or} \qquad t - 5 = 0$$
$$2t = 5 \qquad\qquad\qquad t = 5$$
$$t = \tfrac{5}{2}$$

Since both solutions are positive, they are both feasible solutions to the word problem. Why? (See Figure 4.9.4.)

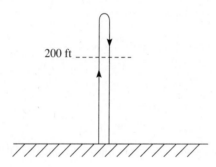

200 ft

Figure 4.9.4

Check: $-16\left(\tfrac{5}{2}\right)^2 + 120\left(\tfrac{5}{2}\right) \overset{?}{=} 200$

$$-16\left(\tfrac{25}{4}\right) + 300 \overset{?}{=} 200$$
$$-100 + 300 = 200$$
$$200 = 200$$
$$-16(5)^2 + 120(5) \overset{?}{=} 200$$
$$-16(25) + 600 \overset{?}{=} 200$$
$$-400 + 600 \overset{?}{=} 200$$
$$200 = 200$$

EXERCISES 4.9

Solve the following problems. The formulas needed are either given in the problem or they can be found in the examples of this section.

1. Find two consecutive integers whose product is 72.

2. Find two consecutive integers whose product is 45 more than three times the larger integer.

3. Find two consecutive even integers whose product is 24 more than eight times the larger integer.

4. Find two consecutive odd integers whose product is 4 less than three times the larger integer.

5. Find two consecutive odd integers whose product is 95 more than twice their sum.

6. Find two consecutive even integers whose product is 10 less than five times their sum.

7. Find two consecutive integers the sum of whose squares is 1 more than twelve times the larger.

8. Find two consecutive integers the difference of whose squares is 131 less than their product.

9. The sum of two numbers is 6. The sum of their squares is 180. What are the two numbers?

10. The sum of two numbers is 9. The sum of their squares is 101. What are the two numbers?

11. The sum of two numbers is 10. The difference of their squares is 4 less than their product. What are the two numbers?

12. The sum of two numbers is 2. The sum of their squares is 49 more than their product. What are the two numbers?

13. The length of a rectangle is 5 ft more than the width. If the area of the rectangle is 36 sq ft, what are the dimensions of the rectangle?

14. The width of a rectangle is 8 in. less than the length. If the area of the rectangle is 33 sq in., what are the dimensions of the rectangle?

15. The length of a rectangle is 3 yd more than twice the width. If the area of the rectangle is 77 sq yd, what are the dimensions of the rectangle?

16. The length of a rectangle is 2 cm less than three times the width. If the area of the rectangle is 56 sq cm, what are the dimensions of the rectangle?

17. The width of a rectangle is 4 m less than half the length. If the area of the rectangle is 90 sq m, what are the dimensions of the rectangle?

18. The width of a rectangle is 1 ft more than half the length. If the area of the rectangle is 60 sq ft, what are the dimensions of the rectangle?

19. The base of a triangle is 5 ft longer than the height. Find the base and the height if the area of the triangle is 42 sq ft.

20. The base of a triangle is 3 in. shorter than the height. Find the base and the height if the area of the triangle is 27 sq in.

21. The base of a triangle is 3 m less than twice the height. Find the base and the height if the area of the triangle is 22 sq m.

22. The base of a triangle is 7 cm less than three times the height. Find the base and the height if the area of the triangle is 13 sq cm.

23. The height of a triangle is 1 in. more than twice the base. Find the base and the height if the area of the triangle is $52\frac{1}{2}$ sq in.

24. The height of a triangle is 7 ft less than twice the base. Find the base and the height if the area of the triangle is $7\frac{1}{2}$ sq ft.

25. The longer leg of a right triangle is 12 in. The hypotenuse is 3 in. less than twice the shorter leg. Find the length of the shorter leg.

26. The hypotenuse of a right triangle is 13 millimeters. The longer leg is 2 millimeters longer than twice the shorter leg. Find the lengths of the two legs.

27. The longer leg of a right triangle is 3 feet longer than three times the shorter leg. The hypotenuse is 3 feet shorter than four times the shorter leg. Find the lengths of the three sides of the right triangle.

28. The shorter leg of a right triangle is 11 meters shorter than twice the longer leg. The hypotenuse is 1 meter longer than the longer leg. Find the lengths of the three sides of the right triangle.

29. The lengths in inches of the three sides of a right triangle are consecutive integers. Find the lengths of the three sides of the right triangle.

30. The longer leg of a right triangle is 4 centimeters longer than four times the shorter leg. The hypotenuse is 1 centimeter longer than the longer leg. Find the lengths of the three sides of the right triangle.

In exercises 31 through 36, an object is moving vertically upward or downward. Its height above ground level is denoted by h, measured in feet, and time is denoted by t, measured in seconds.

31. The height of an object projected upward from ground level is given by $h = -16t^2 + 128t$. When will the object be 240 feet above the ground?

32. The height of an object projected upward from ground level is given by $h = -16t^2 + 160t$. When will the object be 300 feet above the ground?

33. The height of an object projected upward from a 960-foot-tall building is given by $h = -16t^2 + 64t + 960$. When will the object hit the ground? (*Hint:* It reaches the ground when $h = 0$.)

34. The height of an object projected upward from a 480-foot-tall building is given by $h = -16t^2 + 112t + 480$. When will the object hit the ground? (See the hint for exercise 33.)

35. The height of an object dropped from a 100-foot-tall building is given by $h = -16t^2 + 100$. When will the object hit the ground?

36. The height of an object dropped from a 205-foot-tall building is given by $h = -16t^2 + 205$. When will the object be 36 feet above the ground?

In exercises 37 and 38, the kinetic energy of a moving object is given by $E = \frac{1}{2}mv^2$, where E is measured in joules (J), m is the object's mass, measured in kilograms (kg), and v is the object's velocity, measured in meters per second (m/sec).

37. Find the velocity of an object whose mass is 5 kg and kinetic energy is 10 J.

38. Find the velocity of an object whose mass is 4000 kg and kinetic energy is 50,000 J.

39. Tarzan and Jane have a rectangular swimming pool that is 60 m wide and 130 m long. They are going to make a rock border around the pool of uniform width (see Figure 4.9.5). They have enough rocks for 2000 sq m. How wide is the border? [*Hint:* (Area of outer rectangle) − (Area of inner rectangle) = (Area of border).]

40. Homer and Raetta have a rectangular garden that is 20 yd wide and 35 yd long. They are going to make a pebble path of uniform width around the garden. They have enough pebbles for 174 sq yd. How wide is the border? (See hint for exercise 39.)

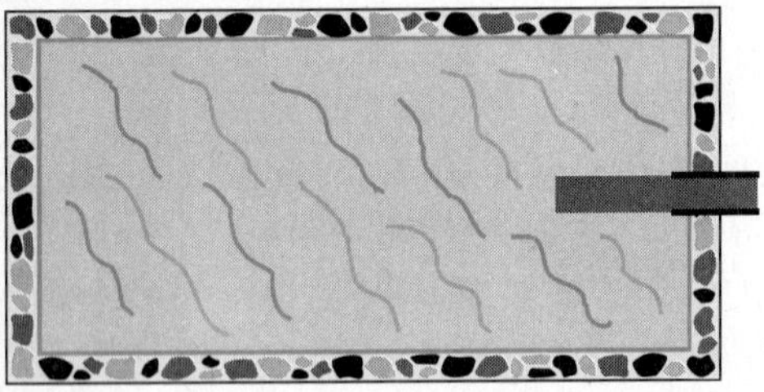

Figure 4.9.5

Chapter 4 Review

Terms to Remember

	Page		Page
Factor	169	**Completely factored**	173
Divisible	169	**Difference of two squares**	176

Formulas

- *Difference of two squares:*

$$a^2 - b^2 = (a + b)(a - b)$$

- *Perfect square trinomials:*

$$a^2 + 2ab + b^2 = (a + b)^2$$
$$a^2 - 2ab + b^2 = (a - b)^2$$

- *Sum of two cubes:*

$$x^3 + y^3 = (x + y)(x^2 - xy + y^2)$$

- *Difference of two cubes:*

$$x^3 - y^3 = (x - y)(x^2 + xy + y^2)$$

Factoring Summary

Given a polynomial to factor:

1. Examine the terms to determine if there is a common factor. If so, factor out the greatest common factor.

2. Consider the number of terms.

Two terms	Difference of squares?
	Difference of cubes?
	Sum of cubes?
Three terms	Perfect square trinomial?
	Leading coefficient of 1?
	Leading coefficient not equal to 1?
Four or more terms	Factor by grouping

3. Examine all of the factors to determine if any factor can be factored further.

Important Facts

- *Zero product theorem:* Let a and b be real numbers. If $a \cdot b = 0$, then $a = 0$ or $b = 0$ or both.

- *Area of a rectangle:* Area = (length)(width)

- *Area of a triangle:* Area = $\frac{1}{2}$(base)(height)

- *Pythagorean theorem:* (Hypotenuse)^2 = (Leg 1)^2 + (Leg 2)^2

Review Exercises

4.1 Find the prime factored form for the following numbers.

1. 231

2. 520

Find the greatest common factor for the following groups of numbers.

3. 56, 42, 70

4. 90, 54, 198

Find the greatest common factor for the following groups of monomials.

5. $20x^3y^2$, $56x^4y^5$

6. $48x^5$, $12x^3$, $36x^2$

7. $15xy^3$, $27x^2y^2$, $33x^3y^2$

8. $16x^3y$, $20xz^2$, $28x^2z$, $12xy^3$

Completely factor the following polynomials.

9. $6x^3 + 12xy$

10. $7a^4b^3 - 15a^3bc^2$

11. $3x^3y - 15xy^2 + 6y$

12. $6x^2y - 10xz + 15y^2z$

13. $-20x^3 + 25xy - 15x^2$

14. $7x^3y + 77x^2y - 21xy^4$

15. $8x^4y + 12x^4y^2 + 4x^3y$

16. $36x^4 - 18x^3 + 45x^2 + 27x$

17. $3x^3y^2 + 9x^2y^3 - 3xy^3 - 12xy^4$

18. $24ax^4 - 30ax^4y + 12ax^3y^2 - 6ax^3$

19. $3x(2x + 5) - y(2x + 5)$

20. $5x^3(x - 4) + 15x(x - 4)$

4.2 Completely factor the following polynomials.

21. $x^2 - 81$

22. $169 - y^2$

23. $16x^2 - y^2$

24. $4x^2 + 9$

25. $36x^2 - 121y^2$

26. $5x^2 - 20$

27. $54x^2 - 24y^2$

28. $16x^2 + 144$

29. $y^4 - 256$

30. $3x^3 - 27x$

31. $16x^4y - 4x^2y$

32. $72x^5y^2 - 50x^3y^4$

33. $y^2 + 12y + 36$

34. $x^2 - 20x + 100$

35. $25x^2 + 10x + 1$

36. $49x^2 + 28xy + 4y^2$

37. $x^2 - 26xy + 169y^2$

38. $9x^2 - 18x + 9$

39. $4x^2 + 10xy + 25y^2$

40. $128x^2 + 96x + 18$

41. $81x^2 + 18xy + y^2$

42. $3x^3 - 36x^2 + 108x$

43. $18x^3y - 24x^2y^2 + 8xy^3$

44. $128x^4y^4 + 576x^3y^5 + 698x^2y^6$

4.3 Completely factor the following polynomials.

45. $x^2 + 6x + 8$

46. $x^2 + 6x + 5$

47. $x^2 - 7x + 10$

48. $x^2 - 9x + 14$

49. $x^2 - 6x - 27$

50. $x^2 - 8x - 20$

51. $x^2 + 7x - 8$

52. $x^2 + 10x - 24$

53. $x^2 - 3x + 4$

54. $x^2 - 8x + 16$

55. $-x^2 + 2x + 35$

56. $-x^2 + 11x - 18$

57. $x^2 + 12x - 13$ **58.** $x^2 - 7x + 8$ **59.** $3x^2 - 36x + 96$

60. $x^3 + 8x^2 + 15x$ **61.** $4x^4 - 16x^3 - 48x^2$ **62.** $6x^3y^2 - 30x^2y^2 + 36xy^2$

63. $x^2 - 11xy + 24y^2$ **64.** $x^2 + 2xy - 63y^2$ **65.** $2x^2 - 14xy - 60y^2$

66. $x^4 + 16x^3y + 15x^2y^2$ **67.** $5x^4y - 45x^3y^2 + 100x^2y^3$ **68.** $-3x^2y^2 - 18xy^3 + 21y^4$

4.4 **Completely factor the following trinomials.**

69. $2x^2 + 5x + 3$ **70.** $2x^2 + 5x + 2$ **71.** $3x^2 - 5x + 2$

72. $3x^2 - x + 4$ **73.** $4x^2 - 11x + 6$ **74.** $5x^2 - 17x + 6$

75. $6x^2 - 11x - 2$ **76.** $5x^2 + 7x - 6$ **77.** $4x^2 + x - 2$

78. $-3x^2 + 2x + 8$ **79.** $-4x^2 - 4x + 3$ **80.** $-6x^2 + 5x + 4$

81. $18x^2 + 18x - 20$ **82.** $6x^2 - 3x - 63$ **83.** $6x^3 - 7x^2 - 10x$

84. $-32x^3 - 72x^2 - 36x$ **85.** $6x^3 + 3x^2 + 9x$ **86.** $48x^4y^2 + 90x^3y^2 - 12x^2y^2$

87. $7x^2 + 23xy + 6y^2$ **88.** $20x^2 + 9xy - 20y^2$ **89.** $16x^2 + 88xy + 40y^2$

90. $-25x^2 + 105xy - 20y^2$ **91.** $-12x^5y + 22x^4y^2 - 6x^3y^3$ **92.** $144x^3y^2 - 132x^2y^3 - 12xy^4$

4.5 **Completely factor the following polynomials.**

93. $6xy + 21x + 10y + 35$ **94.** $6xy + 16x + 9y + 24$ **95.** $8xy + 20x - 14y - 35$

96. $6xy + x - 30y - 5$ **97.** $6x^3 + 14x^2 + 3x + 7$ **98.** $8x^3 + 3x^2 + 40x + 15$

99. $12xy + 18x + 2ay + 3a$ **100.** $4xy + 5x - 16ay - 20a$ **101.** $5x^3 + 2x^2 - 45x - 18$

102. $6x^3 + 11x^2 - 6x - 11$ **103.** $6x^2y + 24x^2 + 27xy + 108x$ **104.** $6xy^2 - 10xy + 30y^2 - 50y$

Completely factor the following trinomials by grouping.

105. $6x^2 + 31x + 40$ **106.** $4x^2 - 17x + 15$ **107.** $10x^2 - x - 2$ **108.** $4x^2 - 8x - 5$

109. $6x^2 - 17xy - 14y^2$ **110.** $6x^2 + 17xy + 10y^2$ **111.** $6x^2 + 11xy - 30y^2$ **112.** $9x^2 - 9xy - 4y^2$

4.6 **Completely factor the following binomials.**

113. $x^3 - 125$ **114.** $27y^3 - 1$ **115.** $8m^3 + 27$ **116.** $64a^3 + 125b^3$

117. $4y^3 - 4$ **118.** $72p^4 - 9pq^3$ **119.** $54x^4y^3 + 2xy^6$ **120.** $5x^5y^2 + 1080x^2y^5$

121. $64x^6 - y^3$ **122.** $8x^6 - 8x^3y^6$

4.8 **Find the solutions of the following equations; also, check your solutions.**

123. $(x - 2)(x + 5) = 0$ **124.** $(3x - 7)(x + 4) = 0$

125. $(2x + 5)(2x - 3)(x - 7) = 0$ **126.** $9x^2 - 12x + 4 = 0$

127. $4x^2 - 49 = 0$ **128.** $6x^2 - 5x - 4 = 0$

129. $x^2 - 13x + 40 = 0$ **130.** $6x^2 - 15x + 5 = 12x + 5$

131. $2x^2 + 5x - 10 = x^2 + 2x + 8$

132. $4x^2 + 4x - 11 = 3(x - 2)$

133. $3x^2 - 11x + 14 = (2x + 1)(x - 2)$

134. $(2x + 5)(2x - 7) = -4x + 1$

135. $(5x + 4)(x - 5) = -24$

136. $(x + 3)(4x + 3) = 13$

137. $(3x + 10)(3x - 2) = (x + 3)(3x + 10)$

138. $(2x + 7)(x + 3) = (3x + 5)(2x + 1)$

139. $4x^3 + 10x^2 - 50x = 0$

140. $15x^3 + 50x^2 + 40x = 0$

141. $(2x + 5)(x^2 + x - 12) = 0$

142. $(x^2 - 1)(3x^2 - 14x - 5) = 0$

4.9 Solve the following problems.

143. Find two consecutive even integers whose product is 10 more than the larger integer.

144. Find two consecutive odd integers whose product is 1 less than their sum.

145. Find two consecutive integers the sum of whose squares is 28 more than the larger integer.

146. The sum of two numbers is 6. The sum of their squares is 68. What are the two numbers?

147. The length of a rectangle is four times the width. If the area of the rectangle is 36 sq ft, what are the dimensions of the rectangle?

148. The length of a rectangle is 3 in. more than twice the width. If the area of the rectangle is 20 sq in., what are the dimensions of the rectangle?

149. The base of a triangle is 4 m less than twice the height. Find the base and the height if the area of the triangle is 35 sq m.

150. The height of a triangle is 1 yd more than twice the base. Find the base and the height if the area of the triangle is 33 sq yd.

151. The hypotenuse of a right triangle is 26 feet. The longer leg is 4 feet longer than twice the shorter leg. Find the lengths of the two legs.

152. The longer leg of a right triangle is 1 inch more than twice the shorter leg. The hypotenuse is 1 inch less than three times the shorter leg. Find the lengths of the three sides of the right triangle.

153. The height of an object projected upward from ground level is given by $h = -16t^2 + 224t$. When will the object be 768 feet above the ground?

154. The height of an object projected upward from a 20-foot-tall tower is given by $h = -16t^2 + 56t + 20$. When will the object be 44 feet above the ground?

Chapter 4   Test   (You should be able to complete this test in 60 minutes.)

Completely factor the following polynomials.

1. $4x^2 - 32x + 55$

2. $3a^4 - 6ab + 3a$

3. $9m^2 + 66m + 121$

4. $x^2 - 25$

5. $x^2 - 2x - 24$

6. $8x^2 + 72$

7. $32x^2 - 2$

8. $x^2 + 12x + 36$

9. $3x^2 - 7x - 26$

10. $12x^2 - 29x - 8$

11. $8x^4y^5 - 36x^3y^6 + 20x^2y^5$

12. $x^2 + 16x + 28$

13. $9y^2 - 49$

14. $6xy - 28x + 3y^2 - 14y$

15. $27x^4 + 27x^3 - 12x^2$

16. $4x^2 - 12xy + 9y^2$ **17.** $y^2 - 14y + 33$ **18.** $2x^2 + 9x + 7$

19. $4x^2y - 10x^2 + 20xy - 50x$ **20.** $2x^3 - 18x^2 + 16x$

Find the solutions of the following equations; also, check your solutions.

21. $3x^2 - 13x + 4 = 0$ **22.** $4x^2 + 5x - 61 = 5(x + 4)$ **23.** $4x^3 - 18x^2 + 18x = 0$

Solve the following problems.

24. The sum of two numbers is 10. The sum of their squares is 122. What are the two numbers?

25. The height of a triangle is 6 ft less than twice the base. Find the base and the height if the area of the triangle is 18 sq ft.

Test Your Memory

These problems review Chapters 1–4.

I. Perform the indicated operations. When necessary, use the rules governing the order of operations.

1. $\dfrac{5^2 - 2^4}{3^2 + 6^2}$

2. $\dfrac{\frac{10}{9}}{15}$

3. $4^{-2} - 4^0$

4. $5^2 \cdot 5^{-5}$

5. $\dfrac{5}{14} - \dfrac{6}{7} + \dfrac{1}{4}$

6. $\left(-\dfrac{2}{5}\right)^{-2}$

7. $(5 \cdot 2^3)^2$

8. $3 + 2[11 - (-6 + 2)]$

II. Find the solutions of the following equations.

9. $4 + 2(x + 3) = 3(2x - 1) + 9x$

10. $5(x + 8) + 3(x - 4) = 0$

11. $4(x + 3) - 7(x + 2) = 3(x + 5)$

12. $\frac{2}{5}(3x + 10) = \frac{3}{2}(x + 4)$

13. $x^2 - 6x + 5 = 0$

14. $3x^2 + 9x + 10 = 2(x + 4)$

15. $(2x + 1)(x - 3) = 30$

16. $8x^3 + 32x^2 + 30x = 0$

III. Find the solution set of each of the following inequalities, and graph the solution set on a number line.

17. $2(4x + 1) \le 5(2x - 1)$

18. $8 - (2x + 5) > 9$

19. $-8 \le 3x + 1 \le -2$

20. $\frac{3}{4}(x + 2) \le \frac{2}{3}(x - 1) + 2$

IV. Simplify each of the following. (Remember to write your answers with only positive exponents.)

21. $\dfrac{12x^{-8}}{20x^4}$

22. $(5x^{-2}y^3)^{-2}$

23. $\left(\dfrac{x^{-3}y^2}{x^4y^{-3}}\right)^{-1}$

24. $\dfrac{(3x^{-2}y^4)^{-2}}{(3x^0y^{-2})^{-3}}$

V. Perform the indicated operations.

25. $3(5x^2 - 4) - 5(x^2 - 4x + 3)$

26. $(3x + 7)^2$

27. $-5x^3y^2(2x^2 + 8xy - 3xy^3)$

28. $(2x^3 - 4x^2 + 9)(4x + 3)$

29. $(9x^4 - 15x^2y^2 + 6xy^3) \div (-12x^2y)$

30. $(12x^3 - 27x^2 - 20x + 45) \div (4x - 9)$

31. $(3x - 5y)(11x - 2y)$

32. $(3x + 8y)(3x - 8y)$

VI. Completely factor the following polynomials.

33. $6x^2 + 7xy + 2y^2$

34. $16xy^2 + 56y^3 + 8y^2$

35. $x^2 - 8$

36. $6x^2 + 15xy - 8x - 20y$

37. $3x^3 + 3x^2 - 60x$

38. $6x^2 - 25x + 4$

39. $49k^2 - 16$

40. $81a^2 + 72ab + 16b^2$

41. $x^2 - 14x + 40$

42. $6y^2 + 40y + 24$

VII. Use algebraic expressions to find the solutions of the following problems.

43. Corye drove her car for 216 miles using $9\frac{3}{5}$ gallons of gasoline. How many miles per gallon did she average?

44. Damian has 3 less than twice as many one-dollar bills as five-dollar bills. He has $46 altogether. How many one-dollar bills and how many five-dollar bills does he have?

45. Bear invested $4500 in two accounts. He invested some in a savings account paying 6% simple interest and the rest in a CD paying 9% simple interest. After 1 year his interest income was $360. How much did Bear invest in each account?

46. The perimeter of a rectangle is 50 m. The length is 13 m more than $\frac{1}{2}$ of the width. What are the dimensions of the rectangle?

47. The sum of two numbers is 9. The sum of their squares is 45. What are the two numbers?

48. The difference of two numbers is 5. The sum of their squares is 73. What are the two numbers?

49. The length of a rectangle is 3 ft less than twice the width. Find the dimensions of the rectangle if it has an area of 54 sq ft.

50. The base of a triangle is 1 ft more than five times the height. Find the base and the height if the area of the triangle is 24 sq ft.

CHAPTER 5

Rational Expressions

5.1 Reducing to Lowest Terms

The basic procedures of arithmetic were introduced in Chapter 1. Many of the most fundamental procedures of algebra are nothing more than generalizations of concepts from arithmetic. For example, in Chapter 1 a rational number was defined to be a fraction (a quotient), p/q, where p and q are integers and q is not zero. In algebra we generalize the concept of a rational number as follows:

Rational Expression

Any expression that can be written as the quotient of two polynomials, where the denominator is not equal to zero, is called a **rational expression**.

REMARKS

1. The following expressions are examples of rational expressions.

$$\frac{9x^2y^3}{12xy^5}, \qquad \frac{8}{7x - 1}, \qquad \frac{2x + 3y}{x - 5y}$$

$$\frac{x^2 - 4}{x^2 - x - 2}, \qquad 3x^2 + 7x + 5$$

Note that $3x^2 + 7x + 5$ is a rational expression since it is equivalent to

$$\frac{3x^2 + 7x + 5}{1}$$

2. In Chapter 1 we saw that the denominator of a rational number could not be zero, because division by zero is undefined. Note that we have the same restriction with rational expressions.

In Chapter 1 we also learned how to reduce a fraction. We recall this procedure in the following example.

$$\frac{18}{30} = \frac{2 \cdot 3 \cdot 3}{2 \cdot 3 \cdot 5} \qquad \text{Factor the numerator and denominator.}$$

$$= \frac{\cancel{2} \cdot \cancel{3} \cdot 3}{\cancel{2} \cdot \cancel{3} \cdot 5} \qquad \text{Divide out the common factors of 2 and 3.}$$

$$= \frac{3}{5}$$

NOTE ▶ When all of the common factors have been divided out, the fraction is said to be in **lowest terms**.

The principle used in reducing fractions is generalized to rational expressions in the following property.

Reduction Property of Fractions
$$\dfrac{ax}{ay} = \dfrac{\cancel{a}x}{\cancel{a}y} = \dfrac{x}{y} \qquad a \neq 0, \ y \neq 0$$

REMARK   The reduction property of fractions says that we can divide out nonzero *factors* (not terms) that are common to both the numerator and denominator.

Let us use the reduction property of fractions to reduce the following rational expression.

$$\frac{9x^2y^3}{12xy^5} = \frac{3 \cdot 3 \cdot x \cdot x \cdot y \cdot y \cdot y}{2 \cdot 2 \cdot 3 \cdot x \cdot y \cdot y \cdot y \cdot y \cdot y} \qquad \text{Completely factor the numerator and denominator.}$$

$$= \frac{\cancel{3} \cdot 3 \cdot \cancel{x} \cdot x \cdot \cancel{y} \cdot \cancel{y} \cdot \cancel{y}}{2 \cdot 2 \cdot \cancel{3} \cdot \cancel{x} \cdot \cancel{y} \cdot \cancel{y} \cdot \cancel{y} \cdot y \cdot y} \qquad \text{Divide out common factors.}$$

$$= \frac{3x}{4y^2} \qquad x \neq 0, \ y \neq 0$$

NOTE ▶ *Although it is important that we note the restrictions $x \neq 0$ and $y \neq 0$, we usually do not write the restrictions. They are just understood.*

Generalizing the previous example, we can use the following two steps when trying to reduce a rational expression to lowest terms.

1. Completely factor both the numerator and denominator of the rational expression.

2. Divide out the factors that are common to both the numerator and denominator.

EXAMPLE 5.1.1

Reduce the following rational expressions to lowest terms.

1. $\dfrac{20x^3y^5}{24x^8y} = \dfrac{2 \cdot 2 \cdot 5y^{5-1}}{2 \cdot 2 \cdot 6x^{8-3}}$ ← The numerator has more factors of y.
                               ← The denominator has more factors of x.

$\qquad\qquad = \dfrac{5y^4}{6x^5}$

NOTE ▶ When both the numerator and denominator are monomials, we can use the laws of exponents developed in Chapter 3 to easily reduce the rational expression.

2. $\dfrac{12x^3y + 6x^2y^2}{15x^4y^3 - 3x^3y^4} = \dfrac{6x^2y(2x + y)}{3x^3y^3(5x - y)}$ Completely factor the numerator and denominator.

$\qquad\qquad = \dfrac{\overset{2}{6}(2x + y)}{\overset{}{3}x^{3-2}y^{3-1}(5x - y)}$

$\qquad\qquad = \dfrac{2(2x + y)}{xy^2(5x - y)}$

NOTE ▶ When reducing a rational expression to lowest terms, all you are really doing is dividing out the greatest common factor of *both* the numerator and denominator.

3. $\dfrac{x^2 + 8x + 15}{x^2 + 13x + 30} = \dfrac{(x + 3)(x + 5)}{(x + 3)(x + 10)}$ Completely factor the numerator and denominator.

$\qquad\qquad = \dfrac{(x + 3)(x + 5)}{(x + 3)(x + 10)}$ Divide out common factors.

$\qquad\qquad = \dfrac{x + 5}{x + 10}$

Try to avoid these mistakes:

Incorrect	Correct
1. $\dfrac{8}{10} = \dfrac{7 + 1}{7 + 3} = \dfrac{1}{3}$ Terms are being divided out.	$\dfrac{8}{10} = \dfrac{2 \cdot 4}{2 \cdot 5} = \dfrac{4}{5}$
2. $\dfrac{x^2 + 8x + \cancel{15}}{x^2 + 13x + \underset{2}{\cancel{30}}} = \dfrac{8x}{13x + 2}$ Terms are being divided out. The reduction property of fractions says that we can divide out only common factors.	$\dfrac{x^2 + 8x + 15}{x^2 + 13x + 30}$ $= \dfrac{(x + 3)(x + 5)}{(x + 3)(x + 10)}$ $= \dfrac{(\cancel{x + 3})(x + 5)}{(\cancel{x + 3})(x + 10)}$ $= \dfrac{x + 5}{x + 10}$

4. $\dfrac{4x^3 - 36x}{4x^5 - 10x^4 - 6x^3} = \dfrac{4x(x^2 - 9)}{2x^3(2x^2 - 5x - 3)}$

$\qquad\qquad = \dfrac{\overset{2}{\cancel{4}}x(x + 3)(\cancel{x - 3})}{\underset{x^2}{\cancel{2x^3}}(2x + 1)(\cancel{x - 3})}$

$\qquad\qquad = \dfrac{2(x + 3)}{x^2(2x + 1)}$

5. $\dfrac{x - 4}{x^2 - 7x + 12} = \dfrac{x - 4}{(x - 4)(x - 3)}$

$\qquad\qquad = \dfrac{\overset{1}{\cancel{(x - 4)}}}{\cancel{(x - 4)}(x - 3)}$ $\qquad$ Don't forget the factor of 1 on top!

$\qquad\qquad = \dfrac{1}{x - 3}$

6. $\dfrac{5x^3 - 10x^2 y}{x - 2y} = \dfrac{5x^2(x - 2y)}{x - 2y}$

$\qquad\qquad = \dfrac{5x^2 \cancel{(x - 2y)}}{\underset{1}{\cancel{(x - 2y)}}}$

$\qquad\qquad = \dfrac{5x^2}{1}$

$\qquad\qquad = 5x^2$

7. $\dfrac{x^2 + x - 6}{2x^2 + 2x} = \dfrac{(x + 3)(x - 2)}{2x(x + 1)}$

After factoring both the numerator and denominator, we found no common factors. The rational expression

$$\dfrac{x^2 + x - 6}{2x^2 + 2x}$$

cannot be reduced.

In Chapter 1 two numbers were called *additive inverses* if their sum was zero. For example, the additive inverse of 5 is -5 since $5 + (-5) = 0$. In algebra we generalize the concept of additive inverses as follows:

Additive Inverses

Two polynomials are called **additive inverses** if their sum is zero.

REMARK   The polynomials $x - 2$ and $2 - x$ are additive inverses since

$$(x - 2) + (2 - x) = x - 2 + 2 - x$$
$$= x - x - 2 + 2$$
$$= 0$$

NOTE ▶ We can find the additive inverse of a polynomial by multiplying the polynomial by -1. Thus, the additive inverse of $x - 2$ is

$$-1 \cdot (x - 2) = -x + 2$$
$$= 2 - x$$

The following example shows how additive inverses are used to reduce some rational expressions.

EXAMPLE 5.1.2   Reduce the following rational expressions to lowest terms.

1. $\dfrac{x - 2}{2 - x}$

There appears to be no common factor. However, $x - 2$ and $2 - x$ are additive inverses. This tells us that if we factor -1 out of the numerator or denominator, but not both, we will generate a common factor:

$$\frac{x - 2}{2 - x} = \frac{-1(-x + 2)}{(2 - x)}$$

$$= \frac{-1(2 - x)}{(2 - x)}$$

$$= \frac{-1}{1}$$

$$= -1$$

2. $\dfrac{x - 5}{25 - x^2} = \dfrac{x - 5}{(5 - x)(5 + x)}$

After completely factoring the numerator and denominator, we obtain factors that are additive inverses of each other:

$$= \frac{-1(-x + 5)}{(5 - x)(5 + x)} \qquad \text{Factor } -1 \text{ out of } x - 5 \text{ in the numerator.}$$

$$= \frac{-1(5 - x)}{(5 - x)(5 + x)} \qquad \text{Divide out the common factor.}$$

$$= \frac{-1}{5 + x}$$

$$\text{or} \qquad = \frac{-1}{x + 5}$$

EXERCISES 5.1

Reduce the following rational expressions to lowest terms.

1. $\dfrac{16x^2y^4}{24x^2y^4}$
2. $\dfrac{20xy^3}{28x^4y}$
3. $\dfrac{27x^4y}{72xy^5}$
4. $\dfrac{27x^5y}{3x^3}$

5. $\dfrac{6x^3y^2}{7x^5y^4}$
6. $\dfrac{20xy^5}{32xy^5}$
7. $\dfrac{8xy^5}{2y^2}$
8. $\dfrac{4xy^4}{9x^4y^5}$

9. $\dfrac{6x^2y^2}{12x^3y^5}$
10. $\dfrac{2y^2}{5x^4}$
11. $\dfrac{18x^2y^4}{12xy^2}$
12. $\dfrac{15x^5}{16x^3y^3}$

13. $\dfrac{14x^3y^4}{18y^8}$
14. $\dfrac{4x^3y}{12x^5y^2}$
15. $\dfrac{3x^3}{4y^2}$
16. $\dfrac{20x^4y^2}{16x^2y}$

17. $\dfrac{6x^3y - 24x^2y}{9x^2y^2 + 27xy^2}$

18. $\dfrac{5x^4y^2 - 15x^3y^3}{10x^6y^3 + 40x^5y^4}$

19. $\dfrac{12x^4y^3 + 24x^3y^3}{32x^2y^5}$

20. $\dfrac{3xy - 6y^2}{10x^3 - 5x^2y}$

21. $\dfrac{90x^4y^2 - 45x^3y^3}{9x^2y^2}$

22. $\dfrac{12x^5y^2 + 48x^4y^2}{54x^3y^3}$

23. $\dfrac{4x^2y}{8x^4y^2 + 32x^3y^2}$

24. $\dfrac{10^2y^2 - 20xy^2}{8x^3y + 8x^2y}$

25. $\dfrac{12x^2 - 4xy}{3y^3 - 9xy^2}$

26. $\dfrac{6xy^2}{18x^3y^3 + 36x^2y^3}$

27. $\dfrac{6x^3y^3 - 24x^2y^4}{18x^4y^5 + 36x^3y^6}$

28. $\dfrac{96x^3y^4 - 32x^2y^5}{8x^2y^3}$

29. $\dfrac{x^2 + 5x + 6}{x^2 + 6x + 8}$

30. $\dfrac{x^2 + 8x + 15}{x^2 + 5x + 6}$

31. $\dfrac{4x^2 - 4x - 15}{6x^2 - 11x - 10}$

32. $\dfrac{5x - 3}{3 - 5x}$

33. $\dfrac{x^2 - 1}{2x^2 - x - 3}$

34. $\dfrac{6x^2 - 11x - 4}{8x^2 + 2x - 3}$

35. $\dfrac{x^2 - 9}{2x^2 - 11x + 15}$

36. $\dfrac{9x^2 + 12x + 4}{3x^2 - 13x - 10}$

37. $\dfrac{3x^2 - 11x + 8}{3x^2 - 14x + 16}$

38. $\dfrac{x^2 + 6x + 9}{x^2 + 9x + 18}$

39. $\dfrac{x^2 - x - 12}{x^2 - x - 2}$

40. $\dfrac{x^2 + 5x + 6}{2x^4 + 10x^3 + 12x^2}$

41. $\dfrac{16 - x^2}{2x^2 - 8x}$

42. $\dfrac{x^2 - x - 12}{4x^2 + 12x}$

43. $\dfrac{x^2 + 8x + 16}{x^2 + 6x + 8}$

44. $\dfrac{24x^2 - 18x}{6x}$

45. $\dfrac{6x^2}{6x^3 - 30x^2}$

46. $\dfrac{4 - x}{x^2 + 2x - 24}$

47. $\dfrac{x^2 + 3x - 10}{2x^2 + 10x}$

48. $\dfrac{x^2 - 1}{3x^2 - 7x + 4}$

49. $\dfrac{2x - 7}{7 - 2x}$

50. $\dfrac{x^2 - x - 20}{x^2 - x - 6}$

51. $\dfrac{4x^2 + 12x + 9}{2x^2 - 5x - 12}$

52. $\dfrac{7x^3y - 21x^2y + 14xy}{x^2 - 3x + 2}$

53. $\dfrac{2 - x}{x^2 + 6x - 16}$

54. $\dfrac{x^2 - 4}{2x^2 + 9x + 10}$

55. $\dfrac{x^2 - 11x + 30}{6 - x}$

56. $\dfrac{12x - 8y}{2y - 3x}$

57. $\dfrac{12x^2 - 8x}{4x}$

58. $\dfrac{12x^3 + 4x^2}{9x^2 + 9x + 2}$

59. $\dfrac{5x^3y - 25x^2y + 20xy}{x^2 - 5x + 4}$

60. $\dfrac{36 - x^2}{4x^2 - 24x}$

61. $\dfrac{12x - 3y}{y - 4x}$

62. $\dfrac{8x^2}{8x^3 - 32x^2}$

63. $\dfrac{x^2 + 3x + 2}{3x^4 + 9x^3 + 6x^2}$

64. $\dfrac{5x^2 - 22x + 21}{5x^2 - 27x + 28}$

65. $\dfrac{6x^3 + 9x^2}{4x^2 + 8x + 3}$

66. $\dfrac{x^2 - 7x + 12}{3 - x}$

67. $\dfrac{15x^4 + 105x^3 + 180x^2}{5x^5 + 20x^4 + 15x^3}$

68. $\dfrac{8x^4 + 56x^3 + 80x^2}{4x^6 + 32x^5 + 48x^4}$

69. $\dfrac{6x^5 - 54x^4 + 120x^3}{3x^5 - 6x^4 - 45x^3}$

70. $\dfrac{15x^5 - 120x^4 + 180x^3}{3x^5 - 9x^4 - 54x^3}$

71. $\dfrac{2x^3y - 4x^2y - 48xy}{8x^4y^3 + 48x^3y^3 + 64x^2y^3}$

72. $\dfrac{3x^3y - 6x^2y - 24xy}{9x^5y^2 + 45x^4y^2 + 54x^3y^2}$

73. $\dfrac{10x^3 - 40x}{6x^4 - 30x^3 + 36x^2}$

74. $\dfrac{6x^3 - 96x}{10x^4 - 50x^3 + 40x^2}$

75. $\dfrac{108x^2 - 12x^4}{9x^3 - 36x^2 + 27x}$

76. $\dfrac{96x^2 - 6x^4}{9x^3 - 54x^2 + 72x}$

77. $\dfrac{2x^4 + 6x^3 + 4x^2}{6x^3y + 3x^2y - 3xy}$

78. $\dfrac{6x^3y + 9x^2y - 6xy}{2x^2y^2 + 10xy^2 + 12y^2}$

79. $\dfrac{ax + 2a + bx + 2b}{ax + 6a + bx + 6b}$

80. $\dfrac{ax + 3a + bx + 3b}{ax + 9a + bx + 9b}$

5.2 Multiplying and Dividing Rational Expressions

In Section 5.1 we saw that the property used to reduce rational expressions was simply a generalization of the technique used to reduce rational numbers. We will now learn how to generalize the techniques of multiplying and dividing rational numbers.

In Chapter 1 we learned how to multiply two rational numbers, as follows:

$$\frac{10}{21} \cdot \frac{7}{8} = \frac{10 \cdot 7}{21 \cdot 8}$$ Write the product of the numerators over the product of the denominators.

$$= \frac{2 \cdot 5 \cdot 7}{3 \cdot 7 \cdot 2 \cdot 2 \cdot 2}$$ Factor the numerator and denominator.

$$= \frac{2 \cdot 5 \cdot \not{7}}{3 \cdot \not{7} \cdot \not{2} \cdot 2 \cdot 2}$$ Divide out common factors.

$$= \frac{5}{12}$$

The principle used in multiplying rational numbers is generalized to rational expressions in the following property.

Multiplication Property of Fractions

$$\frac{c}{d} \cdot \frac{n}{m} = \frac{cn}{dm} \qquad d \neq 0, \, m \neq 0$$

REMARK   To multiply two rational expressions, we simply multiply the numerators and multiply the denominators.

Let's use the multiplication property of fractions to perform the following multiplication.

$$\frac{8x^5}{3y^4} \cdot \frac{9y^2}{10x} = \frac{8x^5 \cdot 9y^2}{3y^4 \cdot 10x}$$ Write the product of the numerators over the product of the denominators.

$$= \frac{8 \cdot 9x^5 y^2}{3 \cdot 10xy^4}$$

$$= \frac{2 \cdot 2 \cdot \not{2} \cdot \not{3} \cdot 3x^{5-1}}{\not{3} \cdot \not{2} \cdot 5y^{4-2}}$$ Divide out common factors.

$$= \frac{12x^4}{5y^2}$$

The following example illustrates how the multiplication property of fractions applies to more complicated rational expressions.

EXAMPLE 5.2.1

Perform the indicated multiplications, and reduce the answers to lowest terms.

1. $\dfrac{6x^5}{10y} \cdot \dfrac{15y^7}{18x^8} = \dfrac{6x^5 \cdot 15y^7}{10y \cdot 18x^8}$

$= \dfrac{6 \cdot 15x^5y^7}{10 \cdot 18x^8y}$

$= \dfrac{2 \cdot 3 \cdot 3 \cdot 5y^{7-1}}{2 \cdot 5 \cdot 2 \cdot 3 \cdot 3x^{8-5}}$

$= \dfrac{y^6}{2x^3}$

2. $\dfrac{5}{x-4} \cdot \dfrac{3x-12}{10x+10} = \dfrac{5}{x-4} \cdot \dfrac{3(x-4)}{10(x+1)}$

Always begin by factoring the numerator and denominator of both fractions, if possible.

$= \dfrac{5 \cdot 3(x-4)}{(x-4) \cdot 10(x+1)}$ Divide out common factors.
$\phantom{= \dfrac{5 \cdot 3(x-4)}{(x-4) \cdot 10}}{}_{2}$

$= \dfrac{3}{2(x+1)}$ The answer is in factored form.
We may stop here or

$= \dfrac{3}{2x+2}$ perform the multiplication.

Later in this chapter we will see the usefulness of leaving an answer in factored form.

3. $\dfrac{x^2+8x+15}{x^2+5x+6} \cdot \dfrac{x^2+3x+2}{x^2+3x-10} = \dfrac{(x+5)(x+3)}{(x+2)(x+3)} \cdot \dfrac{(x+2)(x+1)}{(x+5)(x-2)}$

$= \dfrac{(x+5)(x+3)}{(x+2)(x+3)} \cdot \dfrac{(x+2)(x+1)}{(x+5)(x-2)}$

Observe that we can divide out a common factor from the first fraction's numerator with one in the second fraction's denominator:

$= \dfrac{x+1}{x-2}$

NOTE ▶ When multiplying rational expressions, factoring the numerator and denominator is usually the key step.

4. $\dfrac{4x^2+4x+1}{9x} \cdot \dfrac{6x^2}{2x^2-7x-4} = \dfrac{(2x+1)(2x+1)}{9x} \cdot \dfrac{6x^2}{(2x+1)(x-4)}$

$= \dfrac{(2x+1)(2x+1)}{9x} \cdot \dfrac{6x^2}{(2x+1)(x-4)}$
$\phantom{= \dfrac{(2x+1)(2x+1)}{9}}{}_{3}^{\,2x}$

$= \dfrac{2x(2x+1)}{3(x-4)}$

5. $(x - 2) \cdot \dfrac{x^2 - 3x - 18}{x^2 - 4} = \dfrac{x - 2}{1} \cdot \dfrac{x^2 - 3x - 18}{x^2 - 4}$ Rewrite $x - 2$ as a fraction.

$$= \dfrac{x - 2}{1} \cdot \dfrac{(x - 6)(x + 3)}{(x - 2)(x + 2)}$$

$$= \dfrac{(x - 6)(x + 3)}{x + 2}$$

6. $\dfrac{x - 4}{x^2 + 4x + 3} \cdot \dfrac{x + 1}{2x^2 - 8x} = \dfrac{x - 4}{(x + 3)(x + 1)} \cdot \dfrac{x + 1}{2x(x - 4)}$

$$= \dfrac{1}{2x(x + 3)}$$

Try to avoid this mistake:

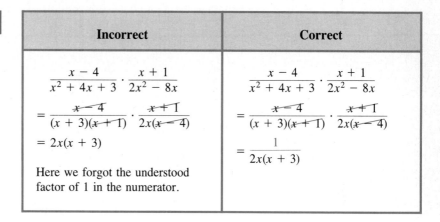

Incorrect	Correct
$\dfrac{x - 4}{x^2 + 4x + 3} \cdot \dfrac{x + 1}{2x^2 - 8x}$	$\dfrac{x - 4}{x^2 + 4x + 3} \cdot \dfrac{x + 1}{2x^2 - 8x}$
$= \dfrac{x - 4}{(x + 3)(x + 1)} \cdot \dfrac{x + 1}{2x(x - 4)}$	$= \dfrac{x - 4}{(x + 3)(x + 1)} \cdot \dfrac{x + 1}{2x(x - 4)}$
$= 2x(x + 3)$	$= \dfrac{1}{2x(x + 3)}$
Here we forgot the understood factor of 1 in the numerator.	

7. $\dfrac{x^2 - x - 6}{4x^2 + 8x} \cdot \dfrac{8x + 24}{9 - x^2} = \dfrac{(x - 3)(x + 2)}{4x(x + 2)} \cdot \dfrac{8(x + 3)}{(3 + x)(3 - x)}$

Factor -1 out of $3 - x$ in the denominator, since we have a factor of $x - 3$ in the numerator.

$$= \dfrac{(x - 3)(x + 2)}{4x(x + 2)} \cdot \dfrac{\overset{2}{8}(x + 3)}{(3 + x)(-1)(x - 3)}$$

$$= \dfrac{2(x + 3)}{(-1)x(3 + x)}$$ Now $x + 3$ and $3 + x$ divide out. Why?

$$= \dfrac{2}{-x}$$

or $\quad = -\dfrac{2}{x}$

The remainder of this section will illustrate how to divide rational expressions. However, we must first reintroduce some terminology.

In Chapter 1 two numbers were called *multiplicative inverses*, or *reciprocals*, if their product was 1. For example, the multiplicative inverse of $\frac{4}{7}$ is $\frac{7}{4}$ since $\frac{4}{7} \cdot \frac{7}{4} = \frac{28}{28} = 1$. In algebra we generalize the concept of multiplicative inverses as follows:

Multiplicative Inverses

Two expressions are called **multiplicative inverses** if their product is 1.

For example, the multiplicative inverse of x/y is y/x since

$$\frac{x}{y} \cdot \frac{y}{x} = \frac{xy}{yx} \qquad \text{provided } x \neq 0, y \neq 0$$

$$= \frac{xy}{yx}$$

$$= 1$$

Suppose that you were asked to find $\frac{2}{3} \div \frac{5}{7}$. From Chapter 1 we know that

$$\frac{2}{3} \div \frac{5}{7} = \frac{2}{3} \cdot \frac{7}{5}$$

$$= \frac{2 \cdot 7}{3 \cdot 5}$$

$$= \frac{14}{15}$$

The principle used in dividing rational numbers is generalized to rational expressions in the following property.

Division Property of Fractions

$$\frac{a}{b} \div \frac{x}{y} = \frac{a}{b} \cdot \frac{y}{x} = \frac{ay}{bx} \qquad b \neq 0, x \neq 0, y \neq 0$$

REMARK   To divide by a rational expression, we invert the divisor (find the multiplicative inverse) and change the operation from division to multiplication.

EXAMPLE 5.2.2

Perform the indicated divisions, and reduce the answers to lowest terms.

1. $\dfrac{8x^2}{5y} \div \dfrac{14x^4}{10y^6} = \dfrac{8x^2}{5y} \cdot \dfrac{10y^6}{14x^4}$ Invert and then multiply.

$\qquad = \dfrac{80x^2y^6}{70x^4y}$

$\qquad = \dfrac{\not{10} \cdot 8y^{6-1}}{\not{10} \cdot 7x^{4-2}}$

$\qquad = \dfrac{8y^5}{7x^2}$

2. $\dfrac{x-3}{2x+2} \div \dfrac{4x-12}{x^2-1} = \dfrac{x-3}{2x+2} \cdot \dfrac{x^2-1}{4x-12}$ Invert and multiply.

$\qquad = \dfrac{x-3}{2(x+1)} \cdot \dfrac{(x+1)(x-1)}{4(x-3)}$ Factor the numerator and denominator.

$\qquad = \dfrac{\cancel{x-3}}{2\cancel{(x+1)}} \cdot \dfrac{\cancel{(x+1)}(x-1)}{4\cancel{(x-3)}}$ Divide out common factors.

$\qquad = \dfrac{x-1}{8}$

3. $\dfrac{x^2-x-12}{x^2+2x-24} \div \dfrac{x^2+x-6}{x^2+4x-12}$

$\qquad = \dfrac{x^2-x-12}{x^2+2x-24} \cdot \dfrac{x^2+4x-12}{x^2+x-6}$ Invert and multiply.

$\qquad = \dfrac{(x-4)(x+3)}{(x+6)(x-4)} \cdot \dfrac{(x+6)(x-2)}{(x+3)(x-2)}$ Factor the numerator and denominator.

$\qquad = \dfrac{\cancel{(x-4)}\cancel{(x+3)}}{\cancel{(x+6)}\cancel{(x-4)}} \cdot \dfrac{\cancel{(x+6)}\cancel{(x-2)}}{\cancel{(x+3)}\cancel{(x-2)}}$ Divide out common factors.

$\qquad = \dfrac{1}{1}$

$\qquad = 1$

4. $\dfrac{10x^2+15x}{x^2+x-20} \div (2x+3)$

$\qquad = \dfrac{10x^2+15x}{x^2+x-20} \div \dfrac{2x+3}{1}$ Rewrite $2x+3$ as a fraction.

$\qquad = \dfrac{10x^2+15x}{x^2+x-20} \cdot \dfrac{1}{2x+3}$ Invert and multiply.

$\qquad = \dfrac{5x(2x+3)}{(x+5)(x-4)} \cdot \dfrac{1}{2x+3}$ Factor the numerator and denominator.

$\qquad = \dfrac{5x\cancel{(2x+3)}}{(x+5)(x-4)} \cdot \dfrac{1}{\cancel{2x+3}}$ Divide out common factors.

$$= \frac{5x}{(x + 5)(x - 4)}$$

EXERCISES 5.2

Perform the indicated operations, and reduce the answers to lowest terms.

1. $\dfrac{4x^3}{5x^2} \cdot \dfrac{3x^2}{2x^2}$

2. $\dfrac{4x}{5x^3} \cdot \dfrac{5x^2}{3x}$

3. $\dfrac{14x}{6x^3} \div \dfrac{2x}{3x^2}$

4. $\dfrac{8x}{2x^4} \div \dfrac{16x}{5x^2}$

5. $\dfrac{5x^5}{2y^4} \cdot \dfrac{2y^2}{3x^4}$

6. $\dfrac{3x^4}{6y^5} \cdot \dfrac{3y^2}{2x^3}$

7. $\dfrac{8x^2y^3}{27xy^4} \div \dfrac{2y^2}{3xy}$

8. $\dfrac{4xy^3}{8x^2} \cdot \dfrac{3x^2y}{2y}$

9. $\dfrac{3x^2y^2}{6xy} \div \dfrac{6x^3y^2}{9x^2y}$

10. $\dfrac{4x^2}{2x^3y} \cdot \dfrac{2xy}{16x^2}$

11. $\dfrac{6xy^2}{5x} \cdot \dfrac{2x^2y^2}{6y^3}$

12. $\dfrac{9xy}{2xy^3} \div \dfrac{10x^2y}{4x^3}$

13. $\dfrac{2xy^2}{5x^2y^3} \div \dfrac{6xy}{18x}$

14. $\dfrac{4x^2y^2}{6x^3y} \div \dfrac{4xy^3}{8y}$

15. $\dfrac{3x^2}{27xy^3} \cdot \dfrac{3y^3}{2xy^2}$

16. $\dfrac{8xy^3}{6x^2y^2} \div \dfrac{4xy^2}{2x^2y}$

17. $\dfrac{7}{2x + 6} \div \dfrac{14x - 28}{8x + 24}$

18. $\dfrac{12x^3y^2}{10x^2 + 5x} \div \dfrac{3xy}{20x^3 + 10x^2}$

19. $\dfrac{x - 3}{3x^2 + 9x} \cdot \dfrac{6x^3 + 6x^2}{5x - 15}$

20. $\dfrac{6}{5x + 10} \div \dfrac{12x - 36}{20x + 40}$

21. $\dfrac{6x^3y^2}{10x - 20} \cdot \dfrac{2x^2 - 4x}{9xy^5}$

22. $\dfrac{x - 2}{4x^2 + 12x} \cdot \dfrac{8x^3 + 48x^2}{7x - 14}$

23. $\dfrac{16x^2y^4}{6x^2 + 9x} \div \dfrac{4xy^2}{18x^3 + 27x^2}$

24. $\dfrac{21x^6y}{8x - 24} \cdot \dfrac{4x^3 - 12x^2}{14x^3y^6}$

25. $\dfrac{x + 3}{4x^2} \cdot \dfrac{2x^5}{2x^2 + x - 15}$

26. $\dfrac{x^2 + 3x - 4}{8x^3} \div \dfrac{5x^3 - 5x^2}{4x^4}$

27. $\dfrac{x^2 - 1}{2x^4 - 8x^3} \div \dfrac{2x^2 - x - 3}{3x^2 - 12x}$

28. $\dfrac{x + 4}{12x^3} \cdot \dfrac{3x^9}{2x^2 + 5x - 12}$

29. $\dfrac{x^2 + 4x - 8}{18x^4} \div \dfrac{3x^4 - 6x^3}{6x^5}$

30. $\dfrac{x^2 - 2x - 15}{4x} \cdot \dfrac{8x^4}{x^2 - 9x + 20}$

31. $\dfrac{x^2 - 4x - 12}{3x^2} \cdot \dfrac{12x^6}{x^2 - 7x + 6}$

32. $\dfrac{x^2 - 9}{5x^5 - 10x^4} \div \dfrac{2x^2 + 5x - 3}{4x^2 - 8x}$

33. $\dfrac{x^2 - x - 12}{x^2 - 3x + 2} \cdot \dfrac{x^2 - 4x + 4}{x^2 + x - 6}$

34. $\dfrac{x^2 + 4x - 5}{x^2 - 9x + 18} \cdot \dfrac{x^2 - 6x + 9}{x^2 + 2x - 15}$

35. $\dfrac{x^2 + 9x + 20}{x^2 + 7x + 12} \div \dfrac{x^2 - x - 30}{x^2 + x - 6}$

36. $\dfrac{x - 4}{2x^2 + 5x + 3} \div \dfrac{x^2 + x - 20}{2x + 3}$

37. $\dfrac{4x^2 + 4x - 3}{6x^2 - 13x + 6} \div \dfrac{4x^2 + 16x + 15}{4x^2 + 4x - 15}$

38. $\dfrac{3x^2 + 11x - 20}{x - 8} \cdot \dfrac{x^2 - 9x + 8}{3x - 4}$

39. $\dfrac{2x^2 - x - 15}{x^2 - 3x + 2} \div \dfrac{4x^2 + 8x - 5}{3x^2 - 7x + 2}$

40. $\dfrac{3x^2 - 10x - 8}{2x^2 - 7x - 4} \cdot \dfrac{2x^2 + 11x + 5}{9x^2 - 4}$

41. $\dfrac{16 - x^2}{x^2 + x - 2} \div \dfrac{x^2 - 2x - 8}{x^2 + 4x + 4}$

42. $\dfrac{2x^2 - 3x - 9}{3x^2 - 7x - 6} \cdot \dfrac{3x^2 - 13x + 4}{2x^2 - 5x - 12}$

43. $\dfrac{x - 1}{2x^2 + 13x + 6} \div \dfrac{x^2 + x - 2}{2x + 1}$

44. $\dfrac{25 - x^2}{x^2 - 8x - 9} \div \dfrac{x^2 - 4x - 5}{x^2 + 2x + 1}$

45. $\dfrac{2x^2 + x - 15}{x - 6} \cdot \dfrac{x^2 - 8x + 12}{2x - 5}$

46. $\dfrac{x^2 - 3x - 18}{x^2 - 2x - 24} \div \dfrac{x^2 - 4x - 21}{x^2 + 3x - 4}$

47. $\dfrac{x^2 - 2x - 15}{x^2 + x - 2} \cdot \dfrac{x^2 - 3x + 2}{x^2 + 2x - 8}$

48. $\dfrac{4x^2 + 8x + 3}{6x^2 + 11x + 3} \div \dfrac{4x^2 - 1}{4x^2 - 8x + 3}$

49. $\dfrac{2x^2 + 3x - 5}{3x^2 - 2x - 1} \cdot \dfrac{3x^2 - 17x + 10}{2x^2 - 5x - 25}$

50. $\dfrac{x^2 - 2x - 8}{x^2 - x - 20} \cdot \dfrac{x^2 - 4x - 5}{x^2 - x - 6}$

51. $\dfrac{x^2 + 3x - 10}{x + 4} \div \dfrac{x + 5}{x^2 + 10x + 24}$

52. $\dfrac{4 - x^2}{2x^2 + 9x + 9} \cdot \dfrac{4x^2 + 20x + 21}{2x^2 + 3x - 14}$

53. $\dfrac{2x^2 + 9x - 5}{3x^2 + 16x + 5} \div \dfrac{2x^2 - 5x + 2}{3x^2 - 5x - 2}$

54. $\dfrac{3x^2 + 11x - 4}{2x^2 + 11x + 12} \div \dfrac{3x^2 - 4x + 1}{2x^2 + x - 3}$

55. $\dfrac{2x^2 - 5x - 3}{2x^2 - x - 15} \cdot \dfrac{2x^2 + 17x + 30}{4x^2 - 1}$

56. $\dfrac{2x^2 - 7x - 15}{2x^2 + 9x + 4} \div \dfrac{2x^2 - 13x + 15}{2x^2 - 5x + 3}$

57. $\dfrac{1 - x^2}{2x^2 + 13x + 20} \cdot \dfrac{6x^2 + 17x + 5}{3x^2 - 2x - 1}$

58. $\dfrac{x^2 + 5x - 6}{x + 3} \div \dfrac{x - 1}{x^2 + 11x + 24}$

59. $\dfrac{x^2 - x - 6}{5x^2 + 30x + 40} \cdot \dfrac{3x^2 + 6x - 24}{x^2 - 5x + 6}$

60. $\dfrac{4x^2 - 8x - 60}{x^2 + 2x - 3} \cdot \dfrac{x^2 + 5x - 6}{5x^2 + 5x - 150}$

61. $(x + 3) \cdot \dfrac{x^2 + 3x - 10}{x^2 + 2x - 3}$

62. $\dfrac{x^2 + 7x + 12}{2x^2 - 9x + 4} \div (x + 4)$

63. $(2x - 3) \cdot \dfrac{x^2 - 11x + 28}{2x^2 - 11x + 12}$

64. $(x - 3) \div \dfrac{9 - x^2}{2x^2 + 11x + 5}$

65. $\dfrac{x^2 + 4x + 4}{3x^2 + 14x - 5} \div (x + 2)$

66. $(x + 2) \cdot \dfrac{x^2 - 5x - 36}{x^2 - x - 6}$

67. $(x - 6) \div \dfrac{36 - x^2}{2x^2 + 5x + 3}$

68. $(3x - 1) \cdot \dfrac{x^2 - 7x + 10}{3x^2 - 7x + 2}$

69. $\dfrac{12x^2 - 30x}{3x^2 - 3x - 6} \div (2x - 5)$

70. $(5x + 3) \cdot \dfrac{4x^2 + 12x}{10x^3 + 6x^2}$

71. $(4x + 3) \cdot \dfrac{9x^2 - 36x}{12x^4 + 9x^3}$

72. $\dfrac{24x^2 - 32x}{2x^2 - 2x - 12} \div (3x - 4)$

73. $\dfrac{4x^3 + 12x^2}{x^2 - 3x - 10} \div \dfrac{8x^3 + 22x^2 - 6x}{4x^2 + 7x - 2}$

74. $\dfrac{4x^3 - 12x^2 - 40x}{2x^2 - 13x + 15} \cdot \dfrac{2x^2 - x - 3}{2x^5 - 2x^3}$

75. $\dfrac{6x^4 + 12x^3 + 6x^2}{5x^2 - 5} \div \dfrac{3x^4 + 9x^3 + 6x^2}{x^2 + x - 2}$

76. $\dfrac{8x^4 - 16x^3 + 8x^2}{5x^2 + 10x - 15} \div \dfrac{2x^3 - 2x}{x^2 + 4x + 3}$

77. $\dfrac{6x^4 + 12x^3 - 18x^2}{2x^3 - 2x^2 - 24x} \cdot \dfrac{2x^2 - 7x - 4}{8x^3 - 4x^2 - 4x}$

78. $\dfrac{6x^3 + 12x^2}{x^2 - x - 12} \div \dfrac{6x^3 + 8x^2 - 8x}{3x^2 + 7x - 6}$

79. $\dfrac{2x^5 + 10x^4 + 8x^3}{3x^2 + 11x - 4} \cdot \dfrac{3x^2 + 5x - 2}{6x^3 - 24x}$

80. $\dfrac{8x^4 + 24x^3 + 16x^2}{3x^3 - 3x^2 - 6x} \cdot \dfrac{2x^2 - x - 6}{8x^3 + 28x^2 + 24x}$

81. $\dfrac{4x^2 + 8x}{x^2 - 8x - 9} \cdot \dfrac{x + 7}{2x^2 - 6x} \cdot \dfrac{x^2 - 2x - 3}{3x^2 + 6x}$

82. $\dfrac{6x^2 + 42x}{x^2 + x - 30} \cdot \dfrac{x + 8}{2x^2 - 8x} \cdot \dfrac{x^2 + 2x - 24}{5x^2 + 35x}$

83. $\dfrac{x^2 + 5x + 6}{3x^2 - 11x - 4} \cdot \dfrac{x^2 - 2x + 1}{2x^2 - 9x + 10} \cdot \dfrac{6x^2 - 13x - 5}{x^2 + 2x - 3}$

84. $\dfrac{x^2 + 8x + 15}{3x^2 + 10x - 8} \cdot \dfrac{x^2 - 4x + 4}{2x^2 + 3x - 2} \cdot \dfrac{6x^2 - 7x + 2}{x^2 + x - 6}$

85. $\dfrac{x + 3}{x^2 + x - 20} \div \dfrac{2x^2 - 5x + 2}{x - 4} \cdot \dfrac{2x^2 + 9x - 5}{2x^2 + 7x + 3}$

86. $\dfrac{x + 5}{x^2 + x - 2} \div \dfrac{3x^2 + 2x - 1}{x - 1} \cdot \dfrac{3x^2 + 5x - 2}{2x^2 + 9x - 5}$

87. $\dfrac{6x^2 + 24x}{x^2 - 2x - 15} \div \dfrac{2x^2 + 3x - 2}{2x^2 + 4x} \div \dfrac{8x^2 + 32x}{2x^2 + 5x - 3}$

88. $\dfrac{3x^2 + 6x}{x^2 - 10x + 24} \div \dfrac{2x^2 + 5x + 3}{3x^2 - 18x} \div \dfrac{12x^2 + 24x}{2x^2 - 5x - 12}$

89. $\dfrac{2ax + 2ay + bx + by}{8x^2 - 24x} \div \dfrac{3ax + 3ay + 2bx + 2by}{4x^2 - 12x}$

90. $\dfrac{ax + ay + 2bx + 2by}{6x^2 - 12x} \div \dfrac{2ax + 2ay + 3bx + 3by}{3x^2 - 6x}$

5.3 Adding and Subtracting Rational Expressions with a Common Denominator

In Chapter 1 we learned how to add two rational numbers with a common denominator, as follows:

$$\frac{2}{9} + \frac{5}{9} = \frac{2 + 5}{9}$$ Add the numerators, and write the sum over the common denominator.

$$= \frac{7}{9}$$

The principle used in adding rational numbers with a common denominator is generalized to rational expressions in the following property.

Addition Property of Fractions
$$\frac{a}{d} + \frac{b}{d} = \frac{a + b}{d} \qquad d \neq 0$$

REMARK   To add two rational expressions with a common denominator, we simply divide the sum of the numerators by the common denominator.

Let's use the addition property of fractions to perform the following additions.

1. $\dfrac{2}{x} + \dfrac{5}{x} = \dfrac{2 + 5}{x}$ Add the numerators. Write the sum over the common denominator.

$\qquad\qquad = \dfrac{7}{x}$

2. $\dfrac{1}{3x} + \dfrac{11}{3x} = \dfrac{1 + 11}{3x}$

$\qquad\qquad = \dfrac{12}{3x}$ Don't stop here.

$\qquad\qquad = \dfrac{\cancel{3} \cdot 4}{\cancel{3} \cdot x}$

$\qquad\qquad = \dfrac{4}{x}$

NOTE ▶ After the addition, there is a common factor of 3 in the numerator and denominator that needs to be divided out. *Always be sure to reduce your answer, if possible.*

We have a similar property for finding the difference of two rational expressions with a common denominator:

Subtraction Property of Fractions

$$\frac{a}{d} - \frac{b}{d} = \frac{a - b}{d} \qquad d \neq 0$$

The following example illustrates how to use both the addition and subtraction properties of fractions.

EXAMPLE 5.3.1

Perform the indicated additions and subtractions.

1. $\dfrac{3x}{5} + \dfrac{4x}{5} = \dfrac{3x + 4x}{5}$ Add the numerators.

$\qquad\qquad = \dfrac{7x}{5}$ Combine like terms.

2. $\dfrac{5}{12y} + \dfrac{1}{12y} = \dfrac{5+1}{12y}$ Add the numerators.

$\qquad\qquad = \dfrac{6}{12y}$ Combine like terms.

$\qquad\qquad = \dfrac{\cancel{6}}{\cancel{6} \cdot 2y}$ Divide out the common factor of 6.

$\qquad\qquad = \dfrac{1}{2y}$

3. $\dfrac{2x+1}{4x} + \dfrac{3x-8}{4x} = \dfrac{2x+1+3x-8}{4x}$ Add the numerators.

$\qquad\qquad = \dfrac{5x-7}{4x}$ Combine like terms.

4. $\dfrac{x^2+3x}{x-2} + \dfrac{2x-14}{x-2} = \dfrac{x^2+3x+2x-14}{x-2}$

$\qquad\qquad = \dfrac{x^2+5x-14}{x-2}$

$\qquad\qquad = \dfrac{(x+7)(\cancel{x-2})}{\cancel{x-2}}$ Factor the numerator, and divide out the common factor of $x-2$.

$\qquad\qquad = x+7$

5. $\dfrac{8x}{x+1} - \dfrac{3x}{x+1} = \dfrac{8x-3x}{x+1}$ Subtract the numerators.

$\qquad\qquad = \dfrac{5x}{x+1}$ Combine like terms.

6. $\dfrac{5x}{8y} - \dfrac{11x}{8y} = \dfrac{5x-11x}{8y}$ Subtract the numerators.

$\qquad\qquad = \dfrac{-6x}{8y}$ Combine like terms.

$\qquad\qquad = \dfrac{\cancel{2} \cdot (-3x)}{\cancel{2} \cdot (4y)}$ Divide out the common factor of 2.

$\qquad\qquad = \dfrac{-3x}{4y}$

7. $\dfrac{4x+5}{x+3} - \dfrac{x+1}{x+3} = \dfrac{4x+5-(x+1)}{x+3}$

$\qquad\qquad = \dfrac{4x+5-x-1}{x+3}$ Eliminate the parentheses by changing the signs.

$\qquad\qquad = \dfrac{3x+4}{x+3}$ Combine like terms.

Try to avoid this mistake:

Incorrect	Correct
$\dfrac{4x + 5}{x + 3} - \dfrac{x + 1}{x + 3}$	$\dfrac{4x + 5}{x + 3} - \dfrac{x + 1}{x + 3}$
$= \dfrac{4x + 5 - x + 1}{x + 3}$	$= \dfrac{4x + 5 - (x + 1)}{x + 3}$
$= \dfrac{3x + 6}{x + 3}$	$= \dfrac{4x + 5 - x - 1}{x + 3}$
	$= \dfrac{3x + 4}{x + 3}$
Here we forgot to change the sign of the 1.	It is a good idea to insert parentheses around the numerator of the fraction we are subtracting. The parentheses help remind us to change the sign of every term in the numerator of the fraction we are subtracting.

8. $\dfrac{3x - 5}{x - 4} - \dfrac{x + 3}{x - 4} = \dfrac{3x - 5 - (x + 3)}{x - 4}$

$= \dfrac{3x - 5 - x - 3}{x - 4}$ Don't forget to change the signs.

$= \dfrac{2x - 8}{x - 4}$

$= \dfrac{2(x - 4)}{x - 4}$ Factor the numerator, and divide out the common factor of $x - 4$.

$= 2$

9. $\dfrac{4x^2 + x - 2}{x^2 + x - 12} - \dfrac{2x^2 + 8x - 5}{x^2 + x - 12} = \dfrac{4x^2 + x - 2 - (2x^2 + 8x - 5)}{x^2 + x - 12}$

$= \dfrac{4x^2 + x - 2 - 2x^2 - 8x + 5}{x^2 + x - 12}$

$= \dfrac{2x^2 - 7x + 3}{x^2 + x - 12}$

$= \dfrac{(2x - 1)(x - 3)}{(x + 4)(x - 3)}$ Factor the numerator and denominator, and divide out the common factor of $x - 3$.

$= \dfrac{2x - 1}{x + 4}$

The following problems illustrate how to extend the addition and subtraction properties of fractions to more than two fractions.

10. $\dfrac{12y}{20x} + \dfrac{7y}{20x} - \dfrac{3y}{20x} = \dfrac{12y + 7y - 3y}{20x}$

$$= \dfrac{16y}{20x}$$

$$= \dfrac{4 \cdot 4y}{4 \cdot 5x}$$

$$= \dfrac{4y}{5x}$$

11. $\dfrac{5x + 1}{x - 9} - \dfrac{x + 8}{x - 9} - \dfrac{2x - 2}{x - 9} = \dfrac{5x + 1 - (x + 8) - (2x - 2)}{x - 9}$

$$= \dfrac{5x + 1 - x - 8 - 2x + 2}{x - 9}$$

$$= \dfrac{2x - 5}{x - 9}$$

The last example of this section will show how to add or subtract two fractions when one denominator is the additive inverse of the other.

EXAMPLE 5.3.2

Perform this addition:

$$\dfrac{8}{x - 7} + \dfrac{3}{7 - x}$$

Since the denominators are not exactly the same, it seems as though we cannot add these fractions. However, the denominators are additive inverses. Multiplying the numerator *and* denominator of one of the fractions by -1 will generate a common denominator.

$$\dfrac{8}{x - 7} + \dfrac{3}{7 - x} = \dfrac{8}{x - 7} + \dfrac{3 \cdot (-1)}{7 - x \cdot (-1)}$$

$$= \dfrac{8}{x - 7} + \dfrac{-3}{-7 + x}$$

$$= \dfrac{8}{x - 7} + \dfrac{-3}{x - 7}$$

$$= \dfrac{8 + (-3)}{x - 7}$$

$$= \dfrac{5}{x - 7}$$

EXERCISES 5.3

Perform the indicated additions and subtractions.

1. $\dfrac{5x}{3} + \dfrac{2x}{3}$

2. $\dfrac{3x}{5} + \dfrac{6x}{5}$

3. $\dfrac{7x}{4} - \dfrac{4x}{4}$

4. $\dfrac{11x}{6} - \dfrac{10x}{6}$

5. $\dfrac{y}{8} + \dfrac{3y}{8}$

6. $\dfrac{3a}{10} - \dfrac{8a}{10}$

7. $\dfrac{2b}{6} - \dfrac{4b}{6}$

8. $\dfrac{5y}{9} + \dfrac{y}{9}$

9. $\dfrac{4}{7x} + \dfrac{2}{7x}$

10. $\dfrac{1}{5x} + \dfrac{3}{5x}$

11. $\dfrac{7y}{4x^2} - \dfrac{2y}{4x^2}$

12. $\dfrac{17x}{9y^2} - \dfrac{4x}{9y^2}$

13. $\dfrac{8}{3xy} - \dfrac{2}{3xy}$

14. $\dfrac{9}{2xy} - \dfrac{3}{2xy}$

15. $\dfrac{4x}{6xy^2} + \dfrac{2x}{6xy^2}$

16. $\dfrac{5y}{8x^2y} + \dfrac{3y}{8x^2y}$

17. $\dfrac{3x+5}{8} + \dfrac{2x+1}{8}$

18. $\dfrac{3x+2}{6} + \dfrac{x+7}{6}$

19. $\dfrac{7x+5}{2y} + \dfrac{3x+1}{2y}$

20. $\dfrac{2x+5}{3y} + \dfrac{4x+7}{3y}$

21. $\dfrac{2x+3}{6} - \dfrac{x}{6}$

22. $\dfrac{3x+4}{5} - \dfrac{x}{5}$

23. $\dfrac{5x-2}{3y} - \dfrac{2x+3}{3y}$

24. $\dfrac{4x-1}{5y} - \dfrac{2x+4}{5y}$

25. $\dfrac{2x+3}{2y} - \dfrac{2x-3}{2y}$

26. $\dfrac{3x+4}{6y} - \dfrac{3x-4}{6y}$

27. $\dfrac{4x}{3x-2y} + \dfrac{8x}{3x-2y}$

28. $\dfrac{5y}{4x-5y} + \dfrac{9y}{4x-5y}$

29. $\dfrac{2x}{x+3} + \dfrac{6}{x+3}$

30. $\dfrac{3x}{x+1} + \dfrac{3}{x+1}$

31. $\dfrac{4x}{2x-5} - \dfrac{2x}{2x-5}$

32. $\dfrac{5x}{3x-4} - \dfrac{3x}{3x-4}$

33. $\dfrac{2x}{3x+2} - \dfrac{x+5}{3x+2}$

34. $\dfrac{5x}{2x+1} - \dfrac{x+7}{2x+1}$

35. $\dfrac{2x^2+17x}{x+6} + \dfrac{x^2-6}{x+6}$

36. $\dfrac{x^2+x}{x+5} + \dfrac{x^2-45}{x+5}$

37. $\dfrac{4x-2}{3x-4} - \dfrac{2x-7}{3x-4}$

38. $\dfrac{7x-1}{5x-3} - \dfrac{4x-8}{5x-3}$

39. $\dfrac{3x+13}{x^2-16} + \dfrac{2x+7}{x^2-16}$

40. $\dfrac{2x+1}{x^2-1} + \dfrac{x+2}{x^2-1}$

41. $\dfrac{2x+5}{x^2-4} - \dfrac{x+3}{x^2-4}$

42. $\dfrac{2x+7}{x^2-9} - \dfrac{x+4}{x^2-9}$

43. $\dfrac{7x-2}{6x+3} + \dfrac{3x+7}{6x+3}$

44. $\dfrac{13x-15}{12x-8} + \dfrac{8x+1}{12x-8}$

45. $\dfrac{7x-3}{21x-28} - \dfrac{x+5}{21x-28}$

46. $\dfrac{18x-11}{20x-15} - \dfrac{6x-2}{20x-15}$

47. $\dfrac{2x+5}{x^2-x-7} + \dfrac{x-12}{x^2-x-7}$

48. $\dfrac{3x-11}{x^2-3x-1} + \dfrac{2x+2}{x^2-3x-1}$

49. $\dfrac{2x^2+1}{x^2-x-12} - \dfrac{x^2-15}{x^2-x-12}$

50. $\dfrac{2x^2+19}{x^2+x-30} - \dfrac{x^2-6}{x^2+x-30}$

51. $\dfrac{5x^2-7}{2x^2+13x+15} - \dfrac{x^2+2}{2x^2+13x+15}$

52. $\dfrac{7x^2+6}{2x^2+13x+6} - \dfrac{3x^2+7}{2x^2+13x+6}$

53. $\dfrac{x-5}{2x^2-5x+3} + \dfrac{x+2}{2x^2-5x+3}$

54. $\dfrac{2x-1}{3x^2-11x-20} + \dfrac{x+5}{3x^2-11x-20}$

55. $\dfrac{2x^2+3x-31}{x^2-2x-15} + \dfrac{x^2-9x-14}{x^2-2x-15}$

56. $\dfrac{2x^2+15x-20}{x^2+3x-4} + \dfrac{2x^2-3x+4}{x^2+3x-4}$

57. $\dfrac{5x^2 - 2x + 8}{x^2 - 4x - 32} - \dfrac{3x^2 - 2x + 14}{x^2 - 4x - 32}$

58. $\dfrac{6x^2 - 7x + 1}{x^2 + 4x - 12} - \dfrac{3x^2 - 7x + 16}{x^2 + 4x - 12}$

59. $\dfrac{x^2 + 7x - 10}{x^2 - 5x - 6} + \dfrac{x^2 - 2x - 2}{x^2 - 5x - 6}$

60. $\dfrac{2x^2 + 19x - 5}{x^2 - 3x - 10} + \dfrac{x^2 - 2x - 1}{x^2 - 3x - 10}$

61. $\dfrac{3x^2 - 4x - 8}{x^2 + 3x - 18} - \dfrac{x^2 - 3x + 7}{x^2 + 3x - 18}$

62. $\dfrac{4x^2 - 7x - 10}{x^2 + 3x - 28} - \dfrac{x^2 + 3x - 2}{x^2 + 3x - 28}$

63. $\dfrac{8}{x - 1} + \dfrac{5}{1 - x}$

64. $\dfrac{9}{x - 3} + \dfrac{2}{3 - x}$

65. $\dfrac{4}{2x - 3} - \dfrac{5}{3 - 2x}$

66. $\dfrac{8}{3x - 2} - \dfrac{7}{2 - 3x}$

67. $\dfrac{4}{5x - 4} + \dfrac{6}{4 - 5x}$

68. $\dfrac{2}{4x - 1} + \dfrac{7}{1 - 4x}$

69. $\dfrac{3x + 4}{5x - 2} + \dfrac{x - 9}{2 - 5x}$

70. $\dfrac{6x + 7}{7x - 4} + \dfrac{2x - 1}{4 - 7x}$

71. $\dfrac{x - 8}{2x - 3} - \dfrac{2x + 5}{3 - 2x}$

72. $\dfrac{x - 9}{3x - 8} - \dfrac{3x + 7}{8 - 3x}$

73. $\dfrac{6x + 9}{2x - 7} + \dfrac{8x + 2}{7 - 2x}$

74. $\dfrac{6x + 7}{5x - 3} + \dfrac{11x + 4}{3 - 5x}$

75. $\dfrac{5x^2 - 2}{9x^2 - 9} + \dfrac{x^2 + 2}{9 - 9x^2}$

76. $\dfrac{5x^2 - 2}{4x^2 - 4} + \dfrac{2x^2 + 1}{4 - 4x^2}$

77. $\dfrac{3x + 7}{x^2 - 9} - \dfrac{x + 5}{9 - x^2}$

78. $\dfrac{x - 7}{x^2 - 4} - \dfrac{2x + 1}{4 - x^2}$

79. $\dfrac{15y}{14x} + \dfrac{11y}{14x} - \dfrac{5y}{14x}$

80. $\dfrac{19y}{15x} + \dfrac{17y}{15x} - \dfrac{11y}{15x}$

81. $\dfrac{5x + 8}{4x + 4} + \dfrac{3x + 5}{4x + 4} + \dfrac{x - 4}{4x + 4}$

82. $\dfrac{4x + 11}{5x + 10} + \dfrac{3x + 13}{5x + 10} + \dfrac{x - 8}{5x + 10}$

83. $\dfrac{x + 3}{2x - 5} - \dfrac{6x + 5}{2x - 5} + \dfrac{x - 1}{2x - 5}$

84. $\dfrac{x + 2}{3x - 7} - \dfrac{8x + 9}{3x - 7} + \dfrac{x - 4}{3x - 7}$

85. $\dfrac{4x^2 - x}{x + 4} - \dfrac{x^2 - 10x}{x + 4} - \dfrac{2x + 20}{x + 4}$

86. $\dfrac{3x^2 - 2x}{x + 5} - \dfrac{x^2 - 6x}{x + 5} - \dfrac{x + 35}{x + 5}$

87. $\dfrac{7x + 3}{x - 2} + \dfrac{x - 5}{2 - x} - \dfrac{3x + 4}{x - 2}$

88. $\dfrac{5x + 2}{x - 1} + \dfrac{2x - 3}{1 - x} - \dfrac{x + 6}{x - 1}$

89. $\dfrac{4x + 1}{3x^2 - 16x - 12} - \dfrac{2x - 5}{3x^2 - 16x - 12} + \dfrac{x - 4}{3x^2 - 16x - 12}$

90. $\dfrac{5x + 2}{2x^2 - 11x - 21} - \dfrac{4x - 9}{2x^2 - 11x - 21} + \dfrac{x - 8}{2x^2 - 11x - 21}$

5.4 Adding and Subtracting Rational Expressions with Different Denominators

The previous section explained how to add and subtract rational expressions with a common denominator. In this section we will extend our work to include adding and subtracting rational expressions with different denominators.

In Chapter 1 we learned how to add and subtract rational numbers with different denominators. We saw that in an addition or a subtraction problem the denominators must be identical. Thus, when the denominators are different, we must first change the fractions to equivalent fractions with a common denominator. This type of problem is illustrated in the following example.

$$\frac{5}{12} + \frac{11}{30} = \frac{5}{2 \cdot 2 \cdot 3} + \frac{11}{2 \cdot 3 \cdot 5}$$

Factor each of the denominators into a product of prime numbers.

$$= \frac{5}{2^2 \cdot 3} + \frac{11}{2 \cdot 3 \cdot 5}$$

Rewrite the factorization using exponential notation.

Now the **least common denominator** (abbreviated **LCD**) is determined by forming a product of all of the distinct factors and raising each factor to its highest power found in any denominator. In this problem, since

$$12 = 2^2 \cdot 3 \qquad \text{and} \qquad 30 = 2 \cdot 3 \cdot 5$$

$$\text{LCD} = 2^2 \cdot 3 \cdot 5$$

$$= 60$$

We now convert the given fractions to equivalent fractions with the LCD, 60, as their denominator.

$$\frac{5}{12} + \frac{11}{30} = \frac{5 \cdot 5}{12 \cdot 5} + \frac{11 \cdot 2}{30 \cdot 2}$$

$$= \frac{25}{60} + \frac{22}{60}$$

$$= \frac{25 + 22}{60}$$

$$= \frac{47}{60}$$

REMARK   To convert $\frac{5}{12}$ to an equivalent fraction with denominator 60, we note that 12 must be multiplied by 5 to obtain 60. Hence, we multiply both the numerator and denominator of $\frac{5}{12}$ by 5. Similarly, to convert $\frac{11}{30}$ to an equivalent fraction with denominator 60, we note that 30 must be multiplied by 2 to obtain 60. So we multiply both the numerator and denominator of $\frac{11}{30}$ by 2.

The following chart shows the similarity between adding fractions with different denominators in arithmetic and algebra.

Arithmetic	Algebra	Procedure
$\dfrac{5}{24} + \dfrac{7}{18}$	$\dfrac{3}{x^2 y} + \dfrac{2}{xy^4}$	
$\dfrac{5}{2 \cdot 2 \cdot 2 \cdot 3} + \dfrac{7}{2 \cdot 3 \cdot 3}$		Factor the denominators.
$\dfrac{5}{2^3 \cdot 3} + \dfrac{7}{2 \cdot 3^2}$		Rewrite the factorization using exponential notation.

		Form the product of all of the distinct factors raised to their highest power.
$LCD = 2^3 \cdot 3^2$ $= 72$	$LCD = x^2y^4$	
$\dfrac{5 \cdot 3}{24 \cdot 3} + \dfrac{7 \cdot 4}{18 \cdot 4}$	$\dfrac{3 \cdot y^3}{x^2y \cdot y^3} + \dfrac{2 \cdot x}{xy^4 \cdot x}$	Convert the fractions to equivalent fractions with the LCD as their denominator. Note that the first fraction's denominator lacks three factors of y and the second fraction's denominator lacks one factor of x.
$\dfrac{15}{72} + \dfrac{28}{72}$	$\dfrac{3y^3}{x^2y^4} + \dfrac{2x}{x^2y^4}$	
$\dfrac{15 + 28}{72}$	$\dfrac{3y^3 + 2x}{x^2y^4}$	Add the numerators.
$\dfrac{43}{72}$	$\dfrac{3y^3 + 2x}{x^2y^4}$	Combine like terms and reduce, if possible.

The following example illustrates how to add and subtract more complicated rational expressions.

EXAMPLE 5.4.1

Perform the indicated additions and subtractions.

1. $\dfrac{5x}{6} + \dfrac{7x}{8} = \dfrac{5x}{2 \cdot 3} + \dfrac{7x}{2^3}$ $\qquad LCD = 2^3 \cdot 3$
$\qquad\qquad\qquad\qquad\qquad\qquad\qquad = 24$

$\qquad\qquad = \dfrac{5x \cdot 4}{6 \cdot 4} + \dfrac{7x \cdot 3}{8 \cdot 3}$

$\qquad\qquad = \dfrac{20x}{24} + \dfrac{21x}{24}$

$\qquad\qquad = \dfrac{20x + 21x}{24}$

$\qquad\qquad = \dfrac{41x}{24}$

2. $\dfrac{3}{2x} - \dfrac{5}{6x} = \dfrac{3}{2 \cdot x} - \dfrac{5}{2 \cdot 3 \cdot x}$ $\qquad LCD = 2 \cdot 3 \cdot x$
$\qquad\qquad\qquad\qquad\qquad\qquad\qquad\qquad = 6x$

$$\frac{3 \cdot 3}{3} - \frac{5}{6x}$$

$$\frac{5}{6x}$$

$$\frac{5}{x}$$

Don't stop here!

$$\frac{\cdot 2}{3x}$$

$$\frac{}{x}$$

reduce your answer, if possible.

$$\frac{y}{y} + \frac{2 \cdot x}{y \cdot x} \qquad \text{LCD} = xy$$

$$+ \frac{2x}{xy}$$

$$\frac{y + 2x}{xy}$$

$$\frac{2}{3y} = \frac{2}{3^2xy^2} - \frac{7}{2 \cdot 3x^3y} \qquad \text{LCD} = 2 \cdot 3^2x^3y^2$$
$$= 18x^3y^2$$

$$= \frac{2}{9xy^2} \cdot \frac{2x^2}{2x^2} - \frac{7}{6x^3y} \cdot \frac{3y}{3y}$$

$$= \frac{4x^2}{18x^3y^2} - \frac{21y}{18x^3y^2}$$

$$= \frac{4x^2 - 21y}{18x^3y^2}$$

5. $\dfrac{3y}{5x} + \dfrac{9y}{4x^2} = \dfrac{3y}{5x} + \dfrac{9y}{2^2x^2} \qquad \text{LCD} = 2^2 \cdot 5 \cdot x^2$
$$= 20x^2$$

$$= \frac{3y \cdot 4x}{5x \cdot 4x} + \frac{9y \cdot 5}{4x^2 \cdot 5}$$

$$= \frac{12xy}{20x^2} + \frac{45y}{20x^2}$$

$$= \frac{12xy + 45y}{20x^2}$$

6. $\dfrac{8}{x + 2} + \dfrac{9}{x - 1} \qquad \text{LCD} = (x + 2)(x - 1)$

$$= \frac{8}{x + 2} \cdot \frac{(x - 1)}{(x - 1)} + \frac{9}{x - 1} \cdot \frac{(x + 2)}{(x + 2)}$$

$$= \frac{8x - 8}{(x + 2)(x - 1)} + \frac{9x + 18}{(x - 1)(x + 2)}$$

$$= \frac{8x - 8 + 9x + 18}{(x + 2)(x - 1)}$$

$$= \frac{17x + 10}{(x + 2)(x - 1)}$$

REMARK The denominator is left in factored form so that we can more easily determine if the answer can be reduced.

7. $\dfrac{6x}{x - 3} - \dfrac{4}{x + 5}$ LCD $= (x - 3)(x + 5)$

$$= \frac{6x \cdot (x + 5)}{x - 3 \cdot (x + 5)} - \frac{4 \cdot (x - 3)}{x + 5 \cdot (x - 3)}$$

$$= \frac{6x^2 + 30x}{(x - 3)(x + 5)} - \frac{4x - 12}{(x + 5)(x - 3)}$$

$$= \frac{6x^2 + 30x - (4x - 12)}{(x - 3)(x + 5)}$$

$$= \frac{6x^2 + 30x - 4x + 12}{(x - 3)(x + 5)}$$ Don't forget to change the signs.

$$= \frac{6x^2 + 26x + 12}{(x - 3)(x + 5)}$$

8. $\dfrac{x}{x + 3} - \dfrac{5}{2x}$ LCD $= 2x(x + 3)$

$$= \frac{x \cdot 2x}{x + 3 \cdot 2x} - \frac{5 \cdot (x + 3)}{2x \cdot (x + 3)}$$

$$= \frac{2x^2}{2x(x + 3)} - \frac{5x + 15}{2x(x + 3)}$$

$$= \frac{2x^2 - (5x + 15)}{2x(x + 3)}$$

$$= \frac{2x^2 - 5x - 15}{2x(x + 3)}$$

9. $\dfrac{2x + 1}{3x - 4} + \dfrac{x - 7}{x + 5}$ LCD $= (3x - 4)(x + 5)$

$$= \frac{(2x + 1) \cdot (x + 5)}{(3x - 4) \cdot (x + 5)} + \frac{(x - 7) \cdot (3x - 4)}{(x + 5) \cdot (3x - 4)}$$

$$= \frac{2x^2 + 11x + 5}{(3x - 4)(x + 5)} + \frac{3x^2 - 25x + 28}{(x + 5)(3x - 4)}$$

$$= \frac{2x^2 + 11x + 5 + 3x^2 - 25x + 28}{(3x - 4)(x + 5)}$$

$$= \frac{5x^2 - 14x + 33}{(3x - 4)(x + 5)}$$

10. $\dfrac{x-4}{2x^2+6x}+\dfrac{7}{x^2-9}$

$=\dfrac{x-4}{2x(x+3)}+\dfrac{7}{(x+3)(x-3)}$ First, factor both
denominators in order
to find the LCD.

$=\dfrac{(x-4)\cdot(x-3)}{2x(x+3)\cdot(x-3)}+\dfrac{7}{(x+3)(x-3)}\cdot\dfrac{\cdot\,2x}{\cdot\,2x}$ $\text{LCD}=2x(x+3)(x-3)$

$=\dfrac{x^2-7x+12}{2x(x+3)(x-3)}+\dfrac{14x}{2x(x+3)(x-3)}$

$=\dfrac{x^2-7x+12+14x}{2x(x+3)(x-3)}$

$=\dfrac{x^2+7x+12}{2x(x+3)(x-3)}$

$=\dfrac{(x+3)(x+4)}{2x(x+3)(x-3)}$ Factor the numerator
and reduce your answer.

$=\dfrac{x+4}{2x(x-3)}$

11. $\dfrac{2x+3}{x^2-4x+4}-\dfrac{x+1}{x^2+3x-10}$

$=\dfrac{2x+3}{(x-2)^2}-\dfrac{x+1}{(x+5)(x-2)}$ Factor each denominator.
$\text{LCD}=(x-2)^2(x+5)$

$=\dfrac{(2x+3)\cdot(x+5)}{(x-2)^2\cdot(x+5)}-\dfrac{(x+1)}{(x+5)(x-2)}\cdot\dfrac{\cdot\,(x-2)}{\cdot\,(x-2)}$

$=\dfrac{2x^2+13x+15}{(x-2)^2(x+5)}-\dfrac{x^2-x-2}{(x-2)^2(x+5)}$

$=\dfrac{2x^2+13x+15-(x^2-x-2)}{(x-2)^2(x+5)}$

$=\dfrac{2x^2+13x+15-x^2+x+2}{(x-2)^2(x+5)}$

$=\dfrac{x^2+14x+17}{(x-2)^2(x+5)}$

12. $\dfrac{x+1}{2x^2-3x-2}+\dfrac{x-5}{x^2+x-6}$

$=\dfrac{x+1}{(2x+1)(x-2)}+\dfrac{x-5}{(x+3)(x-2)}$ Factor each denominator.
$\text{LCD}=(2x+1)(x-2)(x+3)$

$=\dfrac{(x+1)}{(2x+1)(x-2)}\cdot\dfrac{\cdot\,(x+3)}{\cdot\,(x+3)}+\dfrac{(x-5)}{(x+3)(x-2)}\cdot\dfrac{\cdot\,(2x+1)}{\cdot\,(2x+1)}$

$=\dfrac{x^2+4x+3}{(2x+1)(x-2)(x+3)}+\dfrac{2x^2-9x-5}{(2x+1)(x-2)(x+3)}$

$=\dfrac{x^2+4x+3+2x^2-9x-5}{(2x+1)(x-2)(x+3)}$

$$= \frac{3x^2 - 5x - 2}{(2x + 1)(x - 2)(x + 3)}$$

$$= \frac{(3x + 1)(x - 2)}{(2x + 1)(x - 2)(x + 3)} \qquad \text{Factor the numerator}$$
$$\text{and reduce your answer.}$$

$$= \frac{3x + 1}{(2x + 1)(x + 3)}$$

EXERCISES 5.4

Perform the indicated additions and subtractions.

1. $\dfrac{3x}{5} + \dfrac{2x}{7}$
2. $\dfrac{2x}{3} + \dfrac{5x}{4}$
3. $\dfrac{4x}{3} - \dfrac{2x}{9}$
4. $\dfrac{9x}{4} - \dfrac{5x}{12}$

5. $\dfrac{x}{2} + \dfrac{3x}{10}$
6. $\dfrac{7x}{30} - \dfrac{11x}{12}$
7. $\dfrac{7x}{30} - \dfrac{13x}{18}$
8. $\dfrac{x}{6} + \dfrac{7x}{12}$

9. $\dfrac{2}{3x} + \dfrac{5}{2x^2}$
10. $\dfrac{7}{6x} - \dfrac{1}{2x}$
11. $\dfrac{3}{2x} - \dfrac{1}{6x}$
12. $\dfrac{3}{5x} + \dfrac{2}{3x^2}$

13. $\dfrac{3y}{10x} + \dfrac{y}{5x}$
14. $\dfrac{y}{4x} + \dfrac{y}{12x}$
15. $\dfrac{2y}{5x} - \dfrac{4y}{7x^2}$
16. $\dfrac{3y}{4x} - \dfrac{2y}{5x^2}$

17. $\dfrac{2}{x} + \dfrac{4}{y}$
18. $\dfrac{3}{x} + \dfrac{9}{y}$
19. $\dfrac{2y}{3x} - \dfrac{3x}{2y}$
20. $\dfrac{5y}{2x} - \dfrac{2x}{5y}$

21. $\dfrac{5}{x^2y} + \dfrac{6}{xy^3}$
22. $\dfrac{2}{xy^4} + \dfrac{3}{x^2y^2}$
23. $\dfrac{1}{3xy} - \dfrac{5}{9y}$
24. $\dfrac{1}{2xy} - \dfrac{3}{4y}$

25. $\dfrac{1}{x + 3} + \dfrac{4}{2x + 1}$
26. $\dfrac{1}{x + 5} + \dfrac{3}{3x + 2}$
27. $\dfrac{5}{2x - 1} - \dfrac{2}{x - 2}$
28. $\dfrac{7}{3x - 1} - \dfrac{2}{x - 4}$

29. $\dfrac{2}{x - 5} - \dfrac{7}{x + 3}$
30. $\dfrac{3}{x - 3} - \dfrac{6}{x + 5}$
31. $\dfrac{6x}{x + 2} + \dfrac{3}{x - 4}$
32. $\dfrac{2x}{x + 1} + \dfrac{7}{x - 3}$

33. $\dfrac{x}{x - 4} - \dfrac{2}{x - 1}$
34. $\dfrac{x}{x - 3} - \dfrac{4}{x - 2}$
35. $\dfrac{2x}{x - 3} - \dfrac{x}{2x + 1}$
36. $\dfrac{3x}{x - 5} - \dfrac{x}{3x + 2}$

37. $\dfrac{5x}{x + 1} + \dfrac{2}{3x}$
38. $\dfrac{4x}{x + 2} + \dfrac{5}{3x}$
39. $\dfrac{3}{2x - 1} - \dfrac{5}{2x}$
40. $\dfrac{2}{3x - 1} - \dfrac{4}{3x}$

41. $\dfrac{2x}{4x - 1} - \dfrac{2}{5x}$
42. $\dfrac{3x}{3x - 2} - \dfrac{5}{4x}$
43. $\dfrac{x}{4x + 3} + \dfrac{9}{8x}$
44. $\dfrac{x}{3x + 4} + \dfrac{7}{6x}$

45. $\dfrac{2x + 3}{x - 5} + \dfrac{x + 2}{x - 2}$
46. $\dfrac{3x + 1}{x - 4} + \dfrac{x + 3}{x - 1}$
47. $\dfrac{x - 2}{x - 4} - \dfrac{x - 5}{2x - 3}$
48. $\dfrac{x - 1}{x - 6} - \dfrac{x - 3}{2x - 7}$

49. $\dfrac{2x + 1}{3x - 5} - \dfrac{x - 1}{x - 2}$
50. $\dfrac{3x + 4}{2x - 1} - \dfrac{2x - 3}{x - 4}$
51. $\dfrac{3x - 1}{2x + 3} + \dfrac{2x - 5}{3x - 7}$
52. $\dfrac{2x - 3}{3x + 1} + \dfrac{2x - 1}{3x - 5}$

53. $\dfrac{5x}{x^2 - x - 6} + \dfrac{2x}{x^2 + 6x + 8}$

54. $\dfrac{x}{x^2 + 9x + 20} + \dfrac{5x}{x^2 + 3x - 4}$

55. $\dfrac{4}{3x^2 + 12x} - \dfrac{1}{x^2 + 5x + 4}$

56. $\dfrac{2}{x^2 - 6x + 9} - \dfrac{1}{x^2 - 5x + 6}$

57. $\dfrac{3x}{x^2 + x - 2} + \dfrac{2}{x^2 - 4x + 3}$

58. $\dfrac{2x}{x^2 - 4} - \dfrac{5}{x^2 + x - 6}$

59. $\dfrac{3}{x^2 - 2x + 1} - \dfrac{1}{x^2 + 4x - 5}$

60. $\dfrac{3}{2x^2 + 6x} - \dfrac{1}{x^2 + 4x + 3}$

61. $\dfrac{x - 4}{2x^2 + 8x} + \dfrac{7}{x^2 + x - 12}$

62. $\dfrac{x - 3}{3x^2 + 18x} + \dfrac{5}{x^2 + 2x - 24}$

63. $\dfrac{x - 4}{x^2 - 2x - 8} - \dfrac{7}{x^2 + 8x + 12}$

64. $\dfrac{x + 2}{x^2 + 6x + 9} + \dfrac{5}{x^2 - 9}$

65. $\dfrac{x + 1}{x^2 + 8x + 16} + \dfrac{3}{x^2 - 16}$

66. $\dfrac{x - 1}{x^2 + 4x - 5} - \dfrac{8}{x^2 - x - 30}$

67. $\dfrac{x + 7}{x^2 + 2x - 3} - \dfrac{6}{x^2 + x - 2}$

68. $\dfrac{x + 8}{x^2 - 2x - 8} - \dfrac{10}{x^2 - 3x - 4}$

69. $\dfrac{x + 1}{x^2 + 6x + 8} + \dfrac{x - 4}{x^2 - 3x - 10}$

70. $\dfrac{x + 3}{x^2 - 3x - 4} + \dfrac{x + 2}{x^2 - 2x - 3}$

71. $\dfrac{x - 3}{x^2 + 2x - 15} - \dfrac{x - 2}{x^2 + 3x - 10}$

72. $\dfrac{x + 2}{x^2 - 2x - 8} - \dfrac{x + 6}{x^2 + 2x - 24}$

73. $\dfrac{x - 10}{x^2 - 2x - 8} + \dfrac{x + 12}{x^2 - x - 6}$

74. $\dfrac{2x + 14}{x^2 - x - 6} + \dfrac{x - 4}{x^2 + x - 2}$

75. $\dfrac{2x - 8}{x^2 - 2x - 24} - \dfrac{x - 4}{x^2 + 3x - 4}$

76. $\dfrac{2x + 17}{x^2 + 3x - 10} - \dfrac{x + 13}{x^2 + 2x - 15}$

77. $\dfrac{2x}{x - 2} + \dfrac{5}{x + 2} - \dfrac{8x}{x^2 - 4}$

78. $\dfrac{2x}{x + 3} - \dfrac{3}{2x + 1} - \dfrac{11x - 2}{2x^2 + 7x + 3}$

79. $\dfrac{3}{x - 6} + \dfrac{x}{2x + 2} - \dfrac{21}{x^2 - 5x - 6}$

80. $\dfrac{2}{3x^2 + 2x} + \dfrac{1}{4x - 8} - \dfrac{1}{x^2 - 2x}$

5.5 Complex Fractions

This chapter has focused on the study of rational expressions. So far, we have learned how to reduce, multiply and divide, and add and subtract rational expressions. In this section we will learn how to simplify a special type of rational expression.

Complex Fractions
Any fraction that contains a fraction in its numerator and/or denominator is called a **complex fraction**.

REMARK   The following fractions are examples of complex fractions.

$$\dfrac{\dfrac{2}{3}}{\dfrac{7}{5}}, \qquad \dfrac{\dfrac{1}{2}-\dfrac{1}{6}}{\dfrac{1}{3}+\dfrac{1}{4}}, \qquad \dfrac{\dfrac{1}{x}+\dfrac{1}{4}}{x+4}, \qquad \dfrac{x-1}{\dfrac{2}{3x}-\dfrac{2}{3x^2}}, \qquad \dfrac{1+\dfrac{3}{x}-\dfrac{18}{x^2}}{1-\dfrac{8}{x}+\dfrac{15}{x^2}}$$

We are now ready to learn how to convert complex fractions into simple fractions. (A **simple fraction** does not have a fraction in its numerator or denominator.) The following example shows how to simplify complex fractions that contain only one term in the numerator and one term in the denominator.

EXAMPLE 5.5.1

Simplify the following complex fractions.

1. $\dfrac{\dfrac{2}{3}}{\dfrac{7}{5}}$ Since the main fraction bar indicates division, we can easily convert this complex fraction into a division problem.

$= \dfrac{2}{3} \div \dfrac{7}{5}$ Convert into a division problem.

$= \dfrac{2}{3} \cdot \dfrac{5}{7}$ Invert and multiply by the divisor.

$= \dfrac{10}{21}$

2. $\dfrac{\dfrac{4x^2}{3y^4}}{\dfrac{8x^5}{15y^9}} = \dfrac{4x^2}{3y^4} \div \dfrac{8x^5}{15y^9}$

$= \dfrac{4x^2}{3y^4} \cdot \dfrac{15y^9}{8x^5}$

$= \dfrac{4 \cdot 15x^2y^9}{3 \cdot 8x^5y^4}$

$= \dfrac{\cancel{2} \cdot \cancel{2} \cdot \cancel{3} \cdot 5y^{9-4}}{\cancel{3} \cdot \cancel{2} \cdot \cancel{2} \cdot 2x^{5-2}}$

$= \dfrac{5y^5}{2x^3}$

3. $\dfrac{\dfrac{1}{x+3}}{\dfrac{1}{x^2-9}} = \dfrac{1}{x+3} \div \dfrac{1}{x^2-9}$

$$= \frac{1}{x+3} \cdot \frac{x^2 - 9}{1}$$

$$= \frac{1}{\cancel{x+3}} \cdot \frac{\cancel{(x+3)}(x-3)}{1}$$

$$= x - 3$$

4. $\dfrac{\dfrac{15x^2}{2x+7}}{10x} = \dfrac{15x^2}{2x+7} \div 10x$

$$= \frac{\overset{3x}{\cancel{15x^2}}}{2x+7} \cdot \frac{1}{\underset{2}{\cancel{10x}}}$$

$$= \frac{3x}{2(2x+7)}$$

We will investigate two methods for simplifying a complex fraction with more than one term in its numerator or denominator. The first method converts the complex fraction into a complex fraction with just one term in both the numerator and denominator. We then convert the complex fraction into a division problem and simplify as was done in Example 5.5.1.

Method 1

EXAMPLE 5.5.2

Simplify the following complex fractions.

1. $\dfrac{\dfrac{1}{2} - \dfrac{1}{6}}{\dfrac{1}{3} + \dfrac{1}{4}} = \dfrac{\dfrac{1 \cdot 3}{2 \cdot 3} - \dfrac{1}{6}}{\dfrac{1 \cdot 4}{3 \cdot 4} + \dfrac{1 \cdot 3}{4 \cdot 3}}$ Obtain an LCD in both the numerator and denominator of the complex fraction.

$$= \frac{\dfrac{3}{6} - \dfrac{1}{6}}{\dfrac{4}{12} + \dfrac{3}{12}}$$

$$= \frac{\dfrac{2}{6}}{\dfrac{7}{12}}$$ Combine the terms in the numerator and denominator.

$$= \frac{\dfrac{1}{3}}{\dfrac{7}{12}}$$ Reduce the numerator and denominator, if possible.

$$= \frac{1}{3} \div \frac{7}{12}$$

Convert to a division problem.

$$= \frac{1}{\cancel{3}} \cdot \frac{\cancel{12}^{4}}{7}$$

Invert and multiply.

$$= \frac{4}{7}$$

2. $\dfrac{\dfrac{1}{x} + \dfrac{1}{4}}{x + 4} = \dfrac{\dfrac{1 \cdot 4}{x \cdot 4} + \dfrac{1 \cdot x}{4 \cdot x}}{x + 4}$

Obtain an LCD in the numerator.

$$= \frac{\dfrac{4}{4x} + \dfrac{x}{4x}}{x + 4}$$

$$= \frac{\dfrac{4 + x}{4x}}{x + 4}$$

Combine the expressions in the numerator.

$$= \frac{4 + x}{4x} \div (x + 4)$$

Convert to a division problem.

$$= \frac{4 + \cancel{x}}{4x} \cdot \frac{1}{\cancel{x + 4}}$$

Invert and multiply. Divide out common factors.

$$= \frac{1}{4x}$$

3. $\dfrac{\dfrac{x}{2} + 3}{\dfrac{1}{x} + \dfrac{1}{6}} = \dfrac{\dfrac{x}{2} + \dfrac{3 \cdot 2}{1 \cdot 2}}{\dfrac{1 \cdot 6}{x \cdot 6} + \dfrac{1 \cdot x}{6 \cdot x}}$

Obtain an LCD in both the numerator and denominator.

$$= \frac{\dfrac{x}{2} + \dfrac{6}{2}}{\dfrac{6}{6x} + \dfrac{x}{6x}}$$

$$= \frac{\dfrac{x + 6}{2}}{\dfrac{6 + x}{6x}}$$

$$= \frac{x + 6}{2} \div \frac{6 + x}{6x}$$

$$= \frac{\cancel{x + 6}}{\cancel{2}} \cdot \frac{\cancel{6x}^{3}}{\cancel{6 + x}}$$

$$= 3x$$

4. $\dfrac{\dfrac{1}{4} + \dfrac{1}{3x}}{\dfrac{1}{2} - \dfrac{1}{6x}} = \dfrac{\dfrac{1 \cdot 3x}{4 \cdot 3x} + \dfrac{1 \cdot 4}{3x \cdot 4}}{\dfrac{1 \cdot 3x}{2 \cdot 3x} - \dfrac{1}{6x}}$

$= \dfrac{\dfrac{3x}{12x} + \dfrac{4}{12x}}{\dfrac{3x}{6x} - \dfrac{1}{6x}}$

$= \dfrac{\dfrac{3x + 4}{12x}}{\dfrac{3x - 1}{6x}}$

$= \dfrac{3x + 4}{12x} \div \dfrac{3x - 1}{6x}$

$= \dfrac{3x + 4}{\underset{2}{12x}} \cdot \dfrac{6x}{3x - 1}$

$= \dfrac{3x + 4}{2(3x - 1)}$

The second method for simplifying a complex fraction uses the LCD of all of the fractions within the complex fraction. If we multiply the numerator and denominator of the complex fraction by this LCD, the complex fraction will become a simple fraction. The following example illustrates this method for simplifying complex fractions.

Method 2

EXAMPLE 5.5.3

Simplify the following complex fractions.

1. $\dfrac{\dfrac{1}{2} - \dfrac{1}{3}}{\dfrac{1}{4} + \dfrac{1}{6}} = \dfrac{\left(\dfrac{1}{2} - \dfrac{1}{3}\right) \cdot 12}{\left(\dfrac{1}{4} + \dfrac{1}{6}\right) \cdot 12}$ Multiply the numerator and denominator by the LCD, 12.

$= \dfrac{\dfrac{1}{2} \cdot 12 - \dfrac{1}{3} \cdot 12}{\dfrac{1}{4} \cdot 12 + \dfrac{1}{6} \cdot 12}$

$= \dfrac{6 - 4}{3 + 2}$

$= \dfrac{2}{5}$

2.
$$\frac{4 + \dfrac{10}{x}}{x + \dfrac{5}{2}} = \frac{\left(4 + \dfrac{10}{x}\right) \cdot 2x}{\left(x + \dfrac{5}{2}\right) \cdot 2x}$$

Multiply the numerator and denominator by the LCD, $2x$.

$$= \frac{4 \cdot 2x + \dfrac{10}{x} \cdot 2x}{x \cdot 2x + \dfrac{5}{2} \cdot 2x}$$

$$= \frac{8x + 20}{2x^2 + 5x}$$

$$= \frac{4(2x + 5)}{x(2x + 5)}$$

$$= \frac{4}{x}$$

3.
$$\frac{1 + \dfrac{3}{x} - \dfrac{18}{x^2}}{1 - \dfrac{8}{x} + \dfrac{15}{x^2}} = \frac{\left(1 + \dfrac{3}{x} - \dfrac{18}{x^2}\right) \cdot x^2}{\left(1 - \dfrac{8}{x} + \dfrac{15}{x^2}\right) \cdot x^2}$$

$$= \frac{1 \cdot x^2 + \dfrac{3}{x} \cdot x^2 - \dfrac{18}{x^2} \cdot x^2}{1 \cdot x^2 - \dfrac{8}{x} \cdot x^2 + \dfrac{15}{x^2} \cdot x^2}$$

$$= \frac{x^2 + 3x - 18}{x^2 - 8x + 15}$$

$$= \frac{(x + 6)(x - 3)}{(x - 5)(x - 3)}$$

$$= \frac{x + 6}{x - 5}$$

4.
$$\frac{2 - \dfrac{1}{x + 1}}{\dfrac{2}{x + 1} + \dfrac{1}{3}} = \frac{\left(2 - \dfrac{1}{x + 1}\right) \cdot 3(x + 1)}{\left(\dfrac{2}{x + 1} + \dfrac{1}{3}\right) \cdot 3(x + 1)}$$

In this case the LCD is $3(x + 1)$.

$$= \frac{2 \cdot 3(x + 1) - \dfrac{1}{x + 1} \cdot 3(x + 1)}{\dfrac{2}{x + 1} \cdot 3(x + 1) + \dfrac{1}{3} \cdot 3(x + 1)}$$

$$= \frac{6(x + 1) - 3}{6 + (x + 1)}$$

$$= \frac{6x + 6 - 3}{6 + x + 1}$$

$$= \frac{6x + 3}{x + 7}$$

EXERCISES 5.5

Simplify the following complex fractions.

1. $\dfrac{\frac{3}{5}}{\frac{7}{2}}$ 2. $\dfrac{\frac{4}{5}}{\frac{9}{2}}$ 3. $\dfrac{\frac{2}{3}}{\frac{8}{9}}$ 4. $\dfrac{\frac{3}{4}}{\frac{15}{8}}$ 5. $\dfrac{\frac{12}{9}}{\frac{2}{3}}$ 6. $\dfrac{\frac{10}{4}}{\frac{5}{8}}$

7. $\dfrac{\frac{x^3}{y^2}}{\frac{x^4}{y^6}}$ 8. $\dfrac{\frac{x^4}{y^3}}{\frac{x^5}{y^9}}$ 9. $\dfrac{\frac{2x^4}{9y^3}}{\frac{8x^2}{3y}}$ 10. $\dfrac{\frac{3x^5}{10y^7}}{\frac{9x^4}{5y^2}}$ 11. $\dfrac{\frac{5x^2}{3y}}{\frac{7y^6}{2x^3}}$ 12. $\dfrac{\frac{2x^4}{5y^3}}{\frac{9y^4}{7x}}$

13. $\dfrac{\frac{4x^2}{3y^3}}{2}$ 14. $\dfrac{\frac{9x^4}{10y^5}}{3}$ 15. $\dfrac{\frac{5x^4}{9y^3}}{x^2}$ 16. $\dfrac{\frac{6x^5}{7y^2}}{x^3}$ 17. $\dfrac{\frac{9x^4}{10}}{3x^3}$ 18. $\dfrac{\frac{8x^3}{9}}{2x^2}$

19. $\dfrac{\frac{2x^2}{3y^3}}{4x^3}$ 20. $\dfrac{\frac{5x}{7y^2}}{10x^4}$ 21. $\dfrac{\frac{6x^3}{5y}}{\frac{1}{10y^2}}$ 22. $\dfrac{\frac{8x^2}{3y^2}}{\frac{1}{9y^3}}$ 23. $\dfrac{\frac{1}{x-2}}{\frac{1}{x^2-4}}$ 24. $\dfrac{\frac{1}{x-1}}{\frac{1}{x^2-1}}$

25. $\dfrac{\frac{5}{x^2+x-2}}{\frac{10}{x^2+5x+6}}$ 26. $\dfrac{\frac{3}{x^2+3x-4}}{\frac{12}{x^2-6x+5}}$ 27. $\dfrac{\frac{2}{x^2+x-6}}{\frac{6}{x-2}}$ 28. $\dfrac{\frac{4}{x^2+8x+7}}{\frac{16}{x+7}}$

29. $\dfrac{\frac{x^2-9}{7}}{x+3}$ 30. $\dfrac{\frac{x^2-25}{8}}{x+5}$ 31. $\dfrac{\frac{4}{2x+3}}{12}$ 32. $\dfrac{\frac{3}{3x+5}}{9}$

33. $\dfrac{\frac{6x^3}{3x+1}}{8x}$ 34. $\dfrac{\frac{9x^4}{2x+5}}{3x^2}$ 35. $\dfrac{\frac{12x^2}{x+5}}{\frac{15x^4}{x-1}}$ 36. $\dfrac{\frac{18x^5}{x-4}}{\frac{12x^4}{x+3}}$

37. $\dfrac{\frac{2x^2-2x}{x+3}}{\frac{5x^3-5x^2}{x-5}}$ 38. $\dfrac{\frac{3x^3-6x^2}{x+1}}{\frac{4x^2-8x}{x+4}}$ 39. $\dfrac{\frac{5x^2+10x}{2x-6}}{\frac{2x^2+4x}{3x-9}}$ 40. $\dfrac{\frac{4x^2+4x}{6x-12}}{\frac{3x^2+3x}{4x-8}}$

41. $\dfrac{\frac{4x}{3x+9}}{\frac{8x^2-12x}{x+3}}$ 42. $\dfrac{\frac{6x}{5x+10}}{\frac{12x^2-12x}{x+2}}$ 43. $\dfrac{\frac{1}{2}+\frac{1}{4}}{\frac{1}{3}+\frac{1}{6}}$ 44. $\dfrac{\frac{1}{3}+\frac{1}{4}}{\frac{1}{6}+\frac{1}{2}}$

45. $\dfrac{\frac{2}{3}-\frac{1}{2}}{\frac{3}{4}-\frac{1}{6}}$

46. $\dfrac{\frac{3}{4}-\frac{1}{3}}{\frac{1}{2}-\frac{1}{6}}$

47. $\dfrac{\frac{5}{4}-\frac{2}{3}}{\frac{5}{6}+\frac{1}{2}}$

48. $\dfrac{\frac{7}{6}-\frac{1}{2}}{\frac{3}{4}+\frac{5}{3}}$

49. $\dfrac{3+\frac{3}{2x}}{x+\frac{1}{2}}$

50. $\dfrac{2+\frac{3}{2x}}{x+\frac{3}{4}}$

51. $\dfrac{2-\frac{3}{x}}{\frac{10}{3}-\frac{5}{x}}$

52. $\dfrac{4-\frac{4}{3x}}{3-\frac{1}{x}}$

53. $\dfrac{\frac{9}{4}-\frac{1}{x^2}}{\frac{3}{2}+\frac{1}{x}}$

54. $\dfrac{\frac{2}{3}-\frac{3}{2x^2}}{1+\frac{3}{2x}}$

55. $\dfrac{\frac{6}{x}+\frac{3}{2}}{9}$

56. $\dfrac{\frac{2}{x}+\frac{4}{3}}{8}$

57. $\dfrac{\frac{1}{x}+\frac{1}{2}}{x+2}$

58. $\dfrac{\frac{1}{x}+\frac{1}{5}}{x+5}$

59. $\dfrac{x-5}{\frac{1}{25}-\frac{1}{x^2}}$

60. $\dfrac{x-2}{\frac{1}{4}-\frac{1}{x^2}}$

61. $\dfrac{2x+1}{\frac{1}{4}-\frac{3}{x}}$

62. $\dfrac{2x+3}{\frac{1}{2}-\frac{4}{x}}$

63. $\dfrac{1+\frac{1}{x}-\frac{20}{x^2}}{1+\frac{4}{x}-\frac{5}{x^2}}$

64. $\dfrac{1+\frac{1}{x}-\frac{12}{x^2}}{1-\frac{2}{x}-\frac{24}{x^2}}$

65. $\dfrac{1+\frac{5}{x}-\frac{14}{x^2}}{1+\frac{3}{x}-\frac{10}{x^2}}$

66. $\dfrac{1+\frac{7}{x}-\frac{18}{x^2}}{1+\frac{5}{x}-\frac{14}{x^2}}$

67. $\dfrac{1-\frac{4}{x}+\frac{3}{x^2}}{1-\frac{6}{x}+\frac{5}{x^2}}$

68. $\dfrac{1-\frac{5}{x}+\frac{6}{x^2}}{1-\frac{9}{x}+\frac{18}{x^2}}$

69. $\dfrac{3+\frac{1}{x+2}}{\frac{2}{x+2}+\frac{1}{4}}$

70. $\dfrac{2+\frac{1}{x+3}}{\frac{3}{x+3}+\frac{1}{2}}$

71. $\dfrac{5-\frac{3}{x-1}}{\frac{3}{x-1}+\frac{2}{3}}$

72. $\dfrac{4-\frac{3}{x-1}}{\frac{5}{x-1}+\frac{3}{4}}$

73. $\dfrac{\frac{1}{3}+\frac{1}{x-4}}{\frac{2}{3}+\frac{4}{x-4}}$

74. $\dfrac{\frac{1}{2}+\frac{1}{x-6}}{\frac{3}{2}+\frac{5}{x-6}}$

75. $\dfrac{\frac{3}{x}-\frac{1}{x+2}}{\frac{5}{x}-\frac{4}{x+2}}$

76. $\dfrac{\frac{2}{x}-\frac{1}{x+1}}{\frac{6}{x}-\frac{3}{x+1}}$

5.6 Fractional Equations

We have already investigated several types of equations in this text. In particular we have seen how to determine the solutions of both linear and quadratic equations. In this section we will learn how to find the solutions of another type of equation.

Fractional Equation

Any equation that contains at least one fraction is called a **fractional equation**.

REMARK   The following equations are examples of fractional equations.

$$\frac{x}{3} + \frac{2x}{5} = \frac{22}{15}, \qquad \frac{x}{3} - \frac{5}{2} = \frac{2x-1}{2}, \qquad \frac{1}{x} + \frac{3}{2x} = \frac{1}{2}, \qquad \frac{4}{x+1} = \frac{1}{x}$$

$$2 + \frac{4}{x-4} = \frac{x}{x-4}, \qquad \frac{4}{x-2} - \frac{2}{x+3} = \frac{14}{x^2+x-6}$$

A common method for determining the solutions of fractional equations begins with simplifying the equation by eliminating all of the fractions from the equation. We can eliminate the fractions by multiplying by the LCD of all of the fractions on either side of the equation. The following example shows how to find the solutions of fractional equations in which the variable is *not* in any denominator.

EXAMPLE 5.6.1

Solve for *x* in each of the following equations.

1. $\dfrac{x}{3} + \dfrac{2x}{5} = \dfrac{22}{15}$ The LCD is 15. Multiply both sides of the equation by 15.

$$\frac{15}{1}\left(\frac{x}{3} + \frac{2x}{5}\right) = \frac{15}{1} \cdot \frac{22}{15}$$

(It sometimes makes the multiplication easier if the LCD is written as a fraction with a denominator of 1.)

$$\frac{15}{1} \cdot \frac{x}{3} + \frac{15}{1} \cdot \frac{2x}{5} = \frac{15}{1} \cdot \frac{22}{15}$$

$$5x + 6x = 22$$

$$11x = 22$$

$$x = 2$$

Let's check the answer:

$$\frac{2}{3} + \frac{2 \cdot 2}{5} \stackrel{?}{=} \frac{22}{15}$$

$$\frac{2}{3} + \frac{4}{5} \stackrel{?}{=} \frac{22}{15}$$

$$\frac{10}{15} + \frac{12}{15} \stackrel{?}{=} \frac{22}{15}$$

$$\frac{22}{15} = \frac{22}{15}$$

Thus, the solution set is {2}.

2.
$$\frac{x}{3} - \frac{5}{2} = \frac{2x-1}{2}$$ The LCD is 6.

$$\frac{6}{1}\left(\frac{x}{3} - \frac{5}{2}\right) = \frac{6}{1} \cdot \frac{2x-1}{2}$$

$$\frac{6}{1} \cdot \frac{x}{3} - \frac{6}{1} \cdot \frac{5}{2} = \frac{6}{1} \cdot \frac{2x-1}{2}$$

$$2x - 15 = 3(2x - 1)$$

$$2x - 15 = 6x - 3$$

$$-15 = 4x - 3$$

$$-12 = 4x$$

$$-3 = x$$

By checking this solution, the reader can verify that the solution set is $\{-3\}$.

When the LCD contains the variable, multiplying by the LCD *might* cause some problems, so we *must* check our solutions. Consider the following example.

EXAMPLE 5.6.2

Solve for x in the following equations.

1.
$$\frac{1}{x} + \frac{3}{2x} = \frac{1}{2}$$ The LCD is $2x$.

$$\frac{2x}{1}\left(\frac{1}{x} + \frac{3}{2x}\right) = \frac{2x}{1} \cdot \frac{1}{2}$$

$$\frac{2x}{1} \cdot \frac{1}{x} + \frac{2x}{1} \cdot \frac{3}{2x} = \frac{2x}{1} \cdot \frac{1}{2}$$

$$2 + 3 = x$$

$$5 = x$$

Checking the answer:

$$\frac{1}{5} + \frac{3}{2 \cdot 5} \stackrel{?}{=} \frac{1}{2}$$

$$\frac{1}{5} + \frac{3}{10} \stackrel{?}{=} \frac{1}{2}$$

$$\frac{2}{10} + \frac{3}{10} \stackrel{?}{=} \frac{1}{2}$$

$$\frac{5}{10} \stackrel{?}{=} \frac{1}{2}$$

$$\frac{1}{2} = \frac{1}{2}$$

So the solution set is {5}.

2.
$$\frac{4}{x + 1} = \frac{1}{x}$$
The LCD is $x(x + 1)$.

$$\frac{x(x + 1)}{1} \cdot \frac{4}{x + 1} = \frac{x(x + 1)}{1} \cdot \frac{1}{x}$$

$$4x = x + 1$$

$$3x = 1$$

$$x = \frac{1}{3}$$

Checking the answer:

$$\frac{4}{\frac{1}{3} + 1} \overset{?}{=} \frac{1}{\frac{1}{3}}$$

$$\frac{4}{\frac{4}{3}} \overset{?}{=} \frac{1}{\frac{1}{3}}$$

$$4 \div \frac{4}{3} \overset{?}{=} 1 \div \frac{1}{3}$$

$$4 \cdot \frac{3}{4} \overset{?}{=} 1 \cdot \frac{3}{1}$$

$$3 = 3$$

So the solution set is $\{\frac{1}{3}\}$.

3.
$$2 + \frac{4}{x - 4} = \frac{x}{x - 4}$$
The LCD is $x - 4$.

$$\frac{(x - 4)}{1}\left(2 + \frac{4}{x - 4}\right) = \frac{x - 4}{1} \cdot \frac{x}{x - 4}$$

$$\frac{(x - 4)}{1} \cdot 2 + \frac{(x - 4)}{1} \cdot \frac{4}{x - 4} = \frac{x - 4}{1} \cdot \frac{x}{x - 4}$$

$$(x - 4) \cdot 2 + 4 = x$$

$$2x - 8 + 4 = x$$

$$2x - 4 = x$$

$$x - 4 = 0$$

$$x = 4$$

Checking the answer:

$$2 + \frac{4}{4 - 4} \overset{?}{=} \frac{4}{4 - 4}$$

$$2 + \frac{4}{0} \overset{?}{=} \frac{4}{0}$$
  Division by zero

Since division by zero is undefined, 4 is not a solution. Since 4 is the only number we obtained for x and it does not check, the solution set is the empty set, $\emptyset$.

NOTE ▶ Since 4 generates a division by zero, it is not a solution of the equation. It is called an *extraneous solution*. When the LCD of a fractional equation contains the variable, we must always check each solution to verify that it is not an extraneous solution.

4.
$$\frac{4}{x-2} - \frac{2}{x+3} = \frac{14}{x^2+x-6}$$

$$\frac{4}{x-2} - \frac{2}{x+3} = \frac{14}{(x+3)(x-2)}$$ Factor all denominators. The LCD is $(x-2)(x+3)$.

$$\frac{(x-2)(x+3)}{1}\left(\frac{4}{x-2} - \frac{2}{x+3}\right) = \frac{(x-2)(x+3)}{1} \cdot \frac{14}{(x+3)(x-2)}$$

$$\frac{(x-2)(x+3)}{1} \cdot \frac{4}{x-2} - \frac{(x-2)(x+3)}{1} \cdot \frac{2}{x+3} = \frac{(x-2)(x+3)}{1} \cdot \frac{14}{(x+3)(x-2)}$$

$$4(x+3) - 2(x-2) = 14$$

$$4x + 12 - 2x + 4 = 14$$ Watch your signs!

$$2x + 16 = 14$$

$$2x = -2$$

$$x = -1$$

By checking this solution, the reader can verify that the solution set is $\{-1\}$.

5.
$$\frac{x}{x-2} + \frac{1}{x+2} = \frac{-4}{x^2-4}$$

$$\frac{x}{x-2} + \frac{1}{x+2} = \frac{-4}{(x+2)(x-2)}$$ Factor all denominators. The LCD is $(x+2)(x-2)$.

$$\frac{(x+2)(x-2)}{1}\left(\frac{x}{x-2} + \frac{1}{x+2}\right) = \frac{(x+2)(x-2)}{1} \cdot \frac{-4}{(x+2)(x-2)}$$

$$\frac{(x+2)(x-2)}{1} \cdot \frac{x}{x-2} + \frac{(x+2)(x-2)}{1} \cdot \frac{1}{x+2} = \frac{(x+2)(x-2)}{1} \cdot \frac{-4}{(x+2)(x-2)}$$

$$x(x+2) + (x-2) = -4$$

$$x^2 + 2x + x - 2 = -4$$

$$x^2 + 3x - 2 = -4$$

$$x^2 + 3x + 2 = 0$$ A quadratic equation

$$(x+2)(x+1) = 0$$

$$x + 2 = 0 \quad\quad \text{or} \quad\quad x + 1 = 0$$

$$x = -2 \quad\quad\quad\quad\quad x = -1$$

The two *potential solutions* are -2 and -1. When we substitute -2 for x in the original equation, we generate a denominator of zero. Thus, -2 is an extraneous solution. However, when we substitute -1 for x, it is easily shown that it is a solution. Thus, the solution set is $\{-1\}$.

EXERCISES 5.6

Solve for x in the following equations.

1. $\dfrac{2x}{3} + \dfrac{x}{4} = \dfrac{11}{12}$

2. $\dfrac{x}{2} + \dfrac{3x}{7} = \dfrac{13}{7}$

3. $\dfrac{5x}{2} - \dfrac{2x}{3} = -\dfrac{11}{18}$

4. $\dfrac{4x}{3} - \dfrac{2x}{5} = -\dfrac{7}{30}$

5. $\dfrac{x}{10} - 1 = \dfrac{x}{15}$

6. $\dfrac{x}{9} - 2 = \dfrac{x}{6}$

7. $\dfrac{5x}{6} + \dfrac{1}{2} = \dfrac{5x}{4}$

8. $\dfrac{3x}{10} + \dfrac{1}{2} = \dfrac{3x}{5}$

9. $\dfrac{x}{2} - \dfrac{4}{3} = \dfrac{2x - 5}{3}$

10. $\dfrac{x}{3} - \dfrac{3}{2} = \dfrac{2x - 5}{3}$

11. $\dfrac{2x}{5} + 3 = \dfrac{4x + 1}{2}$

12. $\dfrac{3x}{4} + 2 = \dfrac{2x + 1}{3}$

13. $\dfrac{x - 1}{2} - \dfrac{3x + 1}{5} = \dfrac{7}{10}$

14. $\dfrac{x - 4}{3} - \dfrac{4x + 3}{4} = \dfrac{11}{12}$

15. $\dfrac{x + 1}{4} + \dfrac{2x - 5}{2} = \dfrac{x - 3}{8}$

16. $\dfrac{2x + 3}{4} + \dfrac{x - 9}{2} = \dfrac{3x - 5}{8}$

17. $\dfrac{4x}{3} = \dfrac{1}{2}$

18. $\dfrac{2x}{5} = \dfrac{4}{3}$

19. $\dfrac{x - 1}{2} = \dfrac{3x}{4}$

20. $\dfrac{x - 6}{6} = \dfrac{2x}{3}$

21. $\dfrac{x + 1}{6} = \dfrac{2x - 3}{4}$

22. $\dfrac{x + 2}{2} = \dfrac{3x - 1}{3}$

23. $\dfrac{x - 3}{3} = \dfrac{3x + 1}{4}$

24. $\dfrac{2x - 1}{9} = \dfrac{x + 2}{2}$

25. $\dfrac{3}{2x} + \dfrac{1}{x} = \dfrac{3}{4}$

26. $\dfrac{5}{2x} + \dfrac{1}{x} = \dfrac{7}{4}$

27. $\dfrac{4}{3x} - 1 = \dfrac{1}{2x}$

28. $\dfrac{2}{3x} - 1 = \dfrac{3}{2x}$

29. $\dfrac{x}{2} + \dfrac{1}{x} = \dfrac{9}{4}$

30. $\dfrac{2x}{9} + \dfrac{1}{x} = 1$

31. $\dfrac{x}{8} - \dfrac{5}{8x} = -\dfrac{1}{2}$

32. $\dfrac{x}{6} - \dfrac{2}{x} = -\dfrac{2}{3}$

33. $\dfrac{5}{2x - 1} = \dfrac{4}{3x}$

34. $\dfrac{4}{3x + 1} = \dfrac{5}{2x}$

35. $\dfrac{x}{x + 7} = \dfrac{5}{x + 3}$

36. $\dfrac{x}{x + 16} = \dfrac{2}{x - 2}$

37. $\dfrac{x + 1}{x + 2} = \dfrac{x - 5}{x - 3}$

38. $\dfrac{x + 2}{x + 3} = \dfrac{x - 4}{x - 2}$

39. $\dfrac{x + 9}{x + 12} = \dfrac{4x + 3}{3x + 4}$

40. $\dfrac{x - 4}{x - 14} = \dfrac{2x + 1}{5x + 2}$

41. $\dfrac{x}{x - 2} - 3 = \dfrac{2}{x - 2}$

42. $\dfrac{x}{x - 1} - 4 = \dfrac{1}{x - 1}$

43. $\dfrac{x}{x - 3} + 5 = \dfrac{21}{x - 3}$

44. $\dfrac{x}{x - 2} + 3 = \dfrac{10}{x - 2}$

45. $\dfrac{3}{x + 4} + \dfrac{5}{x - 2} = \dfrac{16}{x^2 + 2x - 8}$

46. $\dfrac{2}{x + 3} + \dfrac{7}{x - 1} = \dfrac{25}{x^2 + 2x - 3}$

47. $\dfrac{5}{x + 2} - \dfrac{3}{x + 1} = \dfrac{9}{x^2 + 3x + 2}$

48. $\dfrac{7}{x + 4} - \dfrac{4}{x + 2} = \dfrac{10}{x^2 + 6x + 8}$

49. $\dfrac{4}{5x} + \dfrac{1}{x-2} = \dfrac{1}{5x^2 - 10x}$

50. $\dfrac{5}{3x} + \dfrac{2}{x-1} = \dfrac{17}{3x^2 - 3x}$

51. $\dfrac{3}{2x} + \dfrac{4}{5x+1} = \dfrac{3}{10x^2 + 2x}$

52. $\dfrac{7}{2x} + \dfrac{2}{2x+3} = \dfrac{21}{4x^2 + 6x}$

53. $\dfrac{1}{x-2} - \dfrac{1}{x+2} = \dfrac{4}{5}$

54. $\dfrac{1}{x-3} - \dfrac{1}{x+3} = -\dfrac{6}{5}$

55. $\dfrac{x}{x+4} + \dfrac{3}{x-2} = \dfrac{18}{x^2 + 2x - 8}$

56. $\dfrac{x}{x+3} + \dfrac{6}{x-1} = \dfrac{24}{x^2 + 2x - 3}$

57. $\dfrac{x}{x-2} - \dfrac{2}{x-3} = \dfrac{-2}{x^2 - 5x + 6}$

58. $\dfrac{x}{x-1} - \dfrac{1}{x-4} = \dfrac{-3}{x^2 - 5x + 4}$

59. $\dfrac{2x}{x+3} + \dfrac{17}{x-4} = \dfrac{56}{x^2 - x - 12}$

60. $\dfrac{2x}{x+2} + \dfrac{17}{x-6} = \dfrac{46}{x^2 - 4x - 12}$

61. $\dfrac{2x}{x-5} - \dfrac{x}{x+3} = \dfrac{-24}{x^2 - 2x - 15}$

62. $\dfrac{2x}{x-3} - \dfrac{x}{x+2} = \dfrac{-10}{x^2 - x - 6}$

5.7 Ratios and Proportions

In the previous section we learned how to find the solutions of fractional equations. In this section we will focus on finding the solutions of a special type of fractional equation. However, we must first introduce some new terminology.

Ratio
If x and y are two numbers, then the **ratio** of x to y is the quotient $$\frac{x}{y} \qquad y \neq 0$$

REMARK   The above ratio can be written in three equivalent forms:

1. As a fraction

$$\frac{x}{y}$$

2. With the two numbers separated by the word *to*

x to y

3. With the two numbers separated by a colon (:)

$x : y$

NOTE ▶ The fractional form $\dfrac{x}{y}$ is the most common form of expressing the ratio of x to y.

A ratio is said to be in lowest terms when the two numbers do not have a common factor. The following example illustrates the process of expressing ratios in lowest terms.

EXAMPLE 5.7.1

Write each ratio as a fraction in lowest terms.

1. 3 to 5; as a fraction is $\dfrac{3}{5}$

2. 12 to 9; as a fraction is $\dfrac{12}{9}$

As a fraction in lowest terms: $\dfrac{12}{9} = \dfrac{\cancel{3} \cdot 4}{\cancel{3} \cdot 3} = \dfrac{4}{3}$

3. 8 feet to 12 feet; as a fraction is $\dfrac{8 \text{ feet}}{12 \text{ feet}}$, or simply $\dfrac{8}{12}$

As a fraction in lowest terms: $\dfrac{8}{12} = \dfrac{\cancel{4} \cdot 2}{\cancel{4} \cdot 3} = \dfrac{2}{3}$

4. $7x^2$ to $4x^2$ as a fraction is $\dfrac{7x^2}{4x^2}$ $(x \neq 0)$.

As a fraction in lowest terms: $\dfrac{7\cancel{x^2}}{4\cancel{x^2}} = \dfrac{7}{4}$

5. 4 hours to 2 days

Here we must convert to a common unit of measure. Since 2 days equals 48 hours, the ratio becomes 4 hours to 48 hours; as a fraction is $\dfrac{4 \text{ hours}}{48 \text{ hours}} = \dfrac{4}{48}$.

As a fraction in lowest terms: $\dfrac{4}{48} = \dfrac{\cancel{4} \cdot 1}{\cancel{4} \cdot 12} = \dfrac{1}{12}$

NOTE ▶ If we want to find a ratio of like quantities, the units of measurement must be the same.

Proportion
A **proportion** is an equation that states that two ratios are equal.

REMARK   The equation $a/b = c/d$ ($b \neq 0$, $d \neq 0$) is an example of a proportion.

In the proportion $a/b = c/d$, a and d are called the **extremes** of the proportion, whereas b and c are called the **means** of the proportion. Given a proportion, we can eliminate the fractions by multiplying both sides of the equation by the LCD of the two fractions.

Returning to the given proportion

$$\frac{a}{b} = \frac{c}{d}$$

$$\frac{bd}{1} \cdot \frac{a}{b} = \frac{bd}{1} \cdot \frac{c}{d} \qquad \text{Multiply both sides by the LCD, } bd.$$

$$ad = bc$$

that is, the product of the extremes = product of the means.

NOTE ▶   In every proportion the product of the extremes equals the product of the means. We can quickly find these products with the following schematic approach.

$$\frac{a}{b} \diagdown\!\!\!\!\!\diagup \frac{c}{d}$$

$$ad = bc \qquad \text{Equate the products of the numbers on the diagonals.}$$

The products ad and bc are called **cross products**, for obvious reasons. *In a proportion, cross products are equal.*

In the next example we use cross products to find the solutions of the given proportions.

EXAMPLE 5.7.2   Solve for x in the following proportions.

1. $\dfrac{x}{8} \diagdown\!\!\!\!\!\diagup \dfrac{3}{4}$

$x \cdot 4 = 8 \cdot 3 \qquad \text{Equating the cross products}$

$4x = 24$

$x = 6$

The check is left to the reader. The solution set is $\{6\}$.

2. $\dfrac{2x - 1}{4} \diagdown\!\!\!\!\!\diagup \dfrac{3x}{5}$

$(2x - 1) \cdot 5 = 4 \cdot 3x \qquad \text{Equating the cross products}$

$10x - 5 = 12x$

$-5 = 2x$

$$-\frac{5}{2} = x$$

Again the check is left to the reader. The solution set is $\left\{-\frac{5}{2}\right\}$.

3.
$$\frac{1}{x} \diagdown \frac{3x + 1}{2}$$

$$x \cdot (3x + 1) = 1 \cdot 2 \qquad \text{Equating the cross products}$$

$$3x^2 + x = 2 \qquad \text{We now have a quadratic}$$

$$3x^2 + x - 2 = 0 \qquad \text{equation.}$$

$$(3x - 2)(x + 1) = 0$$

$$3x - 2 = 0 \qquad \text{or} \qquad x + 1 = 0$$

$$3x = 2 \qquad\qquad x = -1$$

$$x = \frac{2}{3}$$

Again, the check is left to the reader. The solution set is $\left\{\frac{2}{3}, -1\right\}$.

4.
$$\frac{5x + 8}{x - 2} \diagdown \frac{x}{x - 2}$$

$$(5x + 8)(x - 2) = x(x - 2) \qquad \text{Equating the cross products}$$

$$5x^2 - 2x - 16 = x^2 - 2x$$

$$4x^2 - 16 = 0$$

$$4(x^2 - 4) = 0$$

$$4(x + 2)(x - 2) = 0$$

$$x + 2 = 0 \qquad \text{or} \qquad x - 2 = 0$$

$$x = -2 \qquad\qquad x = 2$$

The two potential solutions are -2 and 2. However, when we substitute 2 for x in the original proportion we generate a denominator of zero. Thus, 2 is an extraneous solution. However, when we substitute -2 for x, it is easily shown that it is a solution. Thus, $\{-2\}$ is the solution set.

The following example illustrates an application problem using a proportion.

EXAMPLE 5.7.3

"Big D's Miracle Elixir" is guaranteed to cure indigestion. The dosage is 3 tablespoons for every 80 pounds of body weight. Moe, the human cannonball, weighs 200 pounds and has indigestion. How many tablespoons of Big D's Miracle Elixir should Moe take to cure his indigestion?

SOLUTION   Let x = the number of tablespoons Moe should take. Now the ratio of the tablespoons equals the ratio of the weights:

$$\frac{x \text{ tablespoons}}{3 \text{ tablespoons}} \diagdown \hspace{-1.5em} \diagup \frac{200 \text{ pounds}}{80 \text{ pounds}}$$

$$x \cdot 80 = 3 \cdot 200 \qquad \text{Equate the cross products.}$$

$$80x = 600$$

$$x = \frac{600}{80}$$

$$x = 7.5$$

Moe should take 7.5 tablespoons.

NOTE ▶   The numerators in the proportion represented Moe's dosage and weight, whereas the denominators were the given dosage and weight.

EXERCISES 5.7

Write each ratio as a fraction in lowest terms.

1. 4 to 7   **2.** 3 to 11   **3.** 24 to 9   **4.** 10 to 8   **5.** 5 to 30   **6.** 4 to 20

7. 8 miles to 12 miles   **8.** 18 gallons to 24 gallons

9. 49 cups to 21 cups   **10.** 45 years to 10 years

11. $5x^2$ to $2x^2$ **12.** $7x^3$ to $3x^3$ **13.** $6x$ to $27x$ **14.** $6x$ to $16x$

15. $3x^4$ to $2x^3$ **16.** $2x^3$ to $7x^2$ **17.** $20x^3$ to $4x$ **18.** $24x^3$ to $4x$

19. 2 feet to 18 inches   **20.** 3 feet to 21 inches   **21.** 28 hours to 2 days   **22.** 20 hours to 2 days

Solve for x in the following proportions.

23. $\dfrac{x}{6} = \dfrac{2}{3}$ **24.** $\dfrac{x}{10} = \dfrac{3}{5}$ **25.** $\dfrac{2x+1}{5} = \dfrac{7}{2}$ **26.** $\dfrac{3x+2}{4} = \dfrac{5}{3}$

27. $\dfrac{x-4}{5} = \dfrac{2x}{9}$ **28.** $\dfrac{x-3}{4} = \dfrac{3x}{8}$ **29.** $\dfrac{3x-1}{5} = \dfrac{2x-3}{6}$ **30.** $\dfrac{2x-1}{4} = \dfrac{3x-1}{5}$

31. $\dfrac{2}{x} = \dfrac{4}{5}$ **32.** $\dfrac{3}{x} = \dfrac{9}{7}$ **33.** $\dfrac{1}{2x-1} = \dfrac{5}{8}$ **34.** $\dfrac{1}{3x-1} = \dfrac{7}{6}$

35. $\dfrac{2}{3x+2} = \dfrac{4}{3x}$ **36.** $\dfrac{4}{2x-3} = \dfrac{2}{5x}$ **37.** $\dfrac{5}{x+3} = \dfrac{4}{2x+1}$ **38.** $\dfrac{4}{x+2} = \dfrac{5}{3x+1}$

39. $\dfrac{4}{x} = \dfrac{2x+5}{3}$ **40.** $\dfrac{2}{x} = \dfrac{3x+4}{2}$ **41.** $\dfrac{x}{3} = \dfrac{-2}{x-5}$ **42.** $\dfrac{x}{4} = \dfrac{-2}{x-9}$

43. $\dfrac{3x+2}{4x} = \dfrac{5}{3}$ **44.** $\dfrac{2x+1}{5x} = \dfrac{3}{4}$ **45.** $\dfrac{x+1}{x-3} = \dfrac{x-2}{x+4}$ **46.** $\dfrac{x+3}{x-2} = \dfrac{x-2}{x+1}$

47. $\dfrac{2x+1}{x-3} = \dfrac{x}{x-3}$

48. $\dfrac{3x+8}{x-4} = \dfrac{x}{x-4}$

49. $\dfrac{2x+3}{x+2} = \dfrac{6}{x+5}$

50. $\dfrac{2x+5}{x+2} = \dfrac{6}{x+3}$

51. $\dfrac{3}{x+2} = \dfrac{5}{x+2}$

52. $\dfrac{7}{x+4} = \dfrac{2}{x+4}$

53. $\dfrac{2x+3}{x+3} = \dfrac{x-1}{x-3}$

54. $\dfrac{2x-1}{x+4} = \dfrac{x-3}{x-4}$

Use proportions to find the solutions of the following problems.

55. If 5 apples cost $2, what is the cost of 8 apples?

56. If 4 pears cost $3, what is the cost of 10 pears?

57. If 1.6 kilometers = 1 mile, how many miles are in 10 kilometers?

58. If 1.6 kilometers = 1 mile, how many miles are in 24 kilometers?

59. "Dottie's Drinking Diet" is guaranteed to reduce your weight and open your bladder. You must drink 9 12-ounce glasses of pure sparkling water each day for every 160 pounds of body weight. Sharky, the human fish, weighs 200 pounds. How.many glasses of sparkling water must Sharky drink each day?

60. "Bernal's All Tofu Diet" is guaranteed to decrease your appetite and reduce your weight. Each day you can eat only 2 ounces of tofu for every 75 pounds of body weight. Spud, the human potato, weighs 240 pounds. How many ounces of tofu can Spud eat each day?

61. Red's shrimp bisque recipe includes 2 cups of milk for a 6-person serving. How many cups of milk will be required for a recipe serving 26 people?

62. Donna's red bean soup recipe includes 4 cloves of garlic for a 6-person serving. How many cloves of garlic will be required for a recipe serving 26 people?

63. A car uses 2.5 gallons of gasoline on a 45-mile trip. How many gallons of gasoline will it use on a 108-mile trip?

64. A car uses 3.5 gallons of gasoline on an 84-mile trip. How many gallons of gasoline will it use on a 192-mile trip?

65. Fred, of Fred's Famous Fertilizer, recommends 5 pounds of fertilizer for every 120 sq ft of lawn. How many pounds of fertilizer should be used on a lawn that measures 300 sq ft?

66. Mike, of Mike's Manure City, recommends 8 pounds of manure for every 60 sq ft of lawn. How many pounds of manure should be used on a lawn that measures 300 sq ft?

5.8 Applications of Fractional Equations

In the previous section we studied a special type of word problem in which fractional equations must be used. In this section we continue the study of word problems requiring fractional equations. We will consider three types of problems.

Number Problems

EXAMPLE 5.8.1

What number must be added to both the numerator and denominator of $\frac{8}{11}$ to obtain a fraction equivalent to $\frac{2}{3}$?

SOLUTION Let x = the number to be added. Then the numerator becomes $8 + x$ and the denominator becomes $11 + x$.

We then obtain the fraction $\dfrac{8 + x}{11 + x}$, which is equal to $\dfrac{2}{3}$:

$$\frac{8 + x}{11 + x} = \frac{2}{3} \qquad \text{The LCD is } 3(11 + x).$$

$$\frac{3(11 + x)}{1} \cdot \frac{8 + x}{11 + x} = \frac{3(11 + x)}{1} \cdot \frac{2}{3}$$

$$3(8 + x) = 2(11 + x)$$

$$24 + 3x = 22 + 2x$$

$$24 + x = 22$$

$$x = -2$$

If $x = -2$, then

$$\frac{8 + x}{11 + x} = \frac{8 + (-2)}{11 + (-2)}$$

$$= \frac{6}{9}$$

$$= \frac{2 \cdot 3}{3 \cdot 3}$$

$$= \frac{2}{3}$$

Thus, -2 is the number we are looking for.

EXAMPLE 5.8.2

One number is 3 more than another number. The sum of the reciprocals is $\frac{1}{2}$. Find the two numbers.

SOLUTION

If x = smaller number

then $x + 3$ = larger number

and $\dfrac{1}{x}$ = reciprocal of the smaller number

$\dfrac{1}{x + 3}$ = reciprocal of the larger number

Therefore,

$$\frac{1}{x} + \frac{1}{x + 3} = \frac{1}{2} \qquad \text{The LCD is } 2x(x + 3).$$

$$\frac{2x(x + 3)}{1}\left(\frac{1}{x} + \frac{1}{x + 3}\right) = \frac{2x(x + 3)}{1} \cdot \frac{1}{2}$$

$$\frac{2x(x + 3)}{1} \cdot \frac{1}{x} + \frac{2x(x + 3)}{1} \cdot \frac{1}{x + 3} = \frac{2x(x + 3)}{1} \cdot \frac{1}{2}$$

$$2(x + 3) + 2x = x(x + 3)$$

$$2x + 6 + 2x = x^2 + 3x$$

$$4x + 6 = x^2 + 3x$$
$$0 = x^2 - x - 6$$
$$0 = (x - 3)(x + 2)$$

$x - 3 = 0$ or $x + 2 = 0$

$x = 3$ $x = -2$

If $x = 3$, then $x + 3 = 6$, and $\frac{1}{3} + \frac{1}{6} = \frac{2}{6} + \frac{1}{6} = \frac{3}{6} = \frac{1}{2}$. If $x = -2$, then $x + 3 = 1$, and $\frac{1}{-2} + \frac{1}{1} = -\frac{1}{2} + \frac{2}{2} = \frac{1}{2}$. So there are two solutions. The two numbers are 3 and 6, or -2 and 1.

Work Problems

To solve work problems we use this fundamental rule:

> If it takes 2 hr to mow a yard, then in 1 hr $\frac{1}{2}$ of the yard is mowed. If it takes 5 days to paint a house, then in 1 day $\frac{1}{5}$ of the house is painted. In general, if it takes x days (hours or minutes) to complete a job, then in 1 day (hour or minute) $1/x$ of the job is completed.

EXAMPLE 5.8.3

Carlos can eat an entire cake in 12 minutes. Gilbert can eat the same size cake in 60 minutes. Eating together, how long will it take them to consume a cake?

SOLUTION

Let x = number of minutes required to consume a cake eating together.

The following chart describes the part of the cake eaten in 1 minute by Carlos, Gilbert, and both of them.

	Carlos	**Gilbert**	**Both**
$\dfrac{\text{part}}{\text{minute}}$	$\dfrac{1}{12}$	$\dfrac{1}{60}$	$\dfrac{1}{x}$

Then

$$\begin{pmatrix} \text{Part eaten by} \\ \text{Carlos in 1 min} \end{pmatrix} + \begin{pmatrix} \text{Part eaten by} \\ \text{Gilbert in 1 min} \end{pmatrix} = \begin{pmatrix} \text{Part eaten by} \\ \text{both in 1 min} \end{pmatrix}$$

$$\frac{1}{12} \quad + \quad \frac{1}{60} \quad = \quad \frac{1}{x} \qquad \text{The LCD is } 60x.$$

$$\frac{60x}{1}\left(\frac{1}{12} + \frac{1}{60}\right) = \frac{60x}{1} \cdot \frac{1}{x}$$

$$\frac{60x}{1} \cdot \frac{1}{12} + \frac{60x}{1} \cdot \frac{1}{60} = \frac{60x}{1} \cdot \frac{1}{x}$$

$$5x + x = 60$$

$$6x = 60$$

$$x = 10$$

Eating together, they can consume a cake in 10 minutes.

NOTE ▶ Ten minutes is a reasonable answer, since it should take less time for both to eat the cake than it would take Carlos eating alone.

EXAMPLE 5.8.4

With the faucets open all the way, a bathtub can be filled in 8 minutes. When the drain plug is removed, the tub empties in 12 minutes. How long will it take to fill the bathtub if the drain plug is removed?

SOLUTION Let x = number of minutes to fill the tub with the drain plug removed; then

	Faucets open	**Drain open**	**Both**
$\dfrac{\text{part}}{\text{minute}}$	$\dfrac{1}{8}$	$\dfrac{1}{12}$	$\dfrac{1}{x}$

$$\begin{pmatrix}\text{Part of tub filled}\\\text{in 1 min with}\\\text{drain plugged}\end{pmatrix} - \begin{pmatrix}\text{Part of tub}\\\text{drained}\\\text{in 1 min}\end{pmatrix} = \begin{pmatrix}\text{Part of tub filled}\\\text{in 1 min with}\\\text{drain unplugged}\end{pmatrix}$$

$$\frac{1}{8} \quad - \quad \frac{1}{12} \quad = \quad \frac{1}{x} \qquad \text{LCD is } 24x.$$

$$\frac{24x}{1}\left(\frac{1}{8} - \frac{1}{12}\right) = \frac{24x}{1} \cdot \frac{1}{x}$$

$$\frac{24x}{1} \cdot \frac{1}{8} - \frac{24x}{1} \cdot \frac{1}{12} = \frac{24x}{1} \cdot \frac{1}{x}$$

$$3x - 2x = 24$$

$$x = 24$$

It takes 24 minutes to fill the bathtub with the drain unplugged.

Motion Problems

Motion problems require the distance-rate-time formula used in Chapter 2:

$$D = RT$$

For the problems in this section, we need to solve this formula for T:

$$\frac{D}{R} = \frac{RT}{R}$$

$$\frac{D}{R} = T$$

We also will use the charts developed in Chapter 2.

EXAMPLE 5.8.5

Andy's yacht can travel 10 miles down Moon River in the same time that it can travel 8 miles up Moon River. If the speed of the current in Moon River is 4 mph, what is the speed of Andy's yacht in still water?

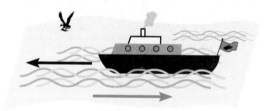

SOLUTION

Let x = speed of Andy's yacht in still water. Traveling down the river will increase his speed by 4 mph. Therefore, his downstream rate is $x + 4$ mph. As he travels up the river, the current will decrease his speed by 4 mph. Therefore, his upstream rate is $x - 4$ mph. So far, the chart for this problem looks like this:

	Distance	Rate	Time
Down Moon River	10	$x + 4$	
Up Moon River	8	$x - 4$	

How do we fill in the time column? Using the formula

$$\frac{D}{R} = T$$

$$\text{Time down} = \frac{\text{Distance down}}{\text{Rate down}} = \frac{10}{x + 4}$$

$$\text{Time up} = \frac{\text{Distance up}}{\text{Rate up}} = \frac{8}{x-4}$$

Thus, the completed chart is as follows:

	Distance	Rate	Time
Down Moon River	10	$x + 4$	$\dfrac{10}{x+4}$
Up Moon River	8	$x - 4$	$\dfrac{8}{x-4}$

We are given that the time down is the same as the time up.

$$\frac{10}{x+4} = \frac{8}{x-4} \qquad \text{The LCD is } (x+4)(x-4).$$

$$\frac{(x+4)(x-4)}{1} \cdot \frac{10}{x+4} = \frac{(x+4)(x-4)}{1} \cdot \frac{8}{x-4}$$

$$10(x - 4) = 8(x + 4)$$
$$10x - 40 = 8x + 32$$
$$2x - 40 = 32$$
$$2x = 72$$
$$x = 36$$

The speed of Andy's yacht in still water is 36 mph.

EXAMPLE 5.8.6

Ramona goes to school during the evening rush hour, averaging 15 mph. After class she goes home along the same route, averaging 45 mph. If she spends a total of 24 min traveling to and from school, what is the distance from Ramona's home to school?

SOLUTION   Let x = distance between home and school.

	Distance	Rate	Time
Home to school	x	15	$\dfrac{x}{15}$
School to home	x	45	$\dfrac{x}{45}$

Since her rates are expressed in miles per *hour*, we need to change her total traveling time from 24 minutes to $\frac{24}{60}$ hour, or $\frac{2}{5}$ hour. Now

$$\left(\begin{array}{c}\text{Time from home}\\ \text{to school}\end{array}\right) + \left(\begin{array}{c}\text{Time from school}\\ \text{to home}\end{array}\right) = \left(\begin{array}{c}\text{Total}\\ \text{time}\end{array}\right)$$

$$\frac{x}{15} \qquad + \qquad \frac{x}{45} \qquad = \qquad \frac{2}{5} \qquad \text{The LCD is 45.}$$

$$\frac{45}{1}\left(\frac{x}{15} + \frac{x}{45}\right) = \frac{45}{1} \cdot \frac{2}{5}$$

$$\frac{45}{1} \cdot \frac{x}{15} + \frac{45}{1} \cdot \frac{x}{45} = \frac{45}{1} \cdot \frac{2}{5}$$

$$3x + x = 18$$

$$4x = 18$$

$$x = 4\frac{1}{2}$$

The distance from Ramona's home to school is $4\frac{1}{2}$ miles.

EXERCISES 5.8

Solve the following problems.

1. What number must be added to both the numerator and denominator of $\frac{5}{17}$ to obtain a fraction equivalent to $\frac{2}{5}$?

2. What number must be added to both the numerator and denominator of $\frac{19}{4}$ to obtain a fraction equivalent to $\frac{7}{2}$?

3. What number must be subtracted from both the numerator and denominator of $\frac{12}{19}$ to obtain a fraction equivalent to $\frac{1}{2}$?

4. What number must be added to the numerator and subtracted from the denominator of $\frac{17}{39}$ to obtain a fraction equivalent to $\frac{3}{4}$?

5. The denominator of a fraction is 4 more than the numerator. If 1 is subtracted from the numerator and 1 is added to the denominator, we obtain a fraction equivalent to $\frac{2}{3}$. Find the original fraction.

6. The denominator of a fraction is 2 less than twice the numerator. If 4 is added to both the numerator and denominator, we obtain a fraction equivalent to $\frac{5}{8}$. Find the original fraction.

7. One number is 8 more than another number. The sum of their reciprocals is $\frac{1}{3}$. Find the two numbers.

8. One number is 15 more than another number. The sum of their reciprocals is $\frac{1}{4}$. Find the two numbers.

9. The sum of two numbers is 10. The sum of their reciprocals is $\frac{5}{8}$. Find the two numbers.

10. The sum of two numbers is 20. The sum of their reciprocals is $\frac{4}{15}$. Find the two numbers.

11. The difference of two numbers is 2. The sum of their reciprocals is $\frac{12}{5}$. Find the two numbers.

12. The difference of two numbers is 4. The sum of their reciprocals is $\frac{12}{7}$. Find the two numbers.

13. Terry can wash his ferry boat in 4 hours. Gerry can wash Terry's ferry in 2 hours. Working together, how long will it take them to wash the ferry?

14. Melissa can clean the bathroom in 1 hour. Michael can clean the bathroom in 3 hours. Working together, how long will it take them to clean the bathroom?

15. Kathy can wash the dishes in 40 min. Melissa takes 60 min to wash the dishes. Working together, how long will it take them to wash the dishes?

16. Necia can wash the church van in 2 hours. When John helps her, they can wash the van in 1 hour 20 minutes. How long does it take John to wash the van if he works alone?

17. Butch can carry all of the merchandise out of the Moss County Pawn Shop in 2 hours. When Bart helps him, they can carry everything out in 1 hour 12 minutes. How long does it take Bart to carry all of the merchandise out if he works alone?

18. Butch can mop the floor of the Moss County Jailhouse mess hall in 15 min less time than it takes Bart. Working together, they can mop the floor in 18 min. How long does it take each of them to mop the floor?

19. The shower room at the Moss County Jailhouse has two drains. After pouring Bernal's All Purpose Drainola into one of the drains, Butch found that it drained the shower twice as fast as the other drain. If together they drain the shower in 4 min, how long would it take each drain to remove the water from the shower?

20. Bart can file through a metal bar three times as fast as Butch can. Together they can file through a metal bar in 9 hours. How long would it take Bart to file through a metal bar by himself?

21. Butch can break down a pile of rocks in 2 hours. Bart can carry rocks to the pile and build it back up in 3 hours. If Butch starts breaking down a pile of rocks and at the same time Bart starts to carry rocks to the pile, how long will it take for the pile of rocks to be broken down?

22. The inlet pipe on Cindy's hot tub can fill the tub in 6 minutes. When the drain is opened, the water drains out in 9 min. One day Cindy's daughter, Larissa, forgot to close the drain before she turned the water on. How long did it take to fill the hot tub with the drain open?

23. Jennifer and Jeremy's pool can be filled in 4 hours. When the drain is opened, the pool empties in 10 hours. How long would it take to fill the pool if the drain were left open?

24. When Juan takes his car to Andele Lube, they first remove the drain plug from the oil pan and drain the oil out in $4\frac{1}{2}$ min. Then they replace the drain plug and pump oil into the crankcase, filling it in $1\frac{1}{2}$ min. However, the last time Juan took his car to Andele Lube, Walker, a new employee, forgot to replace the drain plug before pumping new oil into the crankcase. How long will it take to fill the crankcase with oil?

25. Boudreaux can paddle his mud boat 6 miles up Meloncon Bayou in the same time that it takes him to paddle 10 miles down the bayou. If the speed of the current in Meloncon Bayou is 1 mph, how fast can Boudreaux paddle his mud boat in still water?

26. Jane can swim 1 mile up the Umgawa River in the same time that she can swim 4 miles down the Umgawa River. If the speed of the current in the Umgawa River is 3 mph, how fast can Jane swim in still water?

27. Pat's boat can travel 17 miles down the Strait River in the same time that it can travel 13 miles up the Strait River. If her boat travels 45 mph in still water, what is the speed of the current in the Strait River?

28. Bob's boat can travel 17 km down Mushroom Creek in the same time that it can travel 15 km up Mushroom Creek. If his boat travels 40 km/hr in still water, what is the speed of the current in Mushroom Creek?

29. Ron likes to take his boat 48 miles up Coleman Creek to his favorite fishing hole. The trip up and back takes 5 hours. If the speed of the current in Coleman Creek is 4 mph, what is the speed of Ron's boat in still water?

30. Linda traveled 21 miles up Blue Bayou, then turned around and went 13 miles down Blue Bayou. Her total traveling time was 1 hour 4 minutes. If the speed of her boat in still water is 32 mph, what is the speed of the current in Blue Bayou? (*Hint:* Change 1 hour 4 minutes to hours.)

31. Maria flew to San Antonio averaging 250 mph. She went back home by train averaging 62.5 mph. If she spent a total of 4 hours traveling to and from San Antonio, what is the distance from Maria's home to San Antonio?

32. Bertha flew to Las Vegas averaging 240 mph. She flew home averaging 210 mph. If she spent a total of $12\frac{1}{2}$ hours traveling to and from Las Vegas, what is the distance from Bertha's home to Las Vegas?

33. Maybelline flies her plane 210 miles with the wind. She then flies 165 miles against the wind. Her total flying time is 3 hours. If the wind speed is 15 mph, what is the speed of Maybelline's plane in still air?

34. Petunia, the super pigeon, can fly 75 miles against a 10-mph wind and then turn around and fly back with the wind, spending 4 hours flying. How fast can Petunia fly in still air?

35. Bruce took his horse out at a gallop and covered $10\frac{1}{2}$ miles. He followed the same route back to the stables but traveled 20 mph slower. If he spent a total of 1 hour riding the horse, what was Bruce's speed in each direction?

36. Every morning, Hugo and Janie travel 54 miles into town. In the evening they go back by the same route but travel 15 mph faster. If they spend a total of 3 hours traveling each day, what is their speed in each direction?

Chapter 5 Review

Properties of Fractions

▪ *Reduction property of fractions:*

$$\frac{ax}{ay} = \frac{\cancel{a}x}{\cancel{a}y} = \frac{x}{y} \qquad a \neq 0, \, y \neq 0$$

- *Multiplication property of fractions:*

$$\frac{c}{d} \cdot \frac{n}{m} = \frac{cn}{dm} \qquad d \neq 0, m \neq 0$$

- *Division property of fractions:*

$$\frac{a}{b} \div \frac{x}{y} = \frac{a}{b} \cdot \frac{y}{x} = \frac{ay}{bx} \qquad b \neq 0, x \neq 0, y \neq 0$$

- *Addition property of fractions:*

$$\frac{a}{d} + \frac{b}{d} = \frac{a + b}{d} \qquad d \neq 0$$

- *Subtraction property of fractions:*

$$\frac{a}{d} - \frac{b}{d} = \frac{a - b}{d} \qquad d \neq 0$$

- *Cross-product property of proportions:*

$$\text{If } \frac{a}{b} = \frac{c}{d} \text{ then } \overset{ad}{\frac{a}{b}} \diagup \overset{bc}{\frac{c}{d}} \rightarrow ad = bc \qquad b \neq 0, d \neq 0$$

Review Exercises

5.1 Reduce the following rational expressions to lowest terms.

1. $\dfrac{27x^3y}{18x^2y^4}$
2. $\dfrac{14x^4}{9x^8y^2}$
3. $\dfrac{3x^4 - 6x^3y^2}{24x^9 + 12x^9y^3}$
4. $\dfrac{45x^4y^4 + 15x^3y^5}{5xy^2}$

5. $\dfrac{2x - 6y}{3x^2y - x^3}$
6. $\dfrac{x^2 + 4x + 3}{x^2 - 1}$
7. $\dfrac{2x^3 + 8x^2}{12x^2 + 18x}$
8. $\dfrac{25 - x^2}{x^2 - 10x + 25}$

9. $\dfrac{3x^2 + 5x - 2}{x^2 - 5x - 14}$
10. $\dfrac{6x^2 + xy - 2y^2}{4x^2 - 8xy + 3y^2}$
11. $\dfrac{3 - x}{3x^2 - 5x - 12}$
12. $\dfrac{2x^2 + x - 10}{2 - x}$

13. $\dfrac{6x^3 - 9x^2y - 6xy^2}{6x^2 + 5xy + y^2}$
14. $\dfrac{6x^6y - 6x^5y^2 + 6x^4y^3}{2x^3 - 2x^2y + 2xy^2}$

15. $\dfrac{6x^5 - 22x^4 + 12x^3}{12x^3 + 16x^2 - 16x}$
16. $\dfrac{2ax + 6ay + bx + 3by}{ax + 3ay - 2bx - 6by}$

5.2 Perform the indicated operations, and reduce the answers to lowest terms.

17. $\dfrac{8x^3}{5x} \cdot \dfrac{10x^3}{12x^4}$
18. $\dfrac{7x^2}{4y^5} \cdot \dfrac{6y^3}{49x}$

19. $\dfrac{11y^4}{9x^2} \div \dfrac{2y^8}{3x^6}$
20. $\dfrac{16xy^3}{15x^4} \div \dfrac{12xy^5}{25x^2y}$

21. $\dfrac{5x - 10}{3x - 4} \cdot \dfrac{8 - 6x}{x - 2}$
22. $\dfrac{2x^2 + x}{4x + 8} \cdot \dfrac{2x^2 + 6x}{x^3 + 3x^2}$

23. $\dfrac{3x^2 - 9xy}{6x^2 - 6y^2} \div \dfrac{x^3 - 3x^2y}{xy + y^2}$
24. $\dfrac{5}{3x^2 - 2x} \div \dfrac{10x + 30}{3x^3 - 2x^2}$

25. $\dfrac{x^2 + 10x + 21}{x^2 - 4x + 4} \cdot \dfrac{x^2 - 4}{x^2 + 9x + 14}$

26. $\dfrac{4x^2 - 1}{4x^2 + 22x + 10} \cdot \dfrac{4x^2 + 24x + 20}{2x^2 - 3x + 1}$

27. $\dfrac{2x^2 - 7xy + 3y^2}{2x^2 + xy - y^2} \div \dfrac{3y^2 - 4yx + x^2}{x^2 - xy - 2y^2}$

28. $\dfrac{6x^2 + x - 15}{12x^2 + 17x - 5} \div \dfrac{2x^2 - 11x + 12}{4x^2 - 17x + 4}$

29. $\dfrac{2x^2 + 8x}{2x^2 - 4x - 30} \cdot \dfrac{3x^2 + 7x - 6}{3x^2 - 2x}$

30. $\dfrac{2x^2 - 7xy - 4y^2}{4y^2 + 3yx - x^2} \cdot \dfrac{x^2 - 5xy - 6y^2}{6x^2 + 5xy + y^2}$

31. $\dfrac{x^4 - 6x^3 + 9x^2}{x^2 - x - 6} \div \dfrac{3x^2 - 9x}{9x^2 + 24x + 12}$

32. $\dfrac{12x^2 + 34x + 20}{x^2 - 6x - 16} \div (24x + 20)$

33. $\dfrac{x^2 - 8x - 9}{14x^2 + 63x + 49} \cdot \dfrac{14x^2 + 49x}{x^3 - 5x^2 - 36x} \cdot \dfrac{2x + 8}{2x^2 - 4x - 6}$

34. $\dfrac{x^2 - 16}{3x^2 + 10x + 3} \div \dfrac{x^2 + x - 20}{x^2 + x - 6} \cdot \dfrac{3x^2 - 14x - 5}{2x^2 + 7x - 4}$

35. $\dfrac{x^2 y^2 - 2xy^3 - 3y^4}{2x^2 - xy} \div \dfrac{3x^2 y + 6xy^2 + 3y^3}{2x^2 + 7xy - 4y^2} \div \dfrac{xy + 4y^2}{3x^3 + 33x^2 y + 30xy^2}$

36. $\dfrac{3ax - 12bx + ay - 4by}{6x + 2y} \div \dfrac{2ax - 8bx - 3ay + 12by}{4x^2 - 6xy}$

5.3 Perform the indicated additions and subtractions.

37. $\dfrac{7x}{9} + \dfrac{5x}{9}$

38. $\dfrac{3y}{2x^2} - \dfrac{11y}{2x^2}$

39. $\dfrac{4x + 7}{5y} - \dfrac{2x - 3}{5y}$

40. $\dfrac{4x}{3x + y} + \dfrac{5x}{3x + y}$

41. $\dfrac{6x - 5y}{2x - y} - \dfrac{4x - 4y}{2x - y}$

42. $\dfrac{3x - 7}{x - 4} + \dfrac{11 - 4x}{x - 4}$

43. $\dfrac{2x + 5}{x^2 - x - 12} + \dfrac{x + 4}{x^2 - x - 12}$

44. $\dfrac{2x - 3}{x^2 - 6x + 5} - \dfrac{x + 2}{x^2 - 6x + 5}$

45. $\dfrac{x^2 - 15}{2x - 7} + \dfrac{x^2 - x - 6}{2x - 7}$

46. $\dfrac{x^2 - 5}{6x + 3} - \dfrac{2x + 1}{6x + 3}$

47. $\dfrac{x^2 + 3x - 8}{x^2 - 2x - 15} - \dfrac{2x^2 - 3x + 5}{x^2 - 2x - 15}$

48. $\dfrac{x^2 + 3x + 5}{x^2 + 4x + 4} + \dfrac{x^2 - 3x - 13}{x^2 + 4x + 4}$

49. $\dfrac{6x + 1}{3x - 4} - \dfrac{3x - 13}{4 - 3x}$

50. $\dfrac{4x + 3}{4x^2 - 100} + \dfrac{x - 12}{100 - 4x^2}$

51. $\dfrac{x - 4}{2x + 9} - \dfrac{5x + 2}{2x + 9} + \dfrac{2x - 3}{2x + 9}$

52. $\dfrac{11x + 2}{3x + 3} - \dfrac{2x + 15}{3x + 3} - \dfrac{3x - 1}{3x + 3}$

5.4 Perform the indicated additions and subtractions.

53. $\dfrac{3}{5x} + \dfrac{9}{2x}$

54. $\dfrac{7x}{4y} - \dfrac{5y}{6x}$

55. $\dfrac{3}{8x} - \dfrac{2}{x - 3}$

56. $\dfrac{x}{x - 4} + \dfrac{3x}{x + 2}$

57. $\dfrac{5}{2x + 1} + \dfrac{2x}{x - 3}$

58. $\dfrac{x - 5}{3x} - \dfrac{x}{x + 2}$

59. $\dfrac{3x - 4}{x + 2} - \dfrac{3x + 1}{x - 2}$

60. $\dfrac{5x + 2}{x - 4} + \dfrac{x + 3}{x + 1}$

61. $\dfrac{3}{x^2 + 3x + 2} + \dfrac{2x + 1}{2x^2 + 7x + 6}$

62. $\dfrac{18}{x^2 - 2x - 8} - \dfrac{15}{x^2 - 3x - 4}$

63. $\dfrac{x + 5}{x^2 - 4x + 3} - \dfrac{10 - 2x}{x^2 - 5x + 6}$

64. $\dfrac{2}{3x^2 + 22x + 7} + \dfrac{x}{9x^2 - 1}$

65. $\dfrac{x + 4}{2x^2 - 7x + 5} - \dfrac{3}{x^2 - 2x + 1}$

66. $\dfrac{4x - 3}{x^2 + 3x + 2} - \dfrac{6x - 1}{2x^2 + 5x + 3}$

67. $\dfrac{1}{x - 3} + \dfrac{x}{3x - 12} - \dfrac{2x - 7}{x^2 - 7x + 12}$

68. $\dfrac{x}{2x + 5} - \dfrac{3}{2x - 5} + \dfrac{12x}{4x^2 - 25}$

5.5 Simplify the following complex fractions.

69. $\dfrac{\dfrac{5x^3}{12y^2}}{\dfrac{15x}{8y^8}}$

70. $\dfrac{\dfrac{2y^4}{9x^3}}{6xy^2}$

71. $\dfrac{\dfrac{2}{x - 3}}{\dfrac{4}{x^2 - 9}}$

72. $\dfrac{\dfrac{9}{x^2 - 6x - 16}}{\dfrac{18}{x - 8}}$

73. $\dfrac{\dfrac{4x^2 - 1}{15}}{10x - 5}$

74. $\dfrac{\dfrac{16x^3}{x - 3}}{\dfrac{24x^6}{x + 1}}$

75. $\dfrac{\dfrac{7x^4 - 14x^3}{x + 3}}{\dfrac{21x^3 - 42x^2}{x - 5}}$

76. $\dfrac{\dfrac{3x^2 + 15x}{2x - 12}}{\dfrac{x^3 + 5x^2}{8x - 48}}$

77. $\dfrac{\dfrac{5}{6} - \dfrac{1}{3}}{\dfrac{5}{12} + \dfrac{1}{3}}$

78. $\dfrac{2 - \dfrac{5}{3x}}{x - \dfrac{5}{6}}$

79. $\dfrac{\dfrac{3}{5x} - \dfrac{5x}{3}}{x - \dfrac{3}{5}}$

80. $\dfrac{\dfrac{3}{x - 2} - 1}{x - 5}$

81. $\dfrac{2x + 7}{\dfrac{4}{7} - \dfrac{7}{x^2}}$

82. $\dfrac{1 - \dfrac{14}{x} + \dfrac{48}{x^2}}{1 - \dfrac{3}{x} - \dfrac{18}{x^2}}$

83. $\dfrac{\dfrac{1}{2} - \dfrac{1}{x + 1}}{\dfrac{2}{x + 1} - 1}$

84. $\dfrac{\dfrac{1}{x} - \dfrac{3}{x + 4}}{\dfrac{2}{x} - \dfrac{5}{x + 4}}$

5.6 Solve for x in the following equations.

85. $\dfrac{3x}{5} - \dfrac{x}{2} = \dfrac{7}{10}$

86. $\dfrac{x}{9} + 1 = \dfrac{x}{3}$

87. $\dfrac{3x + 2}{6} - \dfrac{x - 2}{4} = \dfrac{1}{3}$

88. $\dfrac{x + 3}{6} = \dfrac{2x + 5}{8}$

89. $\dfrac{3}{4x} + \dfrac{1}{x} = \dfrac{7}{12}$

90. $x - \dfrac{3}{x} = \dfrac{1}{2}$

91. $\dfrac{x + 4}{x - 1} = \dfrac{6}{x - 3}$

92. $\dfrac{x + 4}{x - 1} = \dfrac{3x - 4}{x + 1}$

93. $\dfrac{x}{x - 3} - 2 = \dfrac{1}{x - 3}$

94. $\dfrac{3}{4x} + \dfrac{2}{x + 3} = \dfrac{9}{4x^2 + 12x}$

95. $\dfrac{x}{x + 1} - \dfrac{5}{x - 3} = \dfrac{-20}{x^2 - 2x - 3}$

96. $\dfrac{2x}{x + 2} - \dfrac{x}{x + 1} = \dfrac{4}{x^2 + 3x + 2}$

5.7 Write each ratio as a fraction in lowest terms.

97. 16 to 12   **98.** $10x^2$ to $14x$ **99.** 24 feet to 30 feet   **100.** 3 years to 8 months

Solve for x in the following proportions.

101. $\dfrac{2x + 5}{7} = \dfrac{1}{3}$ **102.** $\dfrac{5x}{4} = \dfrac{3x + 2}{2}$ **103.** $\dfrac{4}{3x - 2} = \dfrac{2}{5}$ **104.** $\dfrac{5}{x - 2} = \dfrac{3}{2x - 1}$

105. $\dfrac{-1}{2x - 7} = \dfrac{x}{5}$ **106.** $\dfrac{5x - 2}{4x} = \dfrac{3}{2}$ **107.** $\dfrac{x + 3}{x + 2} = \dfrac{x - 6}{x - 10}$ **108.** $\dfrac{x - 1}{x + 2} = \dfrac{2x + 1}{3x + 6}$

Use proportions to find the solutions of the following problems.

109. If 3 pounds of granola cost $4.77, how much do 5 pounds of granola cost?

110. "Bernal's All Purpose Drainola Diet" is guaranteed to reduce your weight and keep you regular. Each day you need to eat 3 ounces of Drainola for every 40 pounds of body weight. Frijole, the human bean, weighs 220 pounds. How many ounces of Drainola does Frijole need to eat each day?

5.8 Solve the following problems.

111. What number must be added to both the numerator and denominator of $\frac{2}{7}$ to obtain a fraction equivalent to $\frac{2}{3}$?

112. The difference of two numbers is 8. The sum of their reciprocals is $\frac{1}{3}$. Find the two numbers.

113. Peggy can paint her living room in 4 hours. If Robin helped her, it would take them 3 hours to paint the living room. How long would it take Robin to paint the living room if he painted alone?

114. Jethro carries water from the Moss County Creek to his washbasin. He can fill it up in 12 minutes. One day he accidentally shoots a hole in his washbasin that allows the water to drain out in 48 minutes. How long does it take Jethro to fill up his washbasin?

115. Running Lizard can paddle his canoe 8 miles up the Tecaboca River in the same time that he can paddle 20 miles down the river. If the speed of the current in the Tecaboca River is 3 mph, how fast can Running Lizard paddle his canoe in still water?

116. Tom ran 4 miles and stopped at a donut shop. He walked home following the same route, but traveled 5 mph slower. If he spent a total of 1 hour 50 minutes running and walking, what was his speed in each direction?

Chapter 5 Test   (You should be able to complete this test in 60 minutes.)

1. Reduce to lowest terms.

$$\frac{5x^2y - 15x^3}{3x^2 + 2xy - y^2}$$

2. Simplify the complex fraction.

$$\frac{\dfrac{3}{x + 1} + 1}{\dfrac{x}{4} - \dfrac{4}{x}}$$

Perform the indicated operations, and reduce the answers to lowest terms.

3. $\dfrac{3x + 10}{x^2 + 11x + 30} - \dfrac{2x + 5}{x^2 + 11x + 30}$

4. $\dfrac{x^3 - 16x}{x^2 - 6x + 8} \cdot \dfrac{6x^2 - 3x - 18}{6x^2 + 9x}$

5. $\dfrac{x + 1}{2x^2 + 9x + 10} + \dfrac{1}{x^2 + 5x + 6}$

6. $\dfrac{3x^2 - 4x + 1}{3x^2 + 8x - 3} \div \dfrac{3x^2 - 18x + 15}{3x - 9}$

Solve for x in the following equations.

7. $\dfrac{1}{6x} - 2 = \dfrac{5}{2x}$

8. $\dfrac{2x}{x + 4} + \dfrac{3}{x + 2} = \dfrac{16}{x^2 + 6x + 8}$

9. Write the following ratio as a fraction in lowest terms.

   42 inches to 66 inches

10. Solve for x in the following proportion.

   $\dfrac{x}{4} = \dfrac{2}{x - 7}$

11. Use a proportion to find the solution of the following problem.

   A car uses 5.5 gallons of gasoline on a 154-mile trip. How many gallons of gasoline will it use on a 175-mile trip?

12. Solve the following problem.

   Janice can fold a basket of clothes twice as fast as her daughter Joy can. Together they can fold a basket of clothes in 20 minutes. How long does it take Janice to fold a basket of clothes by herself?

Test Your Memory

These problems review Chapters 1–5.

I. Perform the indicated operations. When necessary, use the rules governing the order of operations.

1. $\frac{2}{3} + \frac{3}{4} - \frac{5}{6}$

2. $2^0 + 2^{-1} + 2^{-2}$

3. $\left(\frac{1}{2} \cdot \frac{3}{5}\right)^2$

4. $10 - 5[-4 - (2 - 4)]$

II. Completely factor the following polynomials.

5. $x^2 - 13x + 12$

6. $16x^2 - 81y^2$

7. $8x^2 + 6xy - 4x - 3y$

8. $12x^2 + 34xy + 24y^2$

III. Find the solution set of each of the following inequalities, and graph the solution set on a number line.

9. $-1 \le 2x - 3 \le 13$

10. $\frac{1}{3}x + \frac{3}{2} < \frac{3}{4}x - \frac{1}{6}$

IV. Find the solutions of the following equations.

11. $8 - 2(3x + 1) = 5(2 - x) + 7$

12. $2(x + 3) + 4(2x - 1) = 7(x + 3)$

13. $(3x + 1)(x - 2) = 48$

14. $4x^3 - 2x^2 - 42x = 0$

15. $\dfrac{2}{5x} - \dfrac{1}{2x} = \dfrac{3}{10}$

16. $\dfrac{1}{x^2} + \dfrac{1}{4x} = \dfrac{1}{8}$

17. $\dfrac{5}{x} + \dfrac{3}{x + 2} = \dfrac{1}{2x}$

18. $\dfrac{x}{x + 2} - \dfrac{6}{x + 4} = \dfrac{12}{x^2 + 6x + 8}$

19. $\dfrac{4}{x + 1} = \dfrac{2}{3x - 1}$

20. $\dfrac{3}{4} = \dfrac{6x}{2x - 7}$

21. $\dfrac{x}{9} = \dfrac{-2}{x - 9}$

22. $\dfrac{2x - 1}{5} = \dfrac{2}{4x - 3}$

V. Simplify each of the following. Be sure to reduce your answer to lowest terms.

23. $\left(\dfrac{x^{-2}y^{-3}}{x^{-1}y^{-8}}\right)^{-2}$

24. $\dfrac{(2x^2y^4)^{-2}}{(2^{-1}x^3y^{-1})^{-3}}$

25. $\dfrac{3x^3 - 12x^2}{x^2 - 16}$

26. $\dfrac{6x^2 + 7xy - 3y^2}{6x^2 - 11xy + 3y^2}$

27. $\dfrac{\dfrac{3}{x} + \dfrac{1}{2}}{\dfrac{2}{x} + \dfrac{1}{3}}$

28. $\dfrac{\dfrac{x}{2} - 1}{\dfrac{4}{3} - \dfrac{8}{3x}}$

29. $\dfrac{\dfrac{1}{x} + \dfrac{2}{x + 1}}{\dfrac{1}{x} + 3}$

30. $\dfrac{\dfrac{-2}{x + 4} + 1}{\dfrac{x}{3} - \dfrac{4}{3x}}$

VI. Perform the indicated operations.

31. $-2x^3y(5x^2 + 3xy - 4y^4)$

32. $(x^2 + 7x - 4)(x^2 + 3)$

33. $(8x^3 - 24x^2 + 24x - 32) \div (2x - 5)$

34. $(7x + 2y)(7x - 2y)$

35. $\dfrac{x^3 - 4x^2}{2x^2 - 7x + 6} \cdot \dfrac{2x - 3}{5x^2 + 10x}$

36. $\dfrac{x^2 + 2x + 1}{x^2 - x - 6} \cdot \dfrac{2x^4 - 6x^3}{6x^3 + 6x^2}$

37. $\dfrac{9x^2 - 3x}{3x^2 - 7x + 2} \div \dfrac{12x + 9}{4x^2 - 5x - 6}$

38. $\dfrac{x^3 + 5x^2 - 9x - 45}{4x^2 - 8x - 5} \div \dfrac{x^2 + 8x + 15}{2x - 5}$

39. $\dfrac{2x^2 - 5x}{x^2 - 9} + \dfrac{4x - 21}{x^2 - 9}$

40. $\dfrac{3x^2 + 4x - 1}{x^2 + 2x - 8} - \dfrac{2x^2 + 7x - 3}{x^2 + 2x - 8}$

41. $\dfrac{7}{x + 3} - \dfrac{5}{x - 4}$

42. $\dfrac{x + 1}{2x^2 - 3x - 2} + \dfrac{x - 5}{x^2 + x - 6}$

VII. Use algebraic expressions to find the solutions of the following problems.

43. Laverne has $1.40 in nickels and quarters. The number of nickels is 1 more than four times the number of quarters. How many quarters and how many nickels does Laverne have?

44. Lord Thompson deodorant is 44% alcohol. Sir Lancelot deodorant is 35% alcohol. How many ounces of each deodorant must be used to make 45 ounces of deodorant that is 40% alcohol?

45. The perimeter of a rectangle is 30 millimeters. The length is 3 millimeters less than twice the width. What are the dimensions of the rectangle?

46. The base of a triangle is 4 feet more than $\frac{1}{2}$ of the height. Find the base and the height if the area of the triangle is 21 square feet.

47. Luke can eat 15 hard-boiled eggs in 25 minutes. How many hard-boiled eggs can he eat in 1 hour? (We are assuming that Luke eats at a constant rate and that he does not explode.)

48. Nadine's nectarine carrot soup recipe includes $\frac{1}{4}$ teaspoon of orange zest for a 4-person serving. How much orange zest will be required for a recipe serving 14 people?

49. Joe Bob, Jr., takes three times as long as his father Joe Bob, Sr., to wash the dinner dishes. Together they can wash the dinner dishes in 36 minutes. How long does it take Joe Bob, Sr., to wash the dinner dishes by himself?

50. A canal has a current of 2 mph. Find the speed of Reuben's boat in still water if it goes 20 miles downstream in the same time as it goes 16 miles upstream.

CHAPTER  6

Graphs of Linear Equations and Inequalities in Two Variables

6.1 Solutions of Linear Equations in Two Variables

In Chapter 2 we solved linear equations in one variable, such as $5x - 2 = 13$. We discovered that this type of equation has one solution. The solution of this equation is $x = 3$ because substituting 3 for x yields a true statement:

$$5 \cdot 3 - 2 \stackrel{?}{=} 13$$
$$15 - 2 \stackrel{?}{=} 13$$
$$13 = 13$$

In this chapter we will study linear equations in two variables, such as $5x - 2y = 13$. We will find that this type of equation has many solutions. First, let us generalize the type of equation we are talking about.

Linear Equation in Two Variables

Any equation that can be written in the form

$$Ax + By = C$$

where A, B, and C are real numbers and with the property that not both A and B are zero, is called a **linear equation in two variables**.

Returning to the equation $5x - 2y = 13$, we see that a solution must have two numbers, one for x and one for y. For example, if we substitute 7 for x and 11 for y, we get a true statement:

$$5 \cdot 7 - 2 \cdot 11 \stackrel{?}{=} 13$$
$$35 - 22 \stackrel{?}{=} 13$$
$$13 = 13$$

We usually write a solution in parentheses—in this case, $(7, 11)$. This is called an **ordered pair**. The number on the left is called the **first coordinate**, or **x-coordinate**, and is always substituted for x. The number on the right is called the **second coordinate**, or **y-coordinate**, and is always substituted for y. The order in which the pair of numbers is written is therefore important. $(11, 7)$ is a different ordered pair and is not a solution of the above equation, for if we substitute 11 for x and 7 for y the left side of the equation does not equal the right side:

$$5 \cdot 11 - 2 \cdot 7 \stackrel{?}{=} 13$$
$$55 - 14 \stackrel{?}{=} 13$$
$$41 \neq 13$$

NOTE ▶ There are an infinite number of solutions to the equation $5x - 2y = 13$. Some other solutions are listed below.

$$(5, 6) \text{ since } 5 \cdot 5 - 2 \cdot 6 = 25 - 12 = 13$$
$$\left(2, -\tfrac{3}{2}\right) \text{ since } 5 \cdot 2 - 2\left(-\tfrac{3}{2}\right) = 10 + 3 = 13$$
$$(-1, -9) \text{ since } 5(-1) - 2(-9) = -5 + 18 = 13$$

EXAMPLE 6.1.1

For each equation, determine whether the given ordered pair is a solution.

1. $3x + y = 8$, $(3, -1)$

   To determine if $(3, -1)$ is a solution, substitute 3 for x and -1 for y:

   $$3 \cdot 3 + (-1) \stackrel{?}{=} 8$$
   $$9 + (-1) \stackrel{?}{=} 8$$
   $$8 = 8$$

   Since this is a true statement, $(3, -1)$ is a solution.

2. $x - 4y = -12$, $(0, 3)$

   $$0 - 4 \cdot 3 \stackrel{?}{=} -12$$
   $$0 - 12 \stackrel{?}{=} -12$$
   $$-12 = -12$$

   Since this is a true statement, $(0, 3)$ is a solution.

3. $7x + 4y = 9$, $(2, -1)$

   $$7 \cdot 2 + 4(-1) \stackrel{?}{=} 9$$
   $$14 + (-4) \stackrel{?}{=} 9$$
   $$10 \neq 9$$

   Since the left side does not equal the right side, $(2, -1)$ is not a solution.

4. $x = 4$, $(4, -2)$

Since the equation has no y term, how are we going to check the ordered pair? We can rewrite the equation as $x + 0y = 4$. Now we can substitute 4 for x and -2 for y:

$$4 + 0(-2) \overset{?}{=} 4$$
$$4 + 0 \overset{?}{=} 4$$
$$4 = 4$$

Since this is a true statement, $(4, -2)$ is a solution. Actually, any ordered pair with an x-coordinate of 4 is a solution to this equation. This equation says that y can be any number, but x is always 4.

By now you might be wondering how you are supposed to find solutions of linear equations in two variables. We can find a solution by substituting a number for one variable and then solving the resulting equation for the other variable. For example, suppose we substitute 3 for x in the equation $5x + 2y = 7$:

$$5 \cdot 3 + 2y = 7 \qquad \text{We can solve this equation}$$
$$15 + 2y = 7 \qquad \text{for } y.$$
$$2y = -8$$
$$y = -4$$

Thus, when x is 3, y is -4 and $(3, -4)$ is a solution of $5x + 2y = 7$. We can also substitute for y. Let $y = 6$.

$$5x + 2 \cdot 6 = 7 \qquad \text{Solve for } x.$$
$$5x + 12 = 7$$
$$5x = -5$$
$$x = -1$$

Therefore, $(-1, 6)$ is a solution of $5x + 2y = 7$.

EXAMPLE 6.1.2

Find solutions for the equation $4x - 9y = 2$ using the indicated values for x or y.

1. $(-4, \quad)$ Substitute -4 for x and solve for y.

$$4(-4) - 9y = 2$$
$$-16 - 9y = 2$$
$$-9y = 18$$
$$y = -2$$

Therefore, $(-4, -2)$ is a solution.

2. $(\ \ , 2)$ Substitute 2 for y and solve for x.

$$4x - 9 \cdot 2 = 2$$
$$4x - 18 = 2$$
$$4x = 20$$
$$x = 5$$

Therefore, $(5, 2)$ is a solution.

3. $\left(\frac{1}{2}, \ \right)$ Substitute $\frac{1}{2}$ for x and solve for y.

$$4 \cdot \tfrac{1}{2} - 9y = 2$$
$$2 - 9y = 2$$
$$-9y = 0$$
$$y = 0$$

Therefore, $\left(\frac{1}{2}, 0\right)$ is a solution.

Given a linear equation in two variables, it is sometimes useful to solve the equation for one of the variables in terms of the other variable. We studied this process in Section 2.6 where we solved literal equations for a specified variable. Let us see how this can make finding solutions of linear equations easier.

EXAMPLE 6.1.3

Solve each equation for the specified variable, then find solutions using the indicated values for x or y.

1. $5x - 3y = 15$, for y; $x = -6, 0, 3$

We first solve for y in terms of x:

$$5x - 3y = 15$$
$$-3y = -5x + 15$$
$$y = \frac{-5x + 15}{-3}$$

or $y = \frac{5}{3}x - 5$

We now substitute -6, 0, and 3 for x in the bottom equation:

$y = \frac{5}{3}(-6) - 5 = -10 - 5 = -15$; $(-6, -15)$ is a solution.

$y = \left(\frac{5}{3}\right) \cdot 0 - 5 = 0 - 5 = -5$; $(0, -5)$ is a solution.

$y = \left(\frac{5}{3}\right) \cdot 3 - 5 = 5 - 5 = 0$; $(3, 0)$ is a solution.

2. $6x + 5y - 2 = 0$, for x; $y = -2, 0, 4$

We first solve for x in terms of y:

$$6x + 5y - 2 = 0$$
$$6x = -5y + 2$$
$$x = \frac{-5y + 2}{6}$$

or $\qquad x = -\frac{5}{6}y + \frac{1}{3}$

We now substitute -2, 0, and 4 for y in the bottom equation:

$x = -\frac{5}{6}(-2) + \frac{1}{3} = \frac{5}{3} + \frac{1}{3} = \frac{6}{3} = 2$; $(2, -2)$ is a solution.

$x = \left(-\frac{5}{6}\right) \cdot 0 + \frac{1}{3} = 0 + \frac{1}{3} = \frac{1}{3}$; $\left(\frac{1}{3}, 0\right)$ is a solution.

$x = \left(-\frac{5}{6}\right) \cdot 4 + \frac{1}{3} = -\frac{10}{3} + \frac{1}{3} = -\frac{9}{3} = -3$; $(-3, 4)$ is a solution.

There are many applications of equations in two variables, that is, situations where two quantities vary. Let us examine an application of a linear equation in two variables.

EXAMPLE 6.1.4

A farmer wants to enclose a rectangular field with 1600 feet of fence. If the width is 300 feet, what is the length? If the length is 650 feet, what is the width?

SOLUTION

The perimeter of a rectangle is given by the formula $P = 2L + 2W$.

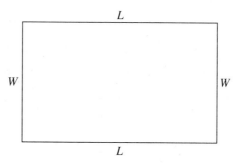

In this case the farmer wants the perimeter to be 1600 feet. Therefore, the equation relating the length and the width is

$$1600 = 2L + 2W$$

When the width is 300 feet, we substitute 300 for W and solve for L:

$$1600 = 2L + 2 \cdot 300$$
$$1600 = 2L + 600$$
$$1000 = 2L$$
$$500 = L$$

When the width is 300 feet, the length is 500 feet. When the length is 650 feet, we substitute 650 for L and solve for W:

$$1600 = 2 \cdot 650 + 2W$$
$$1600 = 1300 + 2W$$
$$300 = 2W$$
$$150 = W$$

When the length is 650 feet, the width is 150 feet.

EXERCISES 6.1

For each equation, determine whether the given ordered pair is a solution.

1. $x + 4y = 7$, (3, 1)

2. $x - 3y = 5$, (11, 2)

3. $5x - y = 3$, (2, 7)

4. $4x + y = 9$, (3, −3)

5. $2x - 7y = 8$, (5, 1)

6. $5x + 4y = 6$, (−2, 1)

7. $7x + 3y = 3$, (6, −13)

8. $4x - 3y = -4$, (14, 20)

9. $12x - 5y + 9 = 0$, (3, 9)

10. $8x + 11y - 5 = 0$, (−2, 1)

11. $3x + 8y + 19 = 0$, (1, −2)

12. $6x - 13y - 8 = 0$, (10, 4)

13. $5x - 2y = 7$, $\left(2, \frac{3}{2}\right)$

14. $7x + 6y = -3$, $\left(-1, \frac{2}{3}\right)$

15. $\frac{7}{2}x - 4y = 2$, (−4, −2)

16. $2x - \frac{5}{3}y = 4$, (4, 3)

17. $\frac{3}{4}x + \frac{5}{6}y = 2$, (−2, 3)

18. $-\frac{7}{10}x + \frac{6}{5}y = -3$, (6, 1)

19. $2.4x - 3.5y = 5$, (5, 2)

20. $1.6x + 4.2y = 20$, (−4, 2)

21. $x = 3$, (3, 7)

22. $x + 5 = 0$, (5, 2)

23. $y - 2 = 0$, (8, −2)

24. $y = -4$, (1, −4)

Find solutions for each equation using the indicated values for x or y.

25. $2x - 5y = 10$; (−5,), (0,), (, 0), (, 2)

26. $3x + 8y = 24$; (−8,), (0,), (, 0), (, −3)

27. $7x + 4y = 14$; (−2,), (0,), (, 0), (, −7)

28. $4x - 5y = 10$; (5,), (0,), (, 0), (, −6)

29. $5x + 3y = -8$; (2,), (−1,), (, 4), (, −11)

30. $2x + 7y = -9$; (−1,), (6,), (, 1), (, −5)

31. $6x - y = -10$; (1,), (−2,), (, −8), (, 4)

32. $8x - y = -4$; (2,), (−3,), (, 12), (, −4)

33. $x + 3y = 5$; $(-1, \quad)$, $(11, \quad)$, $(\quad, 4)$, $(\quad, -1)$

34. $x + 7y = 12$; $(-9, \quad)$, $(5, \quad)$, $(\quad, -1)$, $(\quad, 2)$

35. $x + y + 6 = 0$; $\left(-\frac{5}{2}, \quad\right)$, $\left(\frac{4}{3}, \quad\right)$, $(\quad, 2)$, $\left(\quad, -\frac{7}{4}\right)$

36. $x + y + 9 = 0$; $(3, \quad)$, $\left(-\frac{8}{3}, \quad\right)$, $\left(\quad, -\frac{11}{2}\right)$, $\left(\quad, \frac{3}{4}\right)$

37. $2x - 8y - 5 = 0$; $\left(-\frac{3}{2}, \quad\right)$, $(4, \quad)$, $\left(\quad, -\frac{1}{2}\right)$, $\left(\quad, \frac{1}{3}\right)$

38. $3x - 6y - 8 = 0$; $\left(\frac{2}{3}, \quad\right)$, $(-2, \quad)$; $\left(\quad, \frac{2}{3}\right)$, $(\quad, -2)$

39. $\frac{1}{2}x + y = 5$; $(4, \quad)$, $(-6, \quad)$, $(\quad, 4)$, $(\quad, 10)$

40. $\frac{2}{3}x - y = 4$; $(6, \quad)$, $(-9, \quad)$, $(\quad, -2)$, $(\quad, 2)$

41. $3x + \frac{3}{4}y = 9$; $\left(\frac{3}{2}, \quad\right)$, $(-1, \quad)$, $(\quad, 8)$, $\left(\quad, -\frac{8}{5}\right)$

42. $5x + \frac{3}{2}y = 4$; $\left(-\frac{7}{4}, \quad\right)$, $(2, \quad)$, $(\quad, 6)$, $\left(\quad, \frac{8}{3}\right)$

43. $2.7x - 4y = 5.4$; $(2, \quad)$, $(10, \quad)$, $(\quad, -1.35)$, $(\quad, 6.75)$

44. $3x - 3.6y = 10.8$; $(3, \quad)$, $(7.2, \quad)$, $(\quad, -3)$, $(\quad, 6)$

45. $x = 3$; $(3, \quad)$, $(\quad, 4)$, $(\quad, -2)$, $(\quad, 11)$

46. $x = -7$; $(-7, \quad)$, $(\quad, -5)$, $(\quad, 0)$, $(\quad, 4)$

47. $y = -8$; $(3, \quad)$, $(-5, \quad)$, $(0, \quad)$, $(\quad, -8)$

48. $y = 2$; $\left(\frac{1}{2}, \quad\right)$, $(-7, \quad)$, $(3, \quad)$, $(\quad, 2)$

Solve each equation for y, then find solutions using the indicated values for x.

49. $3x + y = 8$; $x = -2, 0, 3$

50. $4x + y = 5$; $x = -3, 0, 4$

51. $5x - y = -4$; $x = -4, 0, 1$

52. $3x - y = -2$; $x = -5, 0, 3$

53. $2x + 5y = 10$; $x = 5, 0, -10$

54. $5x + 4y = 20$; $x = -4, 0, 8$

55. $-6x + 3y = 5$; $x = -\frac{1}{2}, 2, \frac{1}{6}$

56. $-8x + 2y = 3$; $x = -\frac{3}{4}, -2, \frac{7}{8}$

57. $4x - 7y + 7 = 0$; $x = -3, 7, 14$

58. $3x - 5y + 10 = 0$; $x = 2, -5, 5$

59. $\frac{2}{3}x - \frac{1}{2}y = 5$; $x = 6, \frac{3}{4}, -3$

60. $\frac{3}{4}x - \frac{1}{3}y = 2$; $x = 8, \frac{4}{3}, -4$

61. $3y + 2 = 0$; $x = 5, 0, -2$

62. $4y - 5 = 0$; $x = -4, 0, 7$

Solve each equation for x, then find solutions using the indicated values for y.

63. $x + 5y = 2$; $y = 1, 0, -1$

64. $x + 3y = 4$; $y = 2, 0, -2$

65. $-x + 4y = 3$; $y = -2, 0, 3$

66. $-x + 7y = 5$; $y = -1, 0, 3$

67. $3x + 4y = 12$; $y = 3, 0, -6$

68. $2x + 3y = 6$; $y = 4, 0, -2$

69. $2x - 8y = 5$; $y = \frac{1}{8}, 1, -\frac{3}{4}$

70. $3x - 6y = 7$; $y = -\frac{1}{2}, -2, \frac{1}{3}$

71. $-4x + 9y + 4 = 0$; $y = -1, -4, 4$

72. $-5x + 6y + 5 = 0$; $y = -2, -10, 5$

73. $-\frac{1}{4}x + \frac{3}{2}y = 2$; $y = \frac{2}{3}, 2, -1$

74. $-\frac{1}{5}x + \frac{7}{10}y = 3$; $y = \frac{8}{7}, 6, -2$

75. $3x - 4 = 0$; $y = 8, 0, -\frac{7}{2}$

76. $6x + 2 = 0$; $y = 3, 0, -\frac{2}{5}$

Find a linear equation in two variables for each of problems 77–80, then use that equation to answer the questions.

77. A rancher wants to enclose a rectangular field with 2200 feet of fence. If the width is 430 feet, what is the length? If the length is 850 feet, what is the width?

78. A developer wants to enclose a rectangular lot with 1850 yards of fence. If the width is 320 yards, what is the length? If the length is 775 yards, what is the width?

79. A farmer wants to enclose a rectangular field and divide it in two, using 2500 feet of fence, as shown in Figure 6.1.1. If the width is 350 feet, what is the length? If the length is 830 feet, what is the width?

80. Ray is planning to have a rectangular garden next to his house. He will run 48 feet of fence around three sides, but not along the house, as shown in Figure 6.1.2. If the width is 15 feet, what is the length? If the length is 24 feet, what is the width?

81. The relationship between degrees Fahrenheit and degrees Celsius is given by the equation $5F - 9C = 160$. Find the Celsius equivalent for a temperature of 32°F, the freezing point of water. Find the Fahrenheit equivalent for a temperature of 100°C, the boiling point of water.

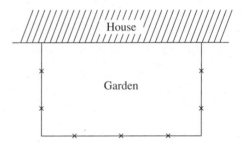

Figure 6.1.2

Figure 6.1.1

6.2 The Cartesian Coordinate System and Graphing Linear Equations

Have you ever had to find a street using a city map? You look up the name of the street in the index and find two coordinates, usually given in the form of a letter and a number. In the portion of the map shown in Figure 6.2.1, Easy Street would have coordinates D2. Letters and numbers are used to distinguish between the horizontal and vertical scales.

In algebra there is a "map" called the **Cartesian coordinate system**. It is named in honor of René Descartes (1596–1650), who formalized the following system. We take a horizontal real number line and lay it across a vertical real number line so that one zero point is over the other. This point of intersection is called the **origin**. The horizontal number line is usually called the **x-axis**, and the vertical number line is called the **y-axis**. (See Figure 6.2.2.)

With this system we can express any point in the plane in terms of its signed distances from the axes, which is given by an ordered pair of real numbers. In Figure 6.2.3 we can locate point A by starting at the origin and going 2 units to the left along the x-axis and 4 units up parallel to the y-axis.

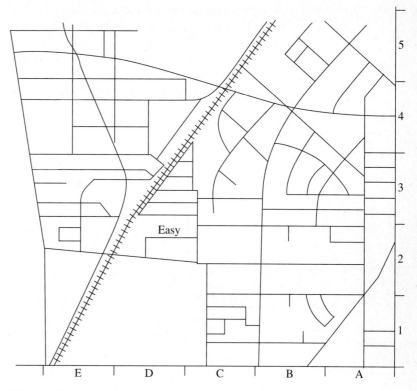

Easy

E   D   C   B   A

Figure 6.2.1

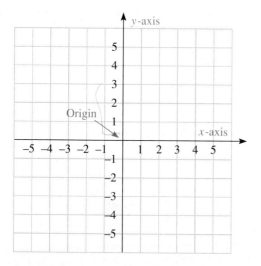

Figure 6.2.2

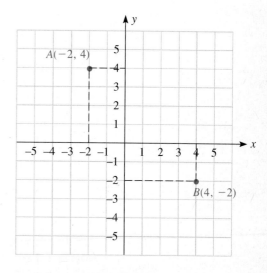

Figure 6.2.3

We associate with point A the ordered pair $(-2, 4)$. We can locate point B by going 4 units to the right along the x-axis and 2 units down parallel to the y-axis. We associate with point B the ordered pair $(4, -2)$. Notice that $(-2, 4)$ and $(4, -2)$ are different ordered pairs and that A and B are different points. When we draw a point associated with an ordered pair (x, y), we say we are **graphing the ordered pair**, or **plotting the point**.

NOTE ▶ The first or x- coordinate is a signed horizontal distance whereas the second or y- coordinate is a signed vertical distance. (See Figure 6.2.4.)

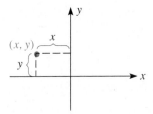

Figure 6.2.4

EXAMPLE 6.2.1

Graph the following ordered pairs.

$$A = (2, 3), \quad B = (0, 4), \quad C = (-1, 2), \quad D = (-2, 0),$$
$$E = (-2, -2), \quad F = (0, -3), \quad G = (2, -3), \quad H = (3, 0)$$

(See Figure 6.2.5.)

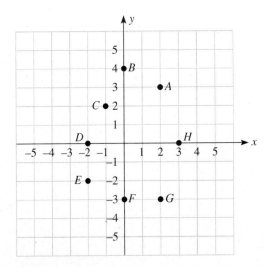

Figure 6.2.5

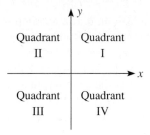

Figure 6.2.6

The *x*-axis and *y*-axis partition the plane into four distinct regions called **quadrants**, which are identified in Figure 6.2.6. From our work in Example 6.2.1, we can see that if a point is in quadrant I both of its coordinates are positive; if it is in quadrant II the first coordinate is negative and the second coordinate is positive; if it is in quadrant III both coordinates are negative; and if it is in quadrant IV the first is positive and the second negative. The points on the axes have one coordinate equal to zero and are not said to be in any quadrant.

To graph a set of ordered pairs, we graph all the ordered pairs in the set.

EXAMPLE 6.2.2

Graph the following set of ordered pairs.

$$\left\{ \left(\tfrac{5}{2}, -4\right), (-3, 1), \left(-\tfrac{4}{3}, 0\right), (-3.2, -2) \right\}$$

(See Figure 6.2.7.)

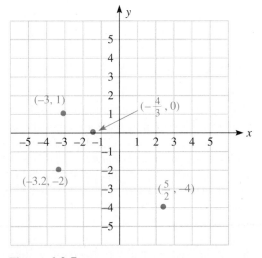

Figure 6.2.7

As we noted in Section 6.1, a solution to an equation in two variables is an ordered pair of real numbers with the property that when the x-coordinate is substituted for x in the equation and the y-coordinate is substituted for y, we obtain a true statement. The next set of ordered pairs that we will graph is one that contains all of the solutions to a linear equation in two variables.

Consider the equation $3x - 2y = 8$. We can write the set of solutions to this equation using set-builder notation:

$$\{(x, y) \qquad | \qquad 3x - 2y = 8\}$$

$\{(x, y)$	$\mid$	$3x - 2y = 8\}$
The set of all ordered pairs of real numbers	such that	The equation the ordered pairs must satisfy in order to be in the set

Let us find some ordered pairs that belong to this set using the techniques developed in Section 6.1:

If $x = 0$,

$3 \cdot 0 - 2y = 8$

$0 - 2y = 8$

$-2y = 8$

$y = -4$

$(0, -4)$ is a solution.

If $x = 2$,

$3 \cdot 2 - 2y = 8$

$6 - 2y = 8$

$-2y = 2$

$y = -1$

$(2, -1)$ is a solution.

If $y = 2$,

$3x - 2 \cdot 2 = 8$

$3x - 4 = 8$

$3x = 12$

$x = 4$

$(4, 2)$ is a solution.

If $y = -6$,

$3x - 2(-6) = 8$

$3x + 12 = 8$

$3x = -4$

$x = -\frac{4}{3}$

$\left(-\frac{4}{3}, -6\right)$ is a solution

These ordered pairs are graphed in Figure 6.2.8(a). We have found four solutions, but we know there are infinitely many. If we plotted more points, we would soon have enough evidence to indicate that we could join the points with a straight line [see Figure 6.2.8(b)] and have a picture that represents all of the ordered pairs that satisfy the equation. This picture is called the **graph of the equation**. The arrows at the end of the line indicate that it continues indefinitely in both directions.

NOTE ▶. It can be shown that the graph of a linear equation in two variables, $Ax + By = C$, is always a straight line.

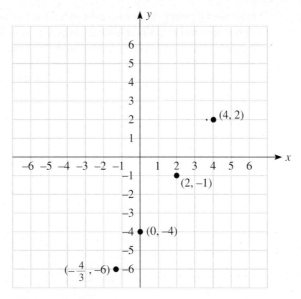

Figure 6.2.8(b)

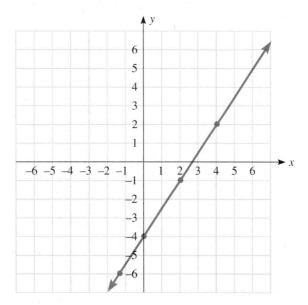

Figure 6.2.8(a)

EXAMPLE 6.2.3

Graph the following equations.

1. $4x + 7y = 2$

If we know beforehand that the graph of an equation is a straight line, how many points do we need to plot? We know from geometry that two points determine a straight line. However, it is usually a good idea to get a third point as a check to make sure no arithmetic errors were made in the calculations of the other two points. Three ordered pair solutions are $(-3, 2)$, $(4, -2)$, and $\left(\frac{1}{2}, 0\right)$. Plotting these points, we obtain the graph in Figure 6.2.9.

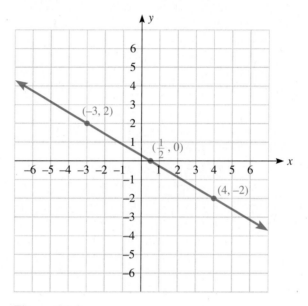

Figure 6.2.9

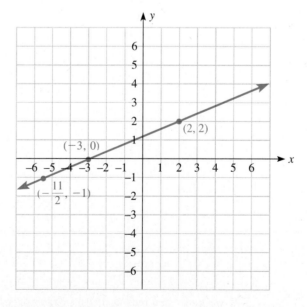

Figure 6.2.10

2. $-2x + 5y - 6 = 0$

Three ordered pair solutions are $(-3, 0)$, $(2, 2)$, and $\left(-\frac{11}{2}, -1\right)$. Plotting these points, we obtain the graph in Figure 6.2.10.

3. $x = 2$

From our work in Section 6.1, we know that any ordered pair with an x-coordinate of 2 is a solution to this equation. Let's use $(2, -4)$, $(2, 0)$, and $(2, 3)$ to graph the vertical line shown in Figure 6.2.11. Any equation of the form $x = a$ generates a vertical line.

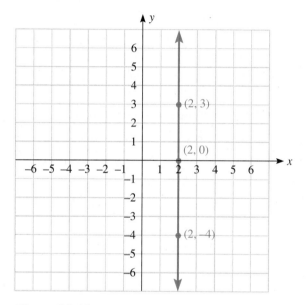

Figure 6.2.11

4. $y = -3$

Any ordered pair with a y-coordinate of -3 is a solution to this equation. Using $(-5, -3)$, $(0, -3)$, and $(2, -3)$, we obtain the horizontal line in Figure 6.2.12. Any equation of the form $y = b$ generates a horizontal line.

The point where a line crosses the y-axis is called the **y-intercept** and is found by substituting 0 for x. The point where a line crosses the x-axis is called the **x-intercept** and is found by substituting 0 for y. For instance, in part 1 of Example 6.2.3, the x-intercept of $\left(\frac{1}{2}, 0\right)$ was found. It is often easiest to graph a linear equation by using the intercepts.

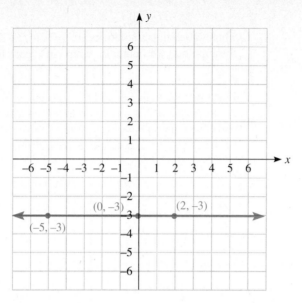

Figure 6.2.12

EXAMPLE 6.2.4

Use the intercepts to graph the following equations.

1. $2x + 3y = -6$

If $x = 0$,	If $y = 0$,
$2 \cdot 0 + 3y = -6$	$2x + 3 \cdot 0 = -6$
$3y = -6$	$2x = -6$
$y = -2$	$x = -3$
$(0, -2)$ is the y-intercept.	$(-3, 0)$ is the x-intercept.

Again, it is a good idea to find a third point; for example, $(3, -4)$ is a solution to this equation. The graph is shown in Figure 6.2.13.

2. $x + 4y = 0$

If $x = 0$,

$0 + 4y = 0$

$4y = 0$

$y = 0$

Therefore, $(0, 0)$ is both the x-intercept and y-intercept. We still need two more ordered pair solutions. Using $(4, -1)$ and $(-4, 1)$, we obtain the graph in Figure 6.2.14. Any equation of the form $Ax + By = 0$ generates a line passing through the origin.

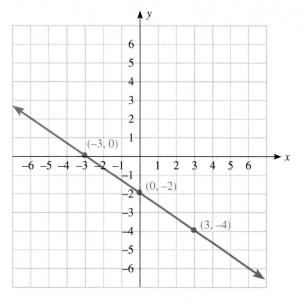

Figure 6.2.13

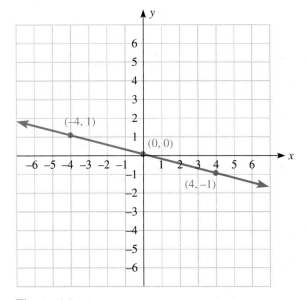

Figure 6.2.14

In Section 6.1 we found solutions for some linear equations by first solving the equation for one variable in terms of the other. In Section 6.4 we will see the importance of solving an equation for *y*.

EXAMPLE 6.2.5

Solve the equation $5x - 2y = 8$ for y, then use the new equation to find ordered pair solutions. Also, graph the equation.

$$5x - 2y = 8$$
$$-2y = -5x + 8$$
$$y = \frac{-5x + 8}{-2}$$

or $\quad y = \frac{5}{2}x - 4$

Now substitute $x = 0$, 2, and 4 into the bottom equation:

$y = \frac{5}{2}(0) - 4 = 0 - 4 = -4;\quad (0, -4)$ is a solution.

$y = \frac{5}{2}(2) - 4 = 5 - 4 = 1;\quad (2, 1)$ is a solution.

$y = \frac{5}{2}(4) - 4 = 10 - 4 = 6;\quad (4, 6)$ is a solution.

Using these three ordered pairs, we obtain the graph in Figure 6.2.15.

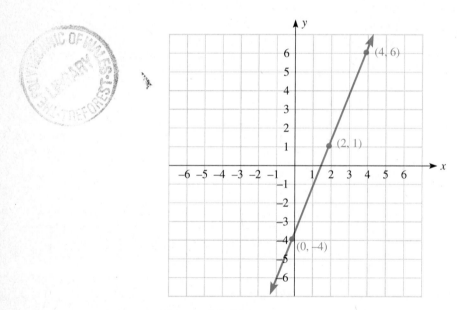

Figure 6.2.15

EXERCISES 6.2

Graph the following sets of ordered pairs.

1. $\{(4, 3), (5, -2), (-1, 6), (-4, -5)\}$

2. $\{(6, 3), (-5, 1), (-4, -6), (1, -1)\}$

3. $\{(3, 0), (-4, 0), (0, 2), (0, -5)\}$

4. $\{(2, 0), (-1, 0), (0, 6), (0, -2)\}$

5. $\left\{\left(3, \frac{3}{2}\right), \left(-\frac{4}{3}, 2\right), \left(-\frac{1}{2}, -5\right), \left(4, -\frac{11}{4}\right)\right\}$

6. $\left\{\left(\frac{5}{2}, 5\right), \left(\frac{6}{5}, -3\right), \left(-4, \frac{2}{3}\right), \left(-6, -\frac{14}{3}\right)\right\}$

7. $\{(-4, -4), (-2, -2), \left(-\frac{1}{2}, -\frac{1}{2}\right), (1, 1), \left(\frac{3}{2}, \frac{3}{2}\right), (3, 3), (5, 5)\}$

8. $\{(-5, 5), (-4, 4), (-2, 2), \left(-\frac{2}{3}, \frac{2}{3}\right), (1, -1), \left(\frac{5}{2}, -\frac{5}{2}\right), (4, -4)\}$

9. $\{(-2, 4), (-1, 1), \left(-\frac{1}{2}, \frac{1}{4}\right), (0, 0), \left(\frac{1}{2}, \frac{1}{4}\right), (1, 1), (2, 4)\}$

10. $\left\{\left(-3, \frac{1}{8}\right), \left(-2, \frac{1}{4}\right), \left(-1, \frac{1}{2}\right), (0, 1), (1, 2), (2, 4)\right\}$

Graph the following equations.

11. $x + y = 5$

12. $x + y = 2$

13. $x - y = 4$

14. $x - y = 3$

15. $x + 2y = 6$

16. $x + 3y = 6$

17. $4x - y = 4$

18. $5x - y = 5$

19. $3x + 5y = -9$

20. $4x + 3y = -8$

21. $7x - 4y = -6$

22. $8x - 5y = -4$

23. $x = \frac{1}{3}y + 2$

24. $x = \frac{1}{2}y - 3$

25. $x = \dfrac{2y - 5}{3}$

26. $x = \dfrac{4y + 1}{5}$

27. $-\frac{3}{4}x + 2y = \frac{1}{2}$

28. $-\frac{2}{3}x + y = \frac{5}{6}$

29. $3.2x + 2.4y = 8$

30. $3x - 1.8y = 4.2$

31. $x = -3$

32. $x = 5$

33. $2x - 5 = 0$

34. $3x + 7 = 0$

35. $y = 4$

36. $y = -2$

37. $4y + 2 = 0$

38. $2y - 3 = 0$

Find the x- and y-intercepts, and use them to graph each equation.

39. $x + y = 3$

40. $x + y = 4$

41. $x - y = -6$

42. $x - y = -5$

43. $x + 4y = -4$

44. $x + 2y = -2$

45. $6x + y = 3$

46. $4x + y = 2$

47. $3x + 4y = 12$

48. $5x + 2y = 10$

49. $2x - 7y = 7$

50. $3x - 8y = 8$

51. $-5x + 9y = 12$

52. $-7x + 3y = 15$

53. $y = \frac{2}{3}x + 4$

54. $y = \frac{3}{4}x - 3$

55. $\frac{2}{5}x - y = \frac{3}{2}$

56. $\frac{5}{2}x - 3y = \frac{11}{4}$

57. $\frac{1}{2}x + \frac{1}{3}y + 2 = 0$

58. $\frac{1}{3}x + \frac{1}{4}y + \frac{3}{2} = 0$

59. $0.6x + 2.7y = -1.8$

60. $1.2x + 3.2y = -4.8$

61. $3x - 5y = 0$

62. $2x - 7y = 0$

63. $4x + 3y = 0$

64. $6x + 5y = 0$

65. $y = x$

66. $y = -x$

Solve each equation for y, then use the new equation to find ordered pair solutions. Also, graph the equations.

67. $3x + y = 4$

68. $2x + y = 5$

69. $4x - y = -6$

70. $5x - y = -4$

71. $-2x + 5y = 10$

72. $-4x + 3y = 9$

73. $6x - 4y = 3$

74. $7x - 2y = -4$

75. $5x + 4y + 8 = 0$

76. $3x + 8y + 4 = 0$

77. $\dfrac{7x - 3y}{5} = 3$

78. $\dfrac{4x - 9y}{7} = 2$

79. $\frac{3}{5}x + \frac{5}{2}y = \frac{7}{4}$

80. $-\frac{5}{4}x + \frac{7}{6}y = \frac{3}{2}$

6.3 The Slope of a Line

An important characteristic of a line is the amount of inclination of the line. The concept of inclination of a line is similar to the pitch of a roof. Some lines are steep and others rise slowly (see Figure 6.3.1). This "steepness" is generally referred to as the slope of the line and requires two numbers to measure it called the **rise** and the **run**.

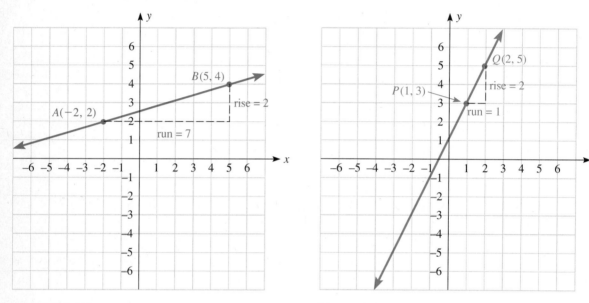

Figure 6.3.1(a)   Figure 6.3.1(b)

In Figure 6.3.1(a), if we start at point A and "run" 7 units to the right and then "rise" up 2 units we arrive at point B (another point on the line). Similarly, in Figure 6.3.1(b) if we start at point P and "run" 1 unit to the right and then "rise" up 2 units we arrive at point Q (another point on the line). The ratio of the rise and the run, that is, $\frac{\text{rise}}{\text{run}}$, is used in the following definition (see Figure 6.3.2).

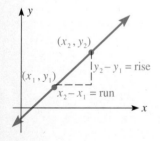

Figure 6.3.2

Slope of a Line

Let (x_1, y_1) and (x_2, y_2) be *any* two distinct points on a line; then the **slope** of that line is defined to be

$$\frac{\text{rise}}{\text{run}} = \frac{y_2 - y_1}{x_2 - x_1}$$

REMARK   It is common practice to let the letter m stand for the slope of a line.

EXAMPLE 6.3.1   Find the slopes of the lines passing through the following pairs of points.

1. $(1, -2)$, $(5, 4)$

Let us choose $(x_1, y_1) = (1, -2)$ and $(x_2, y_2) = (5, 4)$. Then

$$m = \frac{y_2 - y_1}{x_2 - x_1} = \frac{4 - (-2)}{5 - 1} = \frac{6}{4} = \frac{3}{2}$$

But what if we choose $(x_1, y_1) = (5, 4)$ and $(x_2, y_2) = (1, -2)$? Shouldn't we get the same answer for the slope?

$$m = \frac{y_2 - y_1}{x_2 - x_1} = \frac{-2 - 4}{1 - 5} = \frac{-6}{-4} = \frac{3}{2}$$

In either case we obtain the same answer. Therefore, how we choose (x_1, y_1) and (x_2, y_2) is immaterial. The line passing through these two points is shown in Figure 6.3.3. From the graph we can see that each time we start at a point on the line and run 2 units (to the right) and rise (up) 3 units, we arrive at another point on the line.

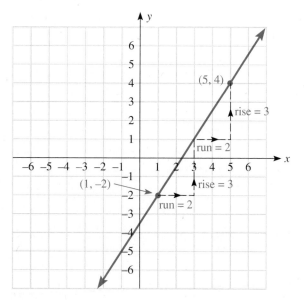

Figure 6.3.3

Try to avoid this mistake:

Incorrect	Correct
$(x_1, y_1) = (-3, -2),$ $(x_2, y_2) = (1, -4)$ $m = \dfrac{y_2 - y_1}{x_1 - x_2} = \dfrac{-4 - (-2)}{-3 - 1}$ $= \dfrac{-2}{-4} = \dfrac{1}{2}$	$(x_1, y_1) = (-3, -2),$ $(x_2, y_2) = (1, -4)$ $m = \dfrac{y_2 - y_1}{x_2 - x_1} = \dfrac{-4 - (-2)}{1 - (-3)}$ $= \dfrac{-2}{4} = -\dfrac{1}{2}$ or if $(x_1, y_1) = (1, -4),$ $(x_2, y_2) = (-3, -2)$ $m = \dfrac{y_2 - y_1}{x_2 - x_1} = \dfrac{-2 - (-4)}{-3 - 1}$ $= \dfrac{2}{-4} = -\dfrac{1}{2}$
The x's were subtracted in a different order than were the y's.	In either case the x's and y's were subtracted in the same order.

2. $(-4, 5), (1, 3)$

Let us choose $(x_1, y_1) = (-4, 5)$ and $(x_2, y_2) = (1, 3)$. Then

$$m = \frac{y_2 - y_1}{x_2 - x_1} = \frac{3 - 5}{1 - (-4)} = \frac{-2}{5}$$

The line passing through these two points is shown in Figure 6.3.4. From the graph we see that we can start at $(-4, 5)$ and run 5 (to the right). The rise is *down* 2, since down is the negative direction on the y-axis. Since $\frac{-2}{5} = \frac{2}{-5}$, we can also start at $(1, 3)$ and run 5 to the *left* (since the negative direction of the x-axis is to the left) and rise (up) 2.

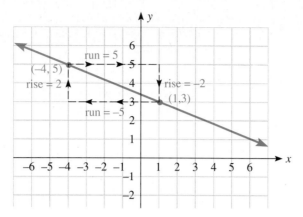

Figure 6.3.4

EXAMPLE 6.3.2

Find the slopes of the lines whose equations are given.

1. $3x - 4y = 8$

To find the slope of the line, we must find two ordered pair solutions of the equation. Let us use $(x_1, y_1) = (0, -2)$ and $(x_2, y_2) = (4, 1)$. Then

$$m = \frac{y_2 - y_1}{x_2 - x_1} = \frac{1 - (-2)}{4 - 0} = \frac{3}{4}$$

What if we use two different points to find the slope? Do we get the same answer? Let us use $(x_1, y_1) = (-4, -5)$ and $(x_2, y_2) = \left(2, -\frac{1}{2}\right)$. Then

$$m = \frac{y_2 - y_1}{x_2 - x_1} = \frac{-\frac{1}{2} - (-5)}{2 - (-4)} = \frac{\frac{9}{2}}{6} = \frac{9}{2} \cdot \frac{1}{6} = \frac{3}{4}$$

NOTE ▶ *Any* two points may be used to find the slope, since the slope is "everywhere the same."

2. $y = 3$

We know from Section 6.1 that any ordered pair with a y-coordinate of 3 will satisfy this equation. Let us use $(x_1, y_1) = (1, 3)$ and $(x_2, y_2) = (5, 3)$. Then

$$m = \frac{y_2 - y_1}{x_2 - x_1} = \frac{3 - 3}{5 - 1} = \frac{0}{4} = 0$$

This says that as we run across a horizontal line there is no rise. *The slope of any horizontal line is zero.* In fact, if a line has zero slope, then it must be a horizontal line.

3. $x = -5$

We need ordered pairs with x-coordinates of -5. Let us use $(x_1, y_1) = (-5, -2)$ and $(x_2, y_2) = (-5, 1)$. Then

$$m = \frac{y_2 - y_1}{x_2 - x_1} = \frac{1 - (-2)}{-5 - (-5)} = \frac{3}{0}$$

which is undefined. *The slope of any vertical line is undefined.* Moreover, if the slope of a line is undefined, then it must be a vertical line. We will avoid the use of the ambiguous expression "no slope."

EXAMPLE 6.3.3

Graph the lines satisfying the following conditions.

1. A line with $m = \frac{1}{2}$ passing through the point $(1, 3)$

We need two points to draw a line. We are given one point, and we can use the slope to find a second point. We start at $(1, 3)$ and run (to the right) 2 units and rise (up) 1 unit, obtaining a second point on the line. Now with two points the line can be drawn as in Figure 6.3.5.

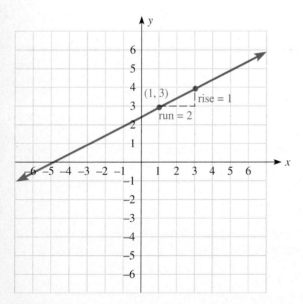

Figure 6.3.5

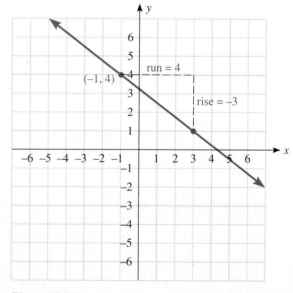

Figure 6.3.6

2. A line with $m = -\dfrac{3}{4}$ passing through the point $(-1, 4)$

   Let us write the slope, $-\dfrac{3}{4}$, as $\dfrac{-3}{4}$. Therefore, we start at $(-1, 4)$ and run (to the right) 4 units and rise (down) 3 units, obtaining a second point on the line. Using these two points, we draw the line shown in Figure 6.3.6.

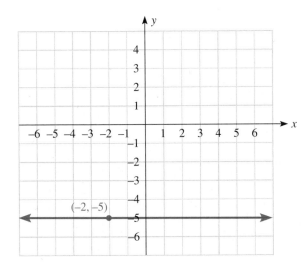

Figure 6.3.7   Figure 6.3.8

3. A line with $m = 0$ passing through the point $(-2, -5)$

   Since the slope is zero, the line is horizontal. Therefore, we simply plot the point $(-2, -5)$ and draw a horizontal line through it, as shown in Figure 6.3.7.

4. A line with undefined slope passing through the point $(4, 2)$

   Since the slope is undefined, the line is vertical. Therefore, we simply plot the point $(4, 2)$ and draw a vertical line through it, as shown in Figure 6.3.8.

EXERCISES 6.3

Find the slopes of the lines passing through the following pairs of points.

1. $(5, 3)$, $(2, 1)$ **2.** $(1, 6)$, $(4, 7)$

3. $(-4, 2)$, $(-2, 6)$ **4.** $(8, -2)$, $(5, 4)$

5. $(2, -3), (-1, -7)$

6. $(-4, -9), (1, -4)$

7. $(4, -3), (2, -3)$

8. $(-1, 5), (3, 5)$

9. $(1, 4), (1, -2)$

10. $(-2, 4), (-2, -3)$

11. $\left(\frac{2}{3}, -1\right), \left(\frac{1}{6}, \frac{3}{2}\right)$

12. $\left(\frac{3}{4}, \frac{5}{2}\right), (2, 0)$

13. $(2.3, 4), (-0.5, 0.5)$

14. $(1.6, -4.2), (2.8, 3)$

15. $(a, 2a), (a + h, 2a + 2h)$

16. $(a, a + 3), (a + h, a + h + 3)$

Find the slopes of the lines whose equations are given.

17. $4x - 3y = 7$ **18.** $5x + 2y = 8$ **19.** $7x + 5y = 10$ **20.** $3x + 8y = 16$

21. $2x - 9y = -5$ **22.** $6x - 4y = -9$ **23.** $x + \frac{2}{3}y = -4$ **24.** $x + \frac{3}{5}y = -6$

25. $y = 2x - 5$ **26.** $y = 3x + 2$ **27.** $y = -\frac{3}{4}x + 5$ **28.** $y = -\frac{5}{2}x - 1$

29. $y = \dfrac{4x - 3}{7}$ **30.** $y = \dfrac{2x + 8}{5}$ **31.** $x = \dfrac{3y + 6}{4}$ **32.** $x = \dfrac{5y - 2}{6}$

33. $y = -4$ **34.** $y - 2 = 0$ **35.** $2x - 3 = 0$ **36.** $x = -1$

Graph the lines satisfying the following conditions.

37. $m = \frac{2}{3}$, passing through $(-2, 1)$

38. $m = \frac{4}{3}$, passing through $(-4, -1)$

39. $m = \frac{1}{4}$, passing through $(1, -3)$

40. $m = \frac{2}{5}$, passing through $(2, 4)$

41. $m = \frac{-3}{2}$, passing through $(3, 1)$

42. $m = \frac{-2}{3}$, passing through $(-5, 6)$

43. $m = \frac{-5}{3}$, passing through $(-1, -1)$

44. $m = \frac{-7}{2}$, passing through $(3, -1)$

45. $m = 3$, passing through $(2, -5)$

46. $m = 2$, passing through $(1, 2)$

47. $m = -4$, passing through $(4, 2)$

48. $m = -3$, passing through $(2, -3)$

49. $m = 1$, passing through $(-2, 2)$

50. $m = -1$, passing through $(-4, 4)$

51. $m = 0$, passing through $(-3, -4)$

52. $m = 0$, passing through $(4, 5)$

53. m is undefined, passing through $(5, 3)$

54. m is undefined, passing through $(-2, -6)$

55. Parallel to the x-axis, passing through $(4, -2)$

56. Parallel to the y-axis, passing through $(-3, 3)$

Find the slopes and the y-intercepts of the lines whose equations are given.
What do the equations tell us about the answers?

57. $y = \frac{2}{3}x + 2$ **58.** $y = \frac{-5}{2}x + 4$ **59.** $y = -3x + 1$ **60.** $y = 2x - 5$

6.4 Equations of Lines

In Section 6.2 we were given equations of lines and we found points on the lines in order to graph them. In Section 6.3 we found slopes of lines, given their equations. In this section we will go in the reverse direction; that is, given information that determines a line, we want to find the equation of the line.

Two common forms make finding equations of lines relatively simple. The first is called the **point-slope form**. This is used when a point on the line and the slope of the line are known. It can also be used when two points on the line are known.

Suppose that we wanted to find the equation of the line with $m = \frac{3}{2}$ and passing through the point $(2, -1)$. If (x, y) is another point on the line, then using $(x_2, y_2) = (x, y)$ and $(x_1, y_1) = (2, -1)$ in the slope formula should yield $\frac{3}{2}$:

$$\frac{y_2 - y_1}{x_2 - x_1} = m$$

$$\frac{y - (-1)}{x - 2} = \frac{3}{2}$$

$$(x - 2) \cdot \frac{y + 1}{x - 2} = \frac{3}{2} \cdot (x - 2) \qquad \text{Multiply both sides by } x - 2.$$

$$y + 1 = \tfrac{3}{2}(x - 2)$$

$$y + 1 = \tfrac{3}{2}x - 3 \qquad \text{Distribute } \tfrac{3}{2} \text{ across the parentheses.}$$

$$y = \tfrac{3}{2}x - 4 \qquad \text{Subtract 1 from both sides.}$$

$$-\tfrac{3}{2}x + y = -4 \qquad \text{Subtract } \tfrac{3}{2}x \text{ from both sides.}$$

$$3x - 2y = 8 \qquad \text{Multiply both sides by } -2.$$

Let us generalize the preceding argument. Let (x_1, y_1) be a particular point on line l, and let (x, y) be any other point (see Figure 6.4.1). Letting $(x_2, y_2) = (x, y)$ in the slope formula, we obtain

$$\frac{y - y_1}{x - x_1} = m$$

$$(x - x_1) \cdot \frac{y - y_1}{x - x_1} = m \cdot (x - x_1) \qquad \text{Multiply both sides by } (x - x_1).$$

$$y - y_1 = m(x - x_1)$$

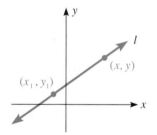

Figure 6.4.1

Point-Slope Form

Given a line with slope m and containing the point (x_1, y_1), the equation of that line is

$$y - y_1 = m(x - x_1)$$

In Section 6.1 we introduced linear equations in two variables as equations that could be written in the form

$$Ax + By = C$$

This is called **standard form**. In the opening example of this section, we wrote our answer, $3x - 2y = 8$, in this form. We use the point-slope form to find the equation, and then we simplify it to standard form.

EXAMPLE 6.4.1

Find the equations of the lines satisfying the given information. Write your answers in standard form.

1. Passing through (4, 1) with $m = \dfrac{-2}{3}$

Let $(x_1, y_1) = (4, 1)$ and substitute into the point-slope form to obtain

$$y - 1 = \frac{-2}{3}(x - 4)$$

Do not stop at this point. Simplify the equation and combine like terms:

$$y - 1 = \frac{-2}{3}x + \frac{8}{3} \qquad \text{Distribute } \frac{-2}{3} \text{ across the parentheses.}$$

$$y = \frac{-2}{3}x + \frac{11}{3} \qquad \text{Add 1 to both sides.}$$

$$\frac{2}{3}x + y = \frac{11}{3} \qquad \text{Add } \tfrac{2}{3}x \text{ to both sides.}$$

$$2x + 3y = 11 \qquad \text{Multiply both sides by 3.}$$

2. Passing through $(-2, -5)$ with $m = 3$

Let $(x_1, y_1) = (-2, -5)$ and substitute into the point-slope form to obtain

$$y - (-5) = 3[x - (-2)]$$
$$y + 5 = 3(x + 2)$$
$$y + 5 = 3x + 6$$
$$y = 3x + 1$$
$$-3x + y = 1 \qquad \text{Multiply both sides by } -1.$$

or $\quad 3x - y = -1$

3. Passing through $(3, -5)$ with $m = 0$

We can find this equation quickly without using the point-slope form. From our work in Sections 6.2 and 6.3, we know that if a line has zero slope, then it is horizontal. We also know that a horizontal line has an equation in the form $y = b$, where b is the y-coordinate of any point on the line. Since $(3, -5)$ is on the line, the equation is $y = -5$.

4. Passing through $(-4, 2)$ with m undefined

This equation *cannot* be found using the point-slope form since there is no number to substitute for the slope. However, from our work in Sections 6.2 and 6.3, we know that if the slope of the line is undefined, then it is vertical. We also know that a vertical line has an equation in the form $x = a$, where a is the x-coordinate of any point on the line. Since $(-4, 2)$ is on the line, the equation is $x = -4$.

5. Passing through $(-4, 3)$ and $(1, -1)$

To find the equation of a line we must know the slope. In this example let $(x_1, y_1) = (-4, 3)$ and $(x_2, y_2) = (1, -1)$. Then the slope is

$$m = \frac{-1 - 3}{1 - (-4)} = \frac{-4}{5}$$

After calculating the slope, we can use the point-slope form. Which of the two given points should we use? It turns out that either point will generate the same equation. Let

$(x_1, y_1) = (-4, 3)$ or $(x_1, y_1) = (1, -1)$

$$y - 3 = \frac{-4}{5}[x - (-4)] \qquad\qquad y - (-1) = \frac{-4}{5}(x - 1)$$

$$y - 3 = \frac{-4}{5}(x + 4) \qquad\qquad y + 1 = \frac{-4}{5}x + \frac{4}{5}$$

$$y - 3 = \frac{-4}{5}x - \frac{16}{5} \qquad\qquad y = \frac{-4}{5}x - \frac{1}{5}$$

$$y = \frac{-4}{5}x - \frac{1}{5} \qquad\qquad \frac{4}{5}x + y = \frac{-1}{5}$$

$$\frac{4}{5}x + y = \frac{-1}{5} \qquad\qquad 4x + 5y = -1$$

$$4x + 5y = -1$$

6. Passing through $\left(-\frac{1}{2}, -3\right)$ and $\left(-4, \frac{9}{4}\right)$

Let $(x_1, y_1) = \left(-\frac{1}{2}, -3\right)$ and $(x_2, y_2) = \left(-4, \frac{9}{4}\right)$. Then

$$m = \frac{\frac{9}{4} - (-3)}{-4 - \left(-\frac{1}{2}\right)} = \frac{\frac{21}{4}}{-\frac{7}{2}} = \left(\frac{21}{4}\right)\left(-\frac{2}{7}\right) = -\frac{3}{2}$$

Now let us substitute $(x_1, y_1) = \left(-\frac{1}{2}, -3\right)$ and $m = \frac{-3}{2}$ into the point-slope form:

$$y - (-3) = \frac{-3}{2}\left[x - \left(-\frac{1}{2}\right)\right]$$

$$y + 3 = \frac{-3}{2}\left(x + \frac{1}{2}\right)$$

$$y + 3 = \frac{-3}{2}x - \frac{3}{4}$$

$$y = \frac{-3}{2}x - \frac{15}{4}$$

$$\frac{3}{2}x + y = \frac{-15}{4}$$

$$6x + 4y = -15$$

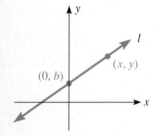

Figure 6.4.2

The second common form for the equation of a line is called the **slope-intercept form**. Assume that line l in Figure 6.4.2. has y-intercept $(0, b)$ and slope m. Then, letting $(x_1, y_1) = (0, b)$ and using the point-slope form, we get

$$y - b = m(x - 0)$$

$$y - b = mx$$

$$y = mx + b$$

Slope-Intercept Form

Given a line l with slope m and y-intercept $(0, b)$, the equation of l is

$$y = mx + b$$

REMARKS

1. Since the x-coordinate of the y-intercept is always zero, we usually call the y-intercept b instead of $(0, b)$.

2. We can use the slope-intercept form to graph a line. We start at the y-intercept and use the slope to determine a second point, as illustrated in the following example.

EXAMPLE 6.4.2

Find the slope and y-intercept of the line $3x + 5y = 10$. Use the slope-intercept form to graph the line.

To find the slope and y-intercept, we simply need to solve the given equation for y. Then the equation is in slope-intercept form, and the coefficient of x is the slope and the constant term is the y-intercept.

$$3x + 5y = 10$$
$$5y = -3x + 10$$
$$y = \frac{-3x + 10}{5}$$
$$y = \frac{-3}{5}x + 2$$

Therefore, the slope is $\dfrac{-3}{5}$ and the y-intercept is 2. To graph the line, start at the y-intercept $(0, 2)$, then use the slope to determine a second point (see Figure 6.4.3).

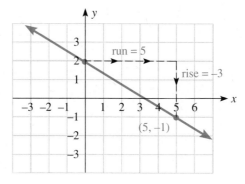

Figure 6.4.3

EXAMPLE 6.4.3

Find the equations of the lines satisfying the given information. Write your answers in slope-intercept form.

1. Slope is 2 and y-intercept is -4.

To find the equation, we substitute 2 for m and -4 for b in the slope-intercept form:

$$y = 2x + (-4) \qquad \text{or} \qquad y = 2x - 4$$

2. Slope is $\frac{1}{2}$ and passing through $(4, -3)$.

Using the point-slope form,

$$y - (-3) = \tfrac{1}{2}(x - 4)$$
$$y + 3 = \tfrac{1}{2}x - 2$$
$$y = \tfrac{1}{2}x - 5$$

If two lines in the same plane never cross, they are called **parallel** [see Figure 6.4.4(a)]. If two lines cross at a right angle, they are called **perpendicular** [see Figure 6.4.4(b)].

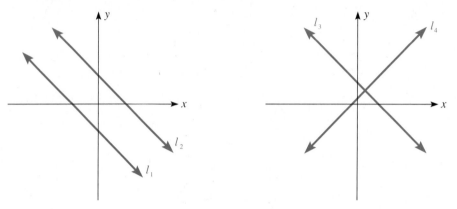

Figure 6.4.4(a)   **Figure 6.4.4(b)**

We can use the slopes of lines to tell us whether a given pair of *nonvertical* lines are parallel or perpendicular. Two distinct nonvertical lines are *parallel* if and only if they have the same slope. In Figure 6.4.4(a), if the slope of line l_1 is m_1 and the slope of line l_2 is m_2, then $m_1 = m_2$. Two nonvertical lines are *perpendicular* if and only if the slopes of the lines are negative reciprocals. (This means their product will be -1.) In Figure 6.4.4(b), if the slope of line l_3 is m_3 and the slope of line l_4 is m_4, then $m_3 = \dfrac{-1}{m_4}$, or $m_3 \cdot m_4 = -1$.

EXAMPLE 6.4.4

Determine whether the following pairs of lines are parallel, perpendicular, or neither.

1. $6x - 4y = 7$, $2x + 3y = 9$

We need to find the slope of each line, which can easily be found by writing the equations in slope-intercept form. Solving each equation for y, we obtain

$$
\begin{aligned}
6x - 4y &= 7 & 2x + 3y &= 9 \\
-4y &= -6x + 7 & 3y &= -2x + 9 \\
y &= \frac{-6x + 7}{-4} & y &= \frac{-2x + 9}{3}
\end{aligned}
$$

$$y = \frac{3}{2}x - \frac{7}{4} \qquad\qquad y = \frac{-2}{3}x + 3$$

$$m = \frac{3}{2} \qquad\qquad m = \frac{-2}{3}$$

Since $\frac{3}{2} \cdot \left(\frac{-2}{3}\right) = -1$, the lines are perpendicular.

2. $x + 2y = 7, \frac{1}{2}x = 4 - y$

Solving each equation for y, we obtain

$$x + 2y = 7 \qquad\qquad \frac{1}{2}x = 4 - y$$

$$2y = -x + 7 \qquad\qquad \frac{1}{2}x + y = 4$$

$$y = \frac{-x + 7}{2} \qquad\qquad y = \frac{-1}{2}x + 4$$

$$y = \frac{-1}{2}x + \frac{7}{2} \qquad\qquad m = \frac{-1}{2}$$

$$m = \frac{-1}{2}$$

Since the slopes are the same, but the y-intercepts are different, these are two distinct parallel lines.

EXERCISES 6.4

Find the slopes and the y-intercepts of the lines whose equations are given.

1. $y = 3x + 2$

2. $y = \frac{-3}{2}x + 4$

3. $4x + 5y = 10$

4. $2x - 3y = 9$

5. $7x - 3y = 5$

6. $5x - 6y = -1$

7. $x + 2y = -3$

8. $-x + 3y = -4$

9. $2x - y + 7 = 0$

10. $6x + y + 3 = 0$

11. $3y + 4 = 0$

12. $4 - 5y = 0$

13. $2x - 7 = 0$

14. $3 - 2x = 0$

Find the equations of the lines satisfying the given information. Write your answers in slope-intercept form.

15. $m = \frac{2}{3}$, y-intercept $= 4$

16. $m = -2$, y-intercept $= \frac{5}{3}$

17. $m = \dfrac{-1}{2}$, passing through $(3, -1)$

18. $m = \frac{4}{3}$, passing through $(2, 1)$

19. $m = -3$, passing through $(-1, -4)$

20. $m = -4$, passing through $(-2, 5)$

21. $m = \frac{3}{4}$, passing through $(4, 2)$

22. $m = \dfrac{-5}{2}$, passing through $(2, -4)$

23. $m = 0$, passing through $(-5, 6)$

24. $m = 0$, passing through $(-3, -5)$

25. $m = 2$, x-intercept $= -3$

26. $m = 3$, x-intercept $= 1$

27. Passing through $(-5, 2)$ and $(-1, -2)$

28. Passing through $(1, -6)$ and $(3, -2)$

29. Passing through $(5, 1)$ and $(2, 3)$

30. Passing through $(-2, -4)$ and $(3, 2)$

31. Passing through $(2, 4)$ and $(-4, 4)$

32. Passing through $(-5, -3)$ and $(-1, -3)$

33. x-intercept $= 4$, y-intercept $= 3$

34. x-intercept $= -5$, y-intercept $= 2$

Find the equations of the lines satisfying the given information. Write your answers in standard form.

35. $m = \dfrac{3}{5}$, passing through $(-1, 2)$

36. $m = \dfrac{1}{4}$, passing through $(-3, -4)$

37. $m = \dfrac{-7}{2}$, passing through $(1, -5)$

38. $m = \dfrac{-8}{3}$, passing through $(-2, 5)$

39. $m = 4$, passing through $(-3, -7)$

40. $m = 3$, passing through $(4, 9)$

41. $m = 0$, passing through $(6, 3)$

42. $m = 0$, passing through $(5, -2)$

43. m is undefined, passing through $(4, -1)$

44. m is undefined, passing through $(-4, -6)$

45. Passing through $(-5, 8)$ and $(-2, 2)$

46. Passing through $(7, 3)$ and $(1, 5)$

47. Passing through $(3, -6)$ and $(5, 1)$

48. Passing through $(6, -2)$ and $(-1, 3)$

49. Passing through $(-4, 3)$ and $(0, 3)$

50. Passing through $(5, 4)$ and $(-2, 4)$

51. Passing through $(-3, 2)$ and $(-3, -6)$

52. Passing through $(4, -4)$ and $(4, 1)$

53. x-intercept $= 3$, y-intercept $= -3$

54. x-intercept $= -8$, y-intercept $= -2$

Determine whether the following pairs of lines are parallel, perpendicular, or neither.

55. $y = 3x - 4$, $y = 3 - 4x$

56. $y = \frac{2}{3}x + 5$, $-y = \frac{2}{3}x - 4$

57. $2x - 4y = 5$, $2x + y = -3$

58. $3x + 5y = 9$, $-5x + 3y = 6$

59. $x - 3y = 4$, $-2x + 6y = 7$

60. $-5x - 10y = 8$, $x + 2y = 3$

61. $5x + 3y = -2$, $\dfrac{-5}{3}x - y = 4$

62. $4x - 7y = 3$, $-2x + \frac{7}{2}y = 5$

63. $4x - 5y + 8 = 0, 2x - 10y - 3 = 0$

64. $x + 3y + 6 = 0, 3x + y - 4 = 0$

65. $3x + 5 = 0, 4y - 7 = 0$

66. $2y + 3 = 0, 5x + 6 = 0$

6.5 Linear Inequalities in Two Variables

In Section 6.1 we defined a linear equation in two variables as an equation that could be written in the form

$$Ax + By = C$$

where A, B, and C are real numbers. We have a similar definition for inequalities.

Linear Inequality in Two Variables

Any inequality that can be written in the form

$$Ax + By \leq C$$

where A, B, and C are real numbers and with the property that not both A and B are zero, is called a **linear inequality in two variables**. The inequality symbol can be $\leq$, $<$, $\geq$, or $>$.

Examples of linear inequalities in two variables are

$$x - 2y \leq 4, \qquad 2x > 3y, \qquad y \leq 3$$

There are an infinite number of ordered pair solutions to any of these inequalities. As with linear equations in two variables, we graph the ordered pair solutions in a Cartesian coordinate system.

EXAMPLE 6.5.1

Graph the inequality $x \geq 3$.

The graph of the equation $x = 3$ is a vertical line that consists of all of the points whose x-coordinates are equal to 3 [see Figure 6.5.1(a)]. The graph of the inequality $x > 3$ consists of all of the points whose x-coordinates are greater than 3. These points lie to the right of the vertical line. Therefore, the graph of $x \geq 3$ is all of the points on or to the right of the line $x = 3$, the shaded region in Figure 6.5.1(b).

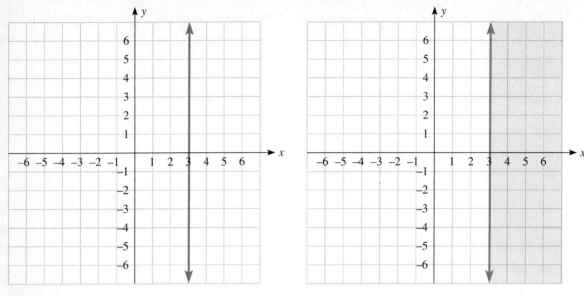

Figure 6.5.1(a)

Figure 6.5.1(b)

EXAMPLE 6.5.2

Graph the inequality $y < 4$.

y is less than 4, so the points on the line $y = 4$ do not satisfy the inequality since their y-coordinates are equal to 4. This fact is indicated by the dashed line in Figure 6.5.2(a).

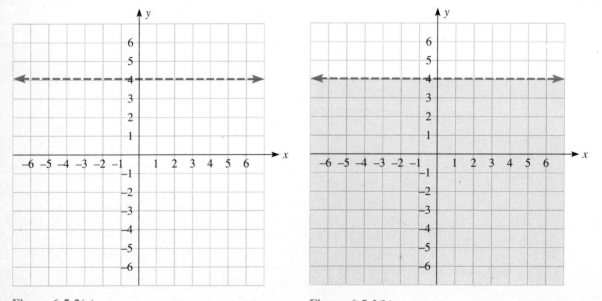

Figure 6.5.2(a)

Figure 6.5.2(b)

The graph of the inequality $y < 4$ consists of all of the points whose y-coordinates are less than 4. These points lie below the horizontal line. Therefore, the graph of $y < 4$ is the shaded region below the horizontal line shown in Figure 6.5.2(b).

The lines $x = 3$ and $y = 4$ in the preceding two examples are called **boundary lines**. Given the linear inequality $Ax + By \leq C$, the boundary line of the inequality is determined by the equation $Ax + By = C$. If the inequality symbol is $\leq$ or $\geq$, the boundary line is solid; if the inequality is $<$ or $>$, the boundary line is dashed.

The boundary line divides the Cartesian plane into two distinct regions called **half-planes**. The points in one of the half-planes satisfy the inequality. When the boundary line is neither vertical nor horizontal, we determine which half-plane satisfies the inequality by testing points from each half-plane in the inequality.

EXAMPLE 6.5.3

Graph the inequality $x - y \geq 2$.

The boundary line is defined by the equation $x - y = 2$. Since the inequality is "greater than or *equal to*," all the points on the boundary line satisfy the given inequality. This fact is indicated by drawing the boundary line solid. In Figure 6.5.3(a) we have drawn the boundary line as well as some points in each half-plane. Let us determine which of these points satisfy the inequality. This work is summarized in the chart on page 324.

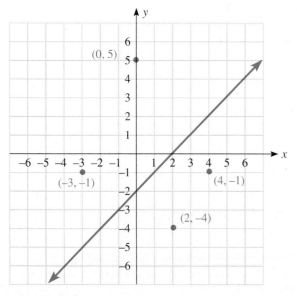

Figure 6.5.3(a)

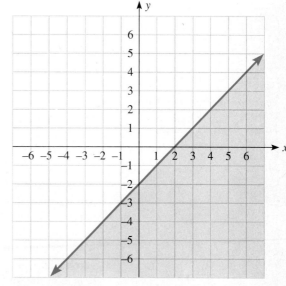

Figure 6.5.3(b)

Point	Substituted into Inequality	True or False?	Point Satisfies Inequality
(0, 5)	$0 - 5 \geq 2$ $-5 \geq 2$	False	No
(−3, −1)	$-3 - (-1) \geq 2$ $-2 \geq 2$	False	No
(4, −1)	$4 - (-1) \geq 2$ $5 \geq 2$	True	Yes
(2, −4)	$2 - (-4) \geq 2$ $6 \geq 2$	True	Yes

Notice in Figure 6.5.3(a) that the points that satisfy the inequality are below the line $x - y = 2$, whereas those that fail are above the line $x - y = 2$. Thus, we conclude that all of the points below and including the line $x - y = 2$ satisfy the inequality $x - y \geq 2$. This is the shaded region in Figure 6.5.3(b).

REMARK   The boundary line divides the Cartesian plane into two half-planes. After you graph the boundary line, pick a point in one half-plane and substitute its coordinates into the given inequality. (Be sure that the point is not on the boundary line.) If that point yields a true statement, then the half-plane that includes this point is the graph of the inequality. If that point does not yield a true statement, then the other half-plane is the graph of the inequality.

EXAMPLE 6.5.4   Graph the following inequalities.

1. $2x + 5y < 10$

We first sketch the boundary line $2x + 5y = 10$. Since the inequality is $<$, all of the points on the boundary line do *not* satisfy the given inequality, and we draw the boundary line dashed. Now we pick some point that is not on the boundary line. The easiest point to use is the origin. Substituting (0, 0) into the given inequality, we obtain $2(0) + 5(0) < 10$, or $0 < 10$. Since this is a true statement, the half-plane that contains (0, 0) is the graph of the inequality. We indicate this by shading the region below the boundary line in Figure 6.5.4.

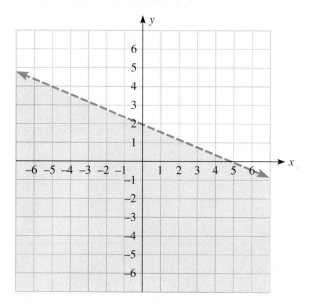

Figure 6.5.4

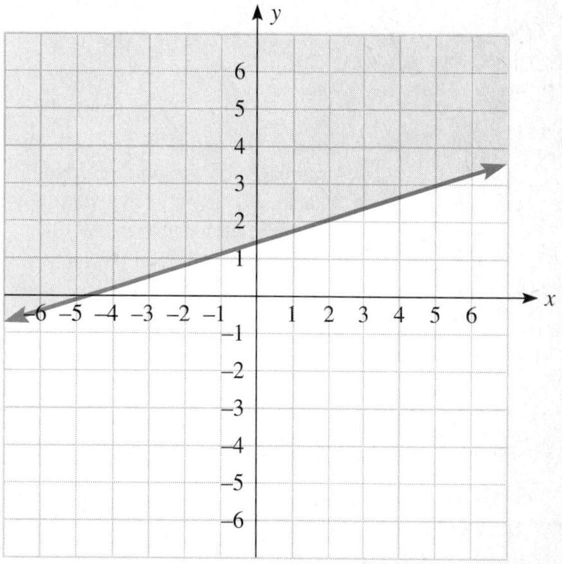

Figure 6.5.5

2. $x - 3y \leq -4$

We first sketch the boundary line $x - 3y = -4$ with a solid line since the inequality is $\leq$. Next we pick some point that is not on the boundary line. Let us use the origin again. Substituting $(0, 0)$ into the given inequality, we obtain $0 - 3(0) \leq -4$, or $0 \leq -4$. Since this is a false statement, the half-plane below the line that contains $(0, 0)$ is *not* the graph of the inequality. Instead we shade the half-plane above the boundary line (see Figure 6.5.5).

EXERCISES 6.5

Graph the following inequalities.

1. $x < 2$

2. $x > 4$

3. $y \geq 3$

4. $y \leq 5$

5. $y < 2x$

6. $y > \frac{3}{2}x$

7. $4x + 3y \geq 0$

8. $2x - 5y \geq 0$

9. $x + y < 4$

10. $x - y \leq 3$

11. $x - y \leq -1$

12. $x + y > 5$

13. $x - 2y > 5$

14. $x - 4y < 4$

15. $3x + y \leq -6$

16. $5x + y \geq -5$

17. $y < \frac{3}{4}x - 2$

18. $y > -\frac{2}{5}x + 3$

19. $4x - 3y \leq 12$

20. $3x - 2y \geq -6$

21. $6x + 4y > 9$

22. $5x + 3y < -9$

23. $\frac{3}{7}x - 2y < 6$

24. $\frac{4}{3}x + 5y > 10$

25. $\frac{2}{5}x + \frac{3}{2}y \le -3$ **26.** $\frac{5}{6}x - \frac{3}{4}y \ge -\frac{5}{2}$ **27.** $\frac{8x - 3y}{5} \ge 2$ **28.** $\frac{4x - 9y}{7} \le 3$

29. $5x + 8y + 10 < 0$ **30.** $3x - 7y + 9 > 0$

31. Judy is starting an exercise program that calls for her to walk and run each day for no more than 30 minutes. Find a linear inequality in two variables, x and y, that gives the relationship between the number of minutes she can walk and the number of minutes she can run each day. Graph the inequality, but note that $x \ge 0$ and $y \ge 0$.

Chapter 6
Review

		Page		*Page*
Terms to Remember	Linear equation in		Graphing an equation	298
	two variables	287	y-intercept	301
	Ordered pair	288	x-intercept	301
	First coordinate	288	Rise	306
	x-coordinate	288	Run	306
	Second coordinate	288	Slope	306
	y-coordinate	288	Point-slope form	313
	Cartesian coordinate		Standard form	314
	system	294	Slope-intercept form	316
	Origin	294	Parallel lines	318
	x-axis	294	Perpendicular lines	318
	y-axis	294	Linear inequality in two	
	Graphing an ordered pair	296	variables	321
	Plotting a point	296	Boundary line	323
	Quadrants	297	Half-plane	323

Formulas and Equations

Slope: The line passing through the points (x_1, y_1) and (x_2, y_2) has a slope

$$m = \frac{y_2 - y_1}{x_2 - x_1}$$

Point-slope form: The equation of the line passing through the point (x_1, y_1) and having slope m is

$$y - y_1 = m(x - x_1)$$

Slope-intercept form: The equation of the line with slope m and y-intercept $(0, b)$ is

$$y = mx + b$$

Standard form: The equation of every line can be written in the form

$$Ax + By = C$$

where A and B are not both zero.

Horizontal line: Every horizontal line has slope $m = 0$. The equation of the horizontal line with y-intercept $(0, b)$ is

$$y = b$$

Vertical line: The slope of every vertical line is undefined. The equation of the vertical line with x-intercept $(a, 0)$ is

$$x = a$$

Review Exercises

6.1 For each equation, determine whether the given ordered pair is a solution.

1. $3x - y = 8$, $(2, -2)$

2. $-4x + 7y + 5 = 0$, $(2, 1)$

3. $y = 2x - 6$, $\left(\frac{3}{2}, 3\right)$

4. $\frac{2}{3}x + 5y = -\frac{7}{6}$, $\left(\frac{3}{4}, -\frac{1}{3}\right)$

5. $2.5x + 3.2y = 4$, $(8, -5)$

6. $y + 5 = 2$, $(-3, 7)$

Find the solutions for each equation using the indicated values for x or y.

7. $6x - 5y = 10$; $(5, \quad)$, $(0, \quad)$, $(\quad, 0)$, $(\quad, -8)$

8. $4x + 7y = 11$; $(3, \quad)$, $(-6, \quad)$, $(\quad, 1)$, $\left(\quad, \frac{3}{4}\right)$

9. $3x - 4y + 9 = 0$; $(1, \quad)$, $\left(\frac{7}{3}, \quad\right)$, $(\quad, -6)$, $\left(\quad, \frac{3}{4}\right)$

10. $\frac{3}{4}x + y = 5$; $(4, \quad)$, $\left(-\frac{8}{3}, \quad\right)$, $(\quad, -1)$, $(\quad, 2)$

11. $1.3x - 2y = 2.6$; $(4, \quad)$, $(-10, \quad)$, $(\quad, 0)$, $(\quad, 0.65)$

12. $x = 5$; $(5, \quad)$, $(\quad, 0)$, $(\quad, -1)$, $(\quad, -8)$

Solve each equation for y, then find solutions using the indicated values for x.

13. $4x - y = 7$; $x = -2, 0, 5$

14. $3x - 5y = 7$; $x = -2, 0, 4$

15. $\frac{1}{4}x + \frac{2}{3}y = 6$; $x = 8, -4, \frac{8}{3}$

Solve each equation for x, then find solutions using the indicated values for y.

16. $x - 7y = 3$; $y = -1, 0, 2$

17. $3x - 5y = 9$; $y = -3, -2, 1$

18. $\frac{1}{5}x + \frac{1}{3}y = 1$; $y = 3, \frac{3}{2}, -6$

Find a linear equation in two variables for each of problems 19 and 20. Then use that equation to answer the questions.

19. A dog owner wants to enclose a rectangular dog run with 72 feet of fence. If the width is 6 feet, what is the length? If the length is 26 feet, what is the width?

20. Buzz drove his sports car up IH 35 at 67 miles per hour. If he drove for $2\frac{1}{2}$ hours, how far did he travel? If he drove 335 miles, how much time did he spend driving?

6.2   Graph the following sets of ordered pairs.

21. $\{(0, 5), (1, 4), (2, 3), (3, 2), (4, 1), (5, 0)\}$

22. $\{(-3, 1), (-2, 4), \left(-\frac{3}{2}, \frac{19}{4}\right), (-1, 5), \left(-\frac{1}{2}, \frac{19}{4}\right), (0, 4), (1, 1)\}$

Graph the following equations.

23. $x + y = -3$
24. $3x - y = 9$
25. $5x - 6y = -10$
26. $4x + 3y = 6$

27. $\frac{4}{5}x + 3y = \frac{6}{5}$
28. $-2x + 1.5y = 3.6$
29. $x = 2$
30. $3y + 5 = 0$

Find the x- and y-intercepts, and use them to graph each equation.

31. $-x + y = -5$
32. $x - 4y = 6$
33. $2x - 6y = -15$
34. $-5x - 7y = 14$

35. $\frac{5}{6}x + \frac{3}{4}y = 5$
36. $2.1x - 3.5y = 2.8$
37. $y = -\frac{5}{8}x + 2$
38. $8x + 5y = 0$

Solve each equation for y, then use the new equation to find ordered pair solutions. Also, graph the equations.

39. $-4x + y = -3$
40. $3x - 5y = 10$
41. $9x + 2y + 6 = 0$
42. $\dfrac{6x - 7y}{5} = -3$

6.3   Find the slopes of the lines passing through the following pairs of points.

43. $(4, -5), (2, -1)$
44. $(-3, 8), (1, 2)$
45. $\left(\frac{1}{4}, -2\right), \left(\frac{7}{4}, 1\right)$
46. $(2.7, 3), (1, 3)$

Find the slopes of the lines whose equations are given.

47. $8x + 5y = 2$
48. $4x - 5y = 8$
49. $x + \frac{7}{4}y = -7$

50. $y = \dfrac{3x + 5}{4}$
51. $x - 3 = 0$
52. $y + 6 = 0$

Graph the lines satisfying the following conditions.

53. $m = \frac{1}{3}$, passing through $(-4, 2)$
54. $m = \dfrac{-4}{5}$, passing through $(2, -1)$

55. $m = -2$, passing through $(-1, -2)$
56. $m = 5$, passing through $(1, 3)$

57. $m = 0$, passing through $(4, -5)$
58. Parallel to the x-axis, passing through $(-3, -5)$

6.4   Find the slopes and y-intercepts of the lines whose equations are given.

59. $y = \frac{2}{5}x - 2$
60. $6x - 5y = 4$
61. $2x + 4y - 7 = 0$
62. $y + 4 = 0$

Find the equations of the lines satisfying the given information. Write your answers in slope-intercept form.

63. $m = \dfrac{-1}{3}$, passing through $(-4, 2)$

64. $m = \frac{5}{4}$, passing through $(-2, -3)$

65. $m = 0$, passing through $(3, -6)$

66. Passing through $(1, 4)$ and $(5, -2)$

67. Passing through $(-1, 5)$ and $(2, 6)$

68. x-intercept $= -3$, y-intercept $= -1$

Find the equations of the lines satisfying the given information. Write your answers in standard form.

69. $m = \frac{6}{5}$, passing through $(-5, -2)$

70. $m = 3$, passing through $(-1, 2)$

71. $m = 0$, passing through $(9, 3)$

72. Passing through $(4, 2)$ and $(4, 10)$

73. Passing through $(-3, 4)$ and $(6, -2)$

74. Passing through $(-7, -4)$ and $(-1, 4)$

Determine whether the following pairs of lines are parallel, perpendicular, or neither.

75. $3x - 4y = 9$, $\dfrac{-3}{8}x + \dfrac{1}{2}y = 5$

76. $5x + 2y = 10$, $\dfrac{-5}{6}x - \dfrac{1}{3}y = 4$

77. $6x + 8y = 7$, $4x - 3y = -3$

78. $-4x + 7y = 8$, $\dfrac{7}{2}x + 2y = -3$

6.5 Graph the following inequalities.

79. $x \le -5$

80. $y > -4$

81. $y < \dfrac{-3}{4}x$

82. $x + y \le 6$

83. $x - 5y > -5$

84. $4x - y < 2$

85. $3x + 8y \le 12$

86. $5x + 6y \ge -15$

Chapter 6 Test   (You should be able to complete this test in 60 minutes.)

1. Find solutions for the equation $4x - 5y = -8$ using the following values for x or y.

$(3, \quad), (0, \quad), (\quad, 0), (\quad -2)$

2. Solve the equation $5x + 6y = 9$ for y, then find solutions using $x = 1, 0$, and 3.

Graph the following equations.

3. $-3x + 7y = 14$

4. $x = -3$

5. $y = 4$

6. $6x - 15y = 20$, using the intercepts

Find the slopes of the lines passing through the following pairs of points.

7. $(-3, 5), (-1, -5)$

8. $(6, 5), (3, 1)$

Find the slopes of the lines whose equations are given.

9. $7x - 3y = 9$

10. $y - 5 = 0$

11. Graph the line with $m = \frac{5}{3}$, passing through $(-2, 1)$.

Find the equations of the lines satisfying the given information. Write your answers in standard form.

12. $m = \dfrac{-2}{5}$, passing through $(10, -3)$

13. m is undefined, passing through $(-6, 4)$

14. Passing through $(-4, -1)$ and $(2, -3)$

15. Passing through $(4, -5)$ and $(1, -5)$

16. Determine whether the lines $4x + 5y = -8$ and $\frac{5}{2}x - 2y = 3$ are parallel, perpendicular, or neither.

Graph the following inequalities.

17. $2x + 5y \leq 0$

18. $6x - 4y < 9$

Test Your Memory

These problems review Chapters 1–6.

I. Find the solution set of each of the following inequalities, and graph the solution set on a number line.

1. $-3 < 2 - x < 6$

2. $\frac{1}{2}x + \frac{3}{4} \leq \frac{1}{3}x + 1$

II. Sketch the graph of the following.

3. $2x - 5y = 6$

4. $x = -3$

5. $y = 4$

6. The line with $m = \frac{3}{4}$, passing through $(1, -2)$

III. Graph the following inequalities.

7. $x - 3y \leq 0$

8. $3x + 6y > 3$

IV. Find the solutions of the following equations.

9. $5(x - 2) + 6(x + 1) = 4(x - 1)$

10. $\frac{1}{2}\left(\frac{5}{3}x + \frac{3}{2}\right) + 2x = \frac{1}{3}(x + 2) - \frac{1}{3}$

11. $(2x + 5)(x + 1) = 2$

12. $(x + 2)(2x - 4) = (x + 4)(x - 1)$

13. $\dfrac{5}{3} = \dfrac{3x}{x - 2}$

14. $\dfrac{x}{2} = \dfrac{2}{3x + 4}$

15. $\dfrac{5}{x + 1} - \dfrac{3}{x} = \dfrac{1}{3x}$

16. $\dfrac{1}{x^2} - \dfrac{1}{6x} = \dfrac{1}{3}$

17. $\dfrac{x - 2}{x + 3} - \dfrac{1}{x + 2} = \dfrac{5}{x^2 + 5x + 6}$

18. $\dfrac{4}{x} + \dfrac{1}{x - 3} = \dfrac{3}{x^2 - 3x}$

V. Simplify each of the following. Be sure to reduce your answer to lowest terms.

19. $\left(\dfrac{x^3 y^{-4}}{x^5 y^{-1}}\right)^{-3}$

20. $\dfrac{(3x^2 y^{-3})^{-2}}{(3^{-1} x^{-2} y^4)^{-1}}$

21. $\dfrac{16x^2 + 24xy + 9y^2}{16x^2 - 9y^2}$

22. $\dfrac{\dfrac{1}{x^2} + \dfrac{1}{2x}}{\dfrac{1}{5x^2} + \dfrac{1}{10x}}$

23. $\dfrac{\dfrac{1}{4} - \dfrac{1}{2x}}{\dfrac{1}{4} - \dfrac{1}{x^2}}$

24. $\dfrac{\dfrac{4}{x} - \dfrac{3}{x + 2}}{\dfrac{1}{x} + 1}$

VI. Perform the indicated operations.

25. $6x^2y^3(2x^4 - 7xy^2 - 3y^3)$

26. $(5x^3 - 7x^2 + 3)(2x - 4)$

27. $(12x^3 - 11x^2 - 10x + 6) \div (3x - 2)$

28. $\dfrac{10x^2 + 5x}{x^2 - 8x + 16} \div \dfrac{6x^3 + 3x^2}{x^2 - 6x + 8}$

29. $\dfrac{2x^2 - 3x}{x^2 + 2x + 1} + \dfrac{2x - 3}{x^2 + 2x + 1}$

30. $\dfrac{2x^2 - 9}{x^2 - x - 12} - \dfrac{x^2 + 7}{x^2 - x - 12}$

31. $\dfrac{3}{4xy} - \dfrac{9}{4x^2}$

32. $\dfrac{2}{x - 1} - \dfrac{5}{3x + 4}$

33. $\dfrac{5}{x^2 + 4x + 4} - \dfrac{1}{x^2 - x - 6}$

34. $\dfrac{2x + 14}{x^2 - x - 6} + \dfrac{x - 6}{x^2 - 4}$

VII. Find the slopes of the following lines.

35. The line passing through $(-3, 8)$ and $(7, -4)$

36. The line passing through $(1, -2)$ and $(-2, -2)$

37. The line whose equation is $2x - 5y = 8$

38. The line whose equation is $x + 3 = 0$

VIII. Find the equations of the lines satisfying the given information. Write your answers in standard form.

39. $m = \frac{3}{4}$, passing through $(-1, 2)$

40. $m = 0$, passing through $(-3, 5)$

41. Passing through $(7, -4)$ and $(-5, 4)$

42. Passing through $(3, -1)$ and $(3, 7)$

IX. Determine whether the following pairs of lines are parallel, perpendicular, or neither.

43. $3x - 2y = -8$ and $2x + 3y = -3$

44. $x - 2y = 2$ and $2x - 4y = 5$

X. Use algebraic expressions to find the solutions of the following problems.

45. Rita has $2.70 in dimes and quarters. She has a total of 18 coins. How many quarters and how many dimes does Rita have?

46. Lamont invested $7000 in two accounts. He invested some in a savings account paying 6% simple interest and the rest in a CD paying 8% simple interest. After 1 year his interest income was $530. How much did Lamont invest in each account?

47. The area of a rectangle is 32 sq ft. The length is 4 ft less than three times the width. What are the dimensions of the rectangle?

48. Paul's pasta with catfish and artichokes recipe includes $\frac{2}{3}$ cup of heavy cream for an 8-person serving. How much heavy cream will be required for a recipe serving 20 people?

49. A swimming pool can be filled by an inlet pipe in 15 hours. The drain can empty the pool in 20 hours. How long will it take to fill the pool if the drain is left open?

50. Eve can row her boat 5 miles up Oak Ridge Creek in the same time that it takes her to row 7 miles down the creek. If Eve can row 3 mph in still water, how fast is the current in Oak Ridge Creek?

Systems of Linear Equations and Inequalities in Two Variables

7.1 Solving a Linear System by Graphing

In Chapter 6, we saw how the solutions to a linear equation in two variables were ordered pairs. If we consider two linear equations together, we form a **linear system of equations**. Examples of linear systems are

$$3x + y = 1 \qquad x + 8y = -1 \qquad x = 3$$
$$x - 2y = -9, \qquad 3x + 4y = 3, \qquad y = 2x - 4$$

The **solutions of a linear system of equations** are the ordered pairs that satisfy *both* equations in the system.

EXAMPLE 7.1.1

For each system determine whether the given ordered pair is a solution.

1. $3x + y = 1$, $(-1, 4)$
   $x - 2y = -9$

In the first equation,   In the second equation,
$$3(-1) + 4 \overset{?}{=} 1 \qquad\qquad -1 - 2 \cdot 4 \overset{?}{=} -9$$
$$-3 + 4 \overset{?}{=} 1 \qquad\qquad -1 - 8 \overset{?}{=} -9$$
$$1 = 1 \qquad\qquad\qquad -9 = -9$$

$(-1, 4)$ is a solution of the system because it satisfies both equations.

2. $x + 8y = -1$, $\left(3, -\frac{1}{2}\right)$
   $3x + 4y = 3$

In the first equation, In the second equation,

$$3 + 8\left(-\tfrac{1}{2}\right) \overset{?}{=} -1 \qquad 3 \cdot 3 + 4\left(-\tfrac{1}{2}\right) \overset{?}{=} 3$$

$$3 + (-4) \overset{?}{=} -1 \qquad 9 + (-2) \overset{?}{=} 3$$

$$-1 = -1 \qquad 7 \neq 3$$

$\left(3, -\tfrac{1}{2}\right)$ is not a solution of the system because it does not satisfy both equations.

In this chapter we will study three methods for finding the solutions of a linear system of equations. In this section we start with the **graphing method**. The graphing method is based on the fact that if we graph both linear equations on one coordinate system and find a point of intersection, the coordinates of this point form an ordered pair that should satisfy both equations.

EXAMPLE 7.1.2

Using the graphing method, determine the solutions of the following linear systems of equations.

1. $x + y = 3$

$x - 2y = -9$

The equations are graphed in Figure 7.1.1, and the point with coordinates $(-1, 4)$ appears to be the point of intersection of the two lines. To be sure that we have found the solution, we must check $(-1, 4)$ in both equations:

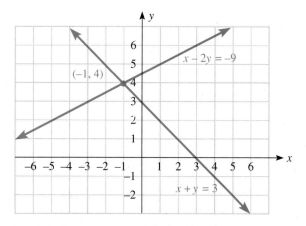

Figure 7.1.1

$$x + y = 3 \qquad x - 2y = -9$$
$$-1 + 4 \overset{?}{=} 3 \qquad -1 - 2(4) \overset{?}{=} -9$$
$$3 = 3 \qquad -1 - 8 \overset{?}{=} -9$$
$$-9 = -9$$

Thus, $(-1, 4)$ is the solution of this linear system of equations.

2. $2x - y = 4$

   $x - 3y = -3$

   From Figure 7.1.2 the solution of the system appears to be the ordered pair $(3, 2)$. Again, we must check $(3, 2)$ in both equations:

$$2x - y = 4 \qquad x - 3y = -3$$
$$2(3) - 2 \overset{?}{=} 4 \qquad 3 - 3(2) \overset{?}{=} -3$$
$$6 - 2 \overset{?}{=} 4 \qquad 3 - 6 \overset{?}{=} -3$$
$$4 = 4 \qquad -3 = -3$$

Thus, $(3, 2)$ is the solution of the linear system of equations.

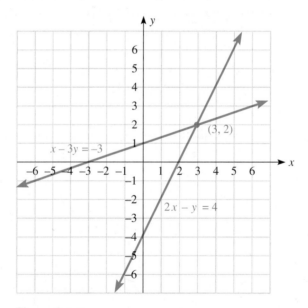

Figure 7.1.2

In the previous example, we considered two systems each of which had only one solution. A system of equations with only one solution is called **independent**. We will see in the next example that there are other types of systems of equations.

EXAMPLE 7.1.3

Using the graphing method, determine the solutions of the following linear systems of equations.

1. $x - 2y = 2$

$\quad\quad 2y = x - 5$

Figure 7.1.3 shows the graphs of the equations. The two lines appear to be parallel. In slope-intercept form, the equations are

$$y = \tfrac{1}{2}x - 1$$

$$y = \tfrac{1}{2}x - \tfrac{5}{2}$$

Since both lines have a slope of $\tfrac{1}{2}$, they are parallel. Therefore, there is no solution to the system of equations.

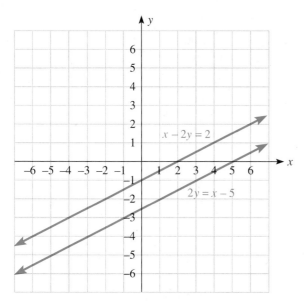

Figure 7.1.3

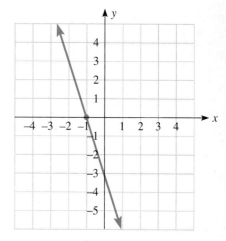

Figure 7.1.4

REMARK A linear system of equations that has no solutions is called **inconsistent**.

2. $\quad 3x + y = -3$

$\quad\quad 6x + 2y = -6$

The two equations have the same graph; that is, they each yield the same line. (See Figure 7.1.4.) Any ordered pair that satisfies one equation also satisfies the other equation. Thus, there are an infinite number of ordered pair solutions.

REMARK A linear system of equations that has infinitely many solutions is called **dependent**.

In summary, the graphing method involves the following steps.

Graphing Method

1. Graph both equations on the same coordinate system.
2. Use the graphs to determine the type of solution.

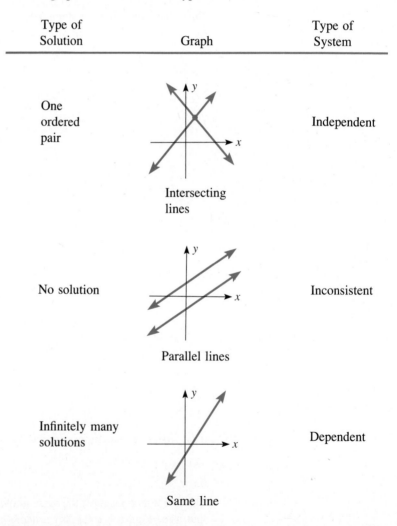

Type of Solution	Graph	Type of System
One ordered pair	Intersecting lines	Independent
No solution	Parallel lines	Inconsistent
Infinitely many solutions	Same line	Dependent

3. In the case of one ordered pair solution, read the coordinates of the ordered pair off the graph.

4. Check the ordered pair in *both* equations.

We end this section with one final comment. The system

$$x - 4y = -8$$
$$2x - 14y = 7$$

has the ordered pair $\left(-23\frac{1}{3}, -3\frac{5}{6}\right)$ as its only solution. You are not likely to be able to find this solution by the graphing method, no matter how carefully you graph your lines! Thus, we need other methods for solving a system of equations that do not rely on graphs. These methods are covered in the next two sections.

EXERCISES 7.1

For each system determine whether the given ordered pair is a solution.

1. $2x - y = 1$, $(4, 7)$
$x - y = -3$

2. $x + 2y = 8$, $(6, 1)$
$x - 3y = 3$

3. $4x + y = 0$, $(-1, 4)$
$3x - y = 7$

4. $3x + 2y = 0$, $(-2, 3)$
$x + y = 1$

5. $x - 5y = -23$, $(-8, -3)$
$x + y = -11$

6. $x + 2y = -13$, $(-5, -4)$
$-x + y = 9$

7. $7x + 2y = -1$, $(3, -11)$
$2x - 3y = 39$

8. $2x + 5y = 26$, $(9, -2)$
$3x - 2y = 23$

9. $6x + y = 1$, $\left(\frac{1}{2}, -2\right)$
$10x + 3y = -1$

10. $2x - 3y = 4$, $\left(1, -\frac{2}{3}\right)$
$5x + 6y = 1$

Using the graphing method, determine the solutions of the following linear systems of equations. If there are no solutions, write *inconsistent*. If there are infinite solutions, write *dependent*.

11. $x + y = 1$
$x + 2y = -2$

12. $x + 3y = 1$
$x - y = -3$

13. $2x - y = 2$
$x - y = 0$

14. $x + y = -5$
$3x - y = 1$

15. $x + 2y = -3$
$2x + 5y = -5$

16. $3y = x + 4$
$x - 3y = 1$

17. $3x - 2y = -2$
$x + y = -9$

18. $2x + y = 6$
$x - y = 3$

19. $y = \frac{1}{2}x + 7$
$y = -x + 4$

20. $y = \frac{3}{2}x - 2$
$y = -\frac{1}{2}x + 6$

21. $y = -2x - 2$
$3x + 2y = -4$

22. $y = x - 6$
$x + 4y = 1$

23. $5x - y = 3$
$-5x + y + 5 = 0$

24. $x + y + 5 = 0$
$2x - y = 5$

25. $3x + 4y - 3 = 0$
$x - 3y + 12 = 0$

26. $3x - 6y - 3 = 0$
$2y - x + 1 = 0$

27. $2x - 3y = 5$
$4x = 6y + 10$

28. $x - 2y = 0$
$x = 3y - 3$

29. $5y + 5 = -3x$
$5y + 25 = x$

30. $3y - 6 = 4x$
$2y - 4 = -3x$

31. $x = \frac{2}{5}y + 3$
     $x = \frac{2}{5}y - 1$

32. $x = -\frac{5}{2}y - 3$
     $x = \frac{1}{2}y + 3$

33. $x = \frac{2y - 4}{7}$

     $x = \frac{3y + 7}{4}$

34. $x = \frac{4y + 5}{-3}$

     $x = \frac{y - 7}{2}$

35. $y = \frac{x - 5}{2}$

     $y = \frac{3x + 5}{2}$

36. $y = \frac{3x - 5}{4}$

     $y = \frac{2x + 8}{-3}$

37. $x - 3 = 0$
     $x - 2y = 1$

38. $x + 4 = 0$
     $x - 3y + 4 = 0$

39. $3x - 5y = 2$
     $y + 1 = 0$

40. $y - 6 = 0$
     $4x - y = 2$

For each system: (a) Determine whether the given ordered pairs are solutions.
(b) From your answers to part (a), determine whether each system is
dependent.

41. $2x - 3y = 3$ $\left(\frac{3}{2}, 0\right), \left(\frac{1}{2}, -\frac{2}{3}\right)$
     $\frac{2x}{3} + 5y = -3$ $(3, -1)$

42. $3x - 2y = 2$ $\left(\frac{2}{3}, 0\right), \left(-\frac{1}{3}, 2\right)$
     $x = \frac{2y + 2}{3}$ $\left(\frac{1}{2}, -\frac{1}{4}\right)$

43. $4x + 12y = -9$ $\left(3, \frac{7}{4}\right), \left(\frac{3}{4}, -1\right)$
     $x + 3y + \frac{9}{4} = 0$ $\left(-\frac{3}{2}, -\frac{1}{4}\right)$

44. $x + 2y = 1$ $\left(\frac{2}{5}, \frac{3}{10}\right), \left(-2, \frac{3}{2}\right)$
     $3x - 4y = 0$ $(4, 3)$

7.2 The Substitution Method

As we mentioned at the end of the previous section, a system with a solution like $\left(-23\frac{1}{3}, -3\frac{5}{6}\right)$ is not solvable by the graphing method. There are two common algebraic methods that precisely find the solutions of a linear system of equations. The first method is called the **substitution method**. We develop this method in the following example.

EXAMPLE 7.2.1

Determine the solutions of the following linear systems of equations.

1. $y = 3$
     $2x + y = 1$

The ordered pair solutions of the first equation must satisfy the condition that their y-coordinate is 3. If we can find an ordered pair solution of the second equation that has a y-coordinate of 3, we will have an ordered pair solution to the system. We can do this by substituting 3 for y in the second equation:

$2x + 3 = 1$

$$2x = -2$$
$$x = -1$$

Thus, $(-1, 3)$ is the ordered pair that satisfies both equations.

REMARK The substitution method relies on the substitution property of real numbers, which states: If $a = b$, then any expression involving a retains the same value if a is replaced by b. Thus, in the above system, since $y = 3$ we could replace y by 3 in the second equation.

2. $y = x + 1$
$$x + 2y = -7$$

The ordered pair solutions of the first equation must satisfy the condition that their y-coordinate is one more than the x-coordinate. If we can find an ordered pair solution of the second equation that satisfies this same condition, we will have an ordered pair solution to the system. We can do this by substituting $x + 1$ for y in the second equation:

$$x + 2(x + 1) = -7$$
$$x + 2x + 2 = -7$$
$$3x + 2 = -7$$
$$3x = -9$$
$$x = -3$$

By substituting $x = -3$ in the first equation we obtain

$$y = -3 + 1$$
$$y = -2$$

Thus, the ordered pair solution of the system is $(-3, -2)$. Let us check the solution:

In the first equation, In the second equation,

$$-2 \overset{?}{=} -3 + 1 \qquad -3 + 2(-2) \overset{?}{=} -7$$
$$-2 = -2 \qquad -3 + (-4) \overset{?}{=} -7$$
$$-7 = -7$$

3. $3x + 4y = 6$
$$x = 2y + 7$$

The ordered pair solutions of the second equation must satisfy the condition that their x-coordinate is 7 more than twice the y-coordinate. We can find an ordered pair solution of the first equation that satisfies this same condition by substituting $2y + 7$ for x in the first equation:

$$3(2y + 7) + 4y = 6$$
$$6y + 21 + 4y = 6$$

$$10y + 21 = 6$$
$$10y = -15$$
$$y = -\frac{3}{2}$$

By substituting $y = -\frac{3}{2}$ in the second equation, we obtain

$$x = 2\left(-\frac{3}{2}\right) + 7$$
$$x = -3 + 7$$
$$x = 4$$

Thus, the ordered pair solution of the system is $\left(4, -\frac{3}{2}\right)$. The check is left to the reader.

In each of the preceding systems, one of the equations was solved for x or solved for y. If neither equation of the system is in this form, then one of the equations must be solved for x or y.

EXAMPLE 7.2.2

Using the substitution method, determine the solutions of the following linear systems of equations.

1. $x - 3y = -1$
   $5x - 7y = 19$

   Let's solve the first equation for x by adding $3y$ to both sides:

   $$x - 3y = -1$$
   $$x = 3y - 1$$

   Substitute $3y - 1$ for x in the second equation:

   $$5(3y - 1) - 7y = 19$$
   $$15y - 5 - 7y = 19$$
   $$8y - 5 = 19$$
   $$8y = 24$$
   $$y = 3$$

   Since $x = 3y - 1$ and $y = 3$,

   $$x = 3(3) - 1$$
   $$x = 9 - 1$$
   $$x = 8$$

Thus, the solution of the system is the ordered pair (8, 3). The reader should check the solution in the two original equations of the system.

2. $2x + 7y = 4$

$4x - y = 3$

Solve the second equation for y:

$4x - y = 3$

$\quad\quad -y = 3 - 4x$ Subtract $4x$ from both sides.

or $\quad\quad y = -3 + 4x$ Multiply both sides by -1.

$\quad\quad y = 4x - 3$

Substitute $4x - 3$ for y in the first equation:

$2x + 7(4x - 3) = 4$

$2x + 28x - 21 = 4$

$30x - 21 = 4$

$30x = 25$

$x = \frac{5}{6}$

Since $y = 4x - 3$ and $x = \frac{5}{6}$,

$y = 4\left(\frac{5}{6}\right) - 3$

$y = \frac{10}{3} - 3$

$y = \frac{1}{3}$

Thus, the solution of the system is the ordered pair $\left(\frac{5}{6}, \frac{1}{3}\right)$. The reader should check the solution in the two original equations of the system.

3. $6x + 2y = 1$

$2x - 3y = -18$

Solve the first equation for y:

$6x + 2y = 1$

$\quad\quad 2y = 1 - 6x$ Subtract $6x$ from both sides.

$\quad\quad y = \frac{1}{2} - 3x$ Multiply both sides by $\frac{1}{2}$.

Substitute $\frac{1}{2} - 3x$ for y in the second equation:

$2x - 3\left(\frac{1}{2} - 3x\right) = -18$

$2x - \frac{3}{2} + 9x = -18$

$11x - \frac{3}{2} = -18$

$11x = \frac{-33}{2}$

$x = \frac{-3}{2}$

Since $y = \frac{1}{2} - 3x$ and $x = \frac{-3}{2}$,

$$y = \frac{1}{2} - 3\left(-\frac{3}{2}\right)$$

$$y = \frac{1}{2} + \frac{9}{2}$$

$$y = 5$$

Thus, the solution of the system is the ordered pair $\left(-\frac{3}{2}, 5\right)$. The reader should check the solution in the two original equations of the system.

4. $\frac{3}{4}x + \frac{2}{3}y = -1$

$\frac{5}{2}x + 6y = 8$

Let us first simplify each equation by multiplying by the LCD of the fractions in each equation:

$$\frac{3}{4}x + \frac{2}{3}y = -1 \xrightarrow[\text{sides by 12.}]{\text{Multiply both}} 9x + 8y = -12$$

$$\frac{5}{2}x + 6y = 8 \xrightarrow[\text{sides by 2.}]{\text{Multiply both}} 5x + 12y = 16$$

Solve the second equation for x:

$$5x + 12y = 16$$

$$5x = 16 - 12y \qquad \text{Subtract } 12y \text{ from both sides.}$$

$$x = \frac{16}{5} - \frac{12}{5}y \qquad \text{Multiply both sides by } \frac{1}{5}.$$

Substitute $\frac{16}{5} - \frac{12}{5}y$ for x in the first equation:

$$9\left(\frac{16}{5} - \frac{12}{5}y\right) + 8y = -12$$

$$\frac{144}{5} - \frac{108}{5}y + 8y = -12$$

$$\frac{144}{5} - \frac{68}{5}y = -12$$

$$-\frac{68}{5}y = -\frac{204}{5}$$

$$y = 3$$

Since $x = \frac{16}{5} - \frac{12}{5}y$ and $y = 3$,

$$x = \frac{16}{5} - \frac{12}{5}(3)$$

$$x = \frac{16}{5} - \frac{36}{5}$$

$$x = \frac{-20}{5} = -4$$

Thus, the solution of the system is the ordered pair $(-4, 3)$. The reader should check the solution in the two original equations of the system.

In the last example of this section, we will see how the substitution method tells us if a system is inconsistent or dependent.

EXAMPLE 7.2.3

Using the substitution method, determine the solutions of the following linear systems of equations.

1. $x - 2y = 2$

$\qquad 2y = x - 5$

Solve the first equation for x:

$x - 2y = 2$

$\qquad x = 2y + 2 \qquad$ Add $2y$ to both sides.

Substitute $2y + 2$ for x in the second equation:

$2y = 2y + 2 - 5$

$2y = 2y - 3$

$\ 0 = -3 \qquad\qquad$ A false statement

This is the same system as in part 1 of Example 7.1.3. The graphing method showed that these equations represented parallel lines and that there was no solution of the system. This observation leads to a conclusion about the substitution method: *If the variable is eliminated leaving a false statement, then the given system has no solution and is inconsistent.*

2. $\quad 3x + y = -3$

$\quad 6x + 2y = -6$

Solve the first equation for y:

$3x + y = -3$

$\qquad y = -3x - 3 \qquad$ Subtract $3x$ from both sides.

Substitute $-3x - 3$ for y in the second equation:

$6x + 2(-3x - 3) = -6$

$\quad 6x - 6x - 6 = -6$

$\qquad\qquad -6 = -6 \qquad$ A true statement

This is the same system as in part 2 of Example 7.1.3. The graphing method showed that these equations represented the same line and that there were infinite solutions to the system. This observation leads to another conclusion about the substitution method: *If the variable is eliminated leaving a true statement, then the given system has infinite solutions and is dependent.*

In summary, the substitution method involves the following steps.

Substitution Method

1. Solve one of the equations for one of the variables.

2. *Substitute* the expression obtained in step 1 in the other equation.

3. You now have an equation with one variable. Solve this equation.

   a. If the equation has one solution, one coordinate of the ordered pair solution has been obtained.

   b. If the variable is eliminated leaving a false statement, the system has no solution and is inconsistent.

   c. If the variable is eliminated leaving a true statement, the system has infinite solutions and is dependent.

4. If one solution was obtained in step 3, substitute this value in the equation of step 1 and solve this equation. You now have both coordinates of the ordered pair solution.

5. (Optional) Check the solution in the two original equations of the system.

EXERCISES 7.2

Using the substitution method, determine the solutions of the following linear systems of equations.

1. $x = 4$
$3x - 2y = 2$

2. $2x + 5y = -1$
$x = -3$

3. $x + 4y = 4$
$y = \frac{3}{2}$

4. $y = 5$
$4x - y = -3$

5. $x = 7y + 1$
$2x - 5y = -7$

6. $x = 2y - 5$
$x - 8y = -2$

7. $3x - 2y = 5$
$x = \frac{2}{3}y + 3$

8. $2x - y = 3$
$x = -\frac{1}{4}y + 3$

9. $y = 2 - 4x$
$9x + 2y = 2$

10. $y = 8x - 7$
$3x - 2y = 14$

11. $5x + 4y = 1$
$y = -\frac{3}{2}x + 1$

12. $7x + 3y = 6$
$y = -\frac{7}{3}x + 2$

13. $x + y = -1$
$5x - y = 5$

14. $x + y = 3$
$-2x - y = 2$

15. $6x + 5y = 23$
$x + y = 5$

16. $2x + 13y = -1$
$x + y = -6$

17. $x - 3y = 0$
$2x - y = 5$

18. $x - 5y = 0$
$-2x + 10y = 3$

19. $y - 4x = 0$
$2x - 3y = \frac{5}{2}$

20. $y - 2x = 0$
$4x + y = 5$

21. $x - y = 5$
$3x - 7y = 3$

22. $x - y = 9$
$2x - 11y = -18$

23. $x - 6y = 3$
$2x + 9y = 13$

24. $x - 8y = 1$
$7x + 4y = -8$

25. $y - 3x = 5$
    $6x - 2y = -10$

26. $y - 7x = 5$
    $8x + 3y = 44$

27. $2x - y = -1$
    $5x - 3y = 2$

28. $9x - y = 11$
    $2x + 3y = 25$

29. $3x + 8y = -29$
    $5x - y = 9$

30. $-12x + 3y = -15$
    $4x - y = 5$

31. $3x - 15y = 2$
    $3y - x = 2$

32. $5x - 12y = 1$
    $6y - x = -2$

33. $-x + 8y = 1$
    $\frac{1}{2}x + 5y = 4$

34. $-x + 10y = -10$
    $\frac{2}{5}x + 11y = -8$

35. $5x + 2y = 27$
    $2x - 3y = 26$

36. $2x - 7y = 4$
    $3x - 4y = -7$

37. $3x - 10y = -14$
    $2x - 25y = 9$

38. $7x - 2y = -1$
    $4x + 3y = 45$

39. $4x - 10y = 9$
    $-2x + 5y = 3$

40. $2x + 5y = -5$
    $4x - 7y = 24$

41. $2x + 7y = 11$
    $13x - 2y = 5$

42. $11x + 8y = -8$
    $3x + 10y = 33$

43. $\frac{3}{4}x - \frac{5}{2}y = 8$
    $\frac{7}{2}x + 2y = 10$

44. $\frac{3}{5}x - \frac{4}{3}y = 17$
    $2x - \frac{1}{5}y = -7$

45. $-\frac{5}{6}x - \frac{7}{3}y = 3$
    $\frac{1}{4}x + \frac{5}{2}y = \frac{9}{2}$

46. $\frac{2}{3}x + \frac{7}{2}y = 18$
    $\frac{7}{8}x + \frac{3}{4}y = \frac{33}{4}$

47. $\frac{2}{9}x + 16y = 3$
    $\frac{11}{15}x - \frac{3}{5}y = 1$

48. $\frac{7}{8}x - \frac{9}{4}y = 15$
    $\frac{7}{12}x - \frac{3}{2}y = 10$

49. $2(5x + 3y) = 1$
    $\frac{1}{3}(2x + 6y) = 7$

50. $3(2x - 5y) = -16$
    $\frac{1}{5}(3x - 11y) = -3$

51. $\frac{1}{4}\left(\frac{3}{2}x + 2y\right) = -2$
    $\frac{1}{10}\left(\frac{5}{6}x + \frac{1}{3}y\right) = \frac{1}{3}$

52. $\frac{3}{2}\left(\frac{2}{3}x + \frac{1}{7}y\right) = -3$
    $\frac{1}{9}\left(\frac{5}{4}x + \frac{3}{2}y\right) = \frac{3}{2}$

53. $6x - 2y + 5 = 3x + 9$
    $x - 4y - 5 = 9 - 4x$

54. $7x - 4y - 7 = 5y + 15$
    $30x + 10y - 13 = 8y - 13x$

55. $4(x - 2y) - 3 = 2(1 - y)$
    $3(y - 3x) + 10 = 7(2 - x)$

56. $4(2x - 2y + 1) = 3(1 - y)$
    $7(x - 2y + 3) = 5(5 - 2y)$

7.3 The Elimination Method

In this section we examine a second algebraic method for finding the solutions of a linear system of equations. It is called the addition or **elimination method**. It is based on the additive property of equality and the substitution property of real numbers:

If $\quad X = Y$ and $A = B$

then $\quad X + A = Y + A \quad$ Additive property of equality

and $\quad X + A = Y + B \quad$ Substitution property of real numbers

We will use the vertical format when applying this method to linear systems of equations, as follows:

$$X = Y$$
$$\underline{A = B}$$
$$X + A = Y + B$$

EXAMPLE 7.3.1

Using the elimination method, determine the solutions of the following linear systems of equations.

1. $2x + y = 1$

$\quad x - y = -10$

As discussed above, we add the equations:

$2x + y = 1$

$\underline{\quad x - y = -10}$

$3x \quad\quad = -9$

We now have an equation with only one variable. The y's were *eliminated* because they had opposite coefficients in the original equations. We now solve this equation:

$3x = -9$

$\quad x = -3$

We have the x-coordinate of the ordered pair solution. By substituting -3 for x in either of the original equations, we can find the y-coordinate. Let's use the first equation:

$2(-3) + y = 1$

$\quad -6 + y = 1$

$\quad\quad\quad y = 7$

Thus, the ordered pair solution of the system is $(-3, 7)$. Let us check the solution:

In the first equation,   In the second equation,

$2(-3) + 7 \overset{?}{=} 1$ $-3 - 7 \overset{?}{=} -10$

$\quad -6 + 7 \overset{?}{=} 1$ $\quad -10 = -10$

$\quad\quad\quad 1 = 1$

2. $-3x + 2y = 4$

$\quad\ 3x + 8y = 1$

By adding the two equations, we eliminate the x's:

$-3x + \ 2y = 4$

$\underline{\ \ 3x + \ 8y = 1}$

$\quad\quad\quad 10y = 5$ Solve this equation.

$\quad\quad\quad\ \ y = \frac{1}{2}$

Don't stop here! We have found only the y-coordinate of the ordered pair solution. Substitute $\frac{1}{2}$ for y in the first equation:

$-3x + 2\left(\frac{1}{2}\right) = 4$

$\quad -3x + 1 = 4$

$$-3x = 3$$
$$x = -1$$

Thus, the ordered pair solution of the system is $\left(-1, \frac{1}{2}\right)$. The check is left to the reader.

In each of the preceding systems, the coefficients of the x's or y's were opposites. Therefore, when we added the equations, one of the variables was eliminated. If this is not the case for a particular system, we must multiply each equation by some nonzero number so that the coefficients of either the x's or y's are opposites.

EXAMPLE 7.3.2

Using the elimination method, determine the solutions of the following linear systems of equations.

1.　$x + 3y = 16$
　　$5x + 4y = 3$

We need only multiply the first equation by -5 to have opposite coefficients on the x's:

$$x + 3y = 16 \xrightarrow[\text{sides by } -5.]{\text{Multiply both}} -5x - 15y = -80$$
$$5x + 4y = 3 \xrightarrow{} \underline{5x + 4y = 3}$$
$$-11y = -77$$
$$y = 7$$

Substitute 7 for y in the first equation:

$$x + 3 \cdot 7 = 16$$
$$x + 21 = 16$$
$$x = -5$$

The ordered pair solution of the system is $(-5, 7)$. The reader should check the solution in the two original equations of the system.

Try to avoid this mistake:

Incorrect	Correct
$x + 3y = 16 \rightarrow -5x - 15y = 16$ $5x + 4y = 3 \rightarrow 5x + 4y = 3$	$x + 3y = 16 \rightarrow -5x - 15y = -80$ $5x + 4y = 3 \rightarrow 5x + 4y = 3$
Here, we forgot to multiply the right side of the first equation.	Remember to multiply *both* sides of the equation.

2. $3x - 5y = 8$

$7x + 4y = 3$

We can choose to eliminate either variable. If the first equation is multiplied by 4 and the second equation is multiplied by 5, the coefficients of the y's will be opposites:

$3x - 5y = 8$ $\xrightarrow[\text{sides by 4.}]{\text{Multiply both}}$ $12x - 20y = 32$

$7x + 4y = 3$ $\xrightarrow[\text{sides by 5.}]{\text{Multiply both}}$ $\underline{35x + 20y = 15}$

$$47x \qquad = 47$$
$$x = 1$$

Substitute 1 for x in the first equation:

$3 \cdot 1 - 5y = 8$

$3 - 5y = 8$

$-5y = 5$

$y = -1$

The ordered pair solution of the system is $(1, -1)$. The reader should check the solution in the two original equations of the system.

3. $\frac{1}{2}x - \frac{2}{3}y = 5$

$-\frac{2}{5}x - \frac{3}{5}y = 13$

Multiply the first equation by 6 and the second equation by 5 to eliminate all of the fractions:

$\frac{1}{2}x - \frac{2}{3}y = 5$ $\xrightarrow[\text{sides by 6.}]{\text{Multiply both}}$ $3x - 4y = 30$

$-\frac{2}{5}x - \frac{3}{5}y = 13$ $\xrightarrow[\text{sides by 5.}]{\text{Multiply both}}$ $-2x - 3y = 65$

Let us eliminate x:

$3x - 4y = 30$ $\xrightarrow[\text{sides by 2.}]{\text{Multiply both}}$ $6x - 8y = 60$

$-2x - 3y = 65$ $\xrightarrow[\text{sides by 3.}]{\text{Multiply both}}$ $\underline{-6x - 9y = 195}$

$$-17y = 255$$
$$y = -15$$

Substitute -15 for y in the equation $3x - 4y = 30$:

$3x - 4(-15) = 30$

$3x + 60 = 30$

$3x = -30$

$x = -10$

The ordered pair solution of the system is $(-10, -15)$. The reader should check the solution in the two *original* equations of the system.

In the last two examples of this section, we will see how the elimination method tells us if a system is inconsistent or dependent.

EXAMPLE 7.3.3

Using the elimination method, determine the solutions of the following linear systems of equations.

1. $6x - 4y = 5$

$-9x + 6y = 7$

Let us eliminate y:

$$6x - 4y = 5 \xrightarrow[\text{sides by 3.}]{\text{Multiply both}} 18x - 12y = 15$$

$$-9x + 6y = 7 \xrightarrow[\text{sides by 2.}]{\text{Multiply both}} \underline{-18x + 12y = 14}$$

$$0 = 29 \quad \text{A false statement}$$

The x's were also eliminated! As we saw in the previous section, *when both variables are eliminated and a false statement results, then the given system has no solution and is inconsistent.*

2. $2x - 5y = 10$

$\frac{1}{2}y = \frac{1}{5}x - 1$

We first write the second equation in the form $ax + by = c$:

$$\frac{1}{2}y = \frac{1}{5}x - 1$$

$$-\frac{1}{5}x + \frac{1}{2}y = -1 \qquad \text{Add } -\frac{1}{5}x \text{ to both sides.}$$

$$10\left(-\frac{1}{5}x\right) + 10\left(\frac{1}{2}y\right) = 10(-1) \qquad \text{Multiply both sides by 10.}$$

$$-2x + 5y = -10$$

We can eliminate x by adding the equations:

$$2x - 5y = 10$$

$$\underline{-2x + 5y = -10}$$

$$0 = 0 \qquad \text{A true statement}$$

The y's were also eliminated! As we saw in the previous section, *when both variables are eliminated and a true statement results, then the given system has infinite solutions and is dependent.*

In summary, the elimination method involves the following steps.

Elimination Method

---

1. Express each equation in the form $ax + by = c$.

2. Multiply each equation by some nonzero number so that the coefficients of either x or y are opposites.

3. Add the equations obtained in step 2.

   a. If you obtain an equation in one variable, solve this equation. You have one coordinate of the ordered pair solution.

   b. If both variables are eliminated leaving a false statement, the system has no solution and is inconsistent.

   c. If both variables are eliminated leaving a true statement, the system has infinite solutions and is dependent.

4. If one coordinate of the ordered pair solution was obtained in step 3, substitute this value into one of the equations that contains both variables and solve this equation. You now have both coordinates of the ordered pair solution.

5. (Optional) Check the solution in the two original equations of the system.

EXERCISES 7.3

Using the elimination method, determine the solutions of the following linear systems of equations.

1. $-x + 2y = -19$
$x - 5y = 40$

2. $x - 4y = -20$
$-x - 2y = 2$

3. $x + y = 11$
$3x - y = 25$

4. $x - y = -1$
$5x + y = 25$

5. $7x + y = 10$
$-y = 3x - 4$

6. $x - 2y = 1$
$-x = y - 2$

7. $3x - 2y = 10$
$2y = 6x - 12$

8. $x - 5y = 3$
$5y = x - 3$

9. $4x + 3y = 5$
$-4x + y = 23$

10. $2x + y = 11$
$-2x + 3y = 41$

11. $2x - 5y = -1$
$3x + 5y = -14$

12. $x - 8y = 9$
$-3x + 8y = 5$

13. $x + 4y = 23$
$-2x + 3y = 42$

14. $-3x - 5y = 11$
$x + 6y = 18$

15. $3x - 9y = 8$
$x - 3y = 5$

16. $x + 5y = -12$
$4x - 3y = -2$

17. $6x + y = 15$
$7x - 5y = 36$

18. $3x - 4y = -12$
$10x + y = -40$

19. $4x - y = 3$
$8x + 3y = -1$

20. $11x - 5y = 2$
$3x - y = 1$

21. $3x + 8y = -46$
$2x - 5y = 21$

22. $7x - 3y = 25$
$4x + 5y = 21$

23. $-2x + 3y = -38$
$5x + 4y = 3$

24. $6x + 9y = 4$
$-4x - 6y = 3$

25. $7x + 3y = -15$
$9x + 2y = -10$

26. $8x + 5y = 13$
$6x + 7y = -13$

27. $-5x + 10y = 25$
$3x - 6y = -15$

28. $-4x + 3y = 19$
$-6x - 5y = 0$

29. $14x + 19y = 15$
$4x - 7y = -4$

30. $6x + 4y = 5$
$22x - 7y = 1$

31. $\frac{3}{4}x + \frac{4}{3}y = -3$
$\frac{5}{2}x + \frac{11}{3}y = -3$

32. $\frac{5}{8}x + \frac{3}{2}y = -5$
$\frac{1}{4}x + \frac{3}{5}y = -2$

33. $\frac{5}{6}x - \frac{4}{9}y = -4$
$\frac{7}{10}x - \frac{4}{15}y = -4$

34. $\frac{6}{7}x - \frac{9}{14}y = 10$
$\frac{2}{9}x - \frac{1}{6}y = 10$

35. $-\frac{x}{5} + \frac{y}{3} = \frac{16}{15}$
$\frac{x}{9} - \frac{y}{4} = -\frac{11}{12}$

36. $\frac{x}{12} - \frac{y}{15} = -\frac{3}{20}$
$-\frac{x}{6} + \frac{y}{8} = \frac{5}{24}$

37. $y = \frac{2}{9}x + \frac{13}{3}$
$3x + 14y = -31$

38. $4x + 3y = 47$
$y = \frac{7}{2}x + 6$

39. $10x - 7y = -8$
$x = \frac{7}{10}y - \frac{4}{5}$

40. $x = \frac{3}{4}y + 1$
$5x - y = 16$

41. $x = \dfrac{11y + 15}{2}$
$y = \dfrac{5x + 3}{14}$

42. $x = \dfrac{7y + 52}{9}$
$y = \dfrac{5x - 20}{3}$

43. $y = \dfrac{5 - 3x}{13}$
$x = \dfrac{22 - 8y}{5}$

44. $y = \dfrac{\frac{9}{2} - 3x}{4}$
$x = \dfrac{3 - \frac{8}{3}y}{2}$

45. $5(x - 2y) + 2x = -3 - 4y$
$7(x - 2y) + 3x = -6 - 5y$

46. $4(x + 3y) - x = 2(13 + y)$
$2(3x + y) - x = -5(1 + y)$

47. $3(2x - 3y - 2) = 2(x - 3y + 4)$
$5(3x - y - 3) = 3(3x + 2y + 2)$

48. $3(5x - 4y) - 2(1 - 5y) = 10$
$5(2x + 3y) - 7(x + 9) = 8y + 6$

49. $\frac{4}{5}\left(\frac{1}{8}x + \frac{1}{6}y\right) = \frac{2}{3}$
$\frac{2}{3}\left(\frac{1}{4}x + \frac{1}{3}y\right) = \frac{1}{2}$

50. $\frac{3}{10}\left(\frac{5}{12}x + \frac{2}{3}y\right) = \frac{1}{20}$
$\frac{15}{14}\left(\frac{7}{5}x + 2y\right) = -\frac{1}{7}$

51. $\frac{2}{3}\left(\frac{1}{6}x + \frac{1}{4}y\right) = \frac{1}{2}$
$\frac{1}{6}\left(\frac{3}{2}x + 2y\right) = \frac{1}{2}$

52. $\frac{7}{10}\left(\frac{2}{7}x - \frac{5}{14}y\right) = -\frac{7}{4}$
$\frac{1}{6}\left(3x - \frac{3}{2}y\right) = -\frac{7}{4}$

53. $\dfrac{3}{x} + \dfrac{5}{y} = \dfrac{1}{2}$
$\dfrac{1}{x} - \dfrac{10}{y} = \dfrac{5}{2}$

54. $\dfrac{5}{x} - \dfrac{1}{y} = 13$
$\dfrac{4}{x} + \dfrac{3}{y} = -1$

7.4 Applications of Linear Systems

Armed with the substitution and elimination methods, we are now prepared to examine application problems in which linear systems of equations can be used. Many problems with two unknowns can be easily solved using the following plan of attack.

1. Read the problem carefully (probably more than once), and determine what two quantities you are looking for.

2. Represent each of the two quantities with a different variable.

3. Find *two* equations involving these two variables.

4. Solve the system formed by these two equations.

5. Determine whether the two values found in step 4 make sense in the statement of the problem.

EXAMPLE 7.4.1

The sum of two numbers is 2. One number is 5 more than twice the other number. Find the two numbers.

SOLUTION

Let x = one number

and y = other number

Then $x + y = 2$ from the first sentence

and $x = 2y + 5$ from the second sentence

Since the second equation is solved for x, substitute $2y + 5$ for x in the first equation:

$$2y + 5 + y = 2$$
$$3y + 5 = 2$$
$$3y = -3$$
$$y = -1$$

Now substitute -1 for y in the second equation:

$$x = 2(-1) + 5$$
$$x = -2 + 5$$
$$x = 3$$

Therefore, the two numbers are 3 and -1. Their sum is 2, and 3 is 5 more than twice -1.

EXAMPLE 7.4.2

The denominator of a fraction is 3 more than the numerator. If 1 is added to the numerator, the resulting fraction is equivalent to $\frac{3}{4}$. Find the original fraction.

SOLUTION

Let n = the numerator

and d = the denominator

The fraction that we are looking for is $\dfrac{n}{d}$. We know that

$$d = n + 3$$ from the first sentence

and $$\frac{n + 1}{d} = \frac{3}{4}$$ from the second sentence

First, simplify the second equation by cross multiplying:

$4(n + 1) = 3d$

$4n + 4 = 3d$

Since the first equation is solved for d, substitute $n + 3$ for d in the simplified second equation:

$4n + 4 = 3(n + 3)$

$4n + 4 = 3n + 9$

$n + 4 = 9$

$n = 5$

Substitute 5 for n in the first equation:

$d = 5 + 3$

$d = 8$

Thus, the fraction is $\dfrac{n}{d} = \dfrac{5}{8}$. The reader should check that this fraction satisfies the conditions of the problem.

EXAMPLE 7.4.3

Benjamin has saved $4.32 in pennies and nickels. If he has a total of 260 coins, how many of each kind of coin does he have?

SOLUTION

Let p = number of pennies

n = number of nickels

NOTE ▶

To convert nickels to cents, multiply the number of nickels by 5. For example,

8 nickels = 5 · 8 cents

= 40 cents

So $1p$ = value of pennies in cents

$5n$ = value of nickels in cents

The following chart organizes the information given in the problem.

	Pennies	**Nickels**	**Total**
Number of coins	p	n	260
Value (in cents)	$1p$	$5n$	432

From the chart, we get the following system.

$p + n = 260$ from the top row

$p + 5n = 432$ from the bottom row

Let us solve the system using the elimination method:

$p + n = 260$ $\xrightarrow[\text{sides by } -1.]{\text{Multiply both}}$ $-p - n = -260$

$p + 5n = 432$ $\xrightarrow{\hspace{2cm}}$ $\underline{p + 5n = \hspace{0.5cm} 432}$

$\hspace{4cm} 4n = \hspace{0.5cm} 172$

$\hspace{4.2cm} n = \hspace{0.5cm} 43$

Substitute 43 for n in the first equation:

$p + 43 = 260$

$\hspace{0.7cm} p = 217$

Benjamin has 43 nickels and 217 pennies. The check is left to the reader.

EXAMPLE 7.4.4

Louisa had $8000 to invest. With part of it she bought a CD paying 9% simple interest. She invested the rest in municipal bonds paying 12% simple interest. After 1 year her interest income was $885. How much did she invest at each rate?

SOLUTION

Let x = amount invested in CD at 9%

$\hspace{0.7cm} y$ = amount invested in bonds at 12%

So $\hspace{0.4cm} 0.09x$ = interest earned from CD at 9%

$\hspace{1.2cm} 0.12y$ = interest earned from bonds at 12%

The following chart organizes the information given in the problem.

	CD	**Bonds**	**Total**
Amount invested	x	y	8000
Interest earned	$0.09x$	$0.12y$	885

From the chart, we get the following system:

$x + y = 8000$ from the top row

$0.09x + 0.12y = 885$ from the bottom row

Using the elimination method,

$$x + y = 8000 \xrightarrow[\text{sides by } -9.]{\text{Multiply both}} -9x - 9y = -72000$$

$$0.09x + 0.12y = 885 \xrightarrow[\text{sides by } 100.]{\text{Multiply both}} \quad \underline{9x + 12y = 88500}$$

$$3y = 16500$$

$$y = 5500$$

Substitute 5500 for y in the first equation:

$$x + 5500 = 8000$$

$$x = 2500$$

Louisa invested $2500 in the CD and $5500 in municipal bonds.

EXAMPLE 7.4.5

Tom's Thirstbuster contains 44% sucrose. Matt's Muscle Energizer contains 2% sucrose. How many liters of each must be used to make 30 liters of Steve's Stimulator, which contains 30% sucrose?

SOLUTION

Let t = number of liters of Tom's Thirstbuster

m = number of liters of Matt's Muscle Energizer

Again, we organize our information with a chart:

	Amount ·	Percent Sucrose =	Amount of Sucrose
Tom's Thirstbuster	t	44% = 0.44	$0.44t$
Matt's Muscle Energizer	m	2% = 0.02	$0.02m$
Steve's Stimulator	30	30% = 0.3	0.3(30) = 9

t liters of Tom's and m liters of Matt's combine to make 30 liters of Steve's (see Figure 7.4.1). So

$$t + m = 30$$

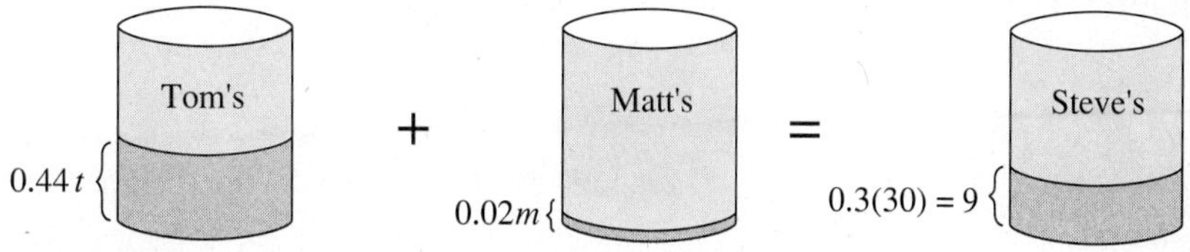

Figure 7.4.1

The sucrose in Tom's and the sucrose in Matt's combine to give the sucrose in Steve's. So

$$0.44t + 0.02m = 9$$

Using the elimination method,

$$t + m = 30 \quad \xrightarrow[\text{sides by } -2.]{\text{Multiply both}} \quad -2t - 2m = -60$$

$$0.44t + 0.02m = 9 \quad \xrightarrow[\text{sides by } 100.]{\text{Multiply both}} \quad \begin{array}{rl} 44t + 2m = & 900 \\ \hline 42t = & 840 \\ t = & 20 \end{array}$$

Substitute 20 for t in the first equation:

$$20 + m = 30$$
$$m = 10$$

Therefore, it takes 20 liters of Tom's Thirstbuster and 10 liters of Matt's Muscle Energizer.

EXAMPLE 7.4.6

At Tukioko's Taco Town, you can buy 3 tacos and 2 burritos for $5.81. If you buy 5 tacos and 1 burrito, it costs you $6.23. What is the cost of each taco and each burrito?

SOLUTION

Let t = cost of 1 taco

$\quad b$ = cost of 1 burrito

Then $\quad 3t + 2b = 5.81 \quad$ from first sentence,

and $\quad 5t + b = 6.23 \quad$ from second sentence.

Using the elimination method,

$$3t + 2b = 5.81 \quad \xrightarrow{} \quad 3t + 2b = 5.81$$

$$5t + b = 6.23 \quad \xrightarrow[\text{sides by } -2.]{\text{Multiply both}} \quad \begin{array}{rl} -10t - 2b = & -12.46 \\ \hline -7t = & -6.65 \\ t = & 0.95 \end{array}$$

Substitute 0.95 for t in the second equation:

$$5(0.95) + b = 6.23$$
$$4.75 + b = 6.23$$
$$b = 1.48$$

Therefore, each taco costs $0.95 and each burrito costs $1.48.

EXERCISES 7.4

Use a linear system of equations to solve each of the following problems.

1. The sum of two numbers is 20. One number is 1 less than twice the other number. Find the two numbers.

2. The sum of two numbers is 30. One number is 2 more than six times the other number. Find the two numbers.

3. The difference of two numbers is 10. The larger number is 12 more than three times the smaller number. Find the two numbers.

4. The difference of two numbers is 40. The larger number is 4 more than four times the smaller number. Find the two numbers.

5. The sum of the digits of a two-digit number is 7. The number is five times the units digit. Find the number. (*Hint:* If t = tens digit and u = units digit, then the number is $10t + u$.)

6. The sum of the digits of a two-digit number is 10. The number is 2 less than eight times the units digit. Find the number. (See the hint for problem 5.)

7. The numerator of a fraction is 3 more than the denominator. If 1 is added to the numerator, the resulting fraction is equivalent to $\frac{3}{2}$. Find the original fraction.

8. The denominator of a fraction is 7 more than the numerator. If 3 is added to the numerator, the resulting fraction is equivalent to $\frac{2}{3}$. Find the original fraction.

9. The denominator of a fraction is 3 more than the numerator. If 4 is added to both the numerator and denominator, the resulting fraction is equivalent to $\frac{2}{3}$. Find the original fraction.

10. The denominator of a fraction is 7 more than the numerator. If 5 is added to both the numerator and denominator, the resulting fraction is equivalent to $\frac{1}{2}$. Find the original fraction.

11. The numerator of a fraction is 2 more than the denominator. If 3 is subtracted from the numerator and 1 is added to the denominator, the resulting fraction is equivalent to $\frac{3}{4}$. Find the original fraction.

12. The denominator of a fraction is 1 more than twice the numerator. If 4 is added to the numerator and 7 is subtracted from the denominator, the resulting fraction is equivalent to $\frac{5}{3}$. Find the original fraction.

13. Mark has $3.55 in nickels and dimes. If he has a total of 42 coins, how many of each kind of coin does he have?

14. Lira has $4.80 in nickels and quarters. If she has a total of 52 coins, how many of each kind of coin does she have?

15. Rupee has $14.80 in dimes and quarters. The number of quarters is 2 less than three times the number of dimes. How many of each kind of coin does she have?

16. Franc has $315 in $5 bills and $20 bills. The number of $5 bills is 3 more than twice the number of $20 bills. How many of each kind of bill does he have?

17. Buck spent $3.01 in 25¢ and 3¢ stamps. He bought a total of 27 stamps. How many of each kind of stamp did he buy?

18. Penny spent $4.51 on 17¢ and 44¢ stamps. She bought a total of 17 stamps. How many of each kind of stamp did she buy?

19. Iphigenia sold 154 tickets to Moss County High School's production of *Snorer in the Class.* Adult tickets cost $8.75 and children's tickets cost $5.50. She collected $1256.50 from the sale of the tickets. How many of each kind of ticket did she sell?

20. "Fingers" Malone sold 87 tickets to Moss County Jail's production of *The Viceman Cometh.* General admission tickets cost $3.50 and inmates' tickets cost $1.25. He collected $113.25 from the sale of the tickets. How many of each kind of ticket did he sell?

21. Amber invested $6000 at her local bank. She invested part of her money in an account paying 8% simple interest, and she invested the rest in a CD paying 10% simple interest. After 1 year her interest income was $550. How much did Amber invest at each rate?

22. Raetta invested $3500 at her credit union. She invested part of her money in a savings account paying 6% simple interest, and she invested the rest in a CD paying 9% simple interest. After 1 year her interest income was $279. How much did Raetta invest at each rate?

23. Heather went to her savings and loan and opened two accounts, one paying 5% and one paying 11%. The amount she invested at 11% was $1600 more than the amount she invested at 5%. After 1 year her interest income was $400. How much did Heather invest at each rate?

24. Gina went to her broker and invested in two mutual funds, one paying 16% and one paying 20%. The amount she invested at 16% was $8500 less than the amount she invested at 20%. After 1 year her interest income was $2780. How much did Gina invest at each rate?

25. Joy went to her broker and invested in two mutual funds, one paying 15% and one paying 18%. The amount she invested at 18% was $1000 more than twice the amount she invested at 15%. After 1 year her interest income was $2475. How much did Joy invest at each rate?

26. Shelley went to her broker and invested in a mutual fund paying 24% and a municipal bond paying 14%. The amount she invested at 24% was 3 times as much as the amount she invested at 14%. After 1 year her interest income was $3612. How much did Shelley invest at each rate?

27. Sun Valley Orange Soda is 15% real fruit juice. Death Valley Orange Soda is 3% fruit juice. How many ounces of each must be used to make 24 ounces of Big Valley Orange Soda, which is 11% fruit juice?

28. Country Girl Trail Mix is 18% fiber. Mountain Man Trail Mix is 10% fiber. How many ounces of each must be used to make 60 ounces of City Slicker Trail Mix, which is 12% fiber?

29. Ethel's Extract is 25% alcohol. Ebenezer's Elixir·is 100% alcohol. How many ounces of each must be used to make 20 ounces of Tommy's Tonic, which is 40% alcohol?

30. Slim's Donut Mix is 4% shortening. Gordo's Donut Mix is 28% shortening. How many pounds of each must be used to make 15 pounds of Generic Donut Mix, which is 12% shortening?

31. At Pat's Plant Palace, you can buy 2 rosebushes and 5 azaleas for $39.75. If you buy 3 rosebushes and 2 azaleas, it costs you $30.75. What is the cost of each rosebush and each azalea?

32. At Mike's Magazine Mart, you can buy 3 back issues of *Guns & Ammo* and 4 back issues of *True Detective Story* for $6.65. If you buy 4 back issues of *Guns & Ammo* and 1 back issue of *True Detective Story*, it costs you $4.10. What is the cost of each issue of *Guns & Ammo* and each issue of *True Detective Story*?

33. Drew and Piper went to Karen's Kookie Kommune to buy some cookies. Drew bought 5 chocolate chip and 8 butter cookies for $4.11. Piper bought 7 chocolate chip and 4 butter cookies for $3.81. What is the cost of each chocolate chip cookie and each butter cookie?

34. Juan and Ernesto went to Holly's Hotdog Hotel to buy hotdogs for lunch. Juan bought 1 chili dog and 1 corny dog for $2.20. Ernesto bought 2 chili dogs and 3 corny dogs for $5.35. What is the cost of each chili dog and each corny dog?

Using a linear system of equations and the slope-intercept form, we can find the equation of a line passing through two given points. If a point on a line has coordinates (4, 7), then we can substitute 4 for x and 7 for y in the equation $y = mx + b$. This leaves an equation in m and b. Doing this twice generates a system of two linear equations, which can be solved for m and b. Substituting these values back into the equation $y = mx + b$ yields the equation of the line.

Use this method to find the equations of the lines passing through the following pairs of points.

35. $(1, -3), (-2, -9)$ **36.** $(2, -5), (4, -11)$ **37.** $(3, 2), (-6, -4)$ **38.** $(-3, -2), (1, -2)$

7.5 Systems of Linear Inequalities in Two Variables

In Section 6.5 we studied linear inequalities in two variables. Recall that the graph of an inequality such as $x + 2y > 4$ is a half-plane indicated by the shaded region in Figure 7.5.1. The boundary line, given by the equation $x + 2y = 4$, is drawn dashed since the ordered pair solutions of the equation do not satisfy the inequality. The boundary line is solid when the inequality symbol is $\leq$ or $\geq$. We determine which side of the boundary line to shade by testing the coordinates of a point that is not on the line.

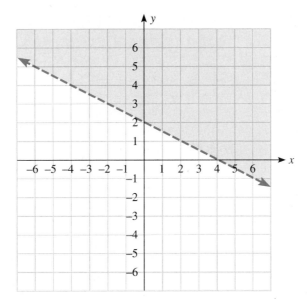

Figure 7.5.1

In this section, when we refer to a **system of linear inequalities**, we are simply considering two inequalities together. The **solution set of a system of inequalities** is the set of ordered pairs that satisfy both of the given inequalities. We can find the graph of the solution set by graphing both inequalities on the same coordinate system. The region where the two graphs overlap contains the points whose ordered pair coordinates satisfy both inequalities, and thus is the graph of the solution set of the system.

EXAMPLE 7.5.1

Graph the solution set of the following system of linear inequalities.

$$x - y < 2$$
$$x > 3$$

We first graph the inequality $x - y < 2$. The boundary line is dashed in Figure 7.5.2. We test the origin by substituting 0 for x and 0 for y in the inequality $x - y < 2$. Since $0 - 0 < 2$ is a true statement, the graph of the inequality is the region above the line $x - y = 2$ in Figure 7.5.2.

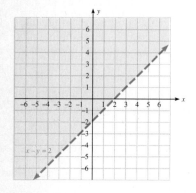

Figure 7.5.2

Figure 7.5.3

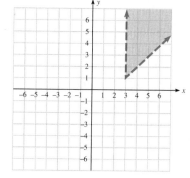

Figure 7.5.4

The graph of $x > 3$ consists of all of those points whose x-coordinates are greater than 3. This is the region to the right of the vertical line $x = 3$, as shown in Figure 7.5.3. Therefore, the graph of the solution set of the system consists of the points that lie above the line $x - y = 2$ and also to the right of the line $x = 3$. These points are indicated by the shaded region in Figure 7.5.4.

EXAMPLE 7.5.2

Graph the solution set of each of the following systems of linear inequalities.

1. $y \geq -2x - 3$
$y \geq \frac{1}{2}x$

Let us first graph the inequality $y \geq -2x - 3$. The boundary line, which in this case is solid, is shown in Figure 7.5.5. We can test the origin by

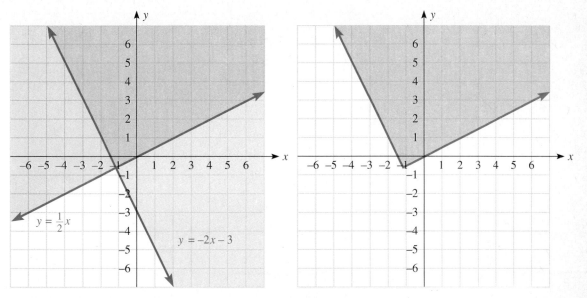

Figure 7.5.5

Figure 7.5.6

substituting 0 for x and for y:

$$0 \overset{?}{\geq} -2 \cdot 0 - 3$$

$$0 \overset{?}{\geq} 0 - 3$$

$$0 \geq -3 \qquad \text{A true statement}$$

Thus, the region above the line $y = -2x - 3$ is shaded. For the inequality $y \geq \frac{1}{2}x$, the boundary line is also solid and is shown in Figure 7.5.5. We can test the point (4, 0) by substituting 4 for x and 0 for y:

$$0 \overset{?}{\geq} \frac{1}{2} \cdot 4$$

$$0 \not\geq 2$$

Since (4, 0), which is below the boundary line, did not satisfy the inequality, the region above the line $y = \frac{1}{2}x$ is shaded. Therefore, the graph of the solution set of the system consists of the points on and above both lines. These points are indicated by the shaded region in Figure 7.5.6.

2. $2x + 5y \geq -12$

$3x + y < -5$

The graph of the first inequality has the boundary line $2x + 5y = -12$, which is the solid line shown in Figure 7.5.7. We can use (0, 0) to test the inequality:

$$2 \cdot 0 + 5 \cdot 0 \overset{?}{\geq} -12$$

$$0 \geq -12 \qquad \text{A true statement}$$

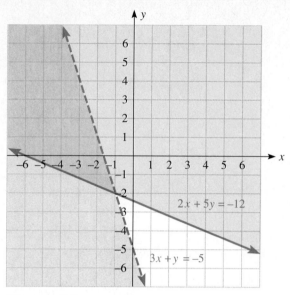

Figure 7.5.7

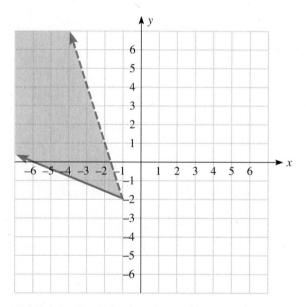

Figure 7.5.8

Therefore, the region above the boundary line is shaded. The graph of the second inequality has the boundary line $3x + y = -5$, which is the dashed line shown in Figure 7.5.7. We can use $(0, 0)$ again as a test point:

$$3 \cdot 0 + 0 \overset{?}{<} -5$$
$$0 \nless -5$$

Since $(0, 0)$, which is above the boundary line, did not satisfy the inequality, the region below the boundary line is shaded. The graph of the solution set of the system consists of those points that are on and above the solid line and below the dashed line. These points are indicated by the shaded region in Figure 7.5.8.

EXERCISES 7.5

Graph the solution set of each of the following systems of linear inequalities.

1. $x + y > 3$
$\quad y < 2$

2. $x - y > -4$
$\quad y > 1$

3. $\quad x \leq 4$
$\quad 2x - y \geq 5$

4. $\quad x \geq -2$
$\quad x - 3y \geq 3$

5. $\quad 2x + 7 > 0$
$\quad 4x + 3y \geq -12$

6. $\quad 3y - 4 \leq 0$
$\quad 5x + 2y > -10$

7. $y \geq 3x$
$\quad y \geq -x - 4$

8. $y \leq -\frac{2}{3}x$
$\quad y \geq \frac{1}{2}x - \frac{7}{2}$

9. $y > 2x + 4$
$\quad y > -2x$

10. $y > 3x + 4$
$\quad y < -x$

11. $3x + 4y < 0$
$\quad x + 3y \geq -5$

12. $\quad 5x - y \leq 0$
$\quad 3x - 2y > 1$

13. $\quad 2x - y > 7$
$\quad 2x - 5y > 11$

14. $2x - 3y > -20$
$\quad 3x - 2y < -15$

15. $x - 3y \geq -11$
$\quad 3x - y \leq 7$

16. $x + 2y \leq -9$
$\quad 2x + y \leq -9$

17. $5x + 6y < 6$
$\quad 3x - 5y \leq -5$

18. $\quad 3x - y \geq -12$
$\quad 3x + 2y < -12$

19. $\quad x - 4y + 36 < 0$
$\quad 2x + 3y - 16 > 0$

20. $2x + 5y + 22 \leq 0$
$\quad x - 6y - 6 \leq 0$

21. $x < \frac{1}{2}y - \frac{5}{2}$
$\quad x < \frac{2}{3}y - \frac{8}{3}$

22. $x > \frac{5}{3}y - 4$
$\quad x > \frac{2}{7}y + \frac{1}{7}$

23. $5x + 4y \leq 6$
$\quad 5x + 4y \geq -4$

24. $x - 3y < 14$
$\quad x - 3y \geq 1$

Find a system of linear inequalities for the shaded regions shown in each of the following graphs.

25.

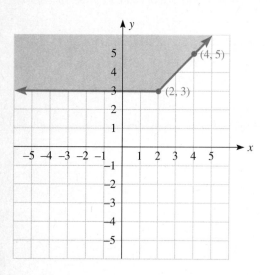

26.

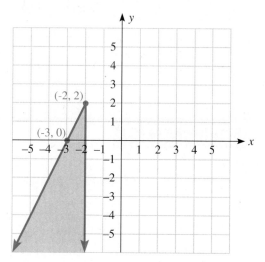

27.

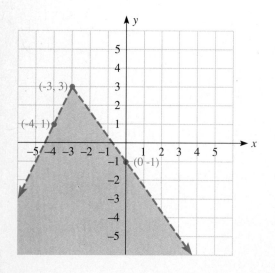

28.

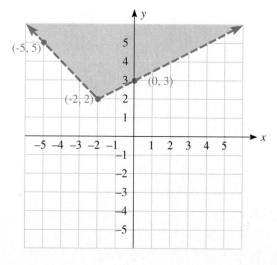

Chapter 7 Review

Terms to Remember

Methods for Finding the Solutions of a Linear System of Equations

• **Graphing Method**

1. Graph both equations on the same coordinate system.

2. Use the graphs to determine the type of solution.

Type of Solution	Graph		Type of System
One ordered pair		Intersecting lines	Independent
No solution		Parallel lines	Inconsistent
Infinite solutions		Same line	Dependent

3. In the case of one ordered pair solution, read the coordinates of the ordered pair off the graph.

4. Check the ordered pair in *both* equations.

▪ Substitution Method

1. Solve one of the equations for one of the variables.

2. Substitute the expression obtained in step 1 in the other equation.

3. You now have an equation with one variable. Solve this equation.
   a. If the equation has one solution, one coordinate of the ordered pair solution has been obtained.
   b. If the variable is eliminated leaving a false statement, the system has no solution and is inconsistent.
   c. If the variable is eliminated leaving a true statement, the system has infinite solutions and is dependent.

4. If one solution was obtained in step 3, substitute this value in the equation of step 1 and solve this equation. You now have both coordinates of the ordered pair solution.

5. (Optional) Check the solution in the two original equations of the system.

▪ Elimination Method

1. Express each equation in the form $ax + by = c$.

2. Multiply each equation by some nonzero number so that the coefficients of either x or y are opposites.

3. Add the equations obtained in step 2.
   a. If you obtain an equation in one variable, solve this equation. You have one coordinate of the ordered pair solution.
   b. If both variables are eliminated leaving a false statement, the system has no solution and is inconsistent.
   c. If both variables are eliminated leaving a true statement, the system has infinite solutions and is dependent.

4. If one solution was obtained in step 3, substitute this value into one of the equations that contains both variables and solve this equation. You now have both coordinates of the ordered pair solution.

5. (Optional) Check the solution in the two original equations of the system.

Which Is the Best Method?

- *Graphing:* Only if an approximate solution is desired.

- *Substitution:* Useful when one coefficient is equal to 1.

- *Elimination:* All other cases.

Review Exercises

7.1 **For each system determine whether the given ordered pair is a solution.**

1. $3x + 5y = 5$, $(5, -2)$
$2x - y = 12$

2. $x + 3y = 6$, $(-6, 4)$
$2x - 3y = 0$

3. $x - 4y = -5$, $\left(-1, -\frac{3}{2}\right)$
$3x + 2y = -6$

4. $2x - 5y = -37$, $(4, 9)$
$-3x + 4y = 24$

Using the graphing method, determine the solutions of the following linear systems of equations. If there are no solutions, write *inconsistent*. If there are infinite solutions, write *dependent*.

5. $x - y = -3$
$2x - y = 0$

6. $x + y = 6$
$3x - y = -2$

7. $3x - y = -1$
$x - 2y = 8$

8. $2x - 3y = 6$
$\frac{3}{2}y = x - 3$

9. $y = \frac{1}{2}x + 2$
$y = -x - 4$

10. $x = 2y + 5$
$x = -y + 2$

11. $3x - 2y + 6 = 0$
$-3x + 2y + 2 = 0$

12. $x + \frac{5}{2}y - 3 = 0$
$3x + 4y - 2 = 0$

7.2 **Using the substitution method, determine the solutions of the following linear systems of equations.**

13. $3x - 7y = 8$
$x = -2$

14. $5x + 6y = -11$
$y = \frac{2}{3}$

15. $x = 3y + 1$
$2x - 5y = 3$

16. $3x - 4y = -22$
$x = y - 5$

17. $7x + 2y = 19$
$y = x - 13$

18. $y = -\frac{3}{2}x - 4$
$3x + 2y = 2$

19. $3x + 8y = 1$
$x + y = 2$

20. $x + y = -11$
$2x - 11y = 4$

21. $2x - 6y = 4$
$-x + 3y = -2$

22. $6x - y = -1$
$3x + 2y = 12$

23. $8x + 3y = 9$
$6x + 5y = 26$

24. $5x + 2y = 2$
$11x + 3y = 17$

25. $\frac{2}{3}x - \frac{1}{5}y = 6$
$\frac{3}{2}x + \frac{5}{4}y = -\frac{7}{2}$

26. $\frac{3}{5}x - \frac{5}{2}y = 7$
$\frac{1}{2}x + \frac{3}{8}y = -4$

7.3 Using the elimination method, determine the solutions of the following linear systems of equations.

27. $x - 3y = 17$
    $-x - 4y = 18$

28. $5x + y = -14$
    $2x - y = -7$

29. $6x - 5y = 14$
    $4x + 5y = -24$

30. $-3x + 7y = -7$
    $3x + 2y = 25$

31. $-2x - y = 4$
    $6x + 3y = 5$

32. $9x + 2y = -50$
    $-3x + 4y = -16$

33. $2x + 3y = 34$
    $5x - 4y = -7$

34. $6x - 4y = 10$
    $-3x + 2y = -5$

35. $8x + 2y = 7$
    $7x + 3y = 3$

36. $11x + 10y = 38$
    $4x + 11y = 31$

37. $\frac{2}{3}x + \frac{7}{15}y = -1$
    $\frac{5}{4}x + \frac{9}{5}y = 12$

38. $\frac{x}{5} - \frac{y}{3} = -1$

    $-\frac{x}{8} + \frac{y}{7} = 2$

39. $y = \frac{3}{4}x + 6$
    $x = \frac{4}{3}y - 8$

40. $2(x + 2y) - 6y = 5(y + 1)$
    $4(x - 2y) - y = 3(x - 1)$

7.4 Use a linear system of equations to solve each of the following problems.

41. The sum of two numbers is 12. One number is 30 more than five times the other number. Find the two numbers.

42. The sum of the digits of a two-digit number is 12. The number is 20 more than eight times the tens digit. Find the number.

43. The numerator of a fraction is 6 more than the denominator. If 2 is added to both the numerator and denominator, the resulting fraction is equivalent to $\frac{5}{3}$. Find the original fraction.

44. Rita has $1150 in $20 bills and $50 bills. The number of $20 bills is 1 less than twice the number of $50 bills. How many of each kind of bill does she have?

45. Ken sold 48 tickets to Dog World of Moss County, featuring Shashu, the killer poodle. Adult tickets cost $24.75 and children's tickets cost $18.50. He collected $925.50 from the sale of the tickets. How many of each kind of ticket did he sell?

46. Tommy and Bettye went to their credit union and opened two accounts. Tommy put his money in an account paying 6% simple interest. Bettye invested $5000 more than Tommy in an account paying 7% simple interest. After 1 year their combined interest income was $1260. How much did they invest at each rate?

47. Ann's Authentic Chili Mix is 55% chili powder. Bennie's Bluff Chili Mix is 7% chili powder. How many ounces of each must be used to make 18 ounces of Fred's Famous Chili Mix, which is 23% chili powder?

48. Ricardo and Suren went to David's Panadería to buy some pan dulce. Ricardo bought 3 empanadas and 1 campechana for $2.96. Suren bought 2 empanadas and 3 campechanas for $3.35. What is the cost of each empanada and each campechana?

7.5 Graph the solution set of each of the following systems of linear inequalities.

49. $y \geq -2$
  $x - y \geq 3$

50. $2x + y > 5$
  $x < 4$

51. $y < \frac{1}{2}x$
  $y > -\frac{1}{2}x + 2$

52. $2x + 3y \leq 0$
  $x - 3y \geq 9$

53. $x - y < -2$
  $3x + y \leq 2$

54. $x - y \leq 3$
  $x - 3y \geq 3$

55. $x + 2y > 1$
  $x + 4y > 4$

56. $4x + 3y \geq -20$
  $3x - 2y \leq -32$

Chapter 7 Test

(You should be able to complete this test in 60 minutes.)

Determine the solutions of the following linear systems of equations using the indicated methods. If there are no solutions, write *inconsistent*. If there are infinite solutions, write *dependent*.

1. $3x + 10y = 62$, substitution
   $-4x + y = -11$

2. $x + 8y = 1$, elimination
   $-3x - 2y = 8$

3. $x + y = 1$, graphing
   $2x + 5y = -4$

4. $\frac{2}{3}x - 5y = 6$, substitution
   $\frac{5}{6}x - \frac{3}{2}y = -2$

5. $3x + 7y = 4$, elimination
   $5x - 2y = -7$

Graph the solution set of each of the following systems of linear inequalities.

6. $y \geq 4$
   $y \geq \frac{2}{3}x + 6$

7. $x - 3y < -5$
   $2x + 5y < 23$

Use a linear system of equations to solve each of the following problems.

8. Walter has $11.60 in nickels and half dollars. If he has a total of 70 coins, how many of each kind of coin does he have?

9. Horace instructed his broker to invest $80,000 for him in the stock market. The broker invested part of his money in public utilities paying an annual dividend of 3%. She invested the rest of his money in Bolivian tin mines paying an annual dividend of 2%. After 1 year his dividend income was $1950. How much did Horace invest at each rate?

Test Your Memory

These problems review Chapters 1–7.

I. Sketch the graph of the lines whose equations are given.

1. $3x + 4y = 8$ **2.** $y = 2x - 3$ **3.** $y = -2$ **4.** $x = 1$

II. Graph the following inequalities.

5. $x + 2y \leq 4$ **6.** $y < 3x - 1$

III. Graph the solution set of each of the following systems of linear inequalities.

7. $x + y < 5$ **8.** $y \geq 3x - 2$
      $x > 2$ $x + 2y \geq 4$

IV. Find the slopes of the following lines.

9. The line passing through $(-5, 8)$ and $(7, -1)$ **10.** The line whose equation is $6x - 4y = 13$

V. Find the equations of the lines satisfying the given information. Write your answers in standard form.

11. $m = \dfrac{-5}{2}$, passing through $(-2, -4)$ **12.** m undefined, passing through $(5, 6)$

13. Passing through $(8, 7)$ and $(-4, 3)$ **14.** Passing through $(1, -4)$ and $(3, -4)$

VI. Determine whether the following pairs of lines are parallel, perpendicular, or neither.

15. $4x - 6y = 1$ and $2x + 3y = 12$ **16.** $x + 4y = 12$ and $4x - y = 2$

VII. Find the solutions of the following equations.

17. $8 - 2(4x + 7) = 3(2x + 2)$

18. $\frac{1}{5}(2x + 3) = \frac{1}{2}\left(\frac{3}{5}x - \frac{1}{2}\right) + 1$

19. $(3x + 3)(x - 5) = -15$

20. $(x + 1)(x - 2) = (x + 1)(2x - 3)$

21. $\dfrac{2}{3x + 1} = \dfrac{3}{5x - 2}$

22. $\dfrac{2}{x + 1} = \dfrac{3x - 1}{16}$

23. $\dfrac{2}{x - 2} - \dfrac{1}{x} = \dfrac{5}{4x}$

24. $\dfrac{x}{x + 3} - \dfrac{1}{x - 4} = 1$

25. $\dfrac{x}{x + 1} - \dfrac{1}{2x - 1} = \dfrac{1}{3}$

26. $\dfrac{x - 1}{x + 3} + \dfrac{2}{x - 8} = \dfrac{2}{x^2 - 5x - 24}$

VIII. Simplify each of the following. Be sure to reduce your answer to lowest terms.

27. $\dfrac{(2x^3 y^{-4})^{-2}}{(2x^{-1} y^{-7})^{-1}}$

28. $\dfrac{2x^2 + 9x + 9}{2x^3 + 3x^2 - 18x - 27}$

29. $\dfrac{\dfrac{1}{6} - \dfrac{1}{2x}}{\dfrac{1}{3x} - \dfrac{1}{x^2}}$

30. $\dfrac{\dfrac{2}{x} + \dfrac{x}{x + 1}}{\dfrac{3}{x} + 1}$

IX. Perform the indicated operations.

31. $(3x^2 + 7x - 4)(8x - 2)$

32. $(16x^3 - 20x^2 - 8x + 5) \div (4x - 1)$

33. $\dfrac{4x^2 - 9y^2}{2y^2 + 11y + 12} \div \dfrac{10xy^2 - 15y^3}{3y^2 + 12y}$

34. $\dfrac{12x^3 + 6x^2}{x^2 + 6x + 5} \cdot \dfrac{x^3 + 5x^2 - x - 5}{6x^3 - 3x^2 - 3x}$

35. $\dfrac{5}{2x} + \dfrac{4}{3y}$

36. $\dfrac{7}{x - 2} - \dfrac{5}{x}$

37. $\dfrac{2x}{x^2 - 5x + 6} - \dfrac{x - 18}{x^2 - 4}$

38. $\dfrac{x + 4}{2x^2 - 5x - 3} + \dfrac{x - 7}{x^2 - 2x - 3}$

X. Determine the solutions of the following linear systems of equations. If there are no solutions, write *inconsistent*. If there are infinite solutions, write *dependent*.

39. $2x + 5y = 1$
   $3x - y = 10$

40. $3x + 9y = 4$
   $2x + 6y = -3$

41. $2x - 8y = 12$
   $3x - 12y = 18$

42. $4x - 3y = 0$
   $2x + 5y = -13$

IX. Use algebraic expressions to find the solutions of the following problems.

43. Mustang Sally has $2.60 in nickels and dimes. The number of dimes is 2 less than three times the number of nickels. How many nickels and how many dimes does Mustang Sally have?

44. Kim has $3.50 in nickels and quarters. He has a total of 22 coins. How many nickels and how many quarters does Kim have? (Use a linear system of equations to solve this problem.)

45. Mrs. Robinson invested $5000 in two accounts. She invested some in a savings account paying 4.5% simple interest and the rest in a CD paying 7.25% simple interest. After 1 year her interest income was $335. How much did Mrs. Robinson invest in each account?

46. Mr. Lime bathroom cleanser is 6% phosphorus. Larry's lime bathroom cleanser is 10% phosphorus. How many ounces of each cleanser must be used to make 20 ounces of bathroom cleanser that is 9% phosphorus? (Use a linear system of equations to solve this problem.)

47. The area of a rectangle is 48 sq m. The length is 5 m more than one-half of the width. What are the dimensions of the rectangle?

48. A car uses 3.5 gallons of gasoline on an 84-mile trip. How many gallons of gasoline will it use on a 288-mile trip?

49. Joe can mow a large yard in 8 hours. If Ray and Joe work together (using two mowers), they can mow the yard in 3 hours. How long would it take Ray working alone to mow the yard?

50. Moe takes his old motorboat to his favorite fishing spot 15 miles downstream. The trip downstream and back takes 4 hours. If the speed of the current is 2 mph, what is the speed of Moe's boat in still water?

CHAPTER 8

Radical Expressions

8.1 Evaluating Radicals

Successfully manipulating radical expressions is one of the most difficult tasks for beginning algebra students. Therefore, we will begin this chapter by extending some of the topics presented earlier in this text.

We know that

$$5^2 = 25$$

and $\quad (-5)^2 = 25$

These equations are saying, "The square of 5 is 25, and the square of −5 is 25."

We can look at these equations from another perspective, as follows:

Square Root of a Number

A number b is called a **square root** of a number a if $b^2 = a$.

From this point of view, the above equations are also saying, "5 and −5 are square roots of 25."

EXAMPLE 8.1.1

Find all of the square roots of the following numbers.

1. 49   The square roots of 49 are 7 and −7, since $7^2 = 49$ and $(-7)^2 = 49$.

2. $\frac{4}{9}$ The square roots of $\frac{4}{9}$ are $\frac{2}{3}$ and $-\frac{2}{3}$, since $\left(\frac{2}{3}\right)^2 = \frac{4}{9}$ and $\left(-\frac{2}{3}\right)^2 = \frac{4}{9}$.

3. 0   Zero has only one square root, 0, since only $0^2 = 0$.

The primary way of indicating square root is with the $\sqrt{}$ symbol, which is called a **radical symbol**, or a **radical sign**. The number or expression inside the radical sign is called the **radicand**. For example, in the expression $\sqrt{25}$, read "the square root of 25," 25 is the radicand. The entire expression $\sqrt{25}$ is called a **radical**, or a **radical expression**.

Principal Square Root
Let a be any real number greater than 0. The **principal square root** of a, denoted $\sqrt{a}$, is the *positive* number b such that $b^2 = a$. If $a = 0$, then $\sqrt{0} = 0$.

REMARKS

1. The symbol $\sqrt{a}$ always indicates the principal square root of a. For example, $\sqrt{25} = 5$.

2. A negative sign in front of the radical sign indicates the negative of the principal square root. For example, $-\sqrt{36} = -6$.

EXAMPLE 8.1.2

Evaluate each of the following square roots.

1. $\sqrt{100} = 10$

2. $\sqrt{\frac{25}{16}} = \frac{5}{4}$

3. $-\sqrt{4} = -2$

4. $\sqrt{0} = 0$

5. $-\sqrt{1} = -1$

6. $-\sqrt{\frac{9}{64}} = -\frac{3}{8}$

Try to avoid this mistake:

Incorrect	Correct
$\sqrt{100} = 10$ or $\sqrt{100} = -10$	$\sqrt{100} = 10$ The symbol $\sqrt{100}$ indicates the principal square root of 100, which is only 10, not -10.

In the previous example, all of the radicands were the square of a rational number. A number that is the square of a rational number is called a **perfect square**. Suppose that you were asked to find the square root of a number that is not a perfect square, for example, $\sqrt{2}$. The square roots of numbers that are not perfect squares are irrational numbers. Thus, $\sqrt{2}$ is an irrational number. In general, any irrational number can be written as a nonterminating and nonrepeating decimal. In practice we can use an approximation for an irrational number. You can use Table 2 in Appendix III or a calculator with a square root key to find an approximation for $\sqrt{2}$:

$$\sqrt{2} \doteq 1.41 \qquad \text{rounded off to two decimal places}$$

EXAMPLE 8.1.3

Use a calculator or Table 2 to find a decimal approximation to two decimal places for the following square roots.

1. $\sqrt{14} \doteq 3.74$ Note: $(3.74)^2 = 13.9876$, so 3.74 is only an approximation for $\sqrt{14}$.

2. $\sqrt{75} \doteq 8.66$

Suppose that you were asked to evaluate $\sqrt{-64}$. This expression does not represent a real number, since no real number squared is -64. In fact, any real number squared is always nonnegative. In the next chapter we will learn how to deal with the square roots of negative numbers.

The following equations illustrate how we can extend some of the concepts presented earlier in this section.

We know that

1. $(-2)^3 = -8$

2. $3^4 = 81$

3. $5^5 = 3125$

Equation 1 illustrates that a **cube root** of -8 is -2. Equation 2 shows that a **fourth root** of 81 is 3. Finally, equation 3 shows that a **fifth root** of 3125 is 5. These ideas are generalized in the following definitions.

Principal *n*th Root of *a* (*n* even)

Let a be any real number greater than zero and n be a positive even integer. The **principal *n*th root** of a, denoted $\sqrt[n]{a}$, is the *positive* real number b such that $b^n = a$. If $a = 0$, then $\sqrt[n]{a} = 0$.

REMARKS

1. The expression $\sqrt[n]{a}$ is read "the *n*th root of *a*."

2. n is called the **index**, or **order**, of the radical.

3. If no index is stated, the index is 2; for example,

$$\sqrt{25} = \sqrt[2]{25}$$

EXAMPLE 8.1.4

Evaluate each of the following radical expressions.

1. $\sqrt{\dfrac{36}{49}} = \dfrac{6}{7}$ because $\dfrac{6}{7} > 0$ and $\left(\dfrac{6}{7}\right)^2 = \dfrac{36}{49}$

2. $\sqrt[4]{625} = 5$ because $5 > 0$ and $5^4 = 625$

NOTE ▶   In order to evaluate these radicals, the student must be familiar with the corresponding powers. See Table 1 in Appendix III.

3. $\sqrt[4]{-81}$

This expression does not represent a real number, since no real number raised to the fourth power is -81. In fact, any real number raised to the fourth power is always nonnegative.

4. $-\sqrt[4]{\dfrac{16}{81}} = -\dfrac{2}{3}$ because $\dfrac{2}{3} > 0$ and $\left(\dfrac{2}{3}\right)^4 = \dfrac{16}{81}$

nth Root of a (n odd)
Let a be any real number and n be a positive odd integer. The **nth root** of a, denoted $\sqrt[n]{a}$, is the real number b such that $b^n = a$.

REMARKS　**1.** When n is odd, the radicand can be any real number.

2. When n is odd, the sign of $\sqrt[n]{a}$ is the same as the sign of a.

EXAMPLE 8.1.5

Evaluate each of the following radical expressions.

1. $\sqrt[3]{64} = 4$　　since $4^3 = 64$

2. $\sqrt[3]{-27} = -3$　　since $(-3)^3 = -27$

3. $-\sqrt[3]{-125} = -(-5)$　　since $(-5)^3 = -125$
$$= 5$$

4. $\sqrt[5]{\dfrac{32}{243}} = \dfrac{2}{3}$　　since $\left(\dfrac{2}{3}\right)^5 = \dfrac{32}{243}$

In Section 8.6 we will need to raise radicals to powers. Consider the next example to see how this is done.

EXAMPLE 8.1.6

Simplify each of the following.

1. $(\sqrt{49})^2 = (7)^2 = 49$

2. $(\sqrt[3]{8})^3 = (2)^3 = 8$

3. $(\sqrt[3]{-8})^3 = (-2)^3 = -8$

4. $(\sqrt[4]{81})^4 = (3)^4 = 81$

We can generalize this example to the following property.

If a is a real number and n is a positive integer, then

$$(\sqrt[n]{a})^n = a$$

EXAMPLE 8.1.7

Simplify each of the following.

1. $(\sqrt{7})^2 = 7$
2. $(\sqrt[3]{5})^3 = 5$
3. $(\sqrt{3x})^2 = 3x$
4. $(\sqrt[4]{y + 2})^4 = y + 2$
5. $(4\sqrt{3})^2 = 4^2(\sqrt{3})^2 = 16 \cdot 3 = 48$
6. $(\sqrt[5]{(6x^3)^2})^5 = (6x^3)^2 = 36x^6$

Suppose that we had exponents on factors in the radicand. What steps do we follow to evaluate those radical expressions? Consider the next example.

EXAMPLE 8.1.8

Evaluate each of the following radical expressions.

1. $\sqrt{5^2} = \sqrt{25} = 5$
2. $\sqrt{(-5)^2} = \sqrt{25} = 5$
3. $\sqrt[3]{2^3} = \sqrt[3]{8} = 2$
4. $\sqrt[3]{(-2)^3} = \sqrt[3]{-8} = -2$

We can generalize this example to the following property.

1. If n is a positive *even* integer and a is a real number, then

$$\sqrt[n]{a^n} = |a|$$

2. If n is a positive *odd* integer and a is a real number, then

$$\sqrt[n]{a^n} = a$$

EXAMPLE 8.1.9

Simplify each of the following.

1. $\sqrt{x^2} = |x|$

2. $\sqrt[3]{y^3} = y$

3. $\sqrt{(6x)^2} = |6x| = 6|x|$

NOTE ▶ For ease of expression, we will assume that the variables represent positive real numbers in the remainder of this section. Consequently,

$$\sqrt{(6x)^2} = 6x$$

4. $\sqrt{49x^2y^2} = \sqrt{(7xy)^2} = 7xy$

5. $\sqrt[3]{-27x^3z^3} = \sqrt[3]{(-3xz)^3} = -3xz$

6. $-\sqrt[3]{64r^3s^3} = -\sqrt[3]{(4rs)^3} = -4rs$

7. $\sqrt{x^2 + 2xy + y^2} = \sqrt{(x+y)^2} = x + y$

EXERCISES 8.1

Evaluate each of the following radical expressions, if possible.

1. $\sqrt{64}$ **2.** $\sqrt{9}$ **3.** $-\sqrt{169}$ **4.** $-\sqrt{144}$

5. $-\sqrt{\dfrac{25}{16}}$ **6.** $-\sqrt{\dfrac{81}{49}}$ **7.** $-\sqrt{\dfrac{25}{121}}$ **8.** $-\sqrt{-4}$

9. $\sqrt{25}$ **10.** $-\sqrt{121}$ **11.** $-\sqrt{\dfrac{121}{144}}$ **12.** $\sqrt{-16}$

13. $\sqrt{1}$ **14.** $\sqrt{81}$ **15.** $\sqrt{\dfrac{169}{64}}$ **16.** $-\sqrt{\dfrac{4}{169}}$

17. $\sqrt{-49}$ **18.** $-\sqrt{\dfrac{1}{36}}$ **19.** $-\sqrt{0}$ **20.** $\sqrt{\dfrac{49}{64}}$

21. $\sqrt{\dfrac{25}{4}}$ **22.** $-\sqrt{\dfrac{25}{144}}$ **23.** $-\sqrt{81}$ **24.** $\sqrt{\dfrac{81}{100}}$

25. $-\sqrt{-1}$ **26.** $\sqrt{\dfrac{169}{100}}$ **27.** $\sqrt{-\dfrac{4}{9}}$ **28.** $\sqrt{196}$

29. $-\sqrt{-\dfrac{9}{121}}$ **30.** $-\sqrt{16}$ **31.** $-\sqrt{\dfrac{1}{100}}$ **32.** $\sqrt{\dfrac{49}{9}}$

33. $\sqrt{\sqrt{81}}$ **34.** $\sqrt{\sqrt{16}}$

Use a calculator or Table 2 in Appendix III to find a decimal approximation to two decimal places for the following.

35. $\sqrt{15}$ **36.** $\sqrt{17}$ **37.** $-\sqrt{23}$ **38.** $-\sqrt{21}$

39. $\sqrt{48}$ **40.** $\sqrt{45}$ **41.** $-\sqrt{60}$ **42.** $-\sqrt{68}$

Evaluate each of the following radical expressions, if possible.

43. $\sqrt[3]{8}$ **44.** $\sqrt[3]{27}$ **45.** $\sqrt[3]{-125}$ **46.** $\sqrt[3]{-64}$

47. $\sqrt{-\dfrac{36}{121}}$ **48.** $\sqrt[5]{32}$ **49.** $\sqrt[3]{-\dfrac{125}{216}}$ **50.** $-\sqrt[4]{16}$

51. $-\sqrt{\dfrac{4}{49}}$ **52.** $\sqrt[3]{\dfrac{8}{27}}$ **53.** $\sqrt[4]{16}$ **54.** $-\sqrt{\dfrac{1}{4}}$

55. $\sqrt[3]{\dfrac{1}{125}}$ **56.** $\sqrt[3]{-\dfrac{8}{125}}$ **57.** $-\sqrt[3]{125}$ **58.** $\sqrt[3]{-\dfrac{64}{27}}$

59. $\sqrt[3]{\dfrac{27}{64}}$ **60.** $\sqrt{\dfrac{81}{4}}$ **61.** $-\sqrt{\dfrac{1}{49}}$ **62.** $-\sqrt[3]{-8}$

63. $-\sqrt[3]{-27}$ **64.** $-\sqrt{\dfrac{9}{100}}$ **65.** $\sqrt[3]{216}$ **66.** $\sqrt[4]{81}$

67. $\sqrt{\dfrac{169}{4}}$ **68.** $\sqrt[3]{1}$ **69.** $-\sqrt[3]{-\dfrac{125}{27}}$ **70.** $-\sqrt{-\dfrac{64}{25}}$

71. $\sqrt[3]{-\dfrac{1}{27}}$ **72.** $\sqrt{\dfrac{169}{9}}$ **73.** $\sqrt[4]{-1}$ **74.** $-\sqrt[3]{64}$

75. $\sqrt[4]{\dfrac{16}{81}}$ **76.** $-\sqrt[3]{\dfrac{1}{8}}$ **77.** $\sqrt[3]{-\dfrac{1}{64}}$ **78.** $\sqrt{\dfrac{121}{16}}$

79. $\sqrt[3]{-\dfrac{216}{125}}$ **80.** $\sqrt[4]{\dfrac{1}{256}}$ **81.** $-\sqrt{-\dfrac{81}{4}}$ **82.** $-\sqrt[5]{\dfrac{1}{243}}$

83. $-\sqrt[3]{\dfrac{64}{125}}$ **84.** $\sqrt[4]{-16}$ **85.** $\sqrt[5]{243}$ **86.** $\sqrt[3]{\dfrac{125}{8}}$

87. $-\sqrt[4]{81}$ **88.** $\sqrt{-\dfrac{64}{81}}$ **89.** $-\sqrt[5]{\dfrac{1}{32}}$ **90.** $-\sqrt[3]{-\dfrac{125}{8}}$

91. $\sqrt{\sqrt[3]{64}}$

92. $\sqrt[3]{\sqrt{64}}$

Simplify each of the following.

93. $(\sqrt{8})^2$

94. $(\sqrt[4]{14})^4$

95. $(\sqrt[3]{-4})^3$

96. $(\sqrt[5]{-10})^5$

97. $(5\sqrt{2})^2$

98. $(2\sqrt[3]{7})^3$

99. $(\sqrt{x^2 + 5x + 7})^2$

100. $(\sqrt[5]{3x + 5})^5$

101. $(\sqrt[3]{(5x^2)^3})^3$

102. $(\sqrt{(-3xy^3)^2})^2$

Simplify each of the following. Assume that the variables represent positive real numbers.

103. $\sqrt{x^2}$

104. $\sqrt{y^2}$

105. $\sqrt{x^2y^2}$

106. $\sqrt{m^2n^2}$

107. $\sqrt{81x^2}$

108. $\sqrt{121y^2}$

109. $\sqrt{4x^2y^2}$

110. $\sqrt{9x^2y^2}$

111. $-\sqrt{25x^2y^2z^2}$

112. $-\sqrt{16x^2y^2z^2}$

113. $\sqrt[3]{x^3}$

114. $\sqrt[3]{y^3}$

115. $\sqrt[3]{27y^3}$

116. $\sqrt[3]{125z^3}$

117. $-\sqrt[3]{125x^3y^3}$

118. $-\sqrt[3]{8m^3n^3}$

119. $-\sqrt[3]{-64p^3q^3}$

120. $-\sqrt[3]{-125j^3k^3}$

121. $\sqrt{x^2 + 2x + 1}$

122. $\sqrt{y^2 + 6y + 9}$

123. $\sqrt{4y^2 + 12y + 9}$

124. $\sqrt{9x^2 + 6x + 1}$

125. $\sqrt{25x^2 + 20xy + 4y^2}$

126. $\sqrt{16x^2 + 24xy + 9y^2}$

8.2 Simplifying Radical Expressions

When we simplify a fraction, we are merely converting the fraction into a more manageable form. In this section we will investigate two properties that will aid us in converting radicals into more manageable forms.

Consider the following examples from arithmetic.

$$\sqrt{4 \cdot 25} = \sqrt{100} = 10$$

and

$$\sqrt{4} \cdot \sqrt{25} = 2 \cdot 5 = 10$$

Thus, $\sqrt{4 \cdot 25} = \sqrt{4} \cdot \sqrt{25}$.

These examples suggest the following property of radicals.

Multiplication Property of Radicals

Let a and b be nonnegative real numbers; then

$$\sqrt{a \cdot b} = \sqrt{a} \cdot \sqrt{b}$$

REMARK   This property states that the square root of the product of two factors is equal to the product of the square roots of each factor.

The following example illustrates one way to use the multiplication property of radicals.

EXAMPLE 8.2.1   Multiply the following radicals. Assume that all variables represent positive real numbers.

1. $\sqrt{2} \cdot \sqrt{7} = \sqrt{2 \cdot 7}$
   $\qquad\qquad = \sqrt{14}$

2. $\sqrt{3} \cdot \sqrt{3} = \sqrt{3 \cdot 3}$
   $\qquad\qquad = \sqrt{9}$
   $\qquad\qquad = 3$

3. $\sqrt{5} \cdot \sqrt{x} = \sqrt{5 \cdot x}$
   $\qquad\qquad = \sqrt{5x}$

4. $\sqrt{2x} \cdot \sqrt{7y} = \sqrt{2x \cdot 7y}$
   $\qquad\qquad\quad = \sqrt{14xy}$

5. $\sqrt{x} \cdot \sqrt{x} = \sqrt{x \cdot x}$
   $\qquad\qquad = \sqrt{x^2}$
   $\qquad\qquad = x \qquad\qquad$ From Section 8.1

6. $\sqrt{3x} \cdot \sqrt{3x} = \sqrt{3x \cdot 3x}$
   $\qquad\qquad\quad = \sqrt{9x^2}$
   $\qquad\qquad\quad = \sqrt{(3x)^2}$
   $\qquad\qquad\quad = 3x$

NOTE ▶   An alternate solution to part 6 would be as follows:

$$\sqrt{3x} \cdot \sqrt{3x} = (\sqrt{3x})^2$$
$$\qquad\quad = 3x \qquad\qquad \text{From Section 8.1}$$

A *square root* (which is free of fractions) *is considered simplified if its radicand contains no perfect square factors* (other than 1). The following example illustrates how to use the multiplication property of radicals to simplify square roots.

$$\sqrt{45} = \sqrt{9 \cdot 5}$$

Since 45 has a perfect square factor of 9, rewrite 45 as a product with 9 as one of the factors.

$$= \sqrt{9} \cdot \sqrt{5}$$ Apply the multiplication property of radicals.

$$= 3 \cdot \sqrt{5}$$

Thus, $45 = 3\sqrt{5}$. Since the remaining radicand contains no perfect square factors, $3\sqrt{5}$ is the simplified form of $\sqrt{45}$.

In general, to simplify a square root (which is free of fractions) we use the following steps.

1. Factor the radicand so that the *largest* perfect square factor within the radicand is one of the factors.

2. Rewrite the square root as a product of the square roots found in step 1 using the multiplication property of radicals.

3. Simplify the square root with the perfect square radicand.

NOTE ▶   It usually is helpful to write down the list of the first eight to ten perfect square integers:

$$1, 4, 9, 16, 25, 36, 49, 64, 81, 100, \ldots$$

Looking at this list sometimes makes it easier to identify the largest perfect square factor within a given radicand.

EXAMPLE 8.2.2   Simplify the following radicals. Assume that all variables represent positive real numbers.

1. $\sqrt{300} = \sqrt{100 \cdot 3}$ Step 1

          $= \sqrt{100} \cdot \sqrt{3}$ Step 2

          $= 10\sqrt{3}$ Step 3

NOTE ▶   It is a good idea to write the perfect square factor first.

2. $\sqrt{98} = \sqrt{49 \cdot 2}$

        $= \sqrt{49}\sqrt{2}$

        $= 7\sqrt{2}$

3. $\sqrt{48} = \sqrt{16 \cdot 3}$

        $= \sqrt{16}\sqrt{3}$

        $= 4\sqrt{3}$

Try to avoid this mistake:

Incorrect	Correct
$\sqrt{48} = \sqrt{4 \cdot 12}$ $= \sqrt{4}\sqrt{12}$ $= 2\sqrt{12}$ Here we did not find the largest perfect square factor of 48. The remaining radicand contains a perfect square factor.	$\sqrt{48} = \sqrt{16 \cdot 3}$ $= \sqrt{16}\sqrt{3}$ $= 4\sqrt{3}$

4. $\sqrt{26}$ Cannot be simplified. 26 has no perfect square factors.

5. $\sqrt{6x^2} = \sqrt{x^2 \cdot 6}$
$= \sqrt{x^2}\sqrt{6}$
$= x\sqrt{6}$

or $= \sqrt{6}x$ Placing the numerical factor first

6. $\sqrt{44x^2} = \sqrt{4x^2 \cdot 11}$ Our largest perfect square factor
$= \sqrt{4x^2}\sqrt{11}$ is $4x^2$ since $4x^2 = (2x)^2$.
$= 2x\sqrt{11}$

NOTE ▶ Here we had to find the largest perfect square factor of the numerical and variable factors.

7. $\sqrt{18x^3} = \sqrt{9x^2 \cdot 2x}$ Our largest perfect square factor
$= \sqrt{9x^2}\sqrt{2x}$ is $9x^2$ since $9x^2 = (3x)^2$.
$= 3x\sqrt{2x}$

8. $\sqrt{x^6} = \sqrt{(x^3)^2}$ x^6 is a perfect square
$= x^3$ since $x^6 = (x^3)^2$.

NOTE ▶ If a factor of the radicand has an even exponent, then it is a perfect square. To find the square root, we divide the exponent by 2. Thus, $\sqrt{x^{10}} = x^5$, $\sqrt{x^{24}} = x^{12}$, and so on.

9. $\sqrt{y^9} = \sqrt{y^8 \cdot y}$
$= \sqrt{y^8}\sqrt{y}$
$= y^4\sqrt{y}$

10. $\sqrt{20x^4y^5} = \sqrt{4x^4y^4 \cdot 5y}$

$\phantom{\textbf{10.} \sqrt{20x^4y^5}} = \sqrt{4x^4y^4}\sqrt{5y}$

$\phantom{\textbf{10.} \sqrt{20x^4y^5}} = 2x^2y^2\sqrt{5y}$

11. $\sqrt{48x^8y^{10}} = \sqrt{16x^8y^{10} \cdot 3}$

$\phantom{\textbf{11.} \sqrt{48x^8y^{10}}} = \sqrt{16x^8y^{10}}\sqrt{3}$

$\phantom{\textbf{11.} \sqrt{48x^8y^{10}}} = 4x^4y^5\sqrt{3}$

12. $\sqrt{75x^3y^7} = \sqrt{25x^2y^6 \cdot 3xy}$

$\phantom{\textbf{12.} \sqrt{75x^3y^7}} = \sqrt{25x^2y^6}\sqrt{3xy}$

$\phantom{\textbf{12.} \sqrt{75x^3y^7}} = 5xy^3\sqrt{3xy}$

Let's consider two other examples from arithmetic:

$$\sqrt{\frac{36}{49}} = \frac{6}{7} \qquad \text{because } \frac{6}{7} > 0 \text{ and } \left(\frac{6}{7}\right)^2 = \frac{36}{49}$$

and

$$\frac{\sqrt{36}}{\sqrt{49}} = \frac{6}{7}$$

Thus, $\sqrt{\dfrac{36}{49}} = \dfrac{\sqrt{36}}{\sqrt{49}}$. These examples suggest another property of radicals.

Division Property of Radicals

Let a and b be nonnegative real numbers with $b \neq 0$; then

$$\sqrt{\frac{a}{b}} = \frac{\sqrt{a}}{\sqrt{b}}$$

REMARK   This property states that the square root of a fraction is equal to the square root of the numerator divided by the square root of the denominator.

The following example illustrates how to apply the division property of radicals.

EXAMPLE 8.2.3

Simplify the following radicals. Assume that all variables represent positive real numbers.

1. $\sqrt{\dfrac{16}{25}} = \dfrac{\sqrt{16}}{\sqrt{25}}$

   $= \dfrac{4}{5}$

2. $\sqrt{\dfrac{17}{9}} = \dfrac{\sqrt{17}}{\sqrt{9}}$

   $= \dfrac{\sqrt{17}}{3}$

3. $\dfrac{\sqrt{5}}{\sqrt{20}} = \sqrt{\dfrac{5}{20}}$

   $= \sqrt{\dfrac{1}{4}}$

   $= \dfrac{\sqrt{1}}{\sqrt{4}}$

   $= \dfrac{1}{2}$

4. $\dfrac{\sqrt{54}}{\sqrt{32}} = \sqrt{\dfrac{54}{32}}$

   $= \sqrt{\dfrac{27}{16}}$

   $= \dfrac{\sqrt{27}}{\sqrt{16}}$

   $= \dfrac{\sqrt{27}}{4}$ Do not stop here. We must simplify the radical in the numerator.

   $= \dfrac{\sqrt{9 \cdot 3}}{4}$

   $= \dfrac{\sqrt{9}\sqrt{3}}{4}$

   $= \dfrac{3\sqrt{3}}{4}$

NOTE ▶ In Section 8.5 we will find another method for simplifying division problems like parts 3 and 4.

5. $\sqrt{\dfrac{49x^2}{100}} = \dfrac{\sqrt{49x^2}}{\sqrt{100}}$

$$= \frac{7x}{10}$$

6. $\sqrt{\dfrac{1}{81y^2}} = \dfrac{\sqrt{1}}{\sqrt{81y^2}}$

$$= \frac{1}{9y}$$

7. $\sqrt{\dfrac{9x^2y^8}{z^4}} = \dfrac{\sqrt{9x^2y^8}}{\sqrt{z^4}}$

$$= \frac{3xy^4}{z^2}$$

Remember, to find the square root of a factor raised to an even exponent, simply divide the exponent by 2.

8. $\sqrt{\dfrac{27x^2}{4}} = \dfrac{\sqrt{27x^2}}{\sqrt{4}}$

$$= \frac{\sqrt{9x^2 \cdot 3}}{\sqrt{4}}$$

$$= \frac{\sqrt{9x^2}\sqrt{3}}{\sqrt{4}}$$

$$= \frac{3x\sqrt{3}}{2}$$

9. $\sqrt{\dfrac{98x^3}{121}} = \dfrac{\sqrt{98x^3}}{\sqrt{121}}$

$$= \frac{\sqrt{49x^2 \cdot 2x}}{\sqrt{121}}$$

$$= \frac{\sqrt{49x^2}\sqrt{2x}}{\sqrt{121}}$$

$$= \frac{7x\sqrt{2x}}{11}$$

10. $\sqrt{\dfrac{80x^5}{81y^{10}}} = \dfrac{\sqrt{80x^5}}{\sqrt{81y^{10}}}$

$$= \frac{\sqrt{16x^4 \cdot 5x}}{\sqrt{81y^{10}}}$$

$$= \frac{\sqrt{16x^4}\sqrt{5x}}{\sqrt{81y^{10}}}$$

$$= \frac{4x^2\sqrt{5x}}{9y^5}$$

The multiplication and division properties of radicals have been applied only to square roots. However, both of these properties can be extended to higher order radicals. The following example illustrates how we can extend these properties to simplify cube roots.

EXAMPLE 8.2.4

Simplify the following radicals. Assume that all variables represent positive real numbers.

1. $\sqrt[3]{24} = \sqrt[3]{8 \cdot 3}$
$$= \sqrt[3]{8}\sqrt[3]{3}$$
$$= 2\sqrt[3]{3}$$

NOTE ▶ Since we have a cube root, we must now look for factors that are perfect cubes.

2. $\sqrt[3]{40x^3} = \sqrt[3]{8x^3 \cdot 5}$
$$= \sqrt[3]{8x^3}\sqrt[3]{5}$$
$$= 2x\sqrt[3]{5}$$

3. $\sqrt[3]{81x^4} = \sqrt[3]{27x^3 \cdot 3x}$
$$= \sqrt[3]{27x^3}\sqrt[3]{3x}$$
$$= 3x\sqrt[3]{3x}$$

4. $\sqrt[3]{\dfrac{8}{27}} = \dfrac{\sqrt[3]{8}}{\sqrt[3]{27}}$
$$= \dfrac{2}{3}$$

5. $\sqrt[3]{\dfrac{x^3}{125}} = \dfrac{\sqrt[3]{x^3}}{\sqrt[3]{125}}$
$$= \dfrac{x}{5}$$

6. $\sqrt[3]{\dfrac{5x^4}{64}} = \dfrac{\sqrt[3]{5x^4}}{\sqrt[3]{64}}$
$$= \dfrac{\sqrt[3]{x^3 \cdot 5x}}{\sqrt[3]{64}}$$
$$= \dfrac{\sqrt[3]{x^3} \cdot \sqrt[3]{5x}}{\sqrt[3]{64}}$$
$$= \dfrac{x\sqrt[3]{5x}}{4}$$

EXERCISES 8.2

Multiply the following radicals. Assume that all variables represent positive real numbers.

1. $\sqrt{3} \cdot \sqrt{10}$ **2.** $\sqrt{2} \cdot \sqrt{11}$ **3.** $\sqrt{7} \cdot \sqrt{2}$ **4.** $\sqrt{6} \cdot \sqrt{7}$

5. $\sqrt{13} \cdot \sqrt{13}$ **6.** $\sqrt{15} \cdot \sqrt{15}$ **7.** $\sqrt{20} \cdot \sqrt{5}$ **8.** $\sqrt{27} \cdot \sqrt{3}$

9. $\sqrt{3} \cdot \sqrt{12}$ **10.** $\sqrt{2} \cdot \sqrt{18}$ **11.** $\sqrt{7} \cdot \sqrt{x}$ **12.** $\sqrt{6} \cdot \sqrt{x}$

13. $\sqrt{2x} \cdot \sqrt{3y}$ **14.** $\sqrt{3x} \cdot \sqrt{5y}$ **15.** $\sqrt{6x} \cdot \sqrt{6x}$ **16.** $\sqrt{7x} \cdot \sqrt{7x}$

17. $\sqrt{2x} \cdot \sqrt{8x}$ **18.** $\sqrt{3x} \cdot \sqrt{12x}$ **19.** $\sqrt{x} \cdot \sqrt{x^5}$ **20.** $\sqrt{y} \cdot \sqrt{y^7}$

21. $\sqrt{5x^3} \cdot \sqrt{5x^5}$ **22.** $\sqrt{3x^5} \cdot \sqrt{3x^7}$ **23.** $\sqrt{32x^7} \cdot \sqrt{2x^3}$ **24.** $\sqrt{20x^5} \cdot \sqrt{5x^9}$

Simplify the following radicals. Assume that all variables represent positive real numbers.

25. $\sqrt{18}$ **26.** $\sqrt{32}$ **27.** $\sqrt{20}$ **28.** $\sqrt{12}$

29. $\sqrt{65}$ **30.** $\sqrt{99}$ **31.** $\sqrt{150}$ **32.** $\sqrt{175}$

33. $\sqrt{40}$ **34.** $\sqrt{33}$ **35.** $\sqrt{54}$ **36.** $\sqrt{72}$

37. $\sqrt{7x^2}$ **38.** $\sqrt{3x^2}$ **39.** $\sqrt{x^{16}}$ **40.** $\sqrt{x^{12}}$

41. $\sqrt{16x^8}$ **42.** $\sqrt{36x^4}$ **43.** $\sqrt{81x^2y^{10}}$ **44.** $\sqrt{25x^4y^8}$

45. $\sqrt{169x^{12}y^{20}}$ **46.** $\sqrt{121x^{18}y^{10}}$ **47.** $\sqrt{121x}$ **48.** $\sqrt{64x}$

49. $\sqrt{x^{11}}$ **50.** $\sqrt{x^9}$ **51.** $\sqrt{64x^5}$ **52.** $\sqrt{9x^7}$

53. $\sqrt{72x^2}$ **54.** $\sqrt{48x^2}$ **55.** $\sqrt{75x^3}$ **56.** $\sqrt{45x^3}$

57. $\sqrt{48x^8y^6}$ **58.** $\sqrt{72x^4y^6}$ **59.** $\sqrt{5x^6y^5}$ **60.** $\sqrt{6x^4y^7}$

61. $\sqrt{32x^9y}$ **62.** $\sqrt{96x^7y}$

Simplify the following radicals. Assume that all variables represent positive real numbers.

63. $\sqrt{\dfrac{49}{4}}$ **64.** $\sqrt{\dfrac{64}{9}}$ **65.** $\sqrt{\dfrac{5}{64}}$ **66.** $\sqrt{\dfrac{7}{81}}$

67. $\dfrac{\sqrt{6}}{\sqrt{96}}$ **68.** $\dfrac{\sqrt{8}}{\sqrt{72}}$ **69.** $\dfrac{\sqrt{98}}{\sqrt{8}}$ **70.** $\dfrac{\sqrt{500}}{\sqrt{45}}$

71. $\dfrac{\sqrt{200}}{\sqrt{8}}$ **72.** $\dfrac{\sqrt{24}}{\sqrt{6}}$ **73.** $\dfrac{\sqrt{48}}{\sqrt{50}}$ **74.** $\dfrac{\sqrt{135}}{\sqrt{48}}$

75. $\dfrac{\sqrt{160}}{\sqrt{45}}$ **76.** $\dfrac{\sqrt{96}}{\sqrt{18}}$ **77.** $\dfrac{\sqrt{24}}{\sqrt{27}}$ **78.** $\sqrt{\dfrac{81x^2}{25}}$

79. $\sqrt{\dfrac{64x^2}{49}}$ **80.** $\sqrt{\dfrac{1}{121x^2}}$ **81.** $\sqrt{\dfrac{1}{100x^2}}$ **82.** $\sqrt{\dfrac{4x^4y^{10}}{z^2}}$

83. $\sqrt{\dfrac{16x^2y^{12}}{z^4}}$ **84.** $\sqrt{\dfrac{12x^2}{25}}$ **85.** $\sqrt{\dfrac{24x^3}{49}}$ **86.** $\sqrt{\dfrac{45x^3}{4}}$

87. $\sqrt{\dfrac{112x}{81}}$ **88.** $\sqrt{\dfrac{162x}{49}}$ **89.** $\sqrt{\dfrac{18x}{25y^2}}$ **90.** $\sqrt{\dfrac{48x}{121y^2}}$

91. $\sqrt{\dfrac{20x^4}{9y^4}}$ **92.** $\sqrt{\dfrac{63x^4}{4y^4}}$ **93.** $\sqrt{\dfrac{32x^3}{25y^8}}$ **94.** $\sqrt{\dfrac{45x^3}{49y^8}}$

95. $\sqrt{\dfrac{75x^7}{16y^6}}$

Simplify the following radicals. Assume that all variables represent positive real numbers.

96. $\sqrt[3]{54}$ **97.** $\sqrt[3]{81}$ **98.** $\sqrt[3]{32}$ **99.** $\sqrt[3]{40}$

100. $\sqrt[3]{375}$ **101.** $\sqrt[3]{250}$ **102.** $\sqrt[3]{24x^3}$ **103.** $\sqrt[3]{135x^3}$

104. $\sqrt[3]{320x^3}$ **105.** $\sqrt[3]{27x^5}$ **106.** $\sqrt[3]{8x^5}$ **107.** $\sqrt[3]{56x^4}$

108. $\sqrt[3]{88x^4}$ **109.** $\sqrt[3]{\dfrac{x^3}{8}}$ **110.** $\sqrt[3]{\dfrac{x^3}{27}}$ **111.** $\sqrt[3]{\dfrac{125x^3}{27y^3}}$

112. $\sqrt[3]{\dfrac{64x^3}{125y^3}}$ **113.** $\sqrt[3]{\dfrac{7x^4}{64}}$ **114.** $\sqrt[3]{\dfrac{9x^4}{125}}$ **115.** $\sqrt[3]{\dfrac{16x^4}{27y^3}}$

116. $\sqrt[3]{\dfrac{81x^4}{8y^3}}$

8.3 Adding and Subtracting Radical Expressions

In Chapter 3 we learned how to simplify polynomials by combining like terms. The like terms were combined by using the distributive property. For example,

$$2x + 5x = (2 + 5)x \quad \text{By the distributive property}$$
$$= 7x$$

Now, suppose that we wanted to simplify the expression

$$2\sqrt{3} + 5\sqrt{3} = (2 + 5)\sqrt{3} \quad \text{By the distributive property}$$
$$= 7\sqrt{3}$$

NOTE ▶ In this case, both of the square roots had the same radicand and the two terms could be combined by using the distributive property.

This example leads us to the following definition.

> ### Like Radicals
>
> **Like radicals** are radicals that have the same order and radicand.

REMARKS

1. The following expressions contain like radicals.

    a. $\sqrt{2}, \quad 3\sqrt{2}, \quad -8\sqrt{2}$

    b. $3\sqrt{x}, \quad -\sqrt{x}, \quad 7\sqrt{x}, \quad 9\sqrt{x}$

    c. $2x\sqrt{y}, \quad 4x\sqrt{y}$

2. The following expressions do *not* contain like radicals.

    a. $5\sqrt{2}, \quad 4\sqrt[3]{2}$

    Here the radicals are not like because the orders of the radicals are not the same.

    b. $3\sqrt{5}, \quad 5\sqrt{3}$

    Here the radicals are not like because the radicands are not the same.

    The following example illustrates how to combine terms with like radicals by using the distributive property.

EXAMPLE 8.3.1

Perform the indicated operations. Simplify your answer by combining like radical terms. Assume that all variables represent positive real numbers.

1. $3\sqrt{6} + 7\sqrt{6} = (3 + 7)\sqrt{6}$ By the distributive property

               $= 10\sqrt{6}$

2. $2\sqrt{5} - 9\sqrt{5} = (2 - 9)\sqrt{5}$

             $= -7\sqrt{5}$

3. $\sqrt{2} + 3\sqrt{2} - 8\sqrt{2} = (1 + 3 - 8)\sqrt{2}$

                      $= -4\sqrt{2}$

4. $3\sqrt{5} + 5\sqrt{3}$

    The terms cannot be combined, since they do not contain like radicals.

5. $2x\sqrt{y} + 4x\sqrt{y} = (2x + 4x)\sqrt{y}$

               $= 6x\sqrt{y}$

6. $\sqrt{2x} + 5\sqrt{2x} + 8\sqrt{y} - 7\sqrt{y}$

    Let us first group the like radical terms together:

    $= (\sqrt{2x} + 5\sqrt{2x}) + (8\sqrt{y} - 7\sqrt{y})$

$$= (1 + 5)\sqrt{2x} + (8 - 7)\sqrt{y}$$
$$= 6\sqrt{2x} + \sqrt{y}$$

This expression cannot be further simplified, since the remaining terms do not contain like radicals.

Suppose that you were asked to simplify the expression $5\sqrt{3} + \sqrt{12}$. Clearly, the radicals are not like radicals. However, the second term, $\sqrt{12}$, can be simplified:

$$\begin{aligned} 5\sqrt{3} + \sqrt{12} &= 5\sqrt{3} + \sqrt{4 \cdot 3} \qquad &&\text{Simplify the second radical.}\\ &= 5\sqrt{3} + \sqrt{4} \cdot \sqrt{3}\\ &= 5\sqrt{3} + 2\sqrt{3}\\ &= 7\sqrt{3} \end{aligned}$$

NOTE ▶ After simplifying the second radical, we have generated like radical terms.

This example illustrates the following rule.

> When adding and subtracting radical expressions, always first simplify each radical expression, if possible.

EXAMPLE 8.3.2

Perform the indicated operations. Assume that all variables represent positive real numbers.

1. $\begin{aligned}[t] 7\sqrt{5} + \sqrt{125} &= 7\sqrt{5} + \sqrt{25 \cdot 5}\\ &= 7\sqrt{5} + \sqrt{25}\sqrt{5}\\ &= 7\sqrt{5} + 5\sqrt{5}\\ &= 12\sqrt{5} \end{aligned}$

2. $\begin{aligned}[t] 2\sqrt{8} - 2\sqrt{50} &= 2\sqrt{4 \cdot 2} - 2\sqrt{25 \cdot 2}\\ &= 2\sqrt{4}\sqrt{2} - 2\sqrt{25}\sqrt{2}\\ &= 2 \cdot 2\sqrt{2} - 2 \cdot 5\sqrt{2}\\ &= 4\sqrt{2} - 10\sqrt{2}\\ &= -6\sqrt{2} \end{aligned}$

3. $\begin{aligned}[t] &3\sqrt{3} - 2\sqrt{54} + \sqrt{27} + 5\sqrt{6}\\ &= 3\sqrt{3} - 2\sqrt{9 \cdot 6} + \sqrt{9 \cdot 3} + 5\sqrt{6} \end{aligned}$

$$= 3\sqrt{3} - 2\sqrt{9}\sqrt{6} + \sqrt{9}\sqrt{3} + 5\sqrt{6}$$
$$= 3\sqrt{3} - 2 \cdot 3\sqrt{6} + 3\sqrt{3} + 5\sqrt{6}$$
$$= 3\sqrt{3} - 6\sqrt{6} + 3\sqrt{3} + 5\sqrt{6}$$
$$= (3\sqrt{3} + 3\sqrt{3}) + (-6\sqrt{6} + 5\sqrt{6})$$

Group the like radical terms together.

$$= 6\sqrt{3} - \sqrt{6}$$

4. $\sqrt{48} + \sqrt{72} = \sqrt{16 \cdot 3} + \sqrt{36 \cdot 2}$
$$= \sqrt{16}\sqrt{3} + \sqrt{36}\sqrt{2}$$
$$= 4\sqrt{3} + 6\sqrt{2}$$

This expression cannot be simplified further, since the remaining terms do not contain like radicals.

5. $3x\sqrt{20} + 7\sqrt{5x^2} = 3x\sqrt{4 \cdot 5} + 7\sqrt{x^2 \cdot 5}$
$$= 3x\sqrt{4}\sqrt{5} + 7\sqrt{x^2}\sqrt{5}$$
$$= 3x \cdot 2\sqrt{5} + 7 \cdot x\sqrt{5}$$
$$= 6x\sqrt{5} + 7x\sqrt{5}$$
$$= (6x + 7x)\sqrt{5}$$
$$= 13x\sqrt{5}$$

6. $2\sqrt{27y} + \sqrt{48y} - 4\sqrt{12y} = 2\sqrt{9 \cdot 3y} + \sqrt{16 \cdot 3y} - 4\sqrt{4 \cdot 3y}$
$$= 2\sqrt{9}\sqrt{3y} + \sqrt{16}\sqrt{3y} - 4\sqrt{4}\sqrt{3y}$$
$$= 2 \cdot 3\sqrt{3y} + 4\sqrt{3y} - 4 \cdot 2\sqrt{3y}$$
$$= 6\sqrt{3y} + 4\sqrt{3y} - 8\sqrt{3y}$$
$$= 2\sqrt{3y}$$

7. $x\sqrt{50x} + 3\sqrt{2x^3} = x\sqrt{25 \cdot 2x} + 3\sqrt{x^2 \cdot 2x}$
$$= x\sqrt{25}\sqrt{2x} + 3\sqrt{x^2}\sqrt{2x}$$
$$= x \cdot 5\sqrt{2x} + 3 \cdot x\sqrt{2x}$$
$$= 5x\sqrt{2x} + 3x\sqrt{2x}$$
$$= (5x + 3x)\sqrt{2x}$$
$$= 8x\sqrt{2x}$$

8. $\quad x\sqrt{45y} - 2x\sqrt{80y} + \sqrt{20x^2y}$
$$= x\sqrt{9 \cdot 5y} - 2x\sqrt{16 \cdot 5y} + \sqrt{4x^2 \cdot 5y}$$
$$= x\sqrt{9}\sqrt{5y} - 2x\sqrt{16}\sqrt{5y} + \sqrt{4x^2}\sqrt{5y}$$
$$= x \cdot 3\sqrt{5y} - 2x \cdot 4\sqrt{5y} + 2x\sqrt{5y}$$
$$= 3x\sqrt{5y} - 8x\sqrt{5y} + 2x\sqrt{5y}$$
$$= (3x - 8x + 2x)\sqrt{5y}$$
$$= -3x\sqrt{5y}$$

9. $\quad 2x\sqrt{75xy} - \sqrt{48x^3y} - \sqrt{3x^3y}$
$$= 2x\sqrt{25 \cdot 3xy} - \sqrt{16x^2 \cdot 3xy} - \sqrt{x^2 \cdot 3xy}$$

$$= 2x\sqrt{25}\sqrt{3xy} - \sqrt{16x^2}\sqrt{3xy} - \sqrt{x^2}\sqrt{3xy}$$
$$= 2x \cdot 5\sqrt{3xy} - 4x\sqrt{3xy} - x\sqrt{3xy}$$
$$= 10x\sqrt{3xy} - 4x\sqrt{3xy} - x\sqrt{3xy}$$
$$= (10x - 4x - x)\sqrt{3xy}$$
$$= 5x\sqrt{3xy}$$

10. $3\sqrt{12x^2y} - 4\sqrt{3x^2y} + y\sqrt{8x} + \sqrt{2xy^2}$
$$= 3\sqrt{4x^2 \cdot 3y} - 4\sqrt{x^2 \cdot 3y} + y\sqrt{4 \cdot 2x} + \sqrt{y^2 \cdot 2x}$$
$$= 3\sqrt{4x^2}\sqrt{3y} - 4\sqrt{x^2}\sqrt{3y} + y\sqrt{4}\sqrt{2x} + \sqrt{y^2}\sqrt{2x}$$
$$= 3 \cdot 2x\sqrt{3y} - 4 \cdot x\sqrt{3y} + y \cdot 2\sqrt{2x} + y\sqrt{2x}$$
$$= 6x\sqrt{3y} - 4x\sqrt{3y} + 2y\sqrt{2x} + y\sqrt{2x}$$
$$= (6x\sqrt{3y} - 4x\sqrt{3y}) + (2y\sqrt{2x} + y\sqrt{2x}) \quad \text{Group the like radical}$$
$$= (6x - 4x)\sqrt{3y} + (2y + y)\sqrt{2x} \quad \text{terms together.}$$
$$= 2x\sqrt{3y} + 3y\sqrt{2x}$$

So far we have combined only like radical terms in which the radicals were square roots. The following example illustrates how to combine like radical terms involving cube roots.

EXAMPLE 8.3.3

Perform the indicated operations. Assume that all variables represent positive real numbers.

1. $\sqrt[3]{24} + \sqrt[3]{81} = \sqrt[3]{8 \cdot 3} + \sqrt[3]{27 \cdot 3}$
$$= \sqrt[3]{8} \cdot \sqrt[3]{3} + \sqrt[3]{27} \cdot \sqrt[3]{3}$$
$$= 2\sqrt[3]{3} + 3\sqrt[3]{3}$$
$$= (2 + 3)\sqrt[3]{3}$$
$$= 5\sqrt[3]{3}$$

NOTE ▶ Remember, since we have a *cube* root, we must now look for factors that are perfect cubes.

2. $3\sqrt[3]{16x^3} - x\sqrt[3]{250} = 3\sqrt[3]{8x^3 \cdot 2} - x\sqrt[3]{125 \cdot 2}$
$$= 3\sqrt[3]{8x^3} \cdot \sqrt[3]{2} - x\sqrt[3]{125} \cdot \sqrt[3]{2}$$
$$= 3 \cdot 2x\sqrt[3]{2} - x \cdot 5\sqrt[3]{2}$$
$$= 6x\sqrt[3]{2} - 5x\sqrt[3]{2}$$
$$= (6x - 5x)\sqrt[3]{2}$$
$$= x\sqrt[3]{2}$$

EXERCISES 8.3

Perform the indicated operations. Assume that all variables represent positive real numbers.

1. $2\sqrt{6} + 11\sqrt{6}$

2. $9\sqrt{2} - 10\sqrt{2}$

3. $5\sqrt{7} + 7\sqrt{5}$

4. $\sqrt{10} - \sqrt{6}$

5. $\sqrt{7} + 3\sqrt{7} - 6\sqrt{7}$

6. $12\sqrt{10} - \sqrt{10} - 4\sqrt{10}$

7. $2x\sqrt{5} + x\sqrt{5}$

8. $4x\sqrt{6} - 8x\sqrt{6}$

9. $3y\sqrt{11} + 2y\sqrt{11} - 11y\sqrt{11}$

10. $5z\sqrt{2} + 4z\sqrt{2} - 6z\sqrt{3}$

11. $7\sqrt{2x} - 6\sqrt{2x}$

12. $4\sqrt{3xy} + 6\sqrt{3xy}$

13. $\sqrt{7x} + 3\sqrt{7x} - 4\sqrt{7x}$

14. $\sqrt{6y} + 3\sqrt{6y} + 2\sqrt{6y}$

15. $5x\sqrt{2y} + 7x\sqrt{2y}$

16. $3x\sqrt{5y} - 7y\sqrt{3x}$

17. $3x\sqrt{5x} + 2x\sqrt{5x} + x\sqrt{5x}$

18. $5x^2\sqrt{3y} - 11x^2\sqrt{3y} - x^2\sqrt{3y}$

19. $5\sqrt{2} + 3\sqrt{2} + 7\sqrt{6} - \sqrt{6}$

20. $6\sqrt{5} - 7\sqrt{6} - 8\sqrt{5} - 2\sqrt{6}$

21. $7\sqrt{2x} - 4\sqrt{2x} + 3\sqrt{3y} - 9\sqrt{3y}$

22. $3\sqrt{5x} - 2\sqrt{5x} - \sqrt{5x} + 5\sqrt{x}$

23. $5x\sqrt{2y} - 3y\sqrt{x} + 9x\sqrt{2y} - y\sqrt{x}$

24. $2\sqrt{6xy} + 5\sqrt{6xy} + 2x\sqrt{5y} - 4x\sqrt{5y}$

25. $\sqrt{48} + \sqrt{12}$

26. $\sqrt{18} + \sqrt{50}$

27. $2\sqrt{45} - 2\sqrt{20}$

28. $3\sqrt{27} - 2\sqrt{12}$

29. $\sqrt{75} - \sqrt{24}$

30. $\sqrt{72} - \sqrt{45}$

31. $3\sqrt{63} + 5\sqrt{7}$

32. $4\sqrt{96} + 2\sqrt{6}$

33. $\sqrt{8} - 2\sqrt{32}$

34. $\sqrt{27} - 3\sqrt{12}$

35. $\sqrt{20} - \sqrt{5} - 3\sqrt{45}$

36. $\sqrt{54} - \sqrt{6} - 2\sqrt{24}$

37. $\sqrt{108} + 3\sqrt{27} + 2\sqrt{3}$

38. $\sqrt{80} + 4\sqrt{20} + 3\sqrt{5}$

39. $\sqrt{12} - 2\sqrt{48} + \sqrt{18}$

40. $3\sqrt{8} - 3\sqrt{18} + 2\sqrt{12}$

41. $4\sqrt{3} + \sqrt{12} + 2\sqrt{18} + 2\sqrt{32}$

42. $\sqrt{96} + 4\sqrt{6} + 2\sqrt{27} + 2\sqrt{12}$

43. $\sqrt{20} + \sqrt{54} - \sqrt{80} - \sqrt{150}$

44. $\sqrt{12} + \sqrt{32} - \sqrt{75} - \sqrt{128}$

45. $\sqrt{5} + \sqrt{45} - \sqrt{96} - 2\sqrt{6}$

46. $\sqrt{6} + \sqrt{96} - 2\sqrt{10} - \sqrt{160}$

47. $3\sqrt{12} + 2\sqrt{8} - \sqrt{75} - \sqrt{32}$

48. $2\sqrt{48} + \sqrt{80} - 3\sqrt{12} - 2\sqrt{20}$

49. $2x\sqrt{45} + 4\sqrt{5x^2}$

50. $3x\sqrt{18} + 8\sqrt{2x^2}$

51. $\sqrt{24x^2} - 3\sqrt{54x^2}$

52. $\sqrt{28x^2} - 2\sqrt{112x^2}$

53. $x\sqrt{27} - x\sqrt{12} - \sqrt{3x^2}$

54. $x\sqrt{50} - x\sqrt{32} - \sqrt{2x^2}$

55. $x\sqrt{10} + 2\sqrt{40x^2} - 2\sqrt{90x^2}$

56. $x\sqrt{6} + 3\sqrt{24x^2} - 3\sqrt{54x^2}$

57. $3\sqrt{20x} - \sqrt{45x}$

58. $4\sqrt{24x} - \sqrt{96x}$

59. $\sqrt{75x} + \sqrt{45x}$

60. $\sqrt{50x} + \sqrt{20x}$

61. $\sqrt{10x} - 2\sqrt{40x} + 3\sqrt{160x}$

62. $\sqrt{10x} - 2\sqrt{90x} + 2\sqrt{490x}$

63. $-\sqrt{54x} - \sqrt{96x} - \sqrt{24x}$

64. $-\sqrt{125x} - \sqrt{80x} - \sqrt{45x}$

65. $2x\sqrt{48x} + 2\sqrt{27x^3}$

66. $3x\sqrt{18x} + 2\sqrt{32x^3}$

67. $-2\sqrt{50x^2y} - x\sqrt{72y}$

68. $-4\sqrt{27x^2y} - 2x\sqrt{48y}$

69. $\sqrt{96x^3} - 3\sqrt{6x^3} + 2x\sqrt{150x}$

70. $\sqrt{180x^3} - 4\sqrt{5x^3} + 5x\sqrt{20x}$

71. $\sqrt{24xy^3} - 2y\sqrt{54xy} + \sqrt{12xy}$

72. $\sqrt{90x^3y} - x\sqrt{250xy} + \sqrt{80xy}$

73. $\sqrt{50x} + 2\sqrt{8x} + 2\sqrt{6x^3} + 2x\sqrt{96x}$

74. $\sqrt{180x} + 2\sqrt{20x} + 2\sqrt{2x^3} + 3x\sqrt{32x}$

75. $2\sqrt{7x^2y} + 3x\sqrt{28y} - \sqrt{54xy^2} - \sqrt{6xy^2}$

76. $x\sqrt{96y} + 3\sqrt{54x^2y} - 2y\sqrt{32x} - \sqrt{2xy^2}$

77. $x\sqrt{20} - 2x\sqrt{63y} + 3\sqrt{45x^2} + 5\sqrt{7x^2y}$

78. $x\sqrt{128} - 2x\sqrt{44y} + 3\sqrt{18x^2} + 3\sqrt{11x^2y}$

79. $3\sqrt{48xy} - 4\sqrt{3xy} - \sqrt{36xy} - \sqrt{81xy}$

80. $3\sqrt{150xy} - 10\sqrt{6xy} - \sqrt{64xy} - \sqrt{16xy}$

81. $x\sqrt{40y^3} + 2y\sqrt{90x^2y} + 3xy\sqrt{160y} + 4\sqrt{10x^2y^3}$

82. $x\sqrt{135y^3} + 3y\sqrt{60x^2y} + 2xy\sqrt{60y} + 8\sqrt{15x^2y^3}$

83. $\sqrt[3]{54} + \sqrt[3]{250}$

84. $\sqrt[3]{250} + \sqrt[3]{16}$

85. $2\sqrt[3]{24} - 2\sqrt[3]{81}$

86. $\sqrt[3]{24} - 2\sqrt[3]{81}$

87. $x\sqrt[3]{500} - 2\sqrt[3]{32x^3}$

88. $2x\sqrt[3]{108} - 5\sqrt[3]{4x^3}$

89. $8\sqrt[3]{3x^4} + x\sqrt[3]{24x}$

90. $6\sqrt[3]{3x^4} + x\sqrt[3]{192x}$

8.4 Multiplying Radical Expressions

In Section 8.2 we multiplied some simple radical expressions by using the multiplication property of radicals. In this section we will examine multiplication problems in which the radical expressions are more complicated.

Suppose that we wanted to perform the following multiplication.

$$4\sqrt{5} \cdot 3\sqrt{10} = 4 \cdot 3 \cdot \sqrt{5} \cdot \sqrt{10} \qquad \text{By the commutative property}$$

$$= (4 \cdot 3) \cdot (\sqrt{5}\sqrt{10}) \qquad \text{By the associative property}$$

$$= 12\sqrt{50} \qquad \begin{array}{l}\text{By the multiplication}\\\text{property of radicals}\end{array}$$

NOTE ▶ Do not stop here! After performing the multiplication, always check the radicand and make sure that it contains no perfect square factors:

$$= 12\sqrt{25}\sqrt{2}$$

$$= 12 \cdot 5\sqrt{2}$$

$$= 60\sqrt{2}$$

The following example illustrates how to multiply one-term radical expressions.

EXAMPLE 8.4.1

Perform the indicated multiplications and simplify your answers. Assume that all variables represent positive real numbers.

1. $2\sqrt{6} \cdot 7\sqrt{15} = 2 \cdot 7 \cdot \sqrt{6} \cdot \sqrt{15}$

$$= (2 \cdot 7) \cdot (\sqrt{6}\sqrt{15})$$
$$= 14 \cdot \sqrt{90}$$
$$= 14 \cdot \sqrt{9}\sqrt{10}$$
$$= 14 \cdot 3\sqrt{10}$$
$$= 42\sqrt{10}$$

2. $5\sqrt{24} \cdot \sqrt{3} = 5 \cdot (\sqrt{24} \cdot \sqrt{3})$

$$= 5 \cdot \sqrt{72}$$
$$= 5\sqrt{36}\sqrt{2}$$
$$= 5 \cdot 6\sqrt{2}$$
$$= 30\sqrt{2}$$

NOTE ▶ We could also perform this multiplication as follows:

$$5\sqrt{24} \cdot \sqrt{3} = 5\sqrt{4}\sqrt{6} \cdot \sqrt{3} \qquad \text{Simplify } \sqrt{24}.$$
$$= 5 \cdot 2\sqrt{6} \cdot \sqrt{3}$$
$$= 10\sqrt{6} \cdot \sqrt{3}$$
$$= 10\sqrt{18}$$
$$= 10\sqrt{9}\sqrt{2}$$
$$= 10 \cdot 3\sqrt{2}$$
$$= 30\sqrt{2}$$

REMARK It is sometimes helpful to simplify the radicals at the start of the multiplication problem, if possible.

3. $5\sqrt{3y} \cdot \sqrt{21x} = 5(\sqrt{3y} \cdot \sqrt{21x})$

$$= 5\sqrt{63xy}$$
$$= 5\sqrt{9}\sqrt{7xy}$$
$$= 5 \cdot 3\sqrt{7xy}$$
$$= 15\sqrt{7xy}$$

4. $2x\sqrt{5x} \cdot 4x\sqrt{15y} = (2x \cdot 4x) \cdot (\sqrt{5x} \cdot \sqrt{15y})$
$$= 8x^2\sqrt{75xy}$$
$$= 8x^2\sqrt{25}\sqrt{3xy}$$
$$= 8x^2 \cdot 5\sqrt{3xy}$$
$$= 40x^2\sqrt{3xy}$$

5. $6y\sqrt{2x} \cdot 3x\sqrt{7xy} = (6y \cdot 3x)(\sqrt{2x} \cdot \sqrt{7xy})$
$$= 18xy\sqrt{14x^2y}$$
$$= 18xy\sqrt{x^2}\sqrt{14y}$$
$$= 18xy \cdot x\sqrt{14y}$$
$$= 18x^2y\sqrt{14y}$$

In the previous section we saw that adding and subtracting radical expressions was similar to adding and subtracting polynomials. The following example shows that multiplying radical expressions is similar to multiplying polynomials.

EXAMPLE 8.4.2

Perform the indicated multiplications and simplify your answers. Assume that all variables represent positive real numbers.

1. $\sqrt{6}(\sqrt{10} + \sqrt{7}) = \sqrt{6} \cdot \sqrt{10} + \sqrt{6} \cdot \sqrt{7}$ By the distributive
$$= \sqrt{60} + \sqrt{42} \qquad \text{property}$$
$$= \sqrt{4}\sqrt{15} + \sqrt{42}$$
$$= 2\sqrt{15} + \sqrt{42}$$

NOTE ▶ Remember to make sure that all of the radical expressions have been simplified.

2. $3\sqrt{5}(2\sqrt{6} - \sqrt{5}) = 3\sqrt{5} \cdot 2\sqrt{6} - 3\sqrt{5} \cdot \sqrt{5}$
$$= 6\sqrt{30} - 3\sqrt{25}$$
$$= 6\sqrt{30} - 3 \cdot 5$$
$$= 6\sqrt{30} - 15$$

3. $2\sqrt{x}(5\sqrt{2} - 8\sqrt{y} - 3\sqrt{x})$
$$= 2\sqrt{x} \cdot 5\sqrt{2} - 2\sqrt{x} \cdot 8\sqrt{y} - 2\sqrt{x} \cdot 3\sqrt{x}$$
$$= 10\sqrt{2x} - 16\sqrt{xy} - 6\sqrt{x^2}$$
$$= 10\sqrt{2x} - 16\sqrt{xy} - 6x$$

4. $(\sqrt{5} + \sqrt{3})(\sqrt{7} - \sqrt{2})$

Here we will use the FOIL method to perform the multiplication:

$$(\sqrt{5} + \sqrt{3})(\sqrt{7} - \sqrt{2})$$

$$= \sqrt{5} \cdot \sqrt{7} - \sqrt{5} \cdot \sqrt{2} + \sqrt{3} \cdot \sqrt{7} - \sqrt{3} \cdot \sqrt{2}$$

$$= \sqrt{35} - \sqrt{10} + \sqrt{21} - \sqrt{6}$$

5. $(2\sqrt{3} + 4\sqrt{5})(6\sqrt{3} - \sqrt{5})$

$$= 2\sqrt{3} \cdot 6\sqrt{3} - 2\sqrt{3} \cdot \sqrt{5} + 4\sqrt{5} \cdot 6\sqrt{3} - 4\sqrt{5} \cdot \sqrt{5}$$
$$= 12\sqrt{9} - 2\sqrt{15} + 24\sqrt{15} - 4\sqrt{25}$$
$$= 12 \cdot 3 - 2\sqrt{15} + 24\sqrt{15} - 4 \cdot 5$$
$$= 36 - 2\sqrt{15} + 24\sqrt{15} - 20$$

Now combine like terms!

$$= 16 + 22\sqrt{15}$$

6. $(\sqrt{x} + 6)(\sqrt{x} - 3) = \sqrt{x} \cdot \sqrt{x} - \sqrt{x} \cdot 3 + 6 \cdot \sqrt{x} - 6 \cdot 3$
$$= \sqrt{x^2} - 3\sqrt{x} + 6\sqrt{x} - 18$$
$$= x - 3\sqrt{x} + 6\sqrt{x} - 18$$

Combine the like terms in the middle:

$$= x + 3\sqrt{x} - 18$$

7. $(\sqrt{x} + \sqrt{y})(\sqrt{x} - \sqrt{y})$
$$= \sqrt{x} \cdot \sqrt{x} - \sqrt{x} \cdot \sqrt{y} + \sqrt{y} \cdot \sqrt{x} - \sqrt{y} \cdot \sqrt{y}$$
$$= \sqrt{x^2} - \sqrt{xy} + \sqrt{xy} - \sqrt{y^2}$$
$$= x - \sqrt{xy} + \sqrt{xy} - y$$
$$= x - y$$

8. $(2\sqrt{x} + 3\sqrt{y})(5\sqrt{x} - 8\sqrt{y})$
$$= 2\sqrt{x} \cdot 5\sqrt{x} - 2\sqrt{x} \cdot 8\sqrt{y} + 3\sqrt{y} \cdot 5\sqrt{x} - 3\sqrt{y} \cdot 8\sqrt{y}$$
$$= 10\sqrt{x^2} - 16\sqrt{xy} + 15\sqrt{xy} - 24\sqrt{y^2}$$
$$= 10x - \sqrt{xy} - 24y$$

9. $(4\sqrt{x} + 3)^2 = (4\sqrt{x} + 3)(4\sqrt{x} + 3)$
$$= 4\sqrt{x} \cdot 4\sqrt{x} + 4\sqrt{x} \cdot 3 + 3 \cdot 4\sqrt{x} + 3 \cdot 3$$
$$= 16\sqrt{x^2} + 12\sqrt{x} + 12\sqrt{x} + 9$$
$$= 16x + 24\sqrt{x} + 9$$

NOTE ▶   This problem could also be worked using the square of a binomial formula from Section 3.5:

$$(4\sqrt{x} + 3)^2 = (4\sqrt{x})^2 + 2(4\sqrt{x} \cdot 3) + 3^2$$

$$= 4^2(\sqrt{x})^2 + 2(12\sqrt{x}) + 9$$
$$= 16x + 24\sqrt{x} + 9$$

Try to avoid this mistake:

Incorrect	Correct
$(4\sqrt{x} + 3)^2 = (4\sqrt{x})^2 + 3^2$ $= 4^2(\sqrt{x})^2 + 9$ $= 16x + 9$ An exponent does not distribute over a sum.	$(4\sqrt{x} + 3)^2$ $= (4\sqrt{x})^2 + 2(4\sqrt{x} \cdot 3) + 3^2$ $= 4^2(\sqrt{x})^2 + 2(12\sqrt{x}) + 9$ $= 16x + 24\sqrt{x} + 9$

10. $(\sqrt{x+1} + 2)^2 = (\sqrt{x+1})^2 + 2(\sqrt{x+1} \cdot 2) + 2^2$
$$= x + 1 + 4\sqrt{x+1} + 4$$
$$= x + 4\sqrt{x+1} + 5$$

EXERCISES 8.4

Perform the indicated multiplications and simplify your answers. Assume that all variables represent positive real numbers.

1. $2\sqrt{5} \cdot \sqrt{7}$

2. $6\sqrt{8} \cdot \sqrt{7}$

3. $9\sqrt{2} \cdot \sqrt{22}$

4. $4\sqrt{20} \cdot \sqrt{3}$

5. $6\sqrt{15} \cdot \sqrt{6}$

6. $6\sqrt{7} \cdot \sqrt{6}$

7. $5\sqrt{3} \cdot 2\sqrt{7}$

8. $6\sqrt{12} \cdot 2\sqrt{10}$

9. $3\sqrt{10} \cdot 5\sqrt{14}$

10. $5\sqrt{3} \cdot 5\sqrt{33}$

11. $3\sqrt{7} \cdot 4\sqrt{6}$

12. $5\sqrt{32} \cdot 2\sqrt{2}$

13. $5\sqrt{3x} \cdot \sqrt{7y}$

14. $9\sqrt{3x} \cdot \sqrt{7x}$

15. $3\sqrt{11x} \cdot \sqrt{2y}$

16. $10\sqrt{10x} \cdot \sqrt{2y}$

17. $5\sqrt{2x} \cdot 7\sqrt{3y}$

18. $3\sqrt{3x} \cdot 5\sqrt{15y}$

19. $8\sqrt{2x} \cdot 2\sqrt{7xy}$

20. $3\sqrt{5x} \cdot 2\sqrt{11y}$

21. $3\sqrt{6x} \cdot 8\sqrt{10y}$

22. $2\sqrt{32x} \cdot 3\sqrt{2xy}$

23. $3\sqrt{50xy} \cdot 7\sqrt{2xy}$

24. $3x\sqrt{14x} \cdot 6x\sqrt{2y}$

25. $2\sqrt{3x} \cdot 4y\sqrt{10xy}$

26. $5x\sqrt{15x} \cdot 3y\sqrt{10y}$

27. $9\sqrt{5xy} \cdot 4\sqrt{2xy}$

28. $3x\sqrt{3x} \cdot 4y\sqrt{24xy}$

29. $xy\sqrt{2xy} \cdot 3x\sqrt{6y}$

30. $2xy\sqrt{12xy} \cdot 4y\sqrt{3xy}$

31. $\sqrt{7}(\sqrt{3} + \sqrt{2})$

32. $\sqrt{6}(\sqrt{7} - \sqrt{18})$

33. $2\sqrt{5}(3\sqrt{6} + 4\sqrt{10})$

34. $\sqrt{x}(\sqrt{6} - \sqrt{y})$

35. $\sqrt{x}(\sqrt{5y} - \sqrt{3x})$

36. $3\sqrt{x}(2\sqrt{x} + 9\sqrt{y})$

37. $\sqrt{2}(\sqrt{3} + \sqrt{5} - \sqrt{7})$

38. $2\sqrt{3}(\sqrt{7} - 3\sqrt{2} - 4\sqrt{6})$

39. $5\sqrt{6}(2\sqrt{8} + 3\sqrt{5} - \sqrt{6})$

40. $\sqrt{x}(\sqrt{5} - \sqrt{7y} + \sqrt{3z})$

41. $2\sqrt{x}(3\sqrt{7} + \sqrt{2y} - 5\sqrt{x})$

42. $3\sqrt{xy}(2\sqrt{6} - \sqrt{5x} + \sqrt{4y})$

43. $(\sqrt{5} + \sqrt{2})(\sqrt{7} + \sqrt{3})$

44. $(\sqrt{7} - 2)(\sqrt{3} + 1)$

45. $(\sqrt{5} - \sqrt{3})(\sqrt{6} - \sqrt{2})$

46. $(\sqrt{6} + 4)(\sqrt{2} - 5)$

47. $(\sqrt{x} + \sqrt{3})(\sqrt{y} + \sqrt{2})$

48. $(\sqrt{x} - 3)(\sqrt{y} - 2)$

49. $(\sqrt{2x} + \sqrt{y})(\sqrt{3x} - \sqrt{5y})$

50. $(\sqrt{x} - \sqrt{2y})(\sqrt{x} + 4)$

51. $(\sqrt{5} + \sqrt{3})(2\sqrt{5} - \sqrt{3})$

52. $(\sqrt{2} - \sqrt{7})(5\sqrt{2} - 9\sqrt{7})$

53. $(2\sqrt{2} + 3\sqrt{3})(\sqrt{2} + 4\sqrt{3})$

54. $(\sqrt{x} + 4)(\sqrt{x} + 2)$

55. $(\sqrt{x} - \sqrt{3})(2\sqrt{x} - 5\sqrt{3})$

56. $(2\sqrt{x} - \sqrt{5})(4\sqrt{x} - 2\sqrt{5})$

57. $(\sqrt{x} + \sqrt{y})(5\sqrt{x} - 9\sqrt{y})$

58. $(2\sqrt{x} - 3\sqrt{y})(7\sqrt{x} - 2\sqrt{y})$

59. $(\sqrt{5} + \sqrt{2})(\sqrt{5} - \sqrt{2})$

60. $(2\sqrt{3} + \sqrt{5})(2\sqrt{3} - \sqrt{5})$

61. $(\sqrt{7} + 2)(\sqrt{7} - 2)$

62. $(3\sqrt{2} + 1)(3\sqrt{2} - 1)$

63. $(\sqrt{x} + 3)(\sqrt{x} - 3)$

64. $(\sqrt{x} + \sqrt{6})(\sqrt{x} - \sqrt{6})$

65. $(\sqrt{x} + \sqrt{2y})(\sqrt{x} - \sqrt{2y})$

66. $(2\sqrt{x} + 3\sqrt{y})(2\sqrt{x} - 3\sqrt{y})$

67. $(\sqrt{3} + \sqrt{2})^2$

68. $(\sqrt{6} - 2)^2$

69. $(3\sqrt{5} + 3)^2$

70. $(\sqrt{x} + 6)^2$

71. $(\sqrt{x} - \sqrt{5})^2$

72. $(2\sqrt{x} + 5)^2$

73. $(4\sqrt{x} - 1)^2$

74. $(\sqrt{x} + \sqrt{y})^2$

75. $(\sqrt{x} + 3\sqrt{y})^2$

76. $(\sqrt{x + 2} + 1)^2$

77. $(\sqrt{x - 4} - 2)^2$

78. $(2\sqrt{x - 1} + 3)^2$

8.5 Dividing Radical Expressions

It is usually easier to manipulate a fraction that does not have a radical in its denominator. For example, to obtain a decimal approximation for $1/\sqrt{3}$ requires division by $\sqrt{3} \doteq 1.7320508$. Without a calculator, $1 \div 1.7320508$ would be tedious. It is easier to obtain an approximation by using the following procedure.

$$\frac{1}{\sqrt{3}} = \frac{1}{\sqrt{3}} \cdot \frac{\sqrt{3}}{\sqrt{3}} \qquad \text{Multiply the numerator and denominator by } \sqrt{3}.$$

$$= \frac{\sqrt{3}}{\sqrt{9}}$$

$$= \frac{\sqrt{3}}{3}$$

Thus,

$$\frac{\sqrt{3}}{3} \doteq \frac{1.7320508}{3} \doteq 0.5773503$$

NOTE ▶ $\sqrt{3}/3$, which is equivalent to the original fraction, has a denominator that is a rational number instead of an irrational number. The process of eliminating radicals from the denominator is called **rationalizing the denominator**.

The following example illustrates how to rationalize the denominator in fractions where the denominator contains only one term.

EXAMPLE 8.5.1

Rationalize the denominators of the following radical expressions. Assume that all variables represent positive real numbers.

1. $\dfrac{2}{\sqrt{5}} = \dfrac{2}{\sqrt{5}} \cdot \dfrac{\cdot \sqrt{5}}{\sqrt{5}}$ Multiply the numerator and denominator by $\sqrt{5}$.

$= \dfrac{2\sqrt{5}}{\sqrt{25}}$

$= \dfrac{2\sqrt{5}}{5}$

2. $\sqrt{\dfrac{3}{7}} = \dfrac{\sqrt{3}}{\sqrt{7}}$

$= \dfrac{\sqrt{3} \cdot \sqrt{7}}{\sqrt{7} \cdot \sqrt{7}}$

$= \dfrac{\sqrt{21}}{\sqrt{49}}$

$= \dfrac{\sqrt{21}}{7}$

3. $\dfrac{5}{2\sqrt{6}} = \dfrac{5}{2\sqrt{6}} \cdot \dfrac{\cdot \sqrt{6}}{\sqrt{6}}$

$= \dfrac{5\sqrt{6}}{2\sqrt{36}}$

$= \dfrac{5\sqrt{6}}{2 \cdot 6}$

$= \dfrac{5\sqrt{6}}{12}$

4. $\dfrac{2}{\sqrt{10}} = \dfrac{2}{\sqrt{10}} \cdot \dfrac{\cdot \sqrt{10}}{\sqrt{10}}$

$= \dfrac{2\sqrt{10}}{\sqrt{100}}$

$= \dfrac{2\sqrt{10}}{10}$ Be sure to reduce your answer.

$= \dfrac{\sqrt{10}}{5}$

5. $\dfrac{7}{\sqrt{12}} = \dfrac{7}{\sqrt{12}} \cdot \dfrac{\sqrt{12}}{\sqrt{12}}$

$\qquad = \dfrac{7\sqrt{12}}{\sqrt{144}}$

$\qquad = \dfrac{7\sqrt{12}}{12}$

Don't stop here! We must simplify the radical in the numerator:

$\qquad = \dfrac{7\sqrt{4}\sqrt{3}}{12}$

$\qquad = \dfrac{7 \cdot 2\sqrt{3}}{12}$

$\qquad = \dfrac{14\sqrt{3}}{12}$ Again, be sure to reduce your answer.

$\qquad = \dfrac{7\sqrt{3}}{6}$

NOTE ▶   We also could rationalize the denominator as follows:

$\qquad \dfrac{7}{\sqrt{12}} = \dfrac{7}{\sqrt{4}\sqrt{3}}$ First, simplify the radical in the denominator.

$\qquad\qquad = \dfrac{7}{2\sqrt{3}}$

$\qquad\qquad = \dfrac{7}{2\sqrt{3}} \cdot \dfrac{\sqrt{3}}{\sqrt{3}}$ Now multiply the numerator and denominator by $\sqrt{3}$.

$\qquad\qquad = \dfrac{7\sqrt{3}}{2\sqrt{9}}$

$\qquad\qquad = \dfrac{7\sqrt{3}}{2 \cdot 3}$

$\qquad\qquad = \dfrac{7\sqrt{3}}{6}$

REMARK   Note that either method yields the same result. However, it is usually helpful to *simplify the radicals at the start of a division problem, if possible.*

6. $\dfrac{\sqrt{15}}{\sqrt{8}} = \dfrac{\sqrt{15}}{\sqrt{4}\sqrt{2}}$ Simplify the radical in the denominator.

$\qquad\quad = \dfrac{\sqrt{15}}{2\sqrt{2}}$

$$= \frac{\sqrt{15} \cdot \sqrt{2}}{2\sqrt{2} \cdot \sqrt{2}}$$

$$= \frac{\sqrt{30}}{2 \cdot \sqrt{4}}$$

$$= \frac{\sqrt{30}}{2 \cdot 2}$$

$$= \frac{\sqrt{30}}{4}$$

7. $\dfrac{3\sqrt{3}}{2\sqrt{5}} = \dfrac{3\sqrt{3} \cdot \sqrt{5}}{2\sqrt{5} \cdot \sqrt{5}}$

$$= \frac{3\sqrt{15}}{2\sqrt{25}}$$

$$= \frac{3\sqrt{15}}{2 \cdot 5}$$

$$= \frac{3\sqrt{15}}{10}$$

8. $\dfrac{2\sqrt{11} - 5\sqrt{7}}{\sqrt{6}} = \dfrac{(2\sqrt{11} - 5\sqrt{7}) \cdot \sqrt{6}}{\sqrt{6} \qquad \cdot \sqrt{6}}$

$$= \frac{2\sqrt{11} \cdot \sqrt{6} - 5\sqrt{7} \cdot \sqrt{6}}{\sqrt{6} \cdot \sqrt{6}}$$

$$= \frac{2\sqrt{66} - 5\sqrt{42}}{\sqrt{36}}$$

$$= \frac{2\sqrt{66} - 5\sqrt{42}}{6}$$

9. $\sqrt{\dfrac{2x}{3y}} = \dfrac{\sqrt{2x}}{\sqrt{3y}}$

$$= \frac{\sqrt{2x} \cdot \sqrt{3y}}{\sqrt{3y} \cdot \sqrt{3y}}$$

$$= \frac{\sqrt{6xy}}{\sqrt{9y^2}}$$

$$= \frac{\sqrt{6xy}}{3y}$$

10. $\dfrac{\sqrt{9x}}{\sqrt{8y}} = \dfrac{\sqrt{9}\sqrt{x}}{\sqrt{4}\sqrt{2y}}$ Simplify the radicals in the numerator and denominator.

$$= \frac{3\sqrt{x}}{2\sqrt{2y}}$$

$$= \frac{3\sqrt{x} \cdot \sqrt{2y}}{2\sqrt{2y} \cdot \sqrt{2y}}$$

$$= \frac{3\sqrt{2xy}}{2\sqrt{4y^2}}$$

$$= \frac{3\sqrt{2xy}}{2 \cdot 2y}$$

$$= \frac{3\sqrt{2xy}}{4y}$$

11. $\sqrt{\dfrac{3}{7x^5}} = \dfrac{\sqrt{3}}{\sqrt{7x^5}}$

$$= \frac{\sqrt{3}}{\sqrt{x^4}\sqrt{7x}} \qquad \text{Simplify the radical in the denominator.}$$

$$= \frac{\sqrt{3}}{x^2\sqrt{7x}}$$

$$= \frac{\sqrt{3} \cdot \sqrt{7x}}{x^2\sqrt{7x} \cdot \sqrt{7x}}$$

$$= \frac{\sqrt{21x}}{x^2\sqrt{49x^2}}$$

$$= \frac{\sqrt{21x}}{x^2 \cdot 7x}$$

$$= \frac{\sqrt{21x}}{7x^3}$$

We can now extend the ideas of simplifying radical expressions that were presented in Section 8.2.

A radical expression containing square roots is in simplified form if all of the following conditions are met.

1. The radicand contains no perfect square factors.

2. The radicand contains no fractions.

3. The denominator contains no radicals.

Suppose that you were asked to simplify the fraction

$$\frac{1}{\sqrt{7} + \sqrt{2}}$$

Clearly, this fraction is not in simplified form since the denominator contains radicals. We must now find a factor to multiply $\sqrt{7} + \sqrt{2}$ by to produce a rational number. In the previous section we saw that

$$(\sqrt{x} + \sqrt{y})(\sqrt{x} - \sqrt{y}) = \sqrt{x}\sqrt{x} - \sqrt{x}\sqrt{y} + \sqrt{y}\sqrt{x} - \sqrt{y}\sqrt{y}$$
$$= \sqrt{x^2} - \sqrt{xy} + \sqrt{xy} - \sqrt{y^2}$$
$$= x - y$$

Note that the product of $(\sqrt{x} + \sqrt{y})(\sqrt{x} - \sqrt{y})$ yields an expression free of radicals. Let us apply this fact to our problem:

$$\frac{1}{\sqrt{7} + \sqrt{2}} = \frac{1}{(\sqrt{7} + \sqrt{2})} \cdot \frac{(\sqrt{7} - \sqrt{2})}{(\sqrt{7} - \sqrt{2})}$$
$$= \frac{\sqrt{7} - \sqrt{2}}{(\sqrt{7} + \sqrt{2})(\sqrt{7} - \sqrt{2})}$$
$$= \frac{\sqrt{7} - \sqrt{2}}{\sqrt{7}\sqrt{7} - \sqrt{7}\sqrt{2} + \sqrt{2}\sqrt{7} - \sqrt{2}\sqrt{2}}$$
$$= \frac{\sqrt{7} - \sqrt{2}}{\sqrt{49} - \sqrt{14} + \sqrt{14} - \sqrt{4}}$$
$$= \frac{\sqrt{7} - \sqrt{2}}{7 - 2}$$
$$= \frac{\sqrt{7} - \sqrt{2}}{5}$$

The denominator has been rationalized.

The product $(\sqrt{7} + \sqrt{2})(\sqrt{7} - \sqrt{2})$ yields a rational number. The numbers $\sqrt{7} + \sqrt{2}$ and $\sqrt{7} - \sqrt{2}$ are called **conjugates** of each other. To find the conjugate of a two-term expression involving only square roots, simply change the sign between the terms. *In general, the product of conjugates yields the difference of the squares of the two terms, provided that the radicals are square roots.*

EXAMPLE 8.5.2

Rationalize the denominators of the following radical expressions. Assume that all variables represent positive real numbers.

1. $\dfrac{5}{\sqrt{10} - 2} = \dfrac{5}{(\sqrt{10} - 2)} \cdot \dfrac{(\sqrt{10} + 2)}{(\sqrt{10} + 2)}$ Multiply by the conjugate of the denominator.

$$= \frac{5(\sqrt{10} + 2)}{(\sqrt{10} - 2)(\sqrt{10} + 2)}$$
$$= \frac{5(\sqrt{10} + 2)}{(\sqrt{10})^2 - 2^2}$$
$$= \frac{5(\sqrt{10} + 2)}{10 - 4}$$

$$= \frac{5(\sqrt{10} + 2)}{6}$$

NOTE ▶ Here we left the numerator in factored form to see if we could reduce our answer.

Try to avoid this mistake:

Incorrect	Correct
$\dfrac{5}{\sqrt{10} - 2} = \dfrac{5}{\sqrt{10} - 2} \cdot \dfrac{\sqrt{10}}{\sqrt{10}}$ $= \dfrac{5\sqrt{10}}{\sqrt{10}\sqrt{10} - 2\sqrt{10}}$ $= \dfrac{5\sqrt{10}}{\sqrt{100} - 2\sqrt{10}}$ $= \dfrac{5\sqrt{10}}{10 - 2\sqrt{10}}$ The denominator *still* contains a radical.	$\dfrac{5}{\sqrt{10} - 2}$ $= \dfrac{5}{(\sqrt{10} - 2)} \cdot \dfrac{(\sqrt{10} + 2)}{(\sqrt{10} + 2)}$ $\vdots$ $= \dfrac{5(\sqrt{10} + 2)}{6}$

2. $\dfrac{14}{\sqrt{6} - \sqrt{2}} = \dfrac{14}{(\sqrt{6} - \sqrt{2})} \cdot \dfrac{(\sqrt{6} + \sqrt{2})}{(\sqrt{6} + \sqrt{2})}$

$= \dfrac{14(\sqrt{6} + \sqrt{2})}{(\sqrt{6} - \sqrt{2})(\sqrt{6} + \sqrt{2})}$

$= \dfrac{14(\sqrt{6} + \sqrt{2})}{(\sqrt{6})^2 - (\sqrt{2})^2}$

$= \dfrac{14(\sqrt{6} + \sqrt{2})}{6 - 2}$

$= \dfrac{14(\sqrt{6} + \sqrt{2})}{4}$ Be sure to reduce your answer.

$= \dfrac{7(\sqrt{6} + \sqrt{2})}{2}$

3. $\dfrac{\sqrt{5} + \sqrt{2}}{\sqrt{5} - \sqrt{2}} = \dfrac{(\sqrt{5} + \sqrt{2}) \cdot (\sqrt{5} + \sqrt{2})}{(\sqrt{5} - \sqrt{2}) \cdot (\sqrt{5} + \sqrt{2})}$

$= \dfrac{(\sqrt{5} + \sqrt{2})(\sqrt{5} + \sqrt{2})}{(\sqrt{5} - \sqrt{2})(\sqrt{5} + \sqrt{2})}$

$$= \frac{\sqrt{25} + \sqrt{10} + \sqrt{10} + \sqrt{4}}{(\sqrt{5})^2 - (\sqrt{2})^2}$$

$$= \frac{5 + \sqrt{10} + \sqrt{10} + 2}{5 - 2}$$

$$= \frac{7 + 2\sqrt{10}}{3}$$

4. $\dfrac{2\sqrt{3} + 1}{5\sqrt{2} + 3} = \dfrac{(2\sqrt{3} + 1) \cdot (5\sqrt{2} - 3)}{(5\sqrt{2} + 3) \cdot (5\sqrt{2} - 3)}$

$$= \frac{(2\sqrt{3} + 1)(5\sqrt{2} - 3)}{(5\sqrt{2} + 3)(5\sqrt{2} - 3)}$$

$$= \frac{10\sqrt{6} - 6\sqrt{3} + 5\sqrt{2} - 3}{(5\sqrt{2})^2 - 3^2}$$

$$= \frac{10\sqrt{6} - 6\sqrt{3} + 5\sqrt{2} - 3}{50 - 9}$$

$$= \frac{10\sqrt{6} - 6\sqrt{3} + 5\sqrt{2} - 3}{41}$$

NOTE ▶   Sometimes after rationalizing the denominator we obtain a bizarre looking expression.

5. $\dfrac{\sqrt{x}}{\sqrt{x} - \sqrt{3}} = \dfrac{\sqrt{x}}{(\sqrt{x} - \sqrt{3})} \cdot \dfrac{(\sqrt{x} + \sqrt{3})}{(\sqrt{x} + \sqrt{3})}$

$$= \frac{\sqrt{x}(\sqrt{x} + \sqrt{3})}{(\sqrt{x} - \sqrt{3})(\sqrt{x} + \sqrt{3})}$$

$$= \frac{\sqrt{x}(\sqrt{x} + \sqrt{3})}{(\sqrt{x})^2 - (\sqrt{3})^2}$$

$$= \frac{x + \sqrt{3x}}{x - 3}$$

6. $\dfrac{\sqrt{x} + 2}{\sqrt{x} - 4} = \dfrac{(\sqrt{x} + 2) \cdot (\sqrt{x} + 4)}{(\sqrt{x} - 4) \cdot (\sqrt{x} + 4)}$

$$= \frac{(\sqrt{x} + 2)(\sqrt{x} + 4)}{(\sqrt{x} - 4)(\sqrt{x} + 4)}$$

$$= \frac{x + 4\sqrt{x} + 2\sqrt{x} + 8}{(\sqrt{x})^2 - (4)^2}$$

$$= \frac{x + 6\sqrt{x} + 8}{x - 16}$$

The following example illustrates how to rationalize the denominator when the denominator contains cube roots. Here we will investigate fractions where the denominator contains only one term.

EXAMPLE 8.5.3

Rationalize the denominators of the following radical expressions. Assume that all variables represent positive real numbers.

1. $\dfrac{1}{\sqrt[3]{4}} = \dfrac{1}{\sqrt[3]{4}} \cdot \dfrac{\sqrt[3]{2}}{\sqrt[3]{2}}$

$\quad = \dfrac{\sqrt[3]{2}}{\sqrt[3]{8}}$

$\quad = \dfrac{\sqrt[3]{2}}{2}$

NOTE ▶ Since we have a cube root, we multiply the numerator and denominator by the smallest quantity that will produce a perfect cube in the radicand in the denominator.

2. $\sqrt[3]{\dfrac{5}{3}} = \dfrac{\sqrt[3]{5}}{\sqrt[3]{3}}$ Multiply the numerator and
denominator by $\sqrt[3]{9}$

$\quad = \dfrac{\sqrt[3]{5} \cdot \sqrt[3]{9}}{\sqrt[3]{3} \cdot \sqrt[3]{9}}$

$\quad = \dfrac{\sqrt[3]{45}}{\sqrt[3]{27}}$

$\quad = \dfrac{\sqrt[3]{45}}{3}$

3. $\dfrac{\sqrt[3]{7y}}{\sqrt[3]{2x}} = \dfrac{\sqrt[3]{7y} \cdot \sqrt[3]{4x^2}}{\sqrt[3]{2x} \cdot \sqrt[3]{4x^2}}$

$\quad = \dfrac{\sqrt[3]{28x^2y}}{\sqrt[3]{8x^3}}$

$\quad = \dfrac{\sqrt[3]{28x^2y}}{2x}$

EXERCISES 8.5

Rationalize the denominators of the following radical expressions. Assume that all variables represent positive real numbers.

1. $\dfrac{1}{\sqrt{6}}$

2. $\dfrac{1}{\sqrt{10}}$

3. $\dfrac{7}{\sqrt{2}}$

4. $\dfrac{2}{\sqrt{10}}$

5. $\dfrac{15}{\sqrt{21}}$

6. $\dfrac{14}{\sqrt{21}}$

7. $\dfrac{\sqrt{6}}{\sqrt{5}}$

8. $\dfrac{\sqrt{6}}{\sqrt{10}}$

9. $\sqrt{\dfrac{2}{3}}$

10. $\sqrt{\dfrac{3}{5}}$

11. $\sqrt{\dfrac{5}{2}}$

12. $\dfrac{1}{6\sqrt{7}}$

13. $\dfrac{7}{2\sqrt{3}}$

14. $\dfrac{9}{5\sqrt{2}}$

15. $\dfrac{8}{7\sqrt{2}}$

16. $\dfrac{7}{\sqrt{27}}$

17. $\dfrac{6}{\sqrt{20}}$

18. $\dfrac{10}{\sqrt{24}}$

19. $\dfrac{9}{\sqrt{12}}$

20. $\dfrac{\sqrt{20}}{\sqrt{3}}$

21. $\dfrac{\sqrt{63}}{\sqrt{72}}$

22. $\dfrac{\sqrt{40}}{\sqrt{300}}$

23. $\dfrac{\sqrt{48}}{\sqrt{45}}$

24. $\dfrac{2\sqrt{3}}{5\sqrt{7}}$

25. $\dfrac{7\sqrt{18}}{5\sqrt{2}}$

26. $\dfrac{6\sqrt{32}}{5\sqrt{2}}$

27. $\dfrac{3\sqrt{18}}{4\sqrt{45}}$

28. $\dfrac{\sqrt{3}+\sqrt{7}}{\sqrt{2}}$

29. $\dfrac{2\sqrt{3}-4\sqrt{2}}{\sqrt{5}}$

30. $\dfrac{3\sqrt{2}-5\sqrt{7}}{\sqrt{3}}$

31. $\dfrac{5\sqrt{2}-4\sqrt{3}}{\sqrt{3}}$

32. $\dfrac{\sqrt{5}+5\sqrt{2}}{\sqrt{10}}$

33. $\sqrt{\dfrac{2}{x}}$

34. $\sqrt{\dfrac{3}{x}}$

35. $\sqrt{\dfrac{2}{7x}}$

36. $\sqrt{\dfrac{6x}{5y}}$

37. $\sqrt{\dfrac{5x}{2y}}$

38. $\sqrt{\dfrac{5x}{11y}}$

39. $\sqrt{\dfrac{11}{20x}}$

40. $\sqrt{\dfrac{25}{8x}}$

41. $\sqrt{\dfrac{9x}{50y}}$

42. $\sqrt{\dfrac{4x}{27y}}$

43. $\sqrt{\dfrac{45x}{44y}}$

44. $\dfrac{\sqrt{48x}}{\sqrt{28y}}$

45. $\sqrt{\dfrac{9}{25x^6}}$

46. $\sqrt{\dfrac{16}{49x^4}}$

47. $\sqrt{\dfrac{7}{2x^3}}$

48. $\sqrt{\dfrac{11}{6x^2}}$

49. $\sqrt{\dfrac{18}{7x^7}}$

50. $\sqrt{\dfrac{20}{3x^7}}$

51. $\sqrt{\dfrac{81x^6}{16y^4}}$

52. $\sqrt{\dfrac{3y}{2x^3}}$

53. $\sqrt{\dfrac{11y}{2x^2}}$

54. $\sqrt{\dfrac{13y}{2x^2}}$

55. $\sqrt{\dfrac{72y^3}{5x^3}}$

56. $\dfrac{5}{\sqrt{7}+1}$

57. $\dfrac{11}{\sqrt{10}-2}$

58. $\dfrac{8}{\sqrt{7}-2}$

59. $\dfrac{15}{\sqrt{14}-2}$

60. $\dfrac{8}{\sqrt{3}+2}$

61. $\dfrac{1}{\sqrt{10}+\sqrt{3}}$

62. $\dfrac{1}{\sqrt{7}+\sqrt{3}}$

63. $\dfrac{2}{\sqrt{5}-\sqrt{10}}$

64. $\dfrac{8}{\sqrt{23}-\sqrt{11}}$

65. $\dfrac{5\sqrt{3}}{\sqrt{10}-3}$

66. $\dfrac{2\sqrt{7}}{\sqrt{5}-2}$

67. $\dfrac{5\sqrt{7}}{\sqrt{7}+1}$

68. $\dfrac{2\sqrt{10}}{\sqrt{10}+2}$

69. $\dfrac{\sqrt{10}+1}{\sqrt{10}+2}$

70. $\dfrac{\sqrt{14}+1}{\sqrt{14}+3}$

71. $\dfrac{\sqrt{10}+2}{\sqrt{10}-3}$

72. $\dfrac{\sqrt{13}-4}{\sqrt{13}-3}$

73. $\dfrac{\sqrt{7}+\sqrt{2}}{\sqrt{7}-\sqrt{2}}$

74. $\dfrac{\sqrt{7}+\sqrt{3}}{\sqrt{7}-\sqrt{3}}$

75. $\dfrac{2\sqrt{10}-3\sqrt{6}}{\sqrt{10}-\sqrt{6}}$

76. $\dfrac{\sqrt{3}+5\sqrt{2}}{2\sqrt{3}+\sqrt{2}}$

77. $\dfrac{2\sqrt{3}+5}{3\sqrt{3}-2}$

78. $\dfrac{5\sqrt{2}+1}{3\sqrt{2}-4}$

79. $\dfrac{5\sqrt{6}-2}{3\sqrt{6}-4}$

80. $\dfrac{2\sqrt{5}+3}{3\sqrt{2}+2}$

81. $\dfrac{3\sqrt{2}+1}{6\sqrt{7}-4}$

82. $\dfrac{7\sqrt{3}+2}{2\sqrt{5}-3}$

83. $\dfrac{\sqrt{x}}{\sqrt{x}-2}$

84. $\dfrac{\sqrt{x}}{2\sqrt{x}+1}$

85. $\dfrac{\sqrt{2x}}{\sqrt{x}-5}$

86. $\dfrac{\sqrt{2x}}{\sqrt{x}-6}$

87. $\dfrac{\sqrt{x}}{\sqrt{x}+\sqrt{6}}$

88. $\dfrac{2\sqrt{x}}{\sqrt{x}-\sqrt{2}}$

89. $\dfrac{5\sqrt{x}}{\sqrt{2x}+\sqrt{7}}$

90. $\dfrac{4\sqrt{x}}{\sqrt{3x}+\sqrt{7}}$

91. $\dfrac{\sqrt{x}+1}{\sqrt{x}+5}$

92. $\dfrac{\sqrt{x}-4}{\sqrt{x}-7}$

93. $\dfrac{2\sqrt{x}+3}{\sqrt{x}-1}$

94. $\dfrac{2\sqrt{x}+5}{\sqrt{x}-3}$

95. $\dfrac{\sqrt{x}-3}{2\sqrt{x}-3}$

96. $\dfrac{\sqrt{x}-\sqrt{2}}{\sqrt{x}+\sqrt{2}}$

97. $\dfrac{\sqrt{x}+2\sqrt{3}}{2\sqrt{x}-\sqrt{3}}$

98. $\dfrac{\sqrt{x}+3\sqrt{2}}{3\sqrt{x}-\sqrt{2}}$

99. $\dfrac{\sqrt{x}-\sqrt{y}}{\sqrt{x}+\sqrt{y}}$

100. $\dfrac{\sqrt{x}-5\sqrt{2y}}{\sqrt{x}-\sqrt{2y}}$

101. $\dfrac{1}{\sqrt[3]{9}}$

102. $\dfrac{5}{\sqrt[3]{3}}$

103. $\sqrt[3]{\dfrac{10}{3}}$

104. $\sqrt[3]{\dfrac{7}{3x}}$

105. $\sqrt[3]{\dfrac{5}{4x^3}}$

106. $\sqrt[3]{\dfrac{5x}{2y}}$

107. $\sqrt[3]{\dfrac{5x}{16y^3}}$

8.6 Radical Equations

The first five sections of this chapter have been devoted to evaluating, simplifying, adding, subtracting, multiplying, and dividing radical expressions. Now that we are familiar with radical expressions, we are able to investigate another type of equation.

Radical Equation

Any equation that contains the variable in at least one radicand is called a **radical equation**.

REMARK   The following equations are examples of radical equations.

$$\sqrt{x} = 5, \qquad \sqrt{2x - 1} = -2, \qquad 4\sqrt{x + 11} = 3\sqrt{x + 18}$$

$$\sqrt{2x - 5} + 4 = x, \qquad \sqrt{x + 7} = \sqrt{x} + 1$$

NOTE ▶   In this text we will consider radical equations involving only square roots. However, the techniques presented in this section can be extended to equations containing higher order radicals.

Suppose that you were asked to find the solutions of the radical equation $\sqrt{x} = 5$. By observation, $x = 25$ is the solution of this equation since $\sqrt{25} = 5$. However, not all radical equations are this simple.

A common method for determining the solutions of radical equations begins with eliminating all of the radicals from the equation. By applying the following property, we will learn how to eliminate the radicals and find the solutions of radical equations involving square roots.

Squaring Property of Equality

If $a = b$, then $a^2 = b^2$.

When we use the squaring property of equality, we must be aware of one risk that this property raises.

Suppose that we were given the equation

$$x = 6$$

Now, $x^2 = 6^2$ By the squaring property of equality

$$x^2 = 36$$

$$x^2 - 36 = 0$$

$$(x + 6)(x - 6) = 0$$

$$x + 6 = 0 \qquad \text{or} \qquad x - 6 = 0$$

Thus, $x = -6$ or $x = 6$

Clearly, -6 is not a solution of our original equation. -6 is called an **extraneous solution**. When both sides of an equation are squared, we run the risk of generating extraneous solutions. Therefore, whenever you square both sides of an equation, *you must always check your solutions*.

The following steps will enable us to find the solution(s) of radical equations involving square roots.

1. Isolate one of the radical terms; that is, place one of the radical terms on one side of the equation and all of the remaining terms on the other side of the equation.

2. Using the squaring property of equality, square both sides of the equation.

3. If no radicals remain, find the solution(s) of the resulting equation. If radicals are still present, repeat steps 1 and 2.

4. *Check your potential solutions in the original equation!*

EXAMPLE 8.6.1

Find the solution set of each radical equation.

1. $\sqrt{x} = 5$

    $(\sqrt{x})^2 = 5^2$ By the squaring property

    $x = 25$ of equality

    Now we must check the answer:

    $\sqrt{25} \overset{?}{=} 5$

    $5 = 5$

    Thus, the solution set is {25}.

2. $\sqrt{3x + 1} = 3$

    $(\sqrt{3x + 1})^2 = 3^2$

    $3x + 1 = 9$

    $3x = 8$

    $x = \frac{8}{3}$

    Checking the answer:

    $\sqrt{3 \cdot \left(\frac{8}{3}\right) + 1} \overset{?}{=} 3$

    $\sqrt{8 + 1} \overset{?}{=} 3$

    $\sqrt{9} \overset{?}{=} 3$

    $3 = 3$

    So the solution set is $\left\{\frac{8}{3}\right\}$.

3. $\sqrt{2x - 1} = -2$

    $(\sqrt{2x - 1})^2 = (-2)^2$

    $2x - 1 = 4$

    $2x = 5$

    $x = \frac{5}{2}$

Checking the answer:

$$\sqrt{2 \cdot \tfrac{5}{2} - 1} \stackrel{?}{=} -2$$
$$\sqrt{5 - 1} \stackrel{?}{=} -2$$
$$\sqrt{4} \stackrel{?}{=} -2$$
$$2 \neq -2$$

Since $\tfrac{5}{2}$ is not a solution, the solution set of the original equation is $\varnothing$, the empty set.

NOTE ▶ In the original equation $\sqrt{2x - 1} = -2$, we have a principal square root equal to -2. However, we know that the principal square root is never negative. Thus, when we first investigate this equation we observe that it has no solution.

4. $\sqrt{x + 15} - 1 = 2$

$$\sqrt{x + 15} = 3 \qquad \text{Isolate the radical term.}$$
$$(\sqrt{x + 15})^2 = 3^2$$
$$x + 15 = 9$$
$$x = -6$$

Checking the answer:

$$\sqrt{-6 + 15} - 1 \stackrel{?}{=} 2$$
$$\sqrt{9} - 1 \stackrel{?}{=} 2$$
$$3 - 1 \stackrel{?}{=} 2$$
$$2 = 2$$

So the solution set is $\{-6\}$.

5.
$$x = \sqrt{x + 20}$$
$$x^2 = (\sqrt{x + 20})^2$$
$$x^2 = x + 20 \qquad \text{We now have a quadratic equation.}$$
$$x^2 - x - 20 = 0$$
$$(x - 5)(x + 4) = 0$$
$$x - 5 = 0 \quad \text{or} \quad x + 4 = 0$$
$$x = 5 \quad \text{or} \quad x = -4$$

When $x = 5$

$$5 \stackrel{?}{=} \sqrt{5 + 20}$$
$$5 \stackrel{?}{=} \sqrt{25}$$
$$5 = 5$$

When $x = -4$

$-4 \overset{?}{=} \sqrt{-4 + 20}$

$-4 \overset{?}{=} \sqrt{16}$

$-4 \neq 4$

-4 is an extraneous solution. Thus, the solution set is $\{5\}$.

6. $\qquad 4\sqrt{x + 11} = 3\sqrt{x + 18}$

$\qquad (4\sqrt{x + 11})^2 = (3\sqrt{x + 18})^2$

$4^2 \cdot (\sqrt{x + 11})^2 = 3^2 \cdot (\sqrt{x + 18})^2$

Remember that when you square a product you must square each factor:

$16(x + 11) = 9(x + 18)$

$16x + 176 = 9x + 162$

$\qquad 7x + 176 = 162$

$\qquad\qquad 7x = -14$

$\qquad\qquad\quad x = -2$

Checking the answer:

$4\sqrt{-2 + 11} \overset{?}{=} 3\sqrt{-2 + 18}$

$\qquad 4\sqrt{9} \overset{?}{=} 3\sqrt{16}$

$\qquad 4 \cdot 3 \overset{?}{=} 3 \cdot 4$

$\qquad\quad 12 = 12$

So the solution set is $\{-2\}$.

7. $\sqrt{2x - 5} + 4 = x$

$\qquad \sqrt{2x - 5} = x - 4$ $\qquad\qquad$ Isolate the radical term.

$\qquad (\sqrt{2x - 5})^2 = (x - 4)^2$

$\qquad\quad 2x - 5 = x^2 - 8x + 16$

$\qquad\qquad\quad 0 = x^2 - 10x + 21$

$\qquad\qquad\quad 0 = (x - 7)(x - 3)$

$x - 7 = 0 \qquad$ or $\qquad x - 3 = 0$

$\quad x = 7 \qquad$ or $\qquad\quad x = 3$

When $x = 7$

$\sqrt{2 \cdot 7 - 5} + 4 \overset{?}{=} 7$

$\sqrt{14 - 5} + 4 \overset{?}{=} 7$

$\sqrt{9} + 4 \overset{?}{=} 7$

$3 + 4 \overset{?}{=} 7$

$7 = 7$

When $x = 3$

$$\sqrt{2 \cdot 3 - 5} + 4 \overset{?}{=} 3$$

$$\sqrt{6 - 5} + 4 \overset{?}{=} 3$$

$$\sqrt{1} + 4 \overset{?}{=} 3$$

$$1 + 4 \overset{?}{=} 3$$

$$5 \neq 3$$

3 is an extraneous solution. Thus, the solution set is $\{7\}$.

Try to avoid this mistake:

Incorrect	Correct
$\sqrt{2x - 5} + 4 = x$ $(\sqrt{2x - 5} + 4)^2 = x^2$ $2x - 5 + 8\sqrt{2x - 5} + 16 = x^2$ Here we did not isolate the radical. When we square the left-hand side of this equation, we obtain an equation that is more complicated than the original equation.	$\sqrt{2x - 5} + 4 = x$ $\sqrt{2x - 5} = x - 4$ $(\sqrt{2x - 5})^2 = (x - 4)^2$ $\vdots$ The solution set is $\{7\}$.

8. $\sqrt{x + 7} = \sqrt{x} + 1$

$(\sqrt{x + 7})^2 = (\sqrt{x} + 1)^2$

$x + 7 = (\sqrt{x})^2 + 2\sqrt{x} + 1$

$x + 7 = x + 2\sqrt{x} + 1$ Isolate the remaining radical term.

$6 = 2\sqrt{x}$ Divide both sides by 2.

$3 = \sqrt{x}$

$3^2 = (\sqrt{x})^2$

$9 = x$

Checking the answer:

$$\sqrt{9 + 7} \overset{?}{=} \sqrt{9} + 1$$

$$\sqrt{16} \overset{?}{=} 3 + 1$$

$$4 = 4$$

Thus, the solution set is $\{9\}$.

9. $\sqrt{2x-7} = \sqrt{x} - 1$

$(\sqrt{2x-7})^2 = (\sqrt{x} - 1)^2$

$2x - 7 = (\sqrt{x})^2 - 2\sqrt{x} + 1$

$2x - 7 = x - 2\sqrt{x} + 1$ Isolate the remaining

$x - 8 = -2\sqrt{x}$ radical term.

$(x-8)^2 = (-2\sqrt{x})^2$

$x^2 - 16x + 64 = 4x$

$x^2 - 20x + 64 = 0$

$(x-16)(x-4) = 0$

$x - 16 = 0$ or $x - 4 = 0$

$x = 16$ or $x = 4$

When $x = 16$

$\sqrt{2 \cdot 16 - 7} \stackrel{?}{=} \sqrt{16} - 1$

$\sqrt{32 - 7} \stackrel{?}{=} 4 - 1$

$\sqrt{25} \stackrel{?}{=} 3$

$5 \neq 3$

16 is an extraneous solution.

When $x = 4$

$\sqrt{2 \cdot 4 - 7} \stackrel{?}{=} \sqrt{4} - 1$

$\sqrt{8 - 7} \stackrel{?}{=} 2 - 1$

$\sqrt{1} \stackrel{?}{=} 1$

$1 = 1$

Thus, the solution set is {4}.

Try to avoid this mistake:

Incorrect	Correct
$\sqrt{2x-7} = \sqrt{x} - 1$ $(\sqrt{2x-7})^2 = (\sqrt{x}-1)^2$ $2x - 7 = x + 1$ An exponent does not distribute over a difference.	$\sqrt{2x-7} = \sqrt{x} - 1$ $(\sqrt{2x-7})^2 = (\sqrt{x}-1)^2$ $2x - 7 = (\sqrt{x})^2 - 2\sqrt{x} + 1$ $2x - 7 = x - 2\sqrt{x} + 1$

EXERCISES 8.6

Find the solution set of each radical equation.

1. $\sqrt{x} = 3$ **2.** $\sqrt{x} = 2$ **3.** $\sqrt{x} = -6$ **4.** $\sqrt{x} = -1$

5. $2\sqrt{x} = 7$ **6.** $3\sqrt{x} = 4$ **7.** $5\sqrt{2x} = 1$ **8.** $3\sqrt{4x} = 5$

9. $\sqrt{x-2} = 3$ **10.** $\sqrt{x-4} = 5$ **11.** $\sqrt{2x+1} = 0$ **12.** $\sqrt{3x+2} = 0$

13. $\sqrt{8x-1} = -4$ **14.** $\sqrt{7x-2} = -3$ **15.** $\sqrt{3x-1} = 4$ **16.** $\sqrt{6x-1} = 5$

17. $\sqrt{2x+7} = 2$ **18.** $\sqrt{5x+4} = 1$ **19.** $\sqrt{5x-2} = 6$ **20.** $\sqrt{4x-1} = 6$

21. $\sqrt{3x-1} = \frac{1}{2}$ **22.** $\sqrt{2x-1} = \frac{1}{3}$ **23.** $\sqrt{2x+3} = \frac{3}{4}$ **24.** $\sqrt{2x+3} = \frac{3}{5}$

25. $3\sqrt{3x+7} = 6$ **26.** $2\sqrt{3x+10} = 4$ **27.** $5\sqrt{2x-1} = 2$ **28.** $3\sqrt{2x-1} = 5$

29. $\sqrt{x}+3 = 9$ **30.** $\sqrt{x}+2 = 5$ **31.** $\sqrt{x}+8 = 3$ **32.** $\sqrt{x}+6 = 2$

33. $2\sqrt{x}-1 = 7$ **34.** $2\sqrt{x}-3 = 9$ **35.** $3\sqrt{x}-2 = 5$ **36.** $3\sqrt{x}-4 = 1$

37. $\sqrt{x+6}+3 = 4$ **38.** $\sqrt{x+7}+5 = 7$ **39.** $\sqrt{x-6}-2 = 1$ **40.** $\sqrt{x-4}-1 = 2$

41. $\sqrt{x-5}+6 = 3$ **42.** $\sqrt{x-6}+7 = 2$ **43.** $\sqrt{2x+3}-1 = 1$ **44.** $\sqrt{2x+6}-1 = 2$

45. $\sqrt{4x+9}+1 = 4$ **46.** $\sqrt{9x+4}+3 = 5$ **47.** $\sqrt{3x+10}+2 = 4$ **48.** $\sqrt{5x+9}+3 = 5$

49. $x = \sqrt{4x+12}$ **50.** $x = \sqrt{2x+15}$ **51.** $x = \sqrt{5x-6}$ **52.** $x = \sqrt{5x-4}$

53. $2x = \sqrt{8x-3}$ **54.** $2x = \sqrt{16x-15}$ **55.** $3x = 2\sqrt{6x-3}$ **56.** $3x = 4\sqrt{3x-3}$

57. $\sqrt{5x-3} = \sqrt{2x+1}$ **58.** $\sqrt{3x+4} = \sqrt{7x-3}$

59. $\sqrt{2x-1} = \sqrt{5x-9}$ **60.** $\sqrt{4x-9} = \sqrt{2x-3}$

61. $\sqrt{2x^2+3x-2} = \sqrt{x^2+3x+2}$ **62.** $\sqrt{2x^2+5x+3} = \sqrt{x^2+5x+4}$

63. $\sqrt{4x^2+4x-1} = \sqrt{2x^2-x+2}$ **64.** $\sqrt{5x^2+7x-6} = \sqrt{2x^2-3x+2}$

65. $2\sqrt{x+2} = 3\sqrt{x-3}$ **66.** $4\sqrt{x+12} = 3\sqrt{x+19}$ **67.** $3\sqrt{2x+12} = 5\sqrt{2x-4}$

68. $2\sqrt{7x+4} = 3\sqrt{3x+3}$ **69.** $\sqrt{x-2}+2 = x$ **70.** $\sqrt{4x+6}+1 = 2x$

71. $\sqrt{5x+19}-1 = x$ **72.** $\sqrt{7x+35}-5 = x$ **73.** $\sqrt{4x-1} = 2x$

74. $\sqrt{6-x}+4 = x$ **75.** $\sqrt{2x+3}+2x = 3$ **76.** $\sqrt{x+18}+2x = 0$

77. $\sqrt{3x+4}-2x = 2$ **78.** $\sqrt{3x-1}-3x = -3$ **79.** $2\sqrt{2x+6}+7 = 3x$

80. $5\sqrt{2x+4}-8 = 2x$ **81.** $\sqrt{x+8} = \sqrt{x}+2$ **82.** $\sqrt{x-5} = \sqrt{x}-1$

83. $\sqrt{2x+7} = \sqrt{x}+2$ **84.** $\sqrt{3x-11} = \sqrt{x}-1$ **85.** $\sqrt{4x+3} = 2\sqrt{x}+1$

86. $\sqrt{9x-3} = 3\sqrt{x}-1$ **87.** $2\sqrt{x-5} = 3\sqrt{x}-5$ **88.** $\sqrt{8x+7} = 4\sqrt{x}+1$

Chapter 8 Review

Properties and Definitions

- *Principal square root*

   If a is a real number such that $a \geq 0$, then $\sqrt{a} = b$ if and only if $a = b^2$ and $b \geq 0$.

- *Principal nth root of* a *(n even)*

   If a is a real number such that $a \geq 0$ and n is a positive even integer, then $\sqrt[n]{a} = b$ if and only if $a = b^n$ and $b \geq 0$.

- *nth root of* a *(n odd)*

   If a is any real number and n is a positive odd integer, then $\sqrt[n]{a} = b$ if and only if $a = b^n$.

- *Powers and roots*

   If a is a real number, then

   1. $(\sqrt[n]{a})^n = a$ for any positive integer n

   2. $\sqrt[n]{a^n} = |a|$ for any positive even integer n

   3. $\sqrt[n]{a^n} = a$ for any positive odd integer n

- *Properties of radicals*

   Let a and b be nonnegative real numbers; then

   1. $\sqrt{a \cdot b} = \sqrt{a} \cdot \sqrt{b}$

   2. $\sqrt{\dfrac{a}{b}} = \dfrac{\sqrt{a}}{\sqrt{b}}$, if $b \neq 0$

- *Simplified radical expressions*

A radical expression containing square roots is in simplified form if all of the following conditions are met.

1. The radicand contains no perfect square factors.

2. The radicand contains no fractions.

3. The denominator contains no radicals.

Review Exercises

8.1 **Evaluate each of the following radical expressions, if possible.**

1. $\sqrt{49}$ **2.** $\sqrt{\frac{100}{9}}$ **3.** $-\sqrt{\frac{1}{25}}$ **4.** $\sqrt{-16}$

5. $\sqrt[3]{-64}$ **6.** $-\sqrt[3]{-8}$ **7.** $\sqrt[3]{\frac{27}{125}}$ **8.** $\sqrt[4]{\frac{16}{81}}$

9. $\sqrt[4]{-16}$ **10.** $-\sqrt[5]{-1}$ **11.** $\sqrt{\sqrt{256}}$ **12.** $\sqrt[4]{\sqrt{256}}$

Simplify each of the following.

13. $(\sqrt{21})^2$ **14.** $(\sqrt[4]{7})^4$ **15.** $(6\sqrt{3})^2$ **16.** $(\sqrt[3]{8x}-3)^3$

Simplify each of the following. Assume that the variables represent positive real numbers.

17. $\sqrt{64x^2}$ **18.** $-\sqrt{100m^2n^2}$ **19.** $\sqrt[3]{-8k^3}$ **20.** $\sqrt{p^2+4p+4}$

8.2 **Multiply the following radicals. Assume that all variables represent positive real numbers.**

21. $\sqrt{7}\cdot\sqrt{28}$ **22.** $\sqrt{5x}\cdot\sqrt{7y}$ **23.** $\sqrt{3y}\cdot\sqrt{3y}$ **24.** $\sqrt{18x}\cdot\sqrt{2x^7}$

Simplify the following radicals. Assume that all variables represent positive real numbers.

25. $\sqrt{63}$ **26.** $\sqrt{96}$ **27.** $\sqrt{245}$ **28.** $\sqrt{128}$

29. $\sqrt{13x^{10}}$ **30.** $\sqrt{121x^8y^6}$ **31.** $\sqrt{x^{15}}$ **32.** $\sqrt{60y^5}$

33. $\sqrt{80x^4y^7}$ **34.** $\sqrt{275x^{11}y^{25}}$

Simplify the following radicals. Assume that all variables represent positive real numbers.

35. $\sqrt{\frac{11}{144}}$ **36.** $\frac{\sqrt{5}}{\sqrt{45}}$ **37.** $\frac{\sqrt{18}}{\sqrt{98}}$ **38.** $\frac{\sqrt{240}}{\sqrt{75}}$

39. $\sqrt{\frac{121x^4}{y^{10}}}$ **40.** $\sqrt{\frac{99x^{16}}{16}}$ **41.** $\sqrt{\frac{90x}{169y^2}}$ **42.** $\sqrt{\frac{150x^{15}}{81y^{18}}}$

Simplify the following radicals. Assume that all variables represent positive real numbers.

43. $\sqrt[3]{200}$ **44.** $\sqrt[3]{135x^5}$ **45.** $\sqrt[3]{\dfrac{x^3}{64y^3}}$ **46.** $\sqrt[3]{\dfrac{56x^4}{27y^3}}$

8.3 Perform the indicated operations. Assume that all variables represent positive real numbers.

47. $3\sqrt{5} - 8\sqrt{5}$

48. $\sqrt{13} - 6\sqrt{13} + 8\sqrt{13}$

49. $2x\sqrt{3} + 5x\sqrt{3} - 9x\sqrt{3}$

50. $3\sqrt{5xy} + 7\sqrt{5xy}$

51. $11x\sqrt{7y} + 6x\sqrt{7y}$

52. $9x^2\sqrt{10y} - x^2\sqrt{10y} - 3x^2\sqrt{10y}$

53. $8\sqrt{2} - 4\sqrt{6} + 2\sqrt{2} - \sqrt{6}$

54. $5x\sqrt{3y} + 11\sqrt{xy} - 11x\sqrt{3y} - 3\sqrt{xy}$

55. $\sqrt{48} - \sqrt{75}$

56. $5\sqrt{20} - 3\sqrt{45}$

57. $7\sqrt{28} - 2\sqrt{7} + \sqrt{175}$

58. $\sqrt{80} + \sqrt{99} - 3\sqrt{20} - 5\sqrt{44}$

59. $15x\sqrt{24} + 3\sqrt{54x^2}$

60. $x\sqrt{15} + 4\sqrt{60x^2} - 9\sqrt{135x^2}$

61. $\sqrt{63x} + \sqrt{147x}$

62. $-\sqrt{32x} - \sqrt{50x} - 3\sqrt{200x}$

63. $\sqrt{72x^3y} - x\sqrt{288xy} + \sqrt{24xy}$

64. $8\sqrt{27x^2y} - 5y\sqrt{20x} - x\sqrt{48y} + \sqrt{45xy^2}$

65. $\sqrt[3]{108} + \sqrt[3]{32}$

66. $16\sqrt[3]{48x^4} - 7x\sqrt[3]{162x}$

8.4 Perform the indicated multiplications and simplify your answers. Assume that all variables represent positive real numbers.

67. $5\sqrt{14} \cdot \sqrt{21}$

68. $12\sqrt{5} \cdot \sqrt{18}$

69. $6\sqrt{6} \cdot 7\sqrt{60}$

70. $3\sqrt{48} \cdot 5\sqrt{3}$

71. $12\sqrt{5x} \cdot \sqrt{3x}$

72. $7\sqrt{8x} \cdot 2\sqrt{6y}$

73. $2\sqrt{7x} \cdot 6\sqrt{28xy}$

74. $4x\sqrt{6x} \cdot 5y\sqrt{3xy}$

75. $3\sqrt{10xy} \cdot 5\sqrt{3xy}$

76. $2xy\sqrt{3xy} \cdot 11x\sqrt{27y}$

77. $\sqrt{14}(\sqrt{6} + \sqrt{5})$

78. $5\sqrt{x}(4\sqrt{x} - 6\sqrt{y})$

79. $2\sqrt{6}(7\sqrt{10} - 3\sqrt{5} + \sqrt{21})$

80. $6\sqrt{xy}(5\sqrt{6} - \sqrt{3x} + \sqrt{4y})$

81. $(\sqrt{5} + 5)(\sqrt{6} - 7)$

82. $(\sqrt{x} + \sqrt{3})(\sqrt{y} + \sqrt{2})$

83. $(\sqrt{5x} + \sqrt{y})(\sqrt{2x} - \sqrt{5})$

84. $(3\sqrt{2} - \sqrt{5})(7\sqrt{2} - 2\sqrt{5})$

85. $(4\sqrt{x} + \sqrt{3})(3\sqrt{x} + 2\sqrt{3})$

86. $(5\sqrt{7} + \sqrt{2})(5\sqrt{7} - \sqrt{2})$

87. $(3\sqrt{11} + 8)(3\sqrt{11} - 8)$

88. $(\sqrt{x} + 2\sqrt{y})(\sqrt{x} - 2\sqrt{y})$

89. $(\sqrt{7} - 4)^2$

90. $(\sqrt{x} + 3)^2$

91. $(\sqrt{x} + 5\sqrt{y})^2$

92. $(\sqrt{x-2} - 4)^2$

8.5 Rationalize the denominators of the following radical expressions. Assume that all variables represent positive real numbers.

93. $\dfrac{2}{\sqrt{7}}$

94. $\dfrac{\sqrt{3}}{\sqrt{11}}$

95. $\sqrt{\dfrac{13}{6}}$

96. $\dfrac{4}{3\sqrt{5}}$

97. $\dfrac{12}{\sqrt{18}}$

98. $\dfrac{\sqrt{80}}{\sqrt{24}}$

99. $\dfrac{3\sqrt{60}}{7\sqrt{50}}$

100. $\dfrac{4\sqrt{5}-5\sqrt{3}}{\sqrt{15}}$

101. $\sqrt{\dfrac{11}{3x}}$

102. $\sqrt{\dfrac{9}{20x}}$

103. $\sqrt{\dfrac{16x}{54y}}$

104. $\sqrt{\dfrac{7}{3x^4}}$

105. $\sqrt{\dfrac{20}{13x^3}}$

106. $\sqrt{\dfrac{48y^5}{5x}}$

107. $\dfrac{5}{\sqrt{11}-3}$

108. $\dfrac{10}{\sqrt{7}+\sqrt{5}}$

109. $\dfrac{7\sqrt{2}}{\sqrt{2}+3}$

110. $\dfrac{\sqrt{6}+4}{\sqrt{6}-2}$

111. $\dfrac{3\sqrt{7}-\sqrt{3}}{\sqrt{7}-2\sqrt{3}}$

112. $\dfrac{4\sqrt{5}+3}{3\sqrt{5}+2}$

113. $\dfrac{\sqrt{x}}{3\sqrt{x}+2}$

114. $\dfrac{5\sqrt{x}}{\sqrt{x}-\sqrt{5}}$

115. $\dfrac{\sqrt{x}+3}{\sqrt{x}+5}$

116. $\dfrac{4\sqrt{x}-3}{\sqrt{x}+4}$

117. $\dfrac{\sqrt{x}-2\sqrt{5y}}{\sqrt{x}-\sqrt{5y}}$

118. $\dfrac{4}{\sqrt[3]{6}}$

119. $\sqrt[3]{\dfrac{5}{4}}$

120. $\sqrt[3]{\dfrac{5}{64x^2}}$

121. $\sqrt[3]{\dfrac{2x}{81y^3}}$

8.6 Find the solution set of each radical equation.

122. $4\sqrt{x}=5$

123. $\sqrt{x-7}=4$

124. $\sqrt{4x-9}=-2$

125. $\sqrt{3x-1}=\frac{1}{3}$

126. $5\sqrt{2x+1}=15$

127. $3\sqrt{x}+8=5$

128. $\sqrt{x+8}-3=4$

129. $\sqrt{5x+1}+1=5$

130. $2x=\sqrt{12x-5}$

131. $\sqrt{3x+5}=\sqrt{7x-3}$

132. $\sqrt{2x^2-x-1}=\sqrt{x^2+2x+3}$

133. $5\sqrt{x+2}=3\sqrt{4x-3}$

134. $\sqrt{4x+10}-1=2x$

135. $\sqrt{3x+33}-5=x$

136. $\sqrt{9-5x}+2x=3$

137. $3\sqrt{4x-3}-1=2x$

138. $\sqrt{x-7}=\sqrt{x}-1$

139. $\sqrt{7x-12}=\sqrt{x}+2$

140. $2\sqrt{x+7}=\sqrt{x}+5$

Chapter 8 Test

(You should be able to complete this test in 60 minutes.)

Simplify the following radicals. Assume that all variables represent positive real numbers.

1. $\sqrt{\dfrac{64}{25}}$

2. $\sqrt[3]{-8}$

3. $-\sqrt{121}$

4. $\sqrt{81x^2}$ **5.** $\sqrt{90}$ **6.** $\sqrt[3]{72}$

7. $\sqrt{48x^6y^{11}}$ **8.** $\sqrt{\dfrac{4x^5}{y^2}}$ **9.** $\sqrt[3]{\dfrac{54}{x^6}}$

Perform the indicated operations and simplify your answers. Assume that all variables represent positive real numbers.

10. $(2\sqrt{5})^2$ **11.** $\sqrt{5x} \cdot \sqrt{10y}$ **12.** $\dfrac{\sqrt{3}}{\sqrt{75}}$

13. $5\sqrt{3} - 3\sqrt{5} - \sqrt{3} - 4\sqrt{5}$ **14.** $7\sqrt{6x} \cdot 5\sqrt{15xy}$

15. $2\sqrt{xy}(3\sqrt{2} + \sqrt{5x} - \sqrt{9y})$ **16.** $\sqrt{12} + \sqrt{108}$

17. $(\sqrt{3x} + \sqrt{y})(\sqrt{x} - \sqrt{3})$ **18.** $(3\sqrt{5} - 2)(3\sqrt{5} + 2)$

19. $\sqrt[3]{81} - \sqrt[3]{375}$ **20.** $7\sqrt{10x^2} - \sqrt{40x} - 3x\sqrt{90}$

21. $8x\sqrt{20y} - \sqrt{45xy^2} - 3\sqrt{5x^2y} + 4y\sqrt{80x}$ **22.** $(\sqrt{x} + 2\sqrt{y})^2$

Rationalize the denominators of the following radical expressions. Assume that all variables represent positive real numbers.

23. $\dfrac{3}{\sqrt{5}}$ **24.** $\dfrac{2}{\sqrt[3]{9}}$ **25.** $\sqrt{\dfrac{50}{12}}$

26. $\sqrt{\dfrac{27y^3}{20x^4}}$ **27.** $\dfrac{\sqrt{x}}{5\sqrt{x} + 3}$ **28.** $\dfrac{4\sqrt{5} + \sqrt{2}}{2\sqrt{5} + 3\sqrt{2}}$

Find the solution set of each radical equation.

29. $\sqrt{3x + 10} - 4 = 3$ **30.** $\sqrt{2x - 5} + 11 = 2x$ **31.** $\sqrt{5x + 4} = \sqrt{x} + 4$

Test Your Memory

These problems review Chapters 1–8.

I. Sketch the graphs of the lines whose equations are given.

1. $2x - 3y = 6$

2. $y = 3$

II. Graph the solution set of each of the following systems of linear inequalities.

3. $y \geq x + 3$
$\quad y \leq 5$

4. $x + 2y < 4$
$\quad y - x < 0$

III. Find the equations of the lines satisfying the given information. Write your answers in standard form.

5. $m = -\frac{1}{4}$, passing through $(-3, 0)$

6. $m = 0$, passing through $(1, 2)$

7. Passing through $(7, 9)$ and $(-5, -1)$

8. Passing through $(3, -6)$ and $(3, 4)$

IV. Find the solution set of each equation.

9. $5 + 3(x - 7) = 10 - 2(3x + 1)$

10. $\frac{1}{4}(3x + 1) = \frac{1}{2} - \frac{1}{3}(1 - 2x)$

11. $(4x - 4)(x + 4) = -25$

12. $(x + 3)(x - 4) = (2x + 1)(x + 3)$

13. $\dfrac{3}{4x - 3} = \dfrac{2}{2x + 1}$

14. $\dfrac{x - 2}{5} = \dfrac{2}{2x - 3}$

15. $\dfrac{3}{x + 1} - \dfrac{2}{x} = \dfrac{1}{3x}$

16. $\dfrac{x + 2}{3x + 2} - \dfrac{2}{x + 1} = \dfrac{x + 3}{3x^2 + 5x + 2}$

17. $\sqrt{2x - 1} = 3$

18. $\sqrt{3x + 1} + 5 = 2$

19. $\sqrt{5x + 1} - x = 1$

20. $\sqrt{x + 2} - 2x = 3$

V. Simplify each of the following. Be sure to reduce your answer to lowest terms.

21. $\dfrac{(5x^3y^{-2})^{-1}}{(5x^{-4}y^{-5})^{-2}}$

22. $\dfrac{\dfrac{2}{x} + \dfrac{3x}{x - 4}}{\dfrac{1}{x} + \dfrac{2}{x - 4}}$

VI. Simplify the following radicals. Assume that all variables represent positive real numbers.

23. $\sqrt{\dfrac{49}{100}}$

24. $\sqrt[3]{-27}$

25. $\sqrt{64x^4}$

26. $\sqrt{80}$

27. $\sqrt{72x^8y^9}$

28. $\sqrt{\dfrac{9x^3}{16y^{10}}}$

VII. Perform the indicated operations and simplify your answers. Assume that all variables represent positive real numbers.

29. $\dfrac{2x^2 + 6x}{12x^3 - 24x^2} \cdot \dfrac{12x^3 - 30x^2 + 12x}{2x^2 + 5x - 3}$

30. $\dfrac{2x^2 + 1}{x^2 - 5x + 6} - \dfrac{x + 16}{x^2 - 5x + 6}$

31. $\dfrac{5x}{x - 3} + \dfrac{2}{x - 1}$

32. $\dfrac{2x + 1}{x^2 + x - 12} + \dfrac{x + 3}{x^2 + 9x + 20}$

33. $\sqrt{6x} \cdot \sqrt{15y}$

34. $3\sqrt{x}(2\sqrt{x} - 7\sqrt{y} + 4)$

35. $(\sqrt{7x} + \sqrt{5y})(\sqrt{7x} - \sqrt{5y})$

36. $(2\sqrt{x} + \sqrt{3y})^2$

37. $\sqrt{48} + 2\sqrt{27} - 4\sqrt{12}$

38. $x\sqrt{72x} - \sqrt{2x^3} - 3x\sqrt{18x}$

VIII. Rationalize the denominators of the following radical expressions. Assume that all variables represent positive real numbers.

39. $\sqrt{\dfrac{2}{3}}$

40. $\sqrt{\dfrac{5x}{2y}}$

41. $\dfrac{3}{\sqrt{5} + 1}$

42. $\dfrac{\sqrt{3} - 2}{\sqrt{6} - \sqrt{2}}$

IX. Determine the solutions of the following linear systems of equations. If there are no solutions, write *inconsistent*. If there are infinite solutions, write *dependent*.

43. $3x - 2y = -4$
   $9x + 4y = 3$

44. $2x - 6y = 5$
   $5x - 15y = -1$

X. Use algebraic expressions to find the solutions of the following problems.

45. Bob has $3.80 in dimes and quarters. He has a total of 23 coins. How many dimes and how many quarters does Bob have? (Use a linear system of equations to solve this problem.)

46. Carol invested $6000 in two accounts. She invested some in a savings account paying 5% simple interest and the rest in a CD paying 8% simple interest. After 1 year her interest income was $405. How much did Carol invest in each account? (Use a linear system of equations to solve this problem.)

47. The base of a triangle is 1 foot less than four times the height. Find the base and the height if the area of the triangle is 30 square feet.

48. José's winning meat loaf recipe includes $\frac{3}{4}$ cup of oats for an 8-person serving. How many cups of oats will be required for a recipe serving 28 people?

49. Charley can sort and deliver the company mail in 30 minutes. Benny can sort and deliver the company mail in 20 minutes. Working together, how long will it take Charley and Benny to sort and deliver the company mail?

50. A creek has a current of 3 mph. Find the speed of Billy Joe's boat in still water if it goes 8 miles downstream in the same time as it goes 5 miles upstream.

CHAPTER 9

Quadratic Equations

9.1 Complex Numbers

At the end of Chapter 4 we learned how to find the solutions of factorable quadratic equations. In this chapter we will learn how to find the solutions of any quadratic equation. However, we must first introduce a new set of numbers.

Suppose that you were asked to find the solutions of the quadratic equation $x^2 = 1$. By observation the solutions are $x = 1$ or $x = -1$, since $1^2 = 1$ and $(-1)^2 = 1$. Another method for finding the solutions of $x^2 = 1$ is as follows:

$$x^2 = 1$$
$$x = \pm\sqrt{1}$$
$$x = \pm 1$$

(The expression $x = \pm 1$ means $x = 1$ or $x = -1$.) Let us apply this method to the equation $x^2 = -1$:

$$x^2 = -1$$
$$x = \pm\sqrt{-1}$$

Clearly, the solutions $x = \sqrt{-1}$ or $x = -\sqrt{-1}$ are not real numbers, since the square of any real number cannot be -1. This problem leads to a new set of numbers based on the following definition.

The Imaginary Unit

The **imaginary unit**, denoted i, is defined as

$$i = \sqrt{-1} \qquad \text{and} \qquad i^2 = -1$$

In Chapter 8 the multiplication property of radicals stated that $\sqrt{a} \cdot \sqrt{b} = \sqrt{ab}$ if $a \geq 0$ and $b \geq 0$. This property can be extended to include the case where a and b are not both negative. With the extension of the multiplication property of radicals, we can express the square root of any negative number in terms of i.

EXAMPLE 9.1.1

Express the following numbers in terms of i and simplify.

1. $\sqrt{-9} = \sqrt{9 \cdot (-1)}$

$\qquad = \sqrt{9} \cdot \sqrt{-1} \qquad$ Multiplication property of radicals

$\qquad = 3 \cdot i$

$\qquad = 3i$

2. $\sqrt{-100} = \sqrt{100 \cdot (-1)}$

$\qquad = \sqrt{100} \cdot \sqrt{-1}$

$\qquad = 10 \cdot i$

$\qquad = 10i$

3. $\sqrt{-48} = \sqrt{48 \cdot (-1)}$

$\qquad = \sqrt{48} \cdot \sqrt{-1}$

$\qquad = \sqrt{16} \cdot \sqrt{3} \cdot \sqrt{-1}$

$\qquad = 4\sqrt{3}i \qquad$ or $\qquad 4i\sqrt{3}$

4. $\sqrt{\frac{-4}{9}} = \sqrt{\frac{4}{9} \cdot (-1)}$

$\qquad = \sqrt{\frac{4}{9}} \cdot \sqrt{-1}$

$\qquad = \frac{2}{3}i$

5. $-\sqrt{-36} = -\sqrt{36 \cdot (-1)}$

$\qquad = -\sqrt{36} \cdot \sqrt{-1}$

$\qquad = -6i$

The following definition illustrates how i becomes a base unit for a new set of numbers.

Complex Numbers
Any number that can be expressed in the form $a + bi$, where a and b are real numbers and $i = \sqrt{-1}$, is called a **complex number**.

REMARK   *a* is called the **real** part and *b* is called the **imaginary** part of the complex number $a + bi$.

Examples of complex numbers are $2 + 5i$, $-\frac{1}{2} + \frac{2}{3}i$, 6, and $10i$.

NOTE ▶   1. 6 is a complex number because $6 = 6 + 0i$; the imaginary part of 6 is 0.

2. $10i$ is a complex number because $10i = 0 + 10i$; the real part of $10i$ is 0.

The commutative, associative, and distributive properties of real numbers extend to complex numbers. With the use of these properties, adding and subtracting complex numbers becomes a matter of combining like terms.

EXAMPLE 9.1.2

Perform the indicated operations. Express all answers in the form $a + bi$.

1. $(5 + 7i) + (3 - 2i) = 5 + 7i + 3 - 2i$

$$= 5 + 3 + 7i - 2i \qquad \text{Commutative property}$$
$$= (5 + 3) + (7i - 2i) \qquad \text{Associative property}$$
$$= 8 + 5i \qquad \text{Combine like terms.}$$

NOTE ▶   The form $a + bi$ is called **standard form**.

2. $\left(\frac{1}{2} - \frac{1}{3}i\right) + \left(\frac{2}{3} - \frac{1}{6}i\right) = \frac{1}{2} - \frac{1}{3}i + \frac{2}{3} - \frac{1}{6}i$

$$= \left(\frac{1}{2} + \frac{2}{3}\right) + \left(-\frac{1}{3}i - \frac{1}{6}i\right)$$
$$= \left(\frac{3}{6} + \frac{4}{6}\right) + \left(-\frac{2}{6}i - \frac{1}{6}i\right)$$
$$= \frac{7}{6} - \frac{1}{2}i$$

3. $(3 - 2i) - (5 - 4i) = 3 - 2i - 5 + 4i$

$$= (3 - 5) + (-2i + 4i)$$
$$= -2 + 2i$$

4. $(-7 - 3i) - 10i = -7 - 3i - 10i$

$$= -7 + (-3i - 10i)$$
$$= -7 - 13i$$

The following example shows that multiplication of complex numbers is similar to multiplication of polynomials.

EXAMPLE 9.1.3

Perform the indicated operations. Express all answers in the form $a + bi$.

1. $2(-3 + 8i) = 2(-3) + 2(8i)$ Distributive property
$$= -6 + 16i$$

2. $5i(7 - 2i) = (5i)7 - (5i)(2i)$
$$= 35i - 10i^2$$
$$= 35i - 10(-1) \quad \text{Remember that } i^2 = -1.$$
$$= 35i + 10$$
$$= 10 + 35i$$

3. $(8 + 3i)(4 - 5i)$

To perform this multiplication we will use the FOIL method:

$(8 + 3i)(4 - 5i) = 8 \cdot 4 + 8(-5i) + (3i)4 + (3i)(-5i)$
$$= 32 - 40i + 12i - 15i^2 \quad \text{Combine like terms.}$$
$$= 32 - 28i - 15i^2$$
$$= 32 - 28i - 15(-1) \quad \text{Again, } i^2 = -1.$$
$$= 32 - 28i + 15$$
$$= 47 - 28i$$

4. $(3 + 2i)(3 - 2i) = 3 \cdot 3 + 3(-2i) + (2i)3 + (2i)(-2i)$
$$= 9 - 6i + 6i - 4i^2$$
$$= 9 - 4i^2$$
$$= 9 - 4(-1)$$
$$= 9 + 4$$
$$= 13$$

NOTE ▶

Here the product of two complex numbers is a real number.

5. $(3 - 4i)^2 = (3 - 4i)(3 - 4i)$
$$= 3 \cdot 3 + 3(-4i) + (-4i)3 + (-4i)(-4i)$$
$$= 9 - 12i - 12i + 16i^2$$
$$= 9 - 24i + 16i^2$$
$$= 9 - 24i + 16(-1)$$
$$= 9 - 24i - 16$$
$$= -7 - 24i$$

Suppose that you were given the division problem

$$\frac{4 + 7i}{3 + 2i}$$

How would you go about expressing your answer in standard form? To determine how to place your answer in standard form, refer to part 4 of the previous example.

$$
\begin{aligned}
\frac{4 + 7i}{3 + 2i} &= \frac{(4 + 7i) \cdot (3 - 2i)}{(3 + 2i) \cdot (3 - 2i)} \\
&= \frac{(4 + 7i)(3 - 2i)}{(3 + 2i)(3 - 2i)} \\
&= \frac{12 - 8i + 21i - 14i^2}{9 - 6i + 6i - 4i^2} \\
&= \frac{12 + 13i - 14i^2}{9 - 4i^2} \\
&= \frac{12 + 13i - 14(-1)}{9 - 4(-1)} \\
&= \frac{12 + 13i + 14}{9 + 4} \\
&= \frac{26 + 13i}{13} \\
&= \frac{26}{13} + \frac{13i}{13} \\
&= 2 + i
\end{aligned}
$$

We have placed the complex number in standard form.

The product $(3 + 2i)(3 - 2i)$ yields a real number. The numbers $3 + 2i$ and $3 - 2i$ are called **complex conjugates** of each other. To find the conjugate of a complex number, change the sign of the imaginary part of the complex number. Note that the process of dividing complex numbers is similar to the process of rationalizing the denominator, which we investigated in Chapter 8.

EXAMPLE 9.1.4

Perform the indicated divisions. Express all answers in the form $a + bi$.

1.
$$
\begin{aligned}
\frac{17 - i}{1 - 3i} &= \frac{(17 - i) \cdot (1 + 3i)}{(1 - 3i) \cdot (1 + 3i)} \\
&= \frac{(17 - i)(1 + 3i)}{(1 - 3i)(1 + 3i)} \\
&= \frac{17 + 51i - i - 3i^2}{1 + 3i - 3i - 9i^2}
\end{aligned}
$$

$$= \frac{17 + 50i - 3i^2}{1 - 9i^2}$$

$$= \frac{17 + 50i - 3(-1)}{1 - 9(-1)}$$

$$= \frac{17 + 50i + 3}{1 + 9}$$

$$= \frac{20 + 50i}{10}$$

$$= \frac{20}{10} + \frac{50i}{10}$$

$$= 2 + 5i$$

2. $\dfrac{5 - i}{1 + 4i} = \dfrac{(5 - i) \cdot (1 - 4i)}{(1 + 4i) \cdot (1 - 4i)}$

$$= \frac{(5 - i)(1 - 4i)}{(1 + 4i)(1 - 4i)}$$

$$= \frac{5 - 20i - i + 4i^2}{1 - 4i + 4i - 16i^2}$$

$$= \frac{5 - 21i + 4i^2}{1 - 16i^2}$$

$$= \frac{5 - 21i + 4(-1)}{1 - 16(-1)}$$

$$= \frac{5 - 21i - 4}{1 + 16}$$

$$= \frac{1 - 21i}{17}$$

$$= \frac{1}{17} - \frac{21}{17}i$$

3. $\dfrac{7 + 2i}{i}$

NOTE ▶ The conjugate of i is $-i$. Remember, in order to find the conjugate of a complex number, change the sign of the imaginary part:

$$\frac{7 + 2i}{i} = \frac{(7 + 2i) \cdot (-i)}{i \quad \cdot (-i)}$$

$$= \frac{(7 + 2i)(-i)}{i(-i)}$$

$$= \frac{-7i - 2i^2}{-i^2}$$

$$= \frac{-7i - 2(-1)}{-(-1)}$$

$$= \frac{-7i + 2}{1}$$
$$= -7i + 2$$
$$= 2 - 7i$$

In this chapter we will examine quadratic equations whose solutions contain square roots of negative numbers. The following example illustrates how to simplify this type of number.

EXAMPLE 9.1.5

Express the following in the form $a + bi$ and simplify.

1. $\dfrac{10 + \sqrt{-36}}{2} = \dfrac{10 + \sqrt{36}\sqrt{-1}}{2}$

$$= \frac{10 + 6i}{2}$$

$$= \frac{10}{2} + \frac{6i}{2}$$

$$= 5 + 3i$$

2. $\dfrac{6 - \sqrt{-81}}{12} = \dfrac{6 - \sqrt{81}\sqrt{-1}}{12}$

$$= \frac{6 - 9i}{12}$$

$$= \frac{6}{12} - \frac{9i}{12}$$

$$= \frac{1}{2} - \frac{3}{4}i$$

EXERCISES 9.1

Express the following numbers in terms of i and simplify.

1. $\sqrt{-25}$ **2.** $\sqrt{-4}$ **3.** $\sqrt{-64}$ **4.** $\sqrt{-49}$

5. $\sqrt{-144}$ **6.** $\sqrt{-121}$ **7.** $\sqrt{-169}$ **8.** $\sqrt{-16}$

9. $\sqrt{-20}$ **10.** $\sqrt{-18}$ **11.** $\sqrt{-27}$ **12.** $\sqrt{-24}$

13. $\sqrt{-75}$ **14.** $\sqrt{-12}$ **15.** $\sqrt{-200}$ **16.** $\sqrt{-50}$

17. $\sqrt{-\frac{1}{9}}$ **18.** $\sqrt{-\frac{1}{4}}$ **19.** $\sqrt{-\frac{25}{36}}$ **20.** $\sqrt{-\frac{9}{49}}$

21. $-\sqrt{-49}$ **22.** $-\sqrt{-25}$ **23.** $-\sqrt{-81}$ **24.** $-\sqrt{-64}$

25. $-\sqrt{-\frac{1}{16}}$ **26.** $-\sqrt{-\frac{1}{81}}$ **27.** $-\sqrt{-\frac{49}{4}}$ **28.** $-\sqrt{-\frac{81}{25}}$

Perform the indicated operations. Express all numbers in the form $a + bi$.

29. $(5 + 2i) + (4 - 3i)$ **30.** $(-3 - i) + (1 - 7i)$ **31.** $\left(\frac{1}{2} - \frac{1}{3}i\right) + \left(\frac{1}{4} + \frac{2}{9}i\right)$

32. $\left(-\frac{3}{10} + \frac{7}{3}i\right) + \left(\frac{3}{2} - \frac{1}{4}i\right)$ **33.** $(2 + 4i) - (5 - i)$ **34.** $(1 - 5i) - (-2 - 3i)$

35. $3 + (2 - 5i)$ **36.** $(6 + 4i) - 7i$ **37.** $8i - (2 - 3i)$

38. $\left(\frac{1}{3} + \frac{3}{4}i\right) - \left(\frac{7}{9} - \frac{3}{8}i\right)$ **39.** $\left(-1 + \frac{2}{5}i\right) - \left(-\frac{1}{3} + \frac{3}{4}i\right)$ **40.** $3(5 - i)$

41. $-6(-2 - 3i)$ **42.** $4i(3 - 5i)$ **43.** $-2i(3 - i)$

44. $(2 + i)(3 + 4i)$ **45.** $(3 - 2i)(5 - i)$ **46.** $(-2 + 3i)(1 - 4i)$

47. $(-1 - 3i)(-2 - 5i)$ **48.** $(4 + i)(4 - i)$ **49.** $(-2 + 7i)(-2 - 7i)$

50. $(4 + 2i)^2$ **51.** $(1 - 3i)^2$ **52.** $\dfrac{10 + 11i}{4 + i}$

53. $\dfrac{13 + 11i}{2 - 5i}$ **54.** $\dfrac{16 - 7i}{-1 + 2i}$ **55.** $\dfrac{5 - i}{2 + i}$ **56.** $\dfrac{2 - 4i}{3 - 2i}$ **57.** $\dfrac{-1 + 6i}{5 + 2i}$

58. $\dfrac{5 - 7i}{-3 - 6i}$ **59.** $\dfrac{6 + 3i}{i}$ **60.** $\dfrac{-4 + i}{i}$ **61.** $\dfrac{2 - 5i}{-i}$ **62.** $\dfrac{-1 - 6i}{-i}$

Express the following in the form $a + bi$ and simplify.

63. $\dfrac{12 + \sqrt{-64}}{4}$ **64.** $\dfrac{10 + \sqrt{-16}}{2}$ **65.** $\dfrac{-18 + \sqrt{-36}}{6}$ **66.** $\dfrac{-20 + \sqrt{-64}}{4}$

67. $\dfrac{-38 - \sqrt{-36}}{2}$ **68.** $\dfrac{-24 - \sqrt{-144}}{6}$ **69.** $\dfrac{8 - \sqrt{-64}}{8}$ **70.** $\dfrac{6 - \sqrt{-36}}{6}$

71. $\dfrac{12 + \sqrt{-100}}{6}$ **72.** $\dfrac{4 + \sqrt{-16}}{8}$ **73.** $\dfrac{8 - \sqrt{-4}}{10}$ **74.** $\dfrac{20 - \sqrt{-36}}{12}$

75. $\dfrac{-15 + \sqrt{-25}}{10}$ **76.** $\dfrac{-21 + \sqrt{-49}}{14}$ **77.** $\dfrac{-6 - \sqrt{-36}}{4}$ **78.** $\dfrac{-8 - \sqrt{-144}}{6}$

9.2 Solving Quadratic Equations by the Extraction of Roots Method

In Chapter 4 we defined a *quadratic equation in the variable x* as any equation that can be written in the form $ax^2 + bx + c = 0$, where a, b, and c are real numbers with $a \neq 0$. We then learned how to find the solutions of a special type of quadratic equation—factorable quadratic equations. By the end of this chapter we will be able to find the solutions of any quadratic equation. However, we must first learn some new properties and algebraic procedures.

Suppose that you were asked to find the solution set of the quadratic equation $x^2 = 9$. Using the procedures developed in Chapter 4:

$$x^2 = 9$$
$$x^2 - 9 = 0 \qquad \text{Place the equation in standard form.}$$
$$(x + 3)(x - 3) = 0 \qquad \text{Factor the left-hand side.}$$
$$x + 3 = 0 \quad \text{or} \quad x - 3 = 0$$
$$x = -3 \quad \text{or} \quad x = 3$$

Thus, the solution set is $\{-3, 3\}$.

Another method for finding the solution set of the equation $x^2 = 9$ is as follows:

$$x^2 = 9$$
$$x = \pm\sqrt{9} \qquad \text{Remember that 9 has two square roots.}$$
$$x = \pm 3$$

Recall that $x = \pm 3$ means $x = 3$ or $x = -3$. Again the solution set is $\{-3, 3\}$.

This line of reasoning leads to the following property.

Extraction of Roots Property

If $x^2 = p$, where p is any real number, then $x = \pm\sqrt{p}$.

REMARK We can write this property in a more general form: If $(mx + n)^2 = p$, where m, n, and p are real numbers, then $mx + n = \pm\sqrt{p}$.

EXAMPLE 9.2.1

Use the extraction of roots property to find the solution set of each quadratic equation.

1. $x^2 = 25$
$$x = \pm\sqrt{25}$$
$$x = \pm 5$$

Checking our solutions:

When $x = 5$ When $x = -5$
$$5^2 \overset{?}{=} 25 \qquad\qquad (-5)^2 \overset{?}{=} 25$$
$$25 = 25 \qquad\qquad\quad 25 = 25$$

So $x = 5$ is a solution.   So $x = -5$ is a solution.

Thus, the solution set is $\{5, -5\}$.

2. $x^2 - 8 = 0$

$$x^2 = 8$$

$$x = \pm\sqrt{8}$$

$$x = \pm2\sqrt{2} \qquad \text{Recall that } \sqrt{8} = \sqrt{4}\sqrt{2} = 2\sqrt{2}.$$

Checking our solutions:

When $x = 2\sqrt{2}$	When $x = -2\sqrt{2}$
$(2\sqrt{2})^2 - 8 \overset{?}{=} 0$	$(-2\sqrt{2})^2 - 8 \overset{?}{=} 0$
$2^2(\sqrt{2})^2 - 8 \overset{?}{=} 0$	$(-2)^2(\sqrt{2})^2 - 8 \overset{?}{=} 0$
$4 \cdot 2 - 8 \overset{?}{=} 0$	$4 \cdot 2 - 8 \overset{?}{=} 0$
$8 - 8 \overset{?}{=} 0$	$8 - 8 \overset{?}{=} 0$
$0 = 0$	$0 = 0$

So $x = 2\sqrt{2}$ is a solution. So $x = -2\sqrt{2}$ is a solution.
Thus, the solution set is $\{-2\sqrt{2}, 2\sqrt{2}\}$.

3. $x^2 = -36$

$$x = \pm\sqrt{-36}$$

$$x = \pm\sqrt{36}\sqrt{-1}$$

$$x = \pm6i \qquad \text{Express the solutions in terms of } i.$$

Checking our solutions:

When $x = 6i$	When $x = -6i$
$(6i)^2 \overset{?}{=} -36$	$(-6i)^2 \overset{?}{=} -36$
$6^2 \cdot i^2 \overset{?}{=} -36$	$(-6)^2 \cdot i^2 \overset{?}{=} -36$
$36(-1) \overset{?}{=} -36$	$36 \cdot (-1) \overset{?}{=} -36$
$-36 = -36$	$-36 = -36$

So $x = 6i$ is a solution. So $x = -6i$ is a solution.
Thus, the solution set is $\{6i, -6i\}$.

4. $4x^2 - 27 = 0$

$$4x^2 = 27$$

$$x^2 = \frac{27}{4}$$

$$x = \pm\sqrt{\frac{27}{4}}$$

$$x = \pm\frac{\sqrt{27}}{\sqrt{4}}$$

$$x = \pm\frac{3\sqrt{3}}{2}$$

The solution set is $\left\{ \dfrac{3\sqrt{3}}{2}, \dfrac{-3\sqrt{3}}{2} \right\}$.

NOTE ▶ When we use the extraction of roots property to find the solutions of a quadratic equation, we do not run the risk of generating extraneous solutions. Thus, checking the solutions will be left to the reader for the remainder of this section.

5. $9x^2 + 4 = 0$

$$9x^2 = -4$$

$$x^2 = -\frac{4}{9}$$

$$x = \pm\sqrt{-\frac{4}{9}}$$

$$x = \pm\sqrt{\frac{4}{9}}\sqrt{-1}$$

$$x = \pm\frac{2}{3}i$$

The solution set is $\left\{ \frac{2}{3}i, -\frac{2}{3}i \right\}$.

6. $(x - 5)^2 = 49$

$$x - 5 = \pm\sqrt{49} \qquad \text{By the extraction of roots property}$$

$$x - 5 = \pm 7$$

$$x = 5 \pm 7$$

So we have two possibilities:

$x = 5 + 7$ or $x = 5 - 7$

$x = 12$ or $x = -2$

The solution set is $\{12, -2\}$.

7. $(x + 3)^2 = 20$

$$x + 3 = \pm\sqrt{20}$$

$$x + 3 = \pm 2\sqrt{5} \qquad \text{Recall that } \sqrt{20} = \sqrt{4}\sqrt{5} = 2\sqrt{5}.$$

$$x = -3 \pm 2\sqrt{5}$$

So we have two possibilities:

$$x = -3 + 2\sqrt{5} \qquad \text{or} \qquad x = -3 - 2\sqrt{5}$$

The solution set is $\{-3 + 2\sqrt{5}, -3 - 2\sqrt{5}\}$.

NOTE ▶ Using the $\pm$ notation, we also can express the solution set as $\{-3 \pm 2\sqrt{5}\}$.

8. $(x - 1)^2 = -4$

$$x - 1 = \pm\sqrt{-4}$$

$$x - 1 = \pm\sqrt{4}\sqrt{-1}$$

$$x - 1 = \pm 2i$$

$$x = 1 \pm 2i$$

The solution set is $\{1 \pm 2i\}$.

9. $(2x - 5)^2 = 13$

$$2x - 5 = \pm\sqrt{13}$$

$$2x = 5 \pm \sqrt{13} \qquad \text{Add 5 to both sides.}$$

$$\frac{2x}{2} = \frac{5 \pm \sqrt{13}}{2} \qquad \text{Divide both sides by 2.}$$

$$x = \frac{5 \pm \sqrt{13}}{2}$$

The solution set is $\left\{ \dfrac{5 \pm \sqrt{13}}{2} \right\}$.

10. $(6x - 3)^2 = 72$

$$6x - 3 = \pm\sqrt{72}$$

$$6x - 3 = \pm 6\sqrt{2} \qquad \text{Recall that } \sqrt{72} = \sqrt{36}\sqrt{2} = 6\sqrt{2}.$$

$$6x = 3 \pm 6\sqrt{2} \qquad \text{Add 3 to both sides.}$$

$$\frac{6x}{6} = \frac{3 \pm 6\sqrt{2}}{6} \qquad \text{Divide both sides by 6.}$$

$$x = \frac{3 \pm 6\sqrt{2}}{6}$$

$$x = \frac{3(1 \pm 2\sqrt{2})}{6} \qquad \text{Factor the numerator.}$$

$$x = \frac{\cancel{3}(1 \pm 2\sqrt{2})}{\cancel{6}_2} \qquad \text{Reduce the fraction.}$$

$$x = \frac{1 \pm 2\sqrt{2}}{2}$$

The solution set is $\left\{ \dfrac{1 \pm 2\sqrt{2}}{2} \right\}$.

Try to avoid this mistake:

Incorrect	Correct
$x = \dfrac{3 \pm 6\sqrt{2}}{6}$	$x = \dfrac{3 \pm 6\sqrt{2}}{6}$
$x = \dfrac{\cancel{3} \pm 6\sqrt{2}}{\cancel{6}_2}$	$x = \dfrac{\cancel{3}(1 \pm 2\sqrt{2})}{\cancel{6}_2}$
$x = \dfrac{1 \pm 6\sqrt{2}}{2}$	$x = \dfrac{1 \pm 2\sqrt{2}}{2}$
Terms, instead of factors, were divided out.	Remember, to reduce a fraction, we can divide out only common factors.

11. $(4x + 6)^2 = -100$

$$4x + 6 = \pm\sqrt{-100}$$
$$4x + 6 = \pm\sqrt{100}\sqrt{-1}$$
$$4x + 6 = \pm 10i$$
$$4x = -6 \pm 10i$$
$$x = \frac{-6 \pm 10i}{4}$$
$$x = \frac{-6}{4} \pm \frac{10i}{4} \qquad \text{Write in standard form}$$
$$\qquad\qquad\qquad\quad\text{and reduce.}$$
$$x = -\frac{3}{2} \pm \frac{5}{2}i$$

The solution set is $\left\{ -\dfrac{3}{2} \pm \dfrac{5}{2}i \right\}$.

12. $(3x + 1)^2 - 5 = 13$

$$(3x + 1)^2 = 18 \qquad \text{Add 5 to both sides}$$
$$\qquad\qquad\qquad\quad\text{to isolate the}$$
$$3x + 1 = \pm\sqrt{18} \qquad \text{squared expression.}$$
$$3x + 1 = \pm 3\sqrt{2}$$
$$3x = -1 \pm 3\sqrt{2}$$
$$x = \frac{-1 \pm 3\sqrt{2}}{3}$$

The solution set is $\left\{ \dfrac{-1 \pm 3\sqrt{2}}{3} \right\}$.

The following example shows how we can use the extraction of roots property in application problems.

EXAMPLE 9.2.2

A bowling ball is dropped from the roof of a 400-foot-tall building. The equation that gives the distance the ball has fallen is $d = 16t^2$, where d is the distance (measured in feet) and t is the time (measured in seconds). How long has the ball been falling when it has traveled a distance of 64 feet? When does it hit the ground?

1. To answer the first question, we must substitute 64 for d in the equation $d = 16t^2$.

$$64 = 16t^2$$
$$\frac{64}{16} = \frac{16t^2}{16}$$
$$4 = t^2$$
$$\pm\sqrt{4} = t$$
$$\pm 2 = t$$

Clearly $t \neq -2$, since it does not make sense for the ball to be falling -2 seconds. Thus, the solution is $t = 2$; that is, the ball has been falling for 2 seconds when it has traveled a distance of 64 feet.

2. To answer the second question, we must substitute 400 for d. This substitution is made because when the ball hits the ground it must have traveled 400 feet.

$$400 = 16t^2$$
$$\frac{400}{16} = \frac{16t^2}{16}$$
$$25 = t^2$$
$$\pm\sqrt{25} = t$$
$$\pm 5 = t$$

We reject $t = -5$. Thus, the ball hits the ground 5 seconds after it is dropped.

EXERCISES 9.2

Use the extraction of roots property to find the solution set of each quadratic equation.

1. $x^2 = 64$ **2.** $x^2 = 100$ **3.** $x^2 = 81$ **4.** $x^2 = 23$ **5.** $x^2 = 22$

6. $x^2 = 20$ **7.** $x^2 = 28$ **8.** $x^2 = 45$ **9.** $x^2 = -49$ **10.** $x^2 = -144$

11. $x^2 = -121$ **12.** $x^2 = -14$ **13.** $x^2 = -32$ **14.** $4x^2 - 9 = 0$ **15.** $4x^2 - 25 = 0$

16. $49x^2 - 9 = 0$ **17.** $9x^2 - 7 = 0$ **18.** $4x^2 - 45 = 0$ **19.** $9x^2 - 20 = 0$ **20.** $49x^2 - 8 = 0$

21. $3x^2 - 1 = 0$ **22.** $5x^2 - 9 = 0$ **23.** $3x^2 - 4 = 0$ **24.** $3x^2 - 7 = 0$ **25.** $25x^2 + 4 = 0$

26. $36x^2 + 1 = 0$ **27.** $4x^2 + 81 = 0$ **28.** $(x - 2)^2 = 25$ **29.** $(x + 5)^2 = 1$ **30.** $\left(x - \frac{1}{2}\right)^2 = \frac{9}{4}$

31. $\left(x - \frac{1}{3}\right)^2 = \frac{25}{9}$ **32.** $\left(x + \frac{2}{5}\right)^2 = \frac{36}{25}$ **33.** $\left(x - \frac{1}{3}\right)^2 = \frac{1}{25}$ **34.** $(x - 6)^2 = 23$ **35.** $(x + 3)^2 = 38$

36. $(x - 5)^2 = 27$ **37.** $(x + 2)^2 = 32$ **38.** $(x + 4)^2 = 48$ **39.** $(x - 9)^2 = -49$

40. $(x + 3)^2 = -16$ **41.** $(3x - 1)^2 = 16$ **42.** $(2x - 3)^2 = 64$ **43.** $(5x + 1)^2 = 100$

44. $(7x + 2)^2 = 10$ **45.** $(2x - 5)^2 = 26$ **46.** $(3x - 2)^2 = 22$ **47.** $(3x - 2)^2 = 12$

48. $(5x + 3)^2 = 40$ **49.** $(6x - 2)^2 = 48$ **50.** $(4x - 2)^2 = 80$ **51.** $(3x + 6)^2 = 63$

52. $(2x + 1)^2 = -36$ **53.** $(7x - 3)^2 = -81$ **54.** $(6x - 5)^2 = -49$ **55.** $(2x - 4)^2 = -100$

56. $(4x + 6)^2 = -64$ **57.** $(4x + 10)^2 = -4$ **58.** $(3x - 4)^2 + 5 = 41$ **59.** $(5x + 2)^2 - 1 = 48$

60. $(3x + 2)^2 + 4 = 18$ **61.** $(4x - 6)^2 - 5 = 23$ **62.** $(2x + 3)^2 + 4 = 3$ **63.** $(6x - 9)^2 - 4 = -13$

Use quadratic equations and the extraction of roots property to find the solutions of the following problems.

64. Six is subtracted from a number and the result squared equals 10. What is the number?

65. Three is subtracted from a number and the result squared equals 30. What is the number?

66. Three is added to twice a number and the result squared equals 16. What is the number?

67. Five is added to twice a number and the result squared equals 36. What is the number?

68. The sum of the areas of two squares is 45 ft². If the length of one side of the larger square equals twice the length of one side of the smaller square, what are the lengths of the sides of each square? (*Hint:* Let x = length of a side of the smaller square; then $2x$ = length of a side of the larger square.)

69. The sum of the areas of two squares is 80 ft². If the length of one side of the larger square equals twice the length of one side of the smaller square, what are the lengths of the sides of each square? (*Hint:* See problem 68.)

70. A pumpkin is dropped from the roof of a 576-foot-tall building. The equation that gives the distance the pumpkin has fallen is $d = 16t^2$, where d is the distance (measured in feet) and t is the time (measured in seconds). How long has the pumpkin been falling when it has traveled 144 feet? When does it hit the ground?

71. An algebra book is dropped from the roof of a 784-foot-tall building. The equation that gives the distance the book has fallen is $d = 16t^2$, where d is the distance (measured in feet) and t is the time (measured in seconds). How long has the book been falling when it has traveled 256 feet? When does it hit the ground?

9.3 Solving Quadratic Equations by Completing the Square

Suppose that you were asked to find the solutions of the quadratic equation $x^2 + 6x + 2 = 0$. The left-hand side of this equation cannot be factored using integer coefficients. Thus, we cannot use the factoring method to find the solutions. We cannot use the extraction of roots property since the equation is not in the form

$$(mx + n)^2 = p$$

In this section we will learn how to transform *any* quadratic equation in the variable x into the form $(x + n)^2 = p$. Once a quadratic equation is in this form, we can then use the extraction of roots property to find its solutions.

First, we must introduce some new terminology. Any trinomial that can be expressed as the square of a binomial is called a **perfect square trinomial**. The following trinomials are examples of perfect square trinomials.

$x^2 + 8x + 16$ since $x^2 + 8x + 16 = (x + 4)^2$
$x^2 + 2x + 1$ since $x^2 + 2x + 1 = (x + 1)^2$
$x^2 - 10x + 25$ since $x^2 - 10x + 25 = (x - 5)^2$
$x^2 - 4x + 4$ since $x^2 - 4x + 4 = (x - 2)^2$

(Do you see the relationship between the constant term and the coefficient of x in these trinomials?)

Let's now determine the number that must be added to $x^2 + 6x$ to form a perfect square trinomial. We want

$x^2 + 6x + \underline{\hspace{1cm}} = (x + n)^2$
$x^2 + 6x + \underline{\hspace{1cm}} = x^2 + 2nx + n^2$

This implies that

$6 = 2n$
$3 = n$

Thus, $9 = n^2$.

It appears that the required number is 9. Let's check:

$x^2 + 6x + 9 = (x + 3)(x + 3)$
$\qquad\qquad\ = (x + 3)^2$

Thus, 9 is the number that must be added to $x^2 + 6x$ to form a perfect square trinomial.

The process of finding the number required to change a polynomial into a perfect square trinomial is called **completing the square**.

Generalizing the above argument, suppose that you wanted to determine the number that must be added to $x^2 + bx$ to form a perfect square trinomial. Again we want

$x^2 + bx + \underline{\hspace{1cm}} = (x + n)^2$
$x^2 + bx + \underline{\hspace{1cm}} = x^2 + 2nx + n^2$

This implies that

$$b = 2n$$
$$\tfrac{1}{2}b = n$$

Thus, $\left(\tfrac{1}{2}b\right)^2 = n^2$.

It appears that the required number is $\left(\tfrac{1}{2}b\right)^2$. Let's check:

$$x^2 + bx + \left(\tfrac{1}{2}b\right)^2 = \left(x + \tfrac{1}{2}b\right)\left(x + \tfrac{1}{2}b\right)$$
$$= \left(x + \tfrac{1}{2}b\right)^2$$

NOTE ▶ The above argument shows that *to complete the square on any polynomial of the form* $x^2 + bx$*, add* $\left(\tfrac{1}{2}b\right)^2 = \left(\tfrac{1}{2} \text{ the coefficient of } x\right)^2$ *to that polynomial.*

EXAMPLE 9.3.1

Determine the number that must be added to the following polynomials to form a perfect square trinomial. Express the resulting trinomial as the square of a binomial.

1. $x^2 + 10x$

Here $b = 10$.

$$\left(\tfrac{1}{2}b\right)^2 = \left(\tfrac{1}{2} \cdot 10\right)^2$$
$$= (5)^2$$
$$= 25$$

So $x^2 + 10x + 25$ should be a perfect square trinomial:

$$x^2 + 10x + 25 = (x + 5)(x + 5)$$
$$= (x + 5)^2$$

2. $x^2 - 12x$

Here $b = -12$.

$$\left(\tfrac{1}{2}b\right)^2 = \left[\tfrac{1}{2} \cdot (-12)\right]^2$$
$$= (-6)^2$$
$$= 36$$

So $x^2 - 12x + 36$ should be a perfect square trinomial:

$$x^2 - 12x + 36 = (x - 6)(x - 6)$$
$$= (x - 6)^2$$

3. $x^2 + 3x$

Here $b = 3$.

$$\left(\tfrac{1}{2}b\right)^2 = \left(\tfrac{1}{2} \cdot 3\right)^2$$
$$= \left(\tfrac{3}{2}\right)^2$$
$$= \tfrac{9}{4}$$

So $x^2 + 3x + \frac{9}{4}$ should be a perfect square trinomial:

$$x^2 + 3x + \tfrac{9}{4} = \left(x + \tfrac{3}{2}\right)\left(x + \tfrac{3}{2}\right)$$
$$= \left(x + \tfrac{3}{2}\right)^2$$

NOTE ▶ The perfect square trinomial $x^2 + bx + \left(\tfrac{1}{2}b\right)^2$ will always factor as follows:

$$x^2 + bx + \left(\tfrac{1}{2}b\right)^2 = \left(x + \tfrac{1}{2}b\right)^2$$

4. $x^2 - 5x$

Here $b = -5$.

$$\left(\tfrac{1}{2}b\right)^2 = \left[\tfrac{1}{2} \cdot (-5)\right]^2$$
$$= \left(-\tfrac{5}{2}\right)^2$$
$$= \tfrac{25}{4}$$

So $x^2 - 5x + \frac{25}{4}$ should be a perfect square trinomial:

$$x^2 - 5x + \tfrac{25}{4} = \left(x - \tfrac{5}{2}\right)\left(x - \tfrac{5}{2}\right)$$
$$= \left(x - \tfrac{5}{2}\right)^2$$

Using the completing the square technique, we are now able to transform any quadratic equation into the form $(x + n)^2 = p$. Let's return to the equation $x^2 + 6x + 2 = 0$:

$$x^2 + 6x + 2 = 0$$

$$x^2 + 6x = -2 \qquad \text{Isolate the } x \text{ terms.}$$

$$x^2 + 6x + \underline{\quad} = -2 + \underline{\quad}$$

The blank will be the number that makes the left-hand side a perfect square trinomial:

$$\underline{\quad} = \left(\tfrac{1}{2} \cdot 6\right)^2 = (3)^2 = 9$$

$$x^2 + 6x + 9 = -2 + 9 \qquad \text{Add 9 to both sides of the}$$
$$(x + 3)^2 = 7 \qquad \text{equation.}$$

Now our equation has been transformed so that we can use the extraction of roots property:

$$x + 3 = \pm\sqrt{7}$$
$$x = -3 \pm \sqrt{7}$$

Thus, the solution set is $\{-3 \pm \sqrt{7}\}$.

EXAMPLE 9.3.2

Find the solution set of each quadratic equation by completing the square.

1. $x^2 + 4x - 12 = 0$

$$x^2 + 4x = 12 \qquad \text{Isolate the } x \text{ terms.}$$

$$x^2 + 4x + \underline{\quad} = 12 + \underline{\quad} \qquad \underline{\quad} = \left(\tfrac{1}{2} \cdot 4\right)^2 = (2)^2 = 4$$

$$x^2 + 4x + 4 = 12 + 4 \qquad \text{Add 4 to both sides of the}$$

$$(x + 2)^2 = 16 \qquad \text{equation.}$$

$$x + 2 = \pm\sqrt{16} \qquad \text{By the extraction of roots}$$

$$x + 2 = \pm 4 \qquad \text{property}$$

$$x = -2 \pm 4$$

So we have two possibilities:

$$x = -2 + 4 \qquad \text{or} \qquad x = -2 - 4$$

$$x = 2 \qquad \text{or} \qquad x = -6$$

The solution set is $\{2, -6\}$.

NOTE ▶ In this problem we could have found the solution set by using the factoring method.

2. $x^2 - 2x - 2 = 0$

$$x^2 - 2x = 2 \qquad \text{Isolate the } x \text{ terms.}$$

$$x^2 - 2x + \underline{\quad} = 2 + \underline{\quad} \qquad \underline{\quad} = \left[\tfrac{1}{2} \cdot (-2)\right]^2 = (-1)^2 = 1$$

$$x^2 - 2x + 1 = 2 + 1 \qquad \text{Add 1 to both sides of the equation.}$$

$$(x - 1)^2 = 3$$

$$x - 1 = \pm\sqrt{3} \qquad \text{By the extraction of roots property}$$

$$x = 1 \pm \sqrt{3}$$

Thus, the solution set is $\{1 \pm \sqrt{3}\}$.

3. $4x^2 + 16x + 5 = 0$

$$4x^2 + 16x = -5 \qquad \text{Isolate the } x \text{ terms.}$$

$$x^2 + 4x = -\tfrac{5}{4} \qquad \begin{array}{l}\text{Make the leading coefficient 1}\\ \text{by dividing both sides by 4.}\end{array}$$

$$x^2 + 4x + \underline{\quad} = -\tfrac{5}{4} + \underline{\quad} \qquad \underline{\quad} = \left(\tfrac{1}{2} \cdot 4\right)^2 = (2)^2 = 4$$

$$x^2 + 4x + 4 = -\tfrac{5}{4} + 4 \qquad \text{Add 4 to both sides of the}$$

$$(x + 2)^2 = -\tfrac{5}{4} + \tfrac{16}{4} \qquad \text{equation.}$$

$$(x + 2)^2 = \tfrac{11}{4}$$

$$x + 2 = \pm\sqrt{\frac{11}{4}} \qquad \begin{array}{l}\text{By the extraction of roots}\\ \text{property}\end{array}$$

$$x + 2 = \pm \frac{\sqrt{11}}{\sqrt{4}}$$

$$x + 2 = \pm \frac{\sqrt{11}}{2}$$

$$x = -2 \pm \frac{\sqrt{11}}{2}$$

or $\quad x = -\dfrac{4}{2} \pm \dfrac{\sqrt{11}}{2}$ Obtain an LCD on the right-hand side, and write the solutions as a single fraction.

$$x = \frac{-4 \pm \sqrt{11}}{2}$$

Thus, the solution set is $\left\{ \dfrac{-4 \pm \sqrt{11}}{2} \right\}$.

4. $2x^2 - 6x + 1 = 0$

$\qquad 2x^2 - 6x = -1$ Isolate the x terms.

$\qquad x^2 - 3x = -\dfrac{1}{2}$ Make the leading coefficient 1 by dividing both sides by 2.

$x^2 - 3x + \underline{\quad} = -\dfrac{1}{2} + \underline{\quad}$ $\underline{\quad} = \left[\frac{1}{2} \cdot (-3)\right]^2 = \left(-\frac{3}{2}\right)^2 = \frac{9}{4}$

$x^2 - 3x + \dfrac{9}{4} = -\dfrac{1}{2} + \dfrac{9}{4}$ Add $\frac{9}{4}$ to both sides of the equation.

$\left(x - \dfrac{3}{2}\right)^2 = -\dfrac{2}{4} + \dfrac{9}{4}$

$\left(x - \dfrac{3}{2}\right)^2 = \dfrac{7}{4}$

$\qquad x - \dfrac{3}{2} = \pm\sqrt{\dfrac{7}{4}}$ By the extraction of roots property

$\qquad x - \dfrac{3}{2} = \pm\dfrac{\sqrt{7}}{\sqrt{4}}$

$\qquad x - \dfrac{3}{2} = \pm\dfrac{\sqrt{7}}{2}$

$\qquad x = \dfrac{3}{2} \pm \dfrac{\sqrt{7}}{2}$

or $\quad x = \dfrac{3 \pm \sqrt{7}}{2}$

Thus, the solution set is $\left\{ \dfrac{3 \pm \sqrt{7}}{2} \right\}$.

5. $9x^2 + 6x + 37 = 0$

$\qquad 9x^2 + 6x = -37$ Isolate the x terms.

$\qquad x^2 + \dfrac{6}{9}x = -\dfrac{37}{9}$ Make the leading coefficient 1 by dividing both sides by 9.

$$x^2 + \frac{2}{3}x = -\frac{37}{9}$$

$$x^2 + \frac{2}{3}x + \underline{\quad} = -\frac{37}{9} + \underline{\quad} \qquad \underline{\quad} = \left(\frac{1}{2} \cdot \frac{2}{3}\right)^2 = \left(\frac{1}{3}\right)^2 = \frac{1}{9}$$

$$x^2 + \frac{2}{3}x + \frac{1}{9} = -\frac{37}{9} + \frac{1}{9} \qquad \text{Add } \frac{1}{9} \text{ to both sides of the equation.}$$

$$\left(x + \frac{1}{3}\right)^2 = -\frac{36}{9}$$

$$\left(x + \frac{1}{3}\right)^2 = -4$$

$$x + \frac{1}{3} = \pm\sqrt{-4} \qquad \text{By the extraction of roots property}$$

$$x + \frac{1}{3} = \pm 2i \qquad \text{Recall that } \sqrt{-4} = \sqrt{4}\sqrt{-1} = 2i.$$

$$x = -\frac{1}{3} \pm 2i$$

Thus, the solution set is $\left\{-\dfrac{1}{3} \pm 2i\right\}$.

The following steps will enable us to find the solution set of *any* quadratic equation (that is, any equation of the form $ax^2 + bx + c = 0$, where $a \neq 0$) by completing the square.

1. Isolate the x terms; that is, rewrite the equation as $ax^2 + bx = -c$.

2. Make the leading coefficient 1 by dividing both sides of the equation by a.

3. Determine the number needed to complete the square on the x terms. [Remember that this number is $\left(\frac{1}{2} \text{ coefficient of } x\right)^2$.]

4. Add the number found in step 3 to both sides of the equation.

5. Rewrite the perfect square trinomial on the left-hand side as a binomial squared, and combine the numbers on the right-hand side. The equation now is in the form $(x + n)^2 = p$.

6. Apply the extraction of roots property and solve for x.

EXERCISES 9.3

Determine the number that must be added to the following polynomials to form a perfect square trinomial. Express the resulting trinomial as the square of a binomial.

1. $x^2 + 14x$ **2.** $x^2 + 20x$ **3.** $x^2 + 9x$ **4.** $x^2 + 7x$ **5.** $x^2 - 6x$ **6.** $x^2 - 16x$

7. $x^2 - x$ **8.** $x^2 - 3x$ **9.** $x^2 + \frac{4}{3}x$ **10.** $x^2 + \frac{2}{5}x$ **11.** $x^2 + \frac{8}{5}x$ **12.** $x^2 + \frac{10}{3}x$

13. $x^2 + \frac{1}{4}x$ **14.** $x^2 + \frac{1}{2}x$ **15.** $x^2 + \frac{3}{7}x$ **16.** $x^2 + \frac{5}{3}x$ **17.** $x^2 - \frac{2}{7}x$ **18.** $x^2 - \frac{4}{9}x$

19. $x^2 - \frac{6}{5}x$ **20.** $x^2 - \frac{8}{3}x$ **21.** $x^2 - \frac{1}{3}x$ **22.** $x^2 - \frac{1}{4}x$ **23.** $x^2 - \frac{7}{2}x$ **24.** $x^2 - \frac{7}{3}x$

Find the solution set of each quadratic equation by completing the square.

25. $x^2 - 5x + 6 = 0$ **26.** $x^2 - 7x + 12 = 0$ **27.** $x^2 + 4x - 5 = 0$

28. $x^2 - 2x - 15 = 0$ **29.** $x^2 - 16 = 0$ **30.** $x^2 - 36 = 0$

31. $x^2 - 2x - 1 = 0$ **32.** $x^2 - 2x - 4 = 0$ **33.** $x^2 + 4x + 1 = 0$

34. $x^2 + 6x + 7 = 0$ **35.** $x^2 + 8x + 14 = 0$ **36.** $x^2 + 4x - 2 = 0$

37. $x^2 - 10x + 18 = 0$ **38.** $x^2 - 14x + 46 = 0$ **39.** $x^2 - 4x + 5 = 0$

40. $x^2 - 2x + 5 = 0$ **41.** $x^2 + 8x + 25 = 0$ **42.** $x^2 + 4x + 29 = 0$

43. $-x^2 + 6x - 3 = 0$ **44.** $-x^2 + 8x - 11 = 0$ **45.** $-x^2 - 8x - 13 = 0$

46. $-x^2 - 12x - 34 = 0$ **47.** $2x^2 + 3x - 9 = 0$ **48.** $2x^2 + 3x - 2 = 0$

49. $3x^2 - 14x + 8 = 0$ **50.** $3x^2 - 19x + 6 = 0$ **51.** $4x^2 - 4x - 3 = 0$

52. $4x^2 - 4x - 15 = 0$ **53.** $9x^2 - 6x - 1 = 0$ **54.** $9x^2 - 12x + 1 = 0$

55. $4x^2 - 8x - 1 = 0$ **56.** $4x^2 - 4x - 1 = 0$ **57.** $4x^2 + 12x + 3 = 0$

58. $2x^2 + 4x - 1 = 0$ **59.** $4x^2 + 4x - 7 = 0$ **60.** $4x^2 + 4x - 17 = 0$

61. $9x^2 - 24x + 4 = 0$ **62.** $3x^2 - 12x + 8 = 0$ **63.** $4x^2 + 24x + 31 = 0$

64. $4x^2 + 32x + 61 = 0$ **65.** $9x^2 - 18x + 10 = 0$ **66.** $9x^2 - 12x + 5 = 0$

67. $2x^2 + 2x + 13 = 0$ **68.** $4x^2 + 4x + 17 = 0$

Use quadratic equations and the completing the square method to find the solutions of the following problems.

69. The square of a number plus four times the number is 7. What is the number?

70. The square of a number plus six times the number is 1. What is the number?

71. Twice the square of a number minus two times the number is 1. What is the number?

72. Twice the square of a number minus two times the number is 7. What is the number?

73. The area of a rectangle is 10 square feet. The length is 2 feet more than four times the width. What is the *width* of the rectangle?

74. The area of a rectangle is 6 square feet. The length is 2 feet more than three times the width. What is the *width* of the rectangle?

75. A ball is thrown vertically upward from ground level. The equation that gives the ball's height above ground level is $d = -16t^2 + 80t$, where d is the distance (measured in feet) and t is the time (measured in seconds). When does the ball hit the ground? (*Hint:* The ball hits the ground when $d = 0$.)

76. A stone is thrown vertically upward from ground level. The equation that gives the stone's height above ground level is $d = -16t^2 + 48t$, where d is the distance (measured in feet) and t is the time (measured in seconds). When does the stone hit the ground? (*Hint:* See problem 75.)

9.4 The Quadratic Formula

The completing the square method enables us to find the solutions of any quadratic equation. As we have seen, the completing the square method is often lengthy. In this section we will derive a formula, using the completing the square method, that will streamline the process of finding the solutions of quadratic equations.

First, recall some terminology from Chapter 4. Given a quadratic equation in the variable x, the form $ax^2 + bx + c = 0$ is called **standard form**. In this form, a is the coefficient of x^2, b is the coefficient of x, and c is the constant term. The only restriction we have on a, b, or c is that $a \neq 0$. Why?

EXAMPLE 9.4.1

Place the following quadratic equations in standard form, and then identify a, b, and c.

1. $x^2 + 5x - 7 = 0$

    This equation already is in standard form. Here $a = 1$, $b = 5$, and $c = -7$.

2. $3x^2 + 4x - 1 = 4x - 6$

    $$3x^2 - 1 = -6 \qquad \text{Subtract } 4x \text{ from both sides.}$$
    $$3x^2 + 5 = 0 \qquad \text{Add 6 to both sides.}$$

    Now the quadratic equation is in standard form, and we can identify a, b, and c. Here $a = 3$, $b = 0$, and $c = 5$.

NOTE ▶ $b = 0$ since $3x^2 + 5 = 0$ is equivalent to $3x^2 + 0x + 5 = 0$.

3. $(2x + 1)(x - 5) = 3$

    First, perform the multiplication on the left-hand side:

    $$2x^2 - 10x + x - 5 = 3$$
    $$2x^2 - 9x - 5 = 3$$
    $$2x^2 - 9x - 8 = 0 \qquad \text{Subtract 3 from both sides.}$$

    In this case, $a = 2$, $b = -9$, and $c = -8$.

We will now use the completing the square method to find the solutions of the general quadratic equation $ax^2 + bx + c = 0$, $a \neq 0$:

$$ax^2 + bx + c = 0 \qquad \text{Be sure that the quadratic equation is in standard form.}$$

$$ax^2 + bx = -c \qquad \text{Isolate the } x \text{ terms.}$$

$$x^2 + \frac{b}{a}x = \frac{-c}{a}$$

Make the leading coefficient 1 by dividing both sides of the equation by a.

$$x^2 + \frac{b}{a}x + \underline{\quad} = \frac{-c}{a} + \underline{\quad}$$

$$\underline{\quad} = \left(\frac{1}{2}\cdot\frac{b}{a}\right)^2 = \left(\frac{b}{2a}\right)^2$$
$$= \frac{b^2}{4a^2}$$

$$x^2 + \frac{b}{a}x + \frac{b^2}{4a^2} = \frac{b^2}{4a^2} - \frac{c}{a}$$

$$\left(x + \frac{b}{2a}\right)^2 = \frac{b^2}{4a^2} - \frac{c}{a}$$

Factor the left-hand side.

$$\left(x + \frac{b}{2a}\right)^2 = \frac{b^2}{4a^2} - \frac{4ac}{4a^2}$$

Obtain an LCD on the right-hand side.

$$\left(x + \frac{b}{2a}\right)^2 = \frac{b^2 - 4ac}{4a^2}$$

$$x + \frac{b}{2a} = \pm\sqrt{\frac{b^2 - 4ac}{4a^2}}$$

By the extraction of roots property

$$x + \frac{b}{2a} = \pm\frac{\sqrt{b^2 - 4ac}}{\sqrt{4a^2}}$$

$$x + \frac{b}{2a} = \pm\frac{\sqrt{b^2 - 4ac}}{2a}$$

$$x = \frac{-b}{2a} \pm \frac{\sqrt{b^2 - 4ac}}{2a}$$

$$x = \frac{-b \pm \sqrt{b^2 - 4ac}}{2a}$$

This important result is summarized as follows:

Quadratic Formula

If $ax^2 + bx + c = 0$, where a, b, and c are real numbers with $a \neq 0$, then the solutions of the equation are

$$x = \frac{-b \pm \sqrt{b^2 - 4ac}}{2a}$$

REMARKS

1. To apply the quadratic formula, we must place the quadratic equation in standard form.

2. We can use the quadratic formula to find the solutions of *any* quadratic equation.

3. The quadratic formula is the most important formula in this book and *should be memorized*.

EXAMPLE 9.4.2

Use the quadratic formula to find the solution set of each quadratic equation.

1. $2x^2 - x - 6 = 0$

To apply the quadratic formula, we must identify a, b, and c. In this case, $a = 2$, $b = -1$, and $c = -6$. Thus,

$$x = \frac{-(-1) \pm \sqrt{(-1)^2 - 4 \cdot 2 \cdot (-6)}}{2 \cdot 2}$$

$$= \frac{1 \pm \sqrt{1 - (-48)}}{4}$$

$$= \frac{1 \pm \sqrt{49}}{4}$$

$$= \frac{1 \pm 7}{4}$$

So $x = \dfrac{1 + 7}{4}$ or $x = \dfrac{1 - 7}{4}$

$\qquad = \dfrac{8}{4} \qquad\qquad\qquad = \dfrac{-6}{4}$

$\qquad = 2 \qquad\qquad\qquad\; = -\dfrac{3}{2}$

The solution set is $\left\{2, -\frac{3}{2}\right\}$.

NOTE ▶

In this problem we could have found the solution set by using the factoring method.

2. $2x^2 + 3x = 1$

Write the equation in standard form:

$$2x^2 + 3x - 1 = 0$$

Now a, b, and c can be identified: $a = 2$, $b = 3$, and $c = -1$. So

$$x = \frac{-3 \pm \sqrt{3^2 - 4 \cdot 2 \cdot (-1)}}{2 \cdot 2}$$

$$= \frac{-3 \pm \sqrt{9 - (-8)}}{4}$$

$$= \frac{-3 \pm \sqrt{17}}{4}$$

The solution set is $\left\{\dfrac{-3 \pm \sqrt{17}}{4}\right\}$.

3. $x^2 - 2x - 4 = 0$

$a = 1, b = -2, c = -4$

$$x = \frac{-(-2) \pm \sqrt{(-2)^2 - 4 \cdot 1 \cdot (-4)}}{2 \cdot 1}$$

$$= \frac{2 \pm \sqrt{4 - (-16)}}{2}$$

$$= \frac{2 \pm \sqrt{20}}{2}$$

$$= \frac{2 \pm \sqrt{4}\sqrt{5}}{2}$$

$$= \frac{2 \pm 2\sqrt{5}}{2}$$

$$= \frac{\cancel{2}(1 \pm \sqrt{5})}{\cancel{2}}$$

$$= 1 \pm \sqrt{5}$$

The solution set is $\{1 \pm \sqrt{5}\}$.

Try to avoid this mistake:

Incorrect	Correct
$x = \dfrac{\cancel{2} \pm 2\sqrt{5}}{\cancel{2}}$ $= 1 \pm 2\sqrt{5}$	$x = \dfrac{2 \pm 2\sqrt{5}}{2}$ $= \dfrac{\cancel{2}(1 \pm \sqrt{5})}{\cancel{2}}$ $= 1 \pm \sqrt{5}$ Remember that you can divide out only common *factors*, not terms.

4. $x^2 - 4x + 13 = 0$

$a = 1, b = -4, c = 13$

$$x = \frac{-(-4) \pm \sqrt{(-4)^2 - 4 \cdot 1 \cdot 13}}{2 \cdot 1}$$

$$= \frac{4 \pm \sqrt{16 - 52}}{2}$$

$$= \frac{4 \pm \sqrt{-36}}{2}$$

$$= \frac{4 \pm 6i}{2} \qquad \text{Remember that } \sqrt{-36} = \sqrt{36}\sqrt{-1} = 6i.$$

$$= \frac{\cancel{2}(2 \pm 3i)}{\cancel{2}}$$

$$= 2 \pm 3i$$

The solution set is $\{2 \pm 3i\}$.

Try to avoid this mistake:

Incorrect	Correct
$x =$ $$-(-4) \pm \frac{\sqrt{(-4)^2 - 4 \cdot 1 \cdot 13}}{2 \cdot 1}$$ $$= 4 \pm \frac{\sqrt{16 - 52}}{2}$$ $$= 4 \pm \frac{\sqrt{-36}}{2}$$ $$= 4 \pm \frac{6i}{2}$$ $$= 4 \pm 3i$$ The fraction line does not go all the way across.	$x =$ $$\frac{-(-4) \pm \sqrt{(-4)^2 - 4 \cdot 1 \cdot 13}}{2 \cdot 1}$$ $$= \frac{4 \pm \sqrt{16 - 52}}{2}$$ $$= \frac{4 \pm \sqrt{-36}}{2}$$ $$= \frac{4 \pm 6i}{2}$$ $$= \frac{\cancel{2}(2 \pm 3i)}{\cancel{2}}$$ $$= 2 \pm 3i$$

5. $(2x + 3)(2x + 5) = 2$

$4x^2 + 10x + 6x + 15 = 2$ Perform the multiplication on the left-hand side.

$4x^2 + 16x + 15 = 2$

$4x^2 + 16x + 13 = 0$ Subtract 2 from both sides.

Now $a = 4$, $b = 16$, and $c = 13$.

$$x = \frac{-16 \pm \sqrt{16^2 - 4 \cdot 4 \cdot 13}}{2 \cdot 4}$$

$$= \frac{-16 \pm \sqrt{256 - 208}}{8}$$

$$= \frac{-16 \pm \sqrt{48}}{8}$$

$$= \frac{-16 \pm \sqrt{16}\sqrt{3}}{8}$$

$$= \frac{-16 \pm 4\sqrt{3}}{8}$$

$$= \frac{\overset{1}{\cancel{4}}(-4 \pm \sqrt{3})}{\underset{2}{\cancel{8}}}$$

$$= \frac{-4 \pm \sqrt{3}}{2}$$

The solution set is $\left\{ \dfrac{-4 \pm \sqrt{3}}{2} \right\}$.

EXERCISES 9.4

Place the following quadratic equations in standard form, and then identify a, b, **and** c.

1. $x^2 + 3x - 5 = 0$

2. $x^2 + 7x - 2 = 0$

3. $5x^2 + 4x = 0$

4. $9x^2 + 16x = 0$

5. $3x^2 + 2x - 11 = x - 4$

6. $6x^2 + 8x - 10 = 2x - 1$

7. $8x^2 + 2x - 1 = 5x^2 + x + 8$

8. $5x^2 + 4x - 2 = 3x^2 + x + 9$

9. $-x^2 + 4x + 8 = 2(2x + 5)$

10. $-x^2 + 6x + 8 = 3(2x + 5)$

11. $5x^2 + 3x + 9 = 3(x + 3)$

12. $7x^2 + 12x + 8 = 4(3x + 2)$

13. $(4x + 1)(2x + 3) = 2$

14. $(3x + 7)(2x + 1) = 4$

15. $(2x + 1)(x - 5) = (x + 1)(x - 2)$

16. $(2x + 3)(x - 4) = (x + 3)(x - 1)$

Use the quadratic formula to find the solution set of each quadratic equation.

17. $3x^2 - 8x - 3 = 0$

18. $3x^2 - x - 2 = 0$

19. $x^2 - 16 = 0$

20. $x^2 - 9 = 0$

21. $2x^2 - 3x + 7 = 2x + 7$

22. $2x^2 + x + 3 = 4x + 3$

23. $3x^2 - 2x - 13 = 2x^2 + x - 9$

24. $3x^2 - x - 12 = 2x^2 + 3x - 7$

25. $(5x + 1)(x - 2) = (x - 1)(3x - 1)$

26. $(5x + 6)(x - 1) = (x + 2)(3x - 2)$

27. $7x^2 - 3 = 0$

28. $5x^2 - 2 = 0$

29. $2x^2 + 7x + 2 = 0$

30. $2x^2 + 9x + 1 = 0$

31. $x^2 + 3x - 9 = 0$

32. $x^2 + 5x - 25 = 0$

33. $2x^2 - 7x + 4 = 4x + 5$

34. $2x^2 - 4x - 1 = 3x + 2$

35. $(3x + 1)(x - 4) = 2x(x - 2)$

36. $(3x + 2)(x - 3) = 2x(x - 2)$

37. $x^2 - 6x + 6 = 0$

38. $x^2 - 8x + 14 = 0$

39. $x^2 - 4x - 4 = 0$

40. $x^2 - 6x + 1 = 0$

41. $4x^2 + 24x + 31 = 0$

42. $4x^2 + 20x + 19 = 0$

43. $9x^2 - 12x - 71 = 0$

44. $9x^2 - 18x - 41 = 0$

45. $4x^2 + 10x + 11 = 3x^2 - 8$

46. $9x^2 + 8x + 2 = 8x^2 - 7$

47. $17x^2 - x - 1 = x^2 + 7x + 10$

48. $17x^2 - 3x - 11 = x^2 + 5x + 6$

49. $(2x - 3)(x + 4) = (x + 3)(x + 2)$

50. $(2x - 1)(x + 5) = (x + 7)(x + 2)$

51. $(3x - 19)(2x + 1) = x(2x - 47)$

52. $(3x - 19)(x + 1) = x(x - 26)$

53. $9x^2 + 25 = 0$

54. $4x^2 + 49 = 0$

55. $x^2 - 2x + 10 = 0$

56. $x^2 - 2x + 5 = 0$

57. $2x^2 + 5x + 28 = x^2 - 5x - 1$

58. $2x^2 + x + 23 = x^2 - 7x - 2$

59. $8x^2 - 12x + 5 = 0$

60. $18x^2 - 6x + 1 = 0$

61. $(16x + 15)(x + 1) = 15x + 2$

62. $(16x + 15)(x + 1) = 7x + 2$

Use quadratic equations and the quadratic formula to find the solutions of the following problems.

63. The square of a number minus four times the number is 3. What is the number?

64. The square of a number minus four times the number is 1. What is the number?

65. Twice the square of a number plus two times the number is -5. What is the number?

66. Twice the square of a number plus two times the number is -13. What is the number?

67. The area of a rectangle is 3 square feet. The length is 1 foot more than three times the width. What is the *width* of the rectangle?

68. The area of a rectangle is 6 square feet. The length is 2 feet more than three times the width. What is the *width* of the rectangle?

69. A ball is thrown vertically upward from the roof of a building. The equation that gives the ball's height above ground level is $d = -16t^2 + 96t + 32$, where d is the distance (measured in feet) and t is the time (measured in seconds). How many seconds after the ball is thrown is it 176 feet high? (*Hint:* The ball is 176 feet high when $d = 176$.)

70. A ball is thrown vertically upward from the roof of a building. The equation that gives the ball's height above ground level is $d = -16t^2 + 64t + 40$, where d is the distance (measured in feet) and t is the time (measured in seconds). How many seconds after the ball is thrown is it 104 feet high? (*Hint:* See problem 69.)

9.5 Quadratic Equations Summary

We have investigated several different methods for finding the solutions of quadratic equations. The following chart gives some hints for matching a quadratic equation with the method that most efficiently finds its solutions.

Form	Method	Example
1. $ax^2 + c = 0$ $a \neq 0, c \neq 0$	Use the extraction of roots property	$9x^2 - 5 = 0$ $9x^2 = 5$ $x^2 = \dfrac{5}{9}$ $x = \pm\sqrt{\dfrac{5}{9}}$ $x = \pm\dfrac{\sqrt{5}}{3}$
2. $ax^2 + bx = 0$ $a \neq 0, b \neq 0$	Use the factoring method	$6x^2 + 12x = 0$ $6x(x + 2) = 0$ $6x = 0 \quad \text{or} \quad x + 2 = 0$ $x = 0 \quad \text{or} \qquad x = -2$
3. $(mx + n)^2 = p$ $m \neq 0$	Use the extraction of roots property	$(2x - 1)^2 = 5$ $2x - 1 = \pm\sqrt{5}$ $2x = 1 \pm \sqrt{5}$ $x = \dfrac{1 \pm \sqrt{5}}{2}$
4. $ax^2 + bx + c = 0$ $a \neq 0, b \neq 0, c \neq 0$	**a.** The left-hand side is factorable. Use the factoring method. **b.** The left-hand side cannot be factored using integer coefficients, or you cannot determine the factorization. Use the quadratic formula.	**a.** $x^2 - x - 12 = 0$ $(x - 4)(x + 3) = 0$ $x - 4 = 0 \quad \text{or} \quad x + 3 = 0$ $x = 4 \quad \text{or} \qquad x = -3$ **b.** $x^2 + x - 3 = 0$ $a = 1, b = 1, c = -3$ $x = \dfrac{-1 \pm \sqrt{1^2 - 4 \cdot 1 \cdot (-3)}}{2 \cdot 1}$ $x = \dfrac{-1 \pm \sqrt{1 - (-12)}}{2}$ $x = \dfrac{-1 \pm \sqrt{13}}{2}$

NOTE ▶ The completing the square method is seldom used for finding solutions of quadratic equations, but it does have important applications. We will see one application in the next section.

EXERCISES 9.5

Using the most efficient method, find the solution set of each quadratic equation.

1. $(2x + 1)(2x - 9) = -16x$

2. $(2x + 1)(4x + 3) = 10$

3. $x^2 - 7x + 2 = 0$

4. $(3x + 1)(x - 2) = 0$

5. $5x^2 - 2 = 0$

6. $x^2 - 12x + 37 = 0$

7. $(2x - 3)^2 - 11 = 14$

8. $(4x + 1)(4x + 3) = 2(8x + 1)$

9. $(4x + 1)(x - 7) = -7x - 36$

10. $(5x + 2)^2 = 10$

11. $(5x + 3)(x - 2) = (2x - 1)(x - 3)$

12. $x^2 - 5x + 3 = 0$

13. $6x^2 - 10x = 0$

14. $(3x - 2)^2 - 5 = 4$

15. $(2x + 5)(x - 3) = 0$

16. $3x^2 - 7 = 0$

17. $(3x + 2)(6x + 1) = 9$

18. $(4x + 1)(x - 3) = x - 16$

19. $(2x + 1)(2x + 5) = 4(3x + 1)$

20. $2x^2 + 5x - 1 = 0$

21. $(4x + 1)^2 = 13$

22. $2x^2 - 10x + 11 = 0$

23. $2x^2 + 3x - 3 = 0$

24. $6x^2 - 21x = 0$

25. $2x^2 - 6x + 3 = 0$

26. $3x^2 + 28x + 9 = 0$

27. $(3x - 1)(3x - 2) = 3x$

28. $(6x + 1)(x - 2) = (4x - 7)(x - 1)$

29. $(2x + 4)(x - 3) = (x + 3)(x - 1)$

30. $(9x - 5)(x + 1) = 4x - 3$

31. $6x^2 + 19x + 8 = 0$

32. $(3x - 2)(x - 1) = x$

33. $(x + 3)^2 + 5 = 1$

34. $(2x + 6)(x - 1) = (x + 1)(x - 4)$

35. $(6x - 9)(x + 2) = -18$

36. $8x^2 + 4x - 1 = 0$

37. $(2x + 1)(x - 1) = (x + 7)(x + 8)$

38. $(2x + 3)(x - 6) = (x + 4)(x + 6)$

39. $16x^2 + 8x - 5 = 0$

40. $(x + 7)^2 + 17 = 1$

41. $(4x - 1)(x + 1) = 3x + 4$

42. $(4x + 2)(x + 1) = 2$

43. $x^2 - 8x + 17 = 0$

44. $(3x + 4)(3x - 2) = 6x - 7$

9.6 Graphing Quadratic Equations

In Chapter 6 a *linear equation in two variables* was defined as any equation that could be written in the form $Ax + By = C$, where A, B, and C are real numbers with the property that not both A and B are zero. We saw that the graph of a linear equation in two variables was a straight line.

In this section we will investigate the graphs of **quadratic equations in two variables** of the form $y = ax^2 + bx + c$, where a, b, and c are real numbers with $a \neq 0$. We will examine some of the characteristics of these graphs and see what conclusions we are able to draw.

EXAMPLE 9.6.1

Sketch the graph of the equation $y = x^2$.

The only graphing technique we have, as of now, is to find points on the curve and then smooth in the graph. Since y is in terms of x, we will pick values for x and then solve for y, which will generate ordered pairs (points on the curve). For example, if $x = -3$, then $y = (-3)^2$, so that $y = 9$. This says that the point $(-3, 9)$ is on the curve. Some of the ordered pairs that satisfy the equation $y = x^2$ are given in the chart. By plotting these points and connecting them with a smooth curve, we generate the curve shown in Figure 9.6.1.

x	y
-3	9
-2	4
-1	1
0	0
1	1
2	4
3	9

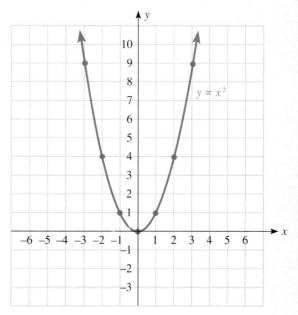

Figure 9.6.1

The curve in Figure 9.6.1 is called a **parabola**. The point $(0, 0)$ is called the **vertex**. In this type of parabola, which opens upward, the vertex is the lowest point on the parabola. The vertex will be the highest point if the parabola opens downward.

EXAMPLE 9.6.2

Sketch the graphs of the following quadratic equations and determine the vertex of each parabola.

1. $y = -x^2$

Some of the ordered pairs that satisfy the equation $y = -x^2$ are given in the chart. Plot these points and smooth in the curve as shown in Figure 9.6.2. In this case the parabola opens downward, so that the vertex is the highest point. Here the vertex is at $(0, 0)$.

x	y
-3	-9
-2	-4
-1	-1
0	0
1	-1
2	-4
3	-9

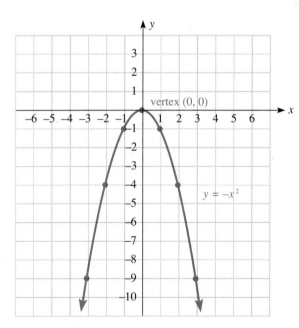

Figure 9.6.2

2. $y = x^2 - 2$

Some of the ordered pairs that satisfy the equation $y = x^2 - 2$ are given in the chart. Plot these points and smooth in the curve as shown in Figure 9.6.3. In this case the parabola opens upward so that the vertex is the lowest point. Here the vertex is at $(0, -2)$.

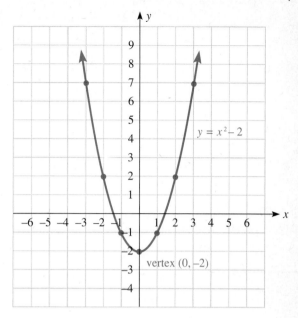

x	y
−3	7
−2	2
−1	−1
0	−2
1	−1
2	2
3	7

Figure 9.6.3

REMARK   These examples aid in illustrating the following facts:

Given a quadratic equation in the form
$$y = ax^2 + bx + c, \ a \neq 0$$

1. The graph of the quadratic equation is a parabola.

2. When $a > 0$, the parabola is ∪-shaped (opens upward) and the vertex is the lowest point on the parabola.

3. When $a < 0$, the parabola is ∩-shaped (opens downward), and the vertex is the highest point on the parabola.

NOTE ▶   In this book we will investigate only the cases in which $a = \pm 1$.

In the next example we examine quadratic equations that are given in a very useful form.

EXAMPLE 9.6.3   Sketch the graphs of the following quadratic equations and determine the vertex of each parabola.

1. $y = (x - 2)^2 + 3$

Some of the ordered pairs that satisfy the equation $y = (x - 2)^2 + 3$ are given in the chart. Plot these points and smooth in the curve as in Figure 9.6.4. Here the vertex is at $(2, 3)$.

x	y
-1	12
0	7
1	4
2	3
3	4
4	7
5	12

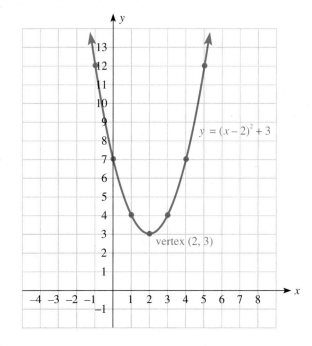

Figure 9.6.4

2. $y = (x + 1)^2 - 2$

Some of the ordered pairs that satisfy the equation $y = (x + 1)^2 - 2$ are given in the chart. Plot these points and smooth in the curve as in Figure 9.6.5. Here the vertex is at $(-1, -2)$.

x	y
-4	7
-3	2
-2	-1
-1	-2
0	-1
1	2
2	7

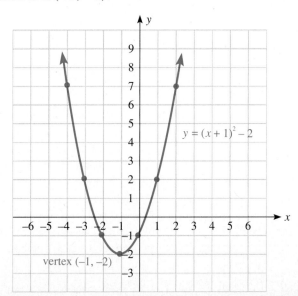

Figure 9.6.5

3. $y = -(x - 1)^2 - 1$

Some of the ordered pairs that satisfy the equation $y = -(x - 1)^2 - 1$ are given in the chart. Plot these points and smooth in the curve as in Figure 9.6.6. Here the vertex is at $(1, -1)$.

x	y
-2	-10
-1	-5
0	-2
1	-1
2	-2
3	-5
4	-10

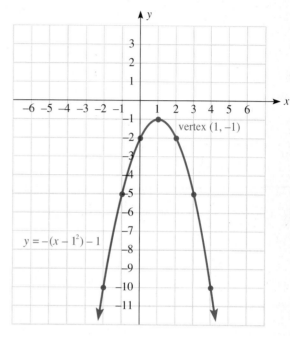

Figure 9.6.6

4. $y = -(x + 2)^2 + 5$

Some of the ordered pairs that satisfy the equation $y = -(x + 2)^2 + 5$ are given in the chart. Plot these points and smooth in the curve as in Figure 9.6.7. Here the vertex is at $(-2, 5)$.

x	y
-5	-4
-4	1
-3	4
-2	5
-1	4
0	1
1	-4

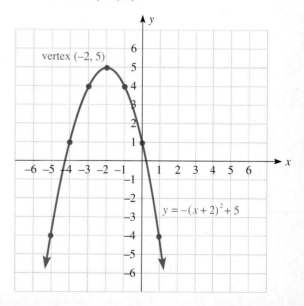

Figure 9.6.7

NOTE ▶   These examples illustrate the following useful fact.

Given a quadratic equation in the form

$$y = a(x - h)^2 + k, \, a \neq 0$$

the coordinates of the vertex are (h, k).

Plotting a lot of points is certainly not the most efficient method for determining the graph of a quadratic equation of the form $y = ax^2 + bx + c$. The following steps will enable us to sketch the graph with a minimum number of points.

Given a quadratic equation in the form

$$y = ax^2 + bx + c, \, a \neq 0$$

1. Use the completing the square process to place the equation in the form $y = a(x - h)^2 + k$.

2. We now know the vertex is at (h, k).

3. Determine some points on either side of the vertex.

4. Smooth in the parabola.

EXAMPLE 9.6.4

Sketch the graph of the following quadratic equation. Find the vertex by completing the square.

$y = x^2 - 6x + 11$

$y = (x^2 - 6x + \underline{\quad}) + 11 - \underline{\quad}$

[The blank is the number that must be added to $x^2 - 6x$ to form a perfect square trinomial. Recall from Section 9.3 that $\underline{\quad} = \left(\frac{1}{2} \text{ coefficient of } x\right)^2$.]

So $\underline{\quad} = \left[\frac{1}{2} \cdot (-6)\right]^2 = (-3)^2 = 9$

$y = (x^2 - 6x + 9) + 11 - 9$

Nine is added inside the parentheses to form a perfect square trinomial. Notice that since 9 is added inside the parentheses, 9 must be subtracted outside the parentheses to preserve the value of the right-hand side of the equation:

$$y = (x^2 - 6x + 9) + 2$$
$$y = (x - 3)^2 + 2$$

Now we know that the vertex is at (3, 2). We choose values of x on either side of $x = 3$ to obtain the ordered pairs in the chart. Plot these points and smooth in the curve as in Figure 9.6.8.

x	y
0	11
1	6
2	3
3	2
4	3
5	6
6	11

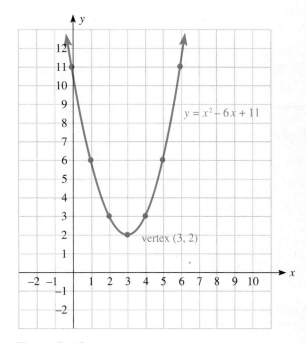

Figure 9.6.8

EXERCISES 9.6

Sketch the graphs of the following quadratic equations and determine the vertex of each parabola.

1. $y = x^2 + 1$　　　　　　　　　**2.** $y = x^2 + 3$

3. $y = x^2 - 5$　　　　　　　　　**4.** $y = x^2 - 1$

5. $y = -x^2 + 2$　　　　　　　　**6.** $y = -x^2 + 4$

7. $y = -x^2 - 3$　　　　　　　　**8.** $y = -x^2 - 6$

9. $y = (x + 4)^2$

10. $y = (x + 1)^2$

11. $y = (x + 3)^2 + 1$

12. $y = (x + 5)^2 + 3$

13. $y = (x - 2)^2 - 2$

14. $y = (x - 6)^2 - 1$

15. $y = (x - 8)^2 + 3$

16. $y = (x - 7)^2 + 4$

17. $y = -(x + 6)^2 - 1$

18. $y = -(x + 2)^2 - 3$

19. $y = -(x + 1)^2 + 2$

20. $y = -(x + 5)^2 + 1$

21. $y = -(x - 3)^2$

22. $y = -(x - 2)^2$

23. $y = -(x - 1)^2 - 2$

24. $y = -(x - 4)^2 - 3$

Sketch the graph of the following quadratic equations. Find the vertex by completing the square.

25. $y = x^2 - 4x + 8$

26. $y = x^2 - 6x + 10$

27. $y = x^2 - 2x - 2$

28. $y = x^2 - 10x + 23$

29. $y = x^2 - 6x + 9$

30. $y = x^2 - 8x + 16$

31. $y = x^2 + 6x + 10$

32. $y = x^2 + 8x + 18$

33. $y = x^2 + 2x - 5$

34. $y = x^2 + 10x + 24$

35. $y = -x^2 + 6x - 5$

36. $y = -x^2 + 4x + 1$

Chapter 9
Review

Arithmetic of Complex Numbers

Let a, b, c, and d be real numbers and $i = \sqrt{-1}$.

$$(a + bi) + (c + di) = (a + c) + (b + d)i$$

$$(a + bi) - (c + di) = (a - c) + (b - d)i$$

$$(a + bi)(c + di) = (ac - bd) + (bc + ad)i$$

$$\frac{a + bi}{c + di} = \frac{(a + bi) \cdot (c - di)}{(c + di) \cdot (c - di)} = \frac{(a + bi)(c - di)}{c^2 + d^2}$$

Solving Quadratic Equations

- *Extraction of roots property*: If $x^2 = p$, where p is any real number, then $x = \pm\sqrt{p}$.

- *Completing the square*: If $ax^2 + bx + c = 0$, where $a \neq 0$, then:

  1. Isolate the x terms; that is, rewrite the equation as $ax^2 + bx = -c$.

  2. Make the leading coefficient 1 by dividing both sides of the equation by a.

  3. Determine the number needed to complete the square on the x terms. [Remember that this number is $\left(\frac{1}{2}\text{ coefficient of } x\right)^2$.]

  4. Add the number found in step 3 to both sides of the equation.

  5. Rewrite the perfect square trinomial on the left-hand side as a binomial squared, and combine the numbers on the right-hand side. The equation now is in the form $(x + n)^2 = p$.

  6. Apply the extraction of roots property and solve for x.

- *Quadratic formula*: If $ax^2 + bx + c = 0$, where a, b, and c are real numbers with $a \neq 0$, then the solutions of the equation are

$$x = \frac{-b \pm \sqrt{b^2 - 4ac}}{2a}$$

- *Summary*: To decide which method is most efficient, consider:

Form	Method
1. $ax^2 + c = 0$ $a \neq 0, c \neq 0$	Use the extraction of roots property.
2. $ax^2 + bx = 0$ $a \neq 0, b \neq 0$	Use the factoring method.
3. $(mx + n)^2 = p$ $m \neq 0$	Use the extraction of roots property.
4. $ax^2 + bx + c = 0$ $a \neq 0, b \neq 0, c \neq 0$	**a.** The left-hand side is factorable. Use the factoring method. **b.** The left-hand side cannot be factored using integer coefficients, or you cannot determine the factorization. Use the quadratic formula.

Quadratic Equation in Two Variables

Given an equation in the form $y = ax^2 + bx + c$, $a \neq 0$, it can be changed to the form $y = a(x - h)^2 + k$ by completing the square in x.

1. The graph of the equation is a parabola.

2. The vertex is the point (h, k).

3. When $a > 0$, the parabola opens upward and the vertex is the lowest point on the parabola.

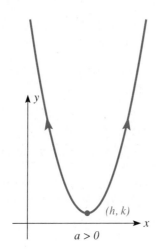

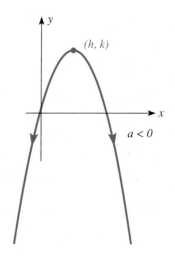

4. When $a < 0$, the parabola opens downward and the vertex is the highest point on the parabola.

Review Exercises

9.1 Express the following numbers in terms of i, and simplify.

1. $\sqrt{-81}$

2. $\sqrt{-60}$

3. $\sqrt{-\dfrac{4}{25}}$

4. $-\sqrt{-\dfrac{1}{36}}$

Perform the indicated operations. Express all numbers in the form $a + bi$.

5. $(-8 + 3i) + (5 + 9i)$

6. $(5 - 4i) - \left(\dfrac{3}{2} + 2i\right)$

7. $4i - (-11 - 7i)$

8. $\left(\dfrac{2}{3} - \dfrac{7}{5}i\right) - \left(\dfrac{1}{6} + \dfrac{11}{10}i\right)$

9. $-4(5 - 9i)$

10. $4i(3 - 8i)$

11. $(5 + 6i)(1 + 4i)$

12. $(-2 + 7i)(3 - i)$

13. $(9 + 2i)(9 - 2i)$

14. $(2 - 5i)^2$

15. $\dfrac{21 + i}{4 + i}$

16. $\dfrac{2 + i}{4 - 3i}$

17. $\dfrac{3 - 11i}{i}$

18. $\dfrac{4 + 13i}{-i}$

Express the following in the form $a + bi$, and simplify.

19. $\dfrac{-12 + \sqrt{-36}}{6}$

20. $\dfrac{10 - \sqrt{-16}}{8}$

9.2 Use the extraction of roots property to find the solution set of each quadratic equation.

21. $x^2 = 16$

22. $x^2 = 40$

23. $x^2 = -50$

24. $64x^2 - 9 = 0$

25. $49x^2 - 20 = 0$

26. $5x^2 - 3 = 0$

27. $9x^2 + 25 = 0$

28. $(x + 8)^2 = 81$

29. $\left(x - \frac{3}{4}\right)^2 = \frac{25}{16}$

30. $(x - 5)^2 = 14$

31. $(x + 1)^2 = 80$

32. $(x - 10)^2 = -25$

33. $(5x + 11)^2 = 36$

34. $(3x - 7)^2 = 60$

35. $(4x - 1)^2 = -9$

36. $(2x + 8)^2 = -4$

37. $(6x - 5)^2 + 11 = 27$

38. $(2x - 7)^2 - 5 = -25$

Use quadratic equations and the extraction of roots property to find the solutions of the following problems.

39. Five is subtracted from a number and the result squared equals 12. What is the number?

40. Nine is added to twice a number and the result squared equals 25. What is the number?

41. The sum of the areas of two squares is 180 square feet. If the length of one side of the larger square equals twice the length of one side of the smaller square, what are the lengths of the sides of each square? (*Hint:* Let x = length of a side of the smaller square; then $2x$ = length of a side of the larger square.)

42. A water balloon is dropped from the roof of a 490-foot-tall building. The equation that gives the distance the balloon has fallen is $d = 16t^2$, where d is the distance (measured in feet) and t is the time (measured in seconds). How long has the balloon been falling when it has traveled 400 feet? When will it hit the head of a 6-foot-tall man who is standing on the ground?

9.3 Determine the number that must be added to the following polynomials to form a perfect square trinomial. Express the resulting trinomial as the square of a binomial.

43. $x^2 - 12x$

44. $x^2 + 5x$

45. $x^2 + \frac{6}{7}x$

46. $x^2 - \frac{5}{4}x$

Find the solution set of each quadratic equation by completing the square.

47. $x^2 - 9x + 14 = 0$

48. $x^2 + 6x + 3 = 0$

49. $x^2 + 2x - 6 = 0$

50. $x^2 + x + 7 = 0$

51. $-x^2 + 10x - 1 = 0$

52. $3x^2 - 8x + 5 = 0$

53. $4x^2 - 12x + 1 = 0$

54. $4x^2 + 40x + 101 = 0$

Use quadratic equations and the completing the square method to find the solutions of the following problems.

55. The square of a number plus eight times the number is 5. What is the number?

56. Twice the square of a number minus two times the number is 9. What is the number?

57. The area of a rectangle is 8 square feet. The length is 2 feet more than five times the width. What is the *width* of the rectangle?

58. An egg is thrown vertically upward from ground level. The equation that gives the egg's height above ground level is $d = -16t^2 + 40t$, where d is the distance (measured in feet) and t is the time (measured in seconds). When does the egg hit the ground? (*Hint:* The egg hits the ground when $d = 0$.)

9.4 Place the following quadratic equations in standard form, and then identify a, b, and c.

59. $4x^2 - 7x = 0$

60. $5x^2 + 3x - 4 = 2x^2 + 8$

61. $(2x - 7)(x + 2) = 5$

Use the quadratic formula to find the solution set of each quadratic equation.

62. $x^2 + 3x - 7 = 5x + 8$

63. $3x(3x - 4) = (x - 3)(x + 1)$

64. $2x^2 + 9x + 8 = 0$

65. $3x^2 - 11 = 0$

66. $4x^2 + 5x - 11 = 3x^2 - 8$

67. $(x + 4)(2x - 5) = (3x - 1)(x - 2)$

68. $2x^2 + x + 8 = 0$

69. $4x^2 - 3x + 8 = 3x^2 + 2x - 5$

Use quadratic equations and the quadratic formula to find the solutions of the following problems.

70. The square of a number minus four times the number is 7. What is the number?

71. Twice the square of a number plus two times the number is -1. What is the number?

72. The area of a rectangle is 8 square feet. The length is 4 feet more than three times the width. What is the *width* of the rectangle?

73. A brick is thrown vertically upward from the roof of a building. The equation that gives the brick's height above ground level is $d = -16t^2 + 128t + 44$, where d is the distance (measured in feet) and t is the time (measured in seconds). How many seconds after the brick is thrown is it 300 feet high? (*Hint:* The brick is 300 feet high when $d = 300$.)

9.5 Using the most efficient method, find the solution set of each quadratic equation.

74. $x^2 + 6x - 2 = 0$

75. $(3x + 2)(x + 2) = (x + 3)(x - 3)$

76. $(2x + 7)^2 = -9$

77. $2x^2 + 7x - 9 = 0$

78. $3x^2 + 8x - 4 = 2x^2 + 4$

79. $3x^2 - 24 = 0$

80. $(2x - 5)(3x + 2) = -10$

81. $5(4x - 1) = (2x + 1)(x + 5)$

9.6 Sketch the graphs of the following quadratic equations and determine the vertex of each parabola.

82. $y = x^2 - 3$ **83.** $y = (x + 3)^2 + 2$ **84.** $y = -(x + 4)^2$ **85.** $y = -(x - 2)^2 + 5$

Sketch the graphs of the following quadratic equations. Find the vertex by completing the square.

Chapter 9 Test   (You should be able to complete this test in 60 minutes.)

I. Perform the indicated operations. Express all numbers in the form $a + bi$.

1. $(5 - 7i) + (-6 - 3i)$ **2.** $(-4 + 3i) - (8 + 4i)$ **3.** $(2 - 5i)(3 + i)$

4. $(4 + 7i)^2$ **5.** $\dfrac{4 - 9i}{i}$ **6.** $\dfrac{17 + i}{2 - 5i}$

II. Find the solution set of each quadratic equation by completing the square.

7. $x^2 - 12x + 28 = 0$ **8.** $5x^2 - 2x - 3 = 0$

III. Using the most efficient method, find the solution set of each quadratic equation.

9. $(x - 5)^2 = 12$ **10.** $(3x - 7)(x + 4) = -30$

11. $x^2 + 8x + 5 = 0$ **12.** $3x^2 - 5x + 5 = x^2 - 3x$

13. $(x + 5)(2x - 1) = (x - 2)(x - 1)$ **14.** $(2x - 5)^2 = 9x$

15. $8x^2 - 14x + 5 = 0$ **16.** $4x^2 + 1 = 0$

IV. Use quadratic equations to find the solutions of the following problems.

17. The sum of the areas of two squares is 153 square feet. If the length of one side of the larger square equals four times the length of one side of the smaller square, what are the lengths of the sides of each square? (*Hint:* Let x = length of a side of the smaller square; then $4x$ = length of a side of the larger square.)

18. Twice the square of a number minus two times the number is 5. What is the number?

V.

19. Sketch the graph and determine the vertex of the quadratic equation $y = (x + 1)^2 - 4$.

20. Sketch the graph of the quadratic equation $y = x^2 - 10x + 26$. Find the vertex by completing the square.

Test Your Memory

These problems review Chapters 1–9.

I. Sketch the graphs of the lines whose equations are given.

1. $y = 2x + 1$

2. $x = -1$

II. Sketch the graphs of the following quadratic equations and determine the vertex of each parabola.

3. $y = (x + 2)^2 + 1$

4. $y = x^2 - 6x + 7$ (*Hint:* Find the vertex by completing the square.)

III. Find the equations of the lines satisfying the given information. Write your answers in standard form.

5. Passing through $(1, 4)$ and $(-2, 4)$

6. Passing through $(9, 1)$ and $(-7, -3)$

IV. Find the solution set of each equation.

7. $3x - 4(2x - 1) = 5 - (x + 4)$

8. $\frac{1}{2} - \frac{1}{4}(x - 3) = 2\left(\frac{1}{4}x + \frac{1}{8}\right)$

9. $(2x + 3)(x + 6) = 18$

10. $(2x + 5)(x + 2) = 2(x + 7)$

11. $\frac{2x + 3}{x + 5} = \frac{7}{5}$

12. $\frac{x + 1}{4} = \frac{3}{2x - 3}$

13. $\frac{1}{x + 1} - \frac{2}{x + 2} = -\frac{1}{6}$

14. $\frac{x - 2}{x - 1} - \frac{4}{2x + 1} = \frac{6}{2x^2 - x - 1}$

15. $\sqrt{5x + 4} = 7$

16. $\sqrt{2x + 3} - 2x = 1$

17. $(x + 3)^2 = 4$
Use the extraction of roots property.

18. $(2x - 1)^2 = 7$
Use the extraction of roots property.

19. $3x^2 + 8x - 3 = 0$
Use the quadratic formula.

20. $x^2 - 2x + 10 = 0$
Use the quadratic formula.

V. Simplify each of the following. Be sure to reduce your answer to lowest terms.

21. $\dfrac{(4^{-1}x^{-4}y^{-3})^{-2}}{(4x^3y^{-4})^{-1}}$

22. $\dfrac{\dfrac{1}{x} + \dfrac{2x}{x-3}}{\dfrac{-2}{x} + \dfrac{6}{x-3}}$

VI. Simplify the following radicals. Assume that all variables represent positive real numbers.

23. $\sqrt[3]{-125}$

24. $\sqrt{98}$

25. $\sqrt{20x^5y^9}$

26. $\sqrt{\dfrac{12x^4}{49y^8}}$

VII. Perform the indicated operations and simplify your answers. Assume that all variables represent positive real numbers.

27. $\dfrac{4x^2 + 4x + 1}{2x^2 - 3x - 2} \div \dfrac{8x^3 + 28x^2 + 12x}{8x^3 - 16x^2}$

28. $\dfrac{x^3 + 3x^2}{x^2 - 4} - \dfrac{4x + 12}{x^2 - 4}$

29. $\dfrac{3}{2x} - \dfrac{4}{3x + 2}$

30. $\dfrac{2x + 9}{3x^2 - 7x - 20} + \dfrac{x - 12}{x^2 - 16}$

31. $5\sqrt{2x}(3\sqrt{6x} + 4\sqrt{y} - 5\sqrt{2y})$

32. $(\sqrt{2x} + \sqrt{5y})(3\sqrt{2x} - \sqrt{5y})$

33. $3\sqrt{80} - \sqrt{45} - 2\sqrt{20}$

34. $2x\sqrt{27x} + \sqrt{12x^3} - 3x\sqrt{48x}$

35. $(3 + 2i) - (4 - 7i)$

36. $(5 + 4i)^2$

37. $\dfrac{3 - 8i}{i}$

38. $\dfrac{9 - 2i}{4 + i}$

VIII. Rationalize the denominators of the following radical expressions. Assume that all variables represent positive real numbers.

39. $\sqrt{\dfrac{7y}{3x}}$

40. $\dfrac{\sqrt{3}}{\sqrt{15} - 2}$

IX. Determine the solutions of the following linear systems of equations. If there are no solutions, write *inconsistent*. If there are infinite solutions, write *dependent*.

41. $6x - 3y = 21$
$4x - 2y = 14$

42. $2x + y = 1$
$8x - 2y = 7$

X. Use algebraic expressions to find the solutions of the following problems.

43. The square of a number minus two times the number is 6. What is the number? (*Hint:* Use the quadratic formula.)

44. Tammy has $400 in tens and twenties. She has a total of 26 bills. How many of each type of bill does she have? (Use a linear system of equations to solve this problem.)

45. Andy's Apple Drink is 12% apple juice. Alan's Apple Delight is 28% apple juice. How many ounces of each drink must be used to make 40 ounces of a drink that is 18% apple juice? (Use a linear system of equations to solve this problem.)

46. The area of a rectangle is 56 square feet. The length is 2 feet more than three times the width. What are the dimensions of the rectangle?

47. Nick's Chinese spareribs recipe includes $\frac{1}{3}$ cup of honey for a 6-person serving. How many cups of honey will be required for a recipe serving 45 people?

48. Henry can wash his father's car in 20 minutes. Henry's sister Sarah can wash the car in 12 minutes. Working together, how long will it take Henry and Sarah to wash their father's car?

49. Maria takes her old motorboat 8 miles downstream. The trip downstream and back takes 2 hours. If the speed of the current is 3 mph, what is the speed of Maria's boat in still water?

50. The sum of the areas of two squares is 160 square feet. If the length of one side of the larger square equals three times the length of one side of the smaller square, what are the lengths of the sides of each square?

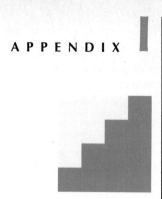

Functions

Fred works at the Moss County Cafeteria earning $6.00 an hour. He works 8 hours per day, for anywhere from one to five days a week. The relationship between the number of hours that he works (here indicated by the letter x) and his earnings (here indicated by the letter y) is represented in the following table.

x	8	16	24	32	40
y	48	96	144	192	240

From the above table we can construct the set {(8, 48), (16, 96), (24, 144), (32, 192), (40, 240)}. The number of hours Fred works and his associated earnings form the first and second coordinates of the ordered pairs in this set. Note that the number of hours that Fred works always is associated with a *unique* amount of earnings.

This example illustrates a very special type of set and leads us to the following definition.

Function

A **function** is a set of ordered pairs such that each first coordinate has one and only one second coordinate.

The set of all first coordinates is called the **domain** of the function. The set of all second coordinates is called the **range** of the function.

In our case the set $\{(8, 48), (16, 96), (24, 144), (32, 192), (40, 240)\}$ is an example of a function, because each first coordinate has one and only one second coordinate. The domain of this function is the set $\{8, 16, 24, 32, 40\}$, and the range is the set $\{48, 96, 144, 192, 240\}$.

EXAMPLE A.I.1

Determine which of the following sets of ordered pairs are functions.

1. $\{(1, 2), (3, 4), (5, 6), (7, 8)\}$

   This set *is* a function because each first coordinate has one and only one second coordinate.

2. $\{(1, 4), (3, 7), (2, 5), (-3, 8), (3, 9)\}$

   This set is *not* a function because the first coordinate 3 has two second coordinates.

3. $\{(1, 2), (2, 2), (3, 2), (4, 5)\}$

   This set *is* a function.

NOTE ▶ Here we have a second coordinate with three first coordinates. However, the definition requires only that each first coordinate have a unique second coordinate.

Quite often sets with an infinite number of ordered pairs are represented by an equation. When this happens it is common practice to let x indicate the first coordinate and y indicate the second coordinate. Whenever a function is represented by an equation, it is understood that *the domain will always be the largest subset of the real numbers that can be substituted for x into the equation and yield a real solution for y.* We will apply this rule in the following example.

EXAMPLE A.I.2

Find the domain of the functions defined by the following equations.

1. $y = \dfrac{9}{x - 3}$

   We can replace x with any real number except 3 and obtain a real value for y. When we substitute 3 for x we obtain

   $$y = \frac{9}{3 - 3}$$

   $$= \frac{9}{0} \qquad \text{Division by zero}$$

Recall that division by zero is undefined. Thus, the domain = {all real numbers except 3}.

2. $y = \sqrt{x}$

For y to be a real number the radicand must be nonnegative, so $x \geq 0$. Thus, the domain = {all real numbers ≥ 0}.

NOTE ▶   If $x = -1$, then $y = \sqrt{-1}$ so $y = i$ (not a real number). Thus, -1 cannot be in the domain.

3. $y = 6x$

The domain = {all real numbers}, since any value for x, when it is multiplied by 6, will yield a real value for y.

Sometimes we use letters to name functions. In our example at the start of this appendix we might choose to let

$$f = \{(8, 48), (16, 96), (24, 144), (32, 192), (40, 240)\}$$

After choosing a letter to name the function, we can use a special notation, called **functional notation**, to designate an element in the range of the function. To see how this notation is applied let us return to our example.

The notation $f(24)$, read "f of 24," is defined to be the second coordinate in the function f when the first coordinate is 24. Thus, in our example $f(24) = 144$.

In general the notation $f(x)$, read "f of x," is used to indicate the second coordinate in the function f when the first coordinate is x, that is, $f(x) = y$. Since $f(x)$ and y both represent the second coordinates in a function, then the equations $y = 6x$ and $f(x) = 6x$ are essentially the same. The $f(x)$ notation simply emphasizes the idea that we are dealing with a function and that the second coordinate depends upon the first coordinate. The following example shows us how to use function notation.

EXAMPLE A.I.3

Let $f(x) = 4x + 3$. Find $f(-4)$, $f(0)$, and $f(6)$.

1. $f(x) = 4x + 3$

       $f(-4) = 4(-4) + 3$ Replace each x with -4.

            $= -16 + 3$

            $= -13$

So $f(-4) = -13$.

This tells us that in function f when the first coordinate is -4 the second coordinate is -13.

2. $f(x) = 4x + 3$

$f(0) = 4 \cdot 0 + 3$ Replace each x with 0.

$= 0 + 3$

$= 3$

So $f(0) = 3$.

This tells us that in function f when the first coordinate is 0 the second coordinate is 3.

3. $f(x) = 4x + 3$

$f(6) = 4 \cdot 6 + 3$ Replace each x with 6.

$= 24 + 3$

$= 27$

So $f(6) = 27$.

This tells us that in function f when the first coordinate is 6 the second coordinate is 27.

In this example we chose the letter f to name our function. In the exercises we will use other letters (f, g, h, and k are common choices) to designate functions.

EXERCISES A.I

Determine which of the following sets of ordered pairs are functions.

1. $\{(0, 1), (-1, 2), (1, 4), (5, 9)\}$

2. $\{(3, 4), (-3, 5), (2, 3), (0, 0)\}$

3. $\{(9, 2), (3, 11), (9, 1)\}$

4. $\{(6, 5), (3, -4), (6, 6)\}$

5. $\{(9, 5), (5, 5), (0, 3), \left(\frac{1}{2}, \frac{1}{3},\right)\}$

6. $\{(1, 1), (2, 1), \left(\frac{1}{4}, \frac{1}{5}\right), (7, 11)\}$

7. $\{(3, -6)\}$

8. $\{(4, -8)\}$

Find the domain of the functions defined by the following equations.

9. $y = 3x$

10. $y = 2x$

11. $y = 8 - x$

12. $y = 7 - 5x$

13. $y = \dfrac{1}{x - 2}$

14. $y = \dfrac{5}{x - 1}$

15. $y = \dfrac{7}{x^2 - 25}$

16. $y = \dfrac{10}{x^2 - 36}$

17. $y = \dfrac{3x}{x^2 - 2x - 8}$

18. $y = \dfrac{2x}{x^2 + 2x - 15}$

19. $y = \sqrt{x - 3}$

20. $y = \sqrt{x - 2}$

21. $y = \sqrt{2x + 1}$

22. $y = \sqrt{2x + 5}$

23. $y = x^2 - 2x + 5$

24. $y = x^2 - 3x + 7$

In exercises 25–44, let $f(x) = 3x - 2$, $g(x) = 5x + 1$, and $k(x) = x^2 - 7$. Find the following.

25. $f(4)$ **26.** $f(2)$ **27.** $f(-3)$ **28.** $f(-7)$

29. $f\left(\frac{5}{2}\right)$ **30.** $f\left(\frac{1}{2}\right)$ **31.** $f(0)$ **32.** $g(0)$

33. $g\left(\frac{1}{10}\right)$ **34.** $g\left(\frac{3}{10}\right)$ **35.** $g\left(-\frac{1}{2}\right)$ **36.** $g\left(-\frac{1}{3}\right)$

37. $g(-6)$ **38.** $g(-4)$ **39.** $k(1)$ **40.** $k(7)$

41. $k(-1)$ **42.** $k(-7)$ **43.** $k(5)$ **44.** $k(2)$

The Pythagorean Theorem

In this appendix we will examine one of the basic theorems of classical geometry. First we must introduce some new terminology. A triangle that contains a 90° angle (a right angle) is called a **right triangle**. In a right triangle the longest side (the side opposite the 90° angle) is called the **hypotenuse**. The remaining sides are called the **legs** of the right triangle (see Figure A.II.1).

hypotenuse

leg

leg

Figure A.II.1

The following theorem is attributed to the Greek mathematician Pythagoras (ca. 580–500 B.C.).

The Pythagorean Theorem

If a and b are the lengths of the legs of a right triangle and c is the length of the hypotenuse (see Figure A.II.2), then

$$a^2 + b^2 = c^2$$

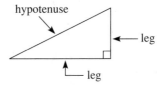

Figure A.II.2

NOTE ▶ The Pythagorean theorem states that in any right triangle the sum of the squares of the legs is equal to the square of the hypotenuse.

Legend has it that when Pythagoras first discovered this theorem he was so excited that he sacrificed 100 oxen to the gods. Since Pythagoras was in all likelihood a vegetarian, this is just an example of one of the many myths that surround him.

In the following examples we will use the Pythagorean theorem and the extraction of roots property to find the length of one side of a right triangle when the lengths of the other two sides are given.

EXAMPLE A.II.1

1. Find the length of the hypotenuse of a right triangle whose legs are 6 ft and 8 ft.

   Draw a sketch:

   Now apply the Pythagorean theorem:

   $$6^2 + 8^2 = c^2$$
   $$36 + 64 = c^2$$
   $$100 = c^2$$
   $$\pm\sqrt{100} = c \qquad \text{By the extraction of roots property}$$
   $$\pm 10 = c$$

   Since c represents the length of the hypotenuse of a right triangle, $c = 10$ ft.

2. Find the length of the hypotenuse of a right triangle whose legs are 1 cm and 7 cm.

   Draw a sketch:

   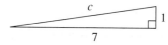

   Now apply the Pythagorean theorem:

   $$1^2 + 7^2 = c^2$$
   $$1 + 49 = c^2$$
   $$50 = c^2$$

$$\pm\sqrt{50} = c$$
$$\pm\sqrt{25}\sqrt{2} = c$$
$$\pm 5\sqrt{2} = c$$

Again, since c represents the length of the hypotenuse of a right triangle, $c = 5\sqrt{2}$ cm.

EXAMPLE A.II.2

In the following problems let a and b represent the lengths of the legs and c represent the length of the hypotenuse in a right triangle.

1. $a = 5$ in., $c = 13$ in., find b

Draw a sketch:

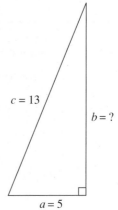

$$a^2 + b^2 = c^2$$
$$5^2 + b^2 = 13^2$$
$$25 + b^2 = 169$$
$$b^2 = 144$$
$$b = \pm\sqrt{144}$$
$$b = \pm 12$$

Since b represents the length of a leg of a right triangle, $b = 12$ in.

2. $b = 3$ m, $c = 6$ m, find a

Draw a sketch:

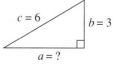

$$a^2 + b^2 = c^2$$
$$a^2 + 3^2 = 6^2$$
$$a^2 + 9 = 36$$
$$a^2 = 27$$
$$a = \pm\sqrt{27}$$
$$a = \pm\sqrt{9}\sqrt{3}$$
$$a = \pm 3\sqrt{3}$$

Since a represents the length of a leg of a right triangle, $a = 3\sqrt{3}$ m.

3. $a = 2$ mm, $c = \sqrt{14}$ mm, find b

Draw a sketch:

$$a^2 + b^2 = c^2$$
$$2^2 + b^2 = (\sqrt{14})^2$$
$$4 + b^2 = 14$$

$$b^2 = 10$$
$$b = \pm\sqrt{10}$$

Since b represents the length of a leg of a right triangle, $b = \sqrt{10}$ mm.

The following example illustrates an application of the Pythagorean theorem.

EXAMPLE A.II.3

A cable 30 ft long is attached to the top of a tower. When the cable is stretched tight and anchored at ground level, it is 18 ft from the base of the tower. How high is the tower?

Draw a sketch:

$$a^2 + b^2 = c^2$$
$$18^2 + b^2 = 30^2$$
$$324 + b^2 = 900$$
$$b^2 = 576$$
$$b = \pm\sqrt{576}$$
$$b = \pm24$$

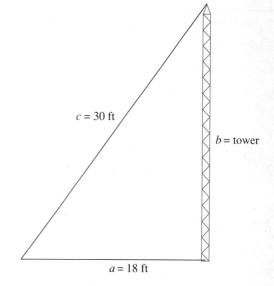

$c = 30$ ft

$b =$ tower

$a = 18$ ft

Thus, the height of the tower is 24 ft.

EXERCISES A.II

In the following exercises let a and b represent the lengths of the legs and c represent the length of the hypotenuse in a right triangle.

1. $a = 3$ in., $b = 4$ in., find c

2. $b = 24$ in., $c = 26$ in., find a

3. $b = 8$ ft, $c = 17$ ft, find a

4. $a = 9$ ft, $b = 12$ ft, find c

5. $a = 10$ yd, $b = 24$ yd, find c

6. $a = 4$ yd, $c = 5$ yd, find b

7. $b = 8$ m, $c = 17$ m, find a

8. $b = 30$ m, $c = 34$ m, find a

9. $a = 12$ cm, $c = 13$ cm, find b

10. $a = 16$ cm, $b = 30$ cm, find c

11. $a = 9$ mm, $b = 12$ mm, find c

12. $a = 5$ mm, $b = 12$ mm, find c

13. $a = 3$ m, $b = 3$ m, find c

14. $a = 2$ m, $b = 2$ m, find c

15. $a = \sqrt{6}$ cm, $c = \sqrt{13}$ cm, find b

16. $a = \sqrt{11}$ cm, $c = \sqrt{13}$ cm, find b

17. $a = \sqrt{7}$ mm, $b = 2\sqrt{5}$ mm, find c

18. $b = 3$ mm, $c = \sqrt{13}$ mm, find a

19. $b = 2$ in., $c = \sqrt{53}$ in., find a

20. $a = \sqrt{10}$ in., $b = 3\sqrt{2}$ in., find c

21. $a = 6$ ft, $c = 6\sqrt{2}$ ft, find b

22. $a = 5$ ft, $c = 5\sqrt{2}$ ft, find b

23. $a = 2\sqrt{3}$ yd, $b = 2$ yd, find c

24. $a = 3\sqrt{3}$ yd, $b = 3$ yd, find c

Use the Pythagorean theorem to find the solution of each of the following problems.

25. Trinh is flying his favorite kite. When 39 ft of string has run out, the kite is caught in the top of "The Kite-Eating Tree." When the string is stretched tight and anchored at ground level, it is 15 ft from the base of the tree. How high is "The Kite-Eating Tree"?

26. Elvira is flying her favorite kite. When 34 ft of string has run out, the kite is caught in the top of Mr. Wilson's oak tree. When the string is stretched tight and anchored at ground level, it is 16 ft from the base of the oak tree. How high is Mr. Wilson's oak tree?

27. Ginny leaves the Moss County Courthouse at 1:00 P.M. in her Model-T, traveling north at a rate of 24 mph. At the same time, Penny leaves the courthouse traveling west on her bicycle at a rate of 10 mph. At 2:00 P.M., what is the distance between Ginny and Penny?

28. Winston leaves school in a car traveling north at a rate of 48 mph. At the same time, Donegal leaves school in a car traveling east at a rate of 36 mph. After 1 hour, what is the distance between Winston and Donegal?

APPENDIX **III**

Tables

Table 1
Powers of Some
Natural Numbers

n	n^3	n^4	n^5	n^6	n^7
1	1	1	1	1	1
2	8	16	32	64	128
3	27	81	243	729	2187
4	64	256	1024	4096	
5	125	625	3125		
6	216	1296			
7	343	2401			
8	512	4096			
9	729	6561			
10	1000	10,000	100,000	1,000,000	10,000,000

Table 2
Squares and Square Roots of Natural Numbers from 1 to 100

n	n^2	$\sqrt{n}$	$\sqrt{10n}$	n	n^2	$\sqrt{n}$	$\sqrt{10n}$
1	1	1.000	3.162	51	2601	7.141	22.583
2	4	1.414	4.472	52	2704	7.211	22.804
3	9	1.732	5.477	53	2809	7.280	23.022
4	16	2.000	6.325	54	2916	7.348	23.238
5	25	2.236	7.071	55	3025	7.416	23.452
6	36	2.449	7.746	56	3136	7.483	23.664
7	49	2.646	8.367	57	3249	7.550	23.875
8	64	2.828	8.944	58	3364	7.616	24.083
9	81	3.000	9.487	59	3481	7.681	24.290
10	100	3.162	10.000	60	3600	7.746	24.495
11	121	3.317	10.488	61	3721	7.810	24.698
12	144	3.464	10.954	62	3844	7.874	24.900
13	169	3.606	11.402	63	3969	7.937	25.100
14	196	3.742	11.832	64	4096	8.000	25.298
15	225	3.873	12.247	65	4225	8.062	25.495
16	256	4.000	12.649	66	4356	8.124	25.690
17	289	4.123	13.038	67	4489	8.185	25.884
18	324	4.243	13.416	68	4624	8.246	26.077
19	361	4.359	13.784	69	4761	8.307	26.268
20	400	4.472	14.142	70	4900	8.367	26.458
21	441	4.583	14.491	71	5041	8.426	26.646
22	484	4.690	14.832	72	5184	8.485	26.833
23	529	4.796	15.166	73	5329	8.544	27.019
24	576	4.899	15.492	74	5476	8.602	27.203
25	625	5.000	15.811	75	5625	8.660	27.386
26	676	5.099	16.125	76	5776	8.718	27.568
27	729	5.196	16.432	77	5929	8.775	27.749
28	784	5.292	16.733	78	6084	8.832	27.928
29	841	5.385	17.029	79	6241	8.888	28.107
30	900	5.477	17.321	80	6400	8.944	28.284
31	961	5.568	17.607	81	6561	9.000	28.460
32	1024	5.657	17.889	82	6724	9.055	28.636
33	1089	5.745	18.166	83	6889	9.110	28.810
34	1156	5.831	18.439	84	7056	9.165	28.983
35	1225	5.916	18.708	85	7225	9.220	29.155
36	1296	6.000	18.974	86	7396	9.274	29.326
37	1369	6.083	19.235	87	7569	9.327	29.496
38	1444	6.164	19.494	88	7744	9.381	29.665
39	1521	6.245	19.748	89	7921	9.434	29.833
40	1600	6.325	20.000	90	8100	9.487	30.000
41	1681	6.403	20.248	91	8281	9.539	30.166
42	1764	6.481	20.494	92	8464	9.592	30.332
43	1849	6.557	20.736	93	8649	9.644	30.496
44	1936	6.633	20.976	94	8836	9.695	30.659
45	2025	6.708	21.213	95	9025	9.747	30.822
46	2116	6.782	21.448	96	9216	9.798	30.984
47	2209	6.856	21.679	97	9409	9.849	31.145
48	2304	6.928	21.909	98	9604	9.899	31.305
49	2401	7.000	22.136	99	9801	9.950	31.464
50	2500	7.071	22.361	100	10000	10.000	31.623

Answers to Odd-Numbered Problems

Chapter 1   EXERCISES 1.1 (p. 6)

1. $\{8\}$ **3.** $\{6, 9\}$ **5.** $\varnothing$ **7.** $\{1, 3, 5, 7, \ldots\}$ **9.** $\{4, 6, 7, 8, 9, 10, 12, 16, 20\}$
11. $\{7, 8, 10, 0, 3, 6, 9, 12, 15, \ldots\}$ **13.** $\{\ldots, -5, -3, -1, 1, 2, 3, 4, \ldots\}$ **15.** $\varnothing$
17. $\{4, 6, 8, 9, 12, 16, 20\}$ **19.** $\{6, 7, 9\}$ **21.** $0.6666\ldots$ **23.** $0.4444\ldots$ **25.** $0.181818\ldots$
27. $1.363636\ldots$ **29.** 0.75 **31.** 1.375 **33.** 0.3125 **35.** $0.\overline{428571}$ **37.** N, W, I, Q, R
39. Q, R **41.** H, R **43.** Q, R **45.** H, R **47.** Belongs to no set (since $\frac{5}{0}$ is undefined)
49. True   **51.** False   **53.** True   **55.** True

EXERCISES 1.2 (p. 9)

1.

3.

5.

7.

9.

11.

13. 6   **15.** π **17.** 21   **19.** 11   **21.** $>$ **23.** $<$ **25.** $<$ **27.** $>$ **29.** $>$ **31.** $>$
33. $>$ **35.** $>$ **37.** $=$ **39.** $=$

EXERCISES 1.3 (p. 19)

1. Commutative property of addition, Associative property of addition, *, *
3. Commutative property of addition, Associative property of addition, *, Inverse property of addition   **5.** $\frac{5}{6}$
7. $\frac{4}{5}$ **9.** $\frac{1}{3}$ **11.** 3   **13.** $-\frac{10}{3}$ **15.** $\frac{4}{3}$ **17.** -3 **19.** 27   **21.** -89 **23.** -33 **25.** -60

27. 6　　**29.** $\frac{8}{11}$　　**31.** $\frac{5}{4}$　　**33.** $\frac{3}{2}$　　**35.** $-\frac{17}{45}$　　**37.** $\frac{61}{24}$　　**39.** $\frac{55}{12}$　　**41.** $\frac{19}{15}$　　**43.** $-\frac{11}{18}$　　**45.** $\frac{35}{132}$
47. $-\frac{4}{9}$　　**49.** $\frac{109}{15}$　　**51.** $\frac{344}{63}$　　**53.** -8.203　　**55.** 57.15　　**57.** -23.863　　**59.** 27.44 cm　　**61.** $91\frac{2}{3}$ ft
63. $15\frac{1}{3}$ yd

EXERCISES 1.4 (p. 24)

1. -53　　**3.** -22　　**5.** 17　　**7.** 17.19　　**9.** -32.22　　**11.** $-\frac{11}{14}$　　**13.** $-\frac{23}{75}$　　**15.** -19　　**17.** -27
19. $\frac{91}{36}$　　**21.** $\frac{17}{30}$　　**23.** $\frac{95}{56}$　　**25.** 51　　**27.** 6.64　　**29.** $\frac{29}{9}$　　**31.** $-\frac{25}{8}$　　**33.** -30　　**35.** $-\frac{93}{34}$　　**37.** $\frac{100}{39}$
39. -61　　**41.** 23.82　　**43.** -5　　**45.** 103.3°　　**47.** 28.92°
49. $-85.31 - 30.00 + 426.50 - 75.00 - 16.38 - 295.99 = -76.18$
Mary is overdrawn again with an account balance of $-\$76.18$.

EXERCISES 1.5 (p. 33)

1. Associative property of multiplication, Inverse property of multiplication, Identity property of multiplication
3. Inverse property of addition, Zero product property
5. Associative property of addition, *, Distributive property, *. *　　**7.** $5 \cdot 4 + 5 \cdot 7$
9. $3 \cdot 12 - 3 \cdot 4 - 3 \cdot 2$　　**11.** $2(7 + 3)$　　**13.** $4(3 + 5 - 2)$　　**15.** $7(11 + 4 + 1)$　　**17.** -80
19. 220　　**21.** -81　　**23.** 18,000　　**25.** $\frac{99}{10}$　　**27.** $-\frac{1}{6}$　　**29.** $-\frac{10}{7}$　　**31.** $-\frac{19}{5}$　　**33.** 20　　**35.** -39.37
37. 12.078　　**39.** $-726,000$　　**41.** -32　　**43.** -16　　**45.** $\frac{1}{18}$　　**47.** $-\frac{149}{42}$　　**49.** $-\frac{1}{10}$　　**51.** $-\frac{3}{4}$
53. -57　　**55.** 88　　**57.** -91　　**59.** -44.819　　**61.** 97.41　　**63.** $-\frac{4}{3}$　　**65.** $\frac{1}{5}$　　**67.** $-\frac{21}{4}$　　**69.** -29
71. $\frac{13}{16}$　　**73.** $28\frac{1}{3}$ feet　　**75.** 22.5675 mm²　　**77.** \$1096.88

EXERCISES 1.6 (p. 40)

1. -7　　**3.** 8　　**5.** -16　　**7.** -6　　**9.** 23　　**11.** 0　　**13.** Indeterminate　　**15.** $-\frac{7}{3}$　　**17.** $-\frac{12}{35}$
19. $\frac{5}{3}$　　**21.** $-\frac{5}{2}$　　**23.** $-\frac{17}{36}$　　**25.** 15　　**27.** $\frac{9}{8}$　　**29.** $-\frac{5}{9}$　　**31.** $\frac{20}{3}$　　**33.** 3　　**35.** -30　　**37.** 5
39. 83　　**41.** $-\frac{38}{15}$　　**43.** 5　　**45.** 3　　**47.** 0　　**49.** $-\frac{2}{5}$　　**51.** $-\frac{65}{22}$　　**53.** 7　　**55.** 4　　**57.** -6
59. $-\frac{3}{10}$　　**61.** $-\frac{1}{15}$　　**63.** $\frac{14}{33}$　　**65.** $25\frac{3}{5}$ mpg　　**67.** $54\frac{6}{11}$ mph　　**69.** $8\frac{1}{4}$ min/mile

EXERCISES 1.7 (p. 43)

1. 216　　**3.** 32　　**5.** 49　　**7.** -64　　**9.** $\frac{4}{25}$　　**11.** 9　　**13.** 5　　**15.** 1　　**17.** 10　　**19.** 6　　**21.** 7
23. 11　　**25.** 15　　**27.** -8　　**29.** -6　　**31.** 0　　**33.** $\frac{1}{108}$　　**35.** $\frac{35}{4}$　　**37.** 0　　**39.** 40　　**41.** -6
43. 3　　**45.** -4　　**47.** $\frac{5}{2}$　　**49.** $-\frac{1}{9}$　　**51.** 24.9π in²

Chapter 1　　REVIEW EXERCISES (p. 44)

1. $\{3, 4, 5, 6, 7, 8, 9, 12, 15\}$　　**3.** $\{3, 6, 9\}$　　**5.** $\{3, 4, 6, 7, 9, 10, 12, 13, 15, 16\}$　　**7.** $1.\overline{3}$　　**9.** 1.625
11. Q, R　　**13.** H, R　　**15.** True　　**17.** True

19. 　　**21.** 　　**23.**

25. 15　　**27.** 8　　**29.** $<$　　**31.** $>$　　**33.** $<$
35. Commutative property of addition, Associative property of addition, Inverse property of addition　　**37.** $\frac{2}{5}$
39. $\frac{1}{6}$　　**41.** -81　　**43.** $\frac{10}{13}$　　**45.** $\frac{19}{15}$　　**47.** $\frac{35}{72}$　　**49.** $\frac{61}{10}$　　**51.** $12\frac{17}{18}$ in　　**53.** -36　　**55.** 90.39
57. $-\frac{47}{30}$　　**59.** 19　　**61.** $-\frac{6}{7}$　　**63.** $-\frac{57}{8}$　　**65.** -86

67. Distributive property, Identity property of multiplication, Inverse property of multiplication, Associative property of addition, Inverse property of addition, Identity property of addition, Identity property of multiplication
69. $5 \cdot 6 - 5 \cdot 8 + 5 \cdot 3$ **71.** $7(2 + 4 - 11)$ **73.** -125 **75.** $\frac{45}{4}$ **77.** $\frac{1}{2}$ **79.** -19 **81.** $\frac{59}{20}$
83. 88 **85.** $-\frac{34}{15}$ **87.** $-\frac{31}{20}$ **89.** $\frac{9}{5}$ **91.** 39.95 m² **93.** 14 **95.** $-\frac{8}{75}$ **97.** -64 **99.** $\frac{14}{225}$
101. 51 **103.** -47 **105.** $-\frac{4}{7}$ **107.** -3 **109.** $-\frac{2}{11}$ **111.** 36 mpg **113.** 8 min/mile
115. 1000 **117.** -1 **119.** 13 **121.** 15 **123.** -32 **125.** $\frac{15}{14}$ **127.** -5 **129.** 4 **131.** $\frac{1}{11}$

Chapter 1   TEST (p. 49)

1. {2, 4} **3.** True **5.** *I, Q, R* **7.** 11 **9.** $>$ **11.** $4(9 + 5)$ **13.** $21\frac{1}{2}$ ft **15.** $6\frac{1}{2}$ min/mile
17. 8 **19.** -46 **21.** -7.22 **23.** $-\frac{5}{4}$ **25.** $\frac{20}{33}$ **27.** -4 **29.** $-\frac{21}{16}$ **31.** -2 **33.** -6.556
35. $-\frac{73}{12}$ **37.** -197 **39.** -4 **41.** 4

Chapter 2   EXERCISES 2.1 (p. 56)

1. 23 **3.** -20 **5.** 14 **7.** -34 **9.** 27 **11.** 0 **13.** 0 **15.** $-\frac{33}{2}$ **17.** -16 **19.** -7
21. 49 **23.** 65 **25.** -28 **27.** 2 **29.** $-\frac{1}{12}$ **31.** $\frac{11}{12}$ **33.** $-\frac{11}{6}$ **35.** 28.119 **37.** $-\frac{1}{4}$
39. 6 **41.** $11x$ **43.** $4x$ **45.** $-10x + 5$ **47.** $-2x - y$ **49.** $6x^2 - 2x + 3z$ **51.** $5x + 2y - 7$
53. $9x^2 - 1$ **55.** $\frac{3}{8}x - 1$ **57.** $-5.698y$ **59.** $\frac{7}{6}x^2 - \frac{1}{4}y$ **61.** $11x - 18$ **63.** $-3x - 4$
65. $9x - 9$ **67.** $x - 17$ **69.** $5x - 16$ **71.** $2x + 2y$ **73.** $2x + 2y + 3a + 3b$ **75.** $-\frac{1}{6}x + \frac{13}{6}$
77. $.19x + .01$ **79.** $8x + 3$ **81.** \$83 **83.** $P = 3x + 7$ **85.** $P = 5x + 5$ **87.** $P = 4x + 18$
89. $P = 6x + 6$

EXERCISES 2.2 (p. 62)

1. {17} **3.** {13} **5.** {-10} **7.** {-2} **9.** {$\frac{7}{8}$} **11.** {-4} **13.** {8.11} **15.** {-4.62}
17. {-5} **19.** {-10} **21.** {10} **23.** {$-\frac{7}{6}$} **25.** {0} **27.** {11} **29.** {-7} **31.** {-8}
33. {3} **35.** {-7} **37.** {1} **39.** {-3} **41.** {-7} **43.** {3.63} **45.** {0} **47.** {9} **49.** {$\frac{1}{6}$}
51. {-5.3} **53.** {2} **55.** {-3} **57.** {$\frac{7}{8}$} **59.** {-9} **61.** {$-\frac{11}{5}$} **63.** {-2} **65.** {9}
67. {31} **69.** {$\frac{5}{6}$} **71.** {-16} **73.** {7} **75.** {8} **77.** {4} **79.** {-28} **81.** {-40} **83.** {-4}
85. {3} **87.** {-12} **89.** {-49}

EXERCISES 2.3 (p. 68)

1. {8} **3.** {15} **5.** {-3} **7.** {-12} **9.** {-18} **11.** {0} **13.** {$\frac{1}{3}$} **15.** {$\frac{3}{2}$} **17.** {$-\frac{5}{28}$}
19. {$\frac{8}{3}$} **21.** {$-\frac{5}{8}$} **23.** {$-\frac{3}{28}$} **25.** {4} **27.** {$\frac{1}{9}$} **29.** {1} **31.** {$-\frac{2}{21}$} **33.** {$-\frac{5}{9}$} **35.** {7}
37. {0} **39.** {$-\frac{5}{8}$} **41.** {$\frac{9}{10}$} **43.** {6} **45.** {3} **47.** {-4} **49.** {3} **51.** {3} **53.** {-2}
55. {-1} **57.** {3} **59.** {$-\frac{3}{2}$} **61.** {-3} **63.** {-6} **65.** {$\frac{5}{3}$} **67.** {3} **69.** {-2} **71.** {-2}
73. {-12} **75.** {3} **77.** {$-\frac{3}{2}$} **79.** {-7} **81.** {$\frac{6}{7}$} **83.** {-24} **85.** {$\frac{3}{7}$} **87.** {3} **89.** {-5}
91. {$\frac{3}{8}$} **93.** {-2} **95.** {$\frac{3}{2}$} **97.** {-20} **99.** 109 miles **101.** \$12,424

EXERCISES 2.4 (p. 75)

1. {-16} **3.** {$\frac{8}{3}$} **5.** {$\frac{2}{5}$} **7.** {$\frac{3}{2}$} **9.** All real numbers **11.** {$-\frac{7}{4}$} **13.** {0} **15.** {-1}
17. {$\frac{11}{10}$} **19.** $\varnothing$ **21.** {$\frac{7}{3}$} **23.** {14} **25.** {3.4} **27.** {-3} **29.** {5} **31.** {$\frac{1}{3}$}
33. All real numbers **35.** {$\frac{8}{5}$} **37.** {$-\frac{2}{3}$} **39.** {-6.1} **41.** {-1} **43.** $\varnothing$ **45.** {14} **47.** {-2}
49. {$\frac{3}{2}$} **51.** {-2} **53.** {-2.1} **55.** {$-\frac{2}{3}$} **57.** {2} **59.** {-4} **61.** $\varnothing$ **63.** {$\frac{8}{3}$} **65.** {5}
67. {-12}

EXERCISES 2.5 (p. 80)

1. $\{-5, 5\}$ **3.** $\{-\frac{2}{3}, \frac{2}{3}\}$ **5.** $\varnothing$ **7.** $\varnothing$ **9.** $\{0\}$ **11.** $\{-5, 5\}$ **13.** $\{-\frac{3}{10}, \frac{3}{10}\}$ **15.** $\{-9, 9\}$
17. $\{-\frac{20}{3}, \frac{20}{3}\}$ **19.** $\{-\frac{2}{3}, \frac{2}{3}\}$ **21.** $\varnothing$ **23.** $\{-\frac{18}{25}, \frac{18}{25}\}$ **25.** $\{-12, 6\}$ **27.** $\{-3, 7\}$ **29.** $\{1, 15\}$
31. $\{\frac{13}{4}, \frac{19}{4}\}$ **33.** $\varnothing$ **35.** $\{-7, 2\}$ **37.** $\{-\frac{13}{4}, \frac{3}{4}\}$ **39.** $\{-\frac{11}{5}, \frac{13}{5}\}$ **41.** $\{\frac{29}{12}, \frac{31}{12}\}$ **43.** $\{-\frac{10}{3}, -\frac{14}{3}\}$
45. $\{\frac{1}{2}, \frac{11}{2}\}$ **47.** $\{-\frac{1}{5}, 1\}$ **49.** $\varnothing$ **51.** $\{\frac{1}{4}\}$ **53.** $\{6, 30\}$ **55.** $\{\frac{11}{2}, \frac{29}{2}\}$ **57.** $\{8\}$ **59.** $\{-8, 2\}$
61. $\varnothing$ **63.** $\{-5, 1\}$ **65.** $\{-13, 1\}$ **67.** $\{-7, 15\}$ **69.** $\{-8\}$ **71.** $\{1, 9\}$ **73.** $\{2, 16\}$
75. $\{3, 13\}$ **77.** $\varnothing$ **79.** $\{-1, 4\}$ **81.** $\{\frac{3}{4}, \frac{9}{4}\}$ **83.** $\{-\frac{11}{2}, \frac{1}{2}\}$ **85.** $\varnothing$ **87.** $\{-2, \frac{6}{5}\}$
89. $\{9, -15\}$ **91.** $\{-4, \frac{28}{3}\}$ **93.** $\{-\frac{1}{6}, \frac{1}{2}\}$ **95.** $\{-\frac{5}{6}, \frac{1}{2}\}$ **97.** $\varnothing$

EXERCISES 2.6 (p. 83)

1. $L = \dfrac{A}{W}$ **3.** $a = \dfrac{F}{M}$ **5.** $r = \dfrac{C}{2\pi}$ **7.** $T = \dfrac{I}{PR}$ **9.** $H = \dfrac{V}{LW}$ **11.** $B = \dfrac{2A}{H}$ **13.** $M = \dfrac{Fr^2}{Gm}$

15. $M = \dfrac{Fr}{v^2}$ **17.** $T = \dfrac{PV}{nR}$ **19.** $m = \dfrac{y - b}{x}$ **21.** $h = \dfrac{S - 2\pi r^2}{2\pi r}$ **23.** $y = \dfrac{7 - 3x}{2}$

25. $y = \dfrac{2x - 12}{7}$ **27.** $y = 2x - z$ **29.** $b = 3a + c$ **31.** $g = \dfrac{2s - 7t}{5}$ **33.** $R = \dfrac{S - \pi rs}{\pi s}$

35. $b = \dfrac{2A - Bh}{h}$ **37.** $a = \dfrac{v - r}{t}$ **39.** $s = \dfrac{v^2 - r^2}{2a}$ **41.** $R = \dfrac{D}{T}, R = 45$ **43.** $s = \dfrac{W}{F}, s = \frac{4}{3}$

45. $B = \dfrac{2A}{H}, B = 24$ **47.** $L = \dfrac{V}{WH}, L = 5$ **49.** $L = \dfrac{P - 2W}{2}, L = 13$ **51.** $x = \dfrac{y - b}{m}, x = -4$

53. $212°$ **55.** $K = \dfrac{F}{x}, K = \frac{7}{3}$ **57.** $b = \dfrac{2A}{h} - B, b = 8$ **59.** $W = \dfrac{2 - 2LH}{2L + 2H}, W = 4$

EXERCISES 2.7 (p. 91)

1. $x + 2$ **3.** $4x$ **5.** $x - 13$ **7.** $\dfrac{x}{4}$ **9.** $x + 5$ **11.** $\frac{1}{2}x$ **13.** $\dfrac{12}{x}$ **15.** $3x + 2$ **17.** $4 - 7x$
19. $5(x + 9)$ **21.** 6 **23.** $\frac{1}{2}$ **25.** -2 **27.** 12, 13, 14 **29.** $-22, -21, -20$ **31.** 29, 31, 33
33. $-12, -10, -8$ **35.** 12, 36, and 16 in **37.** 9, 12, and 16 in **39.** 56, 14, and 28 cm
41. 15 by 22 ft **43.** 9 by 31 ft **45.** 13 by 19 ft **47.** 14 fives, 5 tens **49.** 26 nickels, 13 quarters
51. 11 ones, 17 fives, 7 tens **53.** 12 nickels, 9 dimes, 16 quarters **55.** 6 fives, 13 tens, 2 twenties
57. 22 nickels, 14 quarters **59.** 18 fives, 6 twenties **61.** 25 ones, 8 fives, 15 tens

EXERCISES 2.8 (p. 99)

1. 1.5 hours **3.** 5:00 P.M. **5.** 4.2 hours **7.** 1.4 hours **9.** 4:00 P.M. **11.** 5:30 P.M.
13. Bond–$5,000, Passbook–$3,000 **15.** Retirement–$2,000, Bond–$8,000
17. Certificate–$6,000, Money market–$5,000 **19.** Checking–$2,000, Savings–$5,000
21. Retirement–$9,000, Lifeline–$3,000 **23.** Passbook–$6,000, Retirement–$2,000
25. Big E–12 oz, Big M–6 oz **27.** Bald Away–14 oz, Old Skin Head–9 oz
29. Sweetie–6 oz, This Might be Nectar–10 oz **31.** Bobo's–10 pts, Red's–14 pts
33. Texas–8 pounds, Georgia–6 pounds **35.** West Texas–13 gallons, Moss County–5 gallons

EXERCISES 2.9 (p. 109)

1. $x < 2$ **3.** $x \geq 7$ **5.** $x \leq \frac{1}{15}$

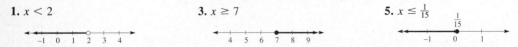

7. $x > 10$

9. $-2 < x < 2$

11. $4 \leq x \leq 10$

13. $-\frac{19}{8} \leq x \leq -\frac{15}{8}$

15. $x > 3$

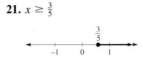

17. $x \leq -\frac{2}{3}$

19. $x < \frac{2}{5}$

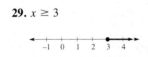

21. $x \geq \frac{3}{5}$

23. $-\frac{4}{3} < x < 2$

25. $-\frac{1}{2} \leq x \leq \frac{5}{4}$

27. $-8 < x < -4$

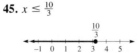

29. $x \geq 3$

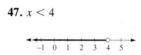

31. $x < 4$

33. $x \geq -\frac{1}{3}$

35. $x > -2$

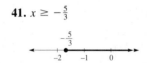

37. $x > \frac{5}{2}$

39. $x \leq \frac{1}{2}$

41. $x \geq -\frac{5}{3}$

43. $x < 4$

45. $x \leq \frac{10}{3}$

47. $x < 4$

49. $x \leq \frac{9}{4}$

51. $x < -1$

53. $x > 0$

55. $x \leq 6$

57. $x \leq 3$

59. $x > -\frac{1}{2}$

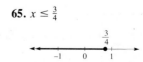

61. $x < -2$

63. $x \geq 2$

65. $x \leq \frac{3}{4}$

67. $x < -5$

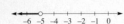

69. $x > \frac{4}{5}$

71. $2 < x < 3$

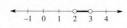

73. $1 \le x \le \frac{12}{5}$

75. $1 \le x \le 6$

77. $-3 < x < 0$

79. $-\frac{3}{2} < x < \frac{1}{2}$

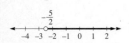

Chapter 2   REVIEW EXERCISES (p. 111)

1. 23   **3.** $-\frac{165}{4}$ **5.** $-\frac{58}{3}$ **7.** $-\frac{3}{5}$ **9.** -8 **11.** $8x - 11$ **13.** $\frac{1}{8}x^2 - \frac{2}{9}y^2$ **15.** $7x + 7$
17. $33x - 9y$ **19.** $\frac{5}{12}x + \frac{17}{12}$ **21.** $\{-5\}$ **23.** $\{-18\}$ **25.** $\{2\}$ **27.** $\{0\}$ **29.** $\{\frac{5}{4}\}$ **31.** $\{8\}$
33. $\{0\}$ **35.** $\{4\}$ **37.** $\{-6\}$ **39.** $\{-5\}$ **41.** $\{-7\}$ **43.** $\{6\}$ **45.** $\{\frac{25}{2}\}$ **47.** $\{12\}$ **49.** $\{\frac{3}{2}\}$
51. $\{\frac{12}{5}\}$ **53.** $\{-2\}$ **55.** $\{3\}$ **57.** $\{-6\}$ **59.** $\{-\frac{27}{8}\}$ **61.** $\{\frac{9}{7}\}$ **63.** $\{\frac{4}{9}\}$ **65.** $\{2\}$ **67.** $\{\frac{1}{3}\}$
69. $\{\frac{4}{5}\}$ **71.** $\{0.5\}$ **73.** $\{0\}$ **75.** $\{\frac{7}{5}\}$ **77.** $\{-7, 7\}$ **79.** $\{-12, 12\}$ **81.** $\{\frac{5}{4}, \frac{11}{4}\}$
83. $\{-\frac{11}{3}, -\frac{1}{3}\}$ **85.** $\varnothing$ **87.** $\{-6, 16\}$ **89.** $\{0, 14\}$ **91.** $\{-\frac{7}{4}, \frac{3}{4}\}$ **93.** $\{-\frac{5}{2}, \frac{13}{2}\}$ **95.** $\{-\frac{1}{4}, \frac{5}{12}\}$

97. $R = \frac{I}{PT}$ **99.** $p = \frac{3n - 5m}{4}$ **101.** $y = \frac{A - 2x^2}{4x}$ or $y = \frac{A}{4x} - \frac{x}{2}$ **103.** $R = \frac{D}{T}$, $R = 42$

105. $L = \frac{P - 2W}{2}$, $L = 12$ **107.** $k = \frac{F}{x}$, $k = \frac{5}{8}$ **109.** -2 **111.** 21, 22, 23

113. 9 in., 19 in., 6 in.   **115.** $W = 7$ m, $L = 15$ m   **117.** 10 nickels, 45 quarters
119. 27 quarters, 12 dimes   **121.** $\frac{3}{4}$ of an hour   **123.** 7:00 P.M.
125. \$1200 in checking account, \$700 in savings account   **127.** \$1350 in 8% CD, \$3650 in 9% CD
129. 5.6 oz of "Real Stuff," 6.4 oz of "Sugar Water"
131. 14 quarts of Moss County honey, 11 quarts of East Connecticut honey
133. $x \le 3$

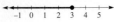

135. $-3 < x < 1$

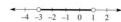

137. $x < -4$

139. $-3 \le x \le \frac{1}{3}$

141. $x > 2$

143. $x \le -\frac{1}{3}$

145. $x > -\frac{5}{2}$

147. $x < -3$

149. $x \le 2$

151. $-3 < x < -1$

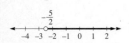

Chapter 2   TEST (p. 114)

1. 22   **3.** $8x^2 - 5x$ **5.** $m = \dfrac{I - V}{P}$ **7.** $\{\frac{13}{5}\}$ **9.** $\{-10\}$ **11.** $\{-\frac{1}{2}, 2\}$

13. $x \le -4$ **15.** $-\frac{3}{4} < x < \frac{3}{4}$

17. 18 fives, 7 twenties

TEST YOUR MEMORY: Chapters 1 and 2 (p. 116)

1.   **3.** $\frac{5}{3}$ **5.** $-\frac{2}{3}$ **7.** -4 **9.** 6.1   **11.** 62   **13.** -27 **15.** $-\frac{7}{4}$ **17.** -2

19. -120 **21.** $\frac{47}{60}$ **23.** 60   **25.** $-\frac{28}{3}$ **27.** $3x^2 - 10x$ **29.** $y = \frac{3}{4}x - 2$ **31.** $\{\frac{11}{3}\}$ **33.** $\{-13\}$
35. $\{5\}$ **37.** $\{4\}$
39. $x \le -3$ **41.** $x > \frac{14}{3}$ **43.** $-4 < x < 5$

45. $18\frac{2}{3}$ ft   **47.** 7×20 ft   **49.** 8 quarters, 24 dimes

Chapter 3   EXERCISES 3.1 (p. 125)

1. Base = 8, exponent = 3   **3.** Base = 2, exponent = 1   **5.** Base = x, exponent = 4
7. Base = 7, exponent = 5   **9.** Base = (-4), exponent = 10   **11.** Base = x, exponent = 8   **13.** 3^5
15. $(-5)^4$ **17.** -6^6 **19.** x^5 **21.** $(2z)^3$ **23.** 64   **25.** 81   **27.** -81 **29.** -625 **31.** 150
33. 28   **35.** 32   **37.** 243   **39.** 256   **41.** 125   **43.** 49   **45.** $\frac{8}{27}$ **47.** 72   **49.** x^{21} **51.** x^{71}
53. $x^5 y^9$ **55.** x^{21} **57.** $-14x^9$ **59.** x^{24} **61.** x^{162} **63.** x^{70} **65.** x^{120} **67.** $36x^2 y^2$
69. $p^{12} t^{84}$ **71.** $x^2 y^6 z^{14}$ **73.** $25x^{14} y^8$ **75.** x^9 **77.** x^{23} **79.** $\dfrac{x^8}{y^2}$ **81.** x^{25} **83.** x^9 **85.** $\dfrac{x^{45}}{y^5}$
87. $\dfrac{x^{32}}{y^8}$ **89.** $\dfrac{36x^{16}}{y^2}$ **91.** $72x^{32}$ **93.** $x^{72} y^{72}$ **95.** $-4x^4 y^8$ **97.** $\dfrac{9x^8}{49y^{16}}$ **99.** $4x^8 y^{16}$
101. $\frac{8}{27}x^3 y^{12}$ **103.** $2x^{13}$ **105.** $32x^6 y^{14}$ **107.** $2x^3 y^{19}$ **109.** $9x^4 y^{18}$

EXERCISES 3.2 (p. 135)

1. 1   **3.** -1 **5.** $\frac{1}{7}$ **7.** $\frac{1}{13}$ **9.** $-\frac{1}{81}$ **11.** $-\frac{1}{125}$ **13.** $-\frac{1}{49}$ **15.** $\frac{25}{36}$ **17.** $-\frac{4}{9}$ **19.** $\frac{2}{9}$
21. $\frac{13}{3}$ **23.** 25   **25.** $-\frac{1}{64}$ **27.** $\frac{4}{9}$ **29.** $\frac{1}{81}$ **31.** $\frac{1}{49}$ **33.** 64   **35.** 48   **37.** x^3 **39.** $\dfrac{1}{x^3}$
41. $\dfrac{1}{x}$ **43.** $\dfrac{1}{x^6}$ **45.** x^{15} **47.** $\dfrac{1}{x^{12}}$ **49.** $x^4 y^2$ **51.** $\dfrac{1}{x^{10} y^{20}}$ **53.** x^7 **55.** 1   **57.** $\dfrac{1}{x^8}$ **59.** x^{14}
61. $\dfrac{y^3}{x^8}$ **63.** $\dfrac{6}{x^2 y^3}$ **65.** $\dfrac{3}{4x^8}$ **67.** $4x^2$ **69.** $\dfrac{6y^6}{5x^2}$ **71.** $\dfrac{9y^4}{25x^{10}}$ **73.** $125y^{12}$ **75.** $\dfrac{8y^{27}}{27x^{21}}$
77. $\dfrac{16}{27x^{16} y^3}$ **79.** $\dfrac{128x^5}{y^{22}}$ **81.** $\dfrac{20}{9x^8 y^4}$ **83.** 5.376×10^6 **85.** 7.51×10^7 **87.** 1.023×10^3
89. 1.01×10^8 **91.** 7.9×10^{-4} **93.** 6.05×10^{-7} **95.** 2.4×10^{-2} **97.** 4.009×10^{-3}
99. 1.4×10^3 **101.** 1.23×10^{-2} **103.** 6×10^{-4} **105.** 1.5×10^2 **107.** 2.4×10^5 **109.** 2.5
111. Approximately 5.12×10^{26} molecules   **113.** Approximately 2.52×10^{13} miles
115. Approximately 3.468×10^6 hares

EXERCISES 3.3 (p. 143)

1. $-7x^2 + 5x + 11$, 2nd degree, trinomial   **3.** $13x - 8$, 1st degree, binomial

5. $4x^2 + 4xy + y^3 - 7$, 3rd degree   **7.** 9, 0 degree, monomial   **9.** $x^4 - 4y^2$, 4th degree, binomial

11. $\frac{2}{3}x^6y^2z$, 9th degree, monomial   **13.** $-4x^2 + 5x - 12$, 2nd degree, trinomial

15. $\frac{7}{6}x^2 + \frac{2}{9}x + \frac{6}{5}$, 2nd degree, trinomial   **17.** $-7.69x^3 + 3.238z^2$, 3rd degree, binomial

19. $9x - 11y + 4z + 3$, 1st degree   **21.** $-\frac{5}{8}x^3y - \frac{1}{6}x^4y^2$, 6th degree, binomial

23. $\frac{7}{8}xz^2$, 3rd degree, monomial   **25.** $5x^2 - 2x - 7$, leading coefficient 5

27. $-x^9 - 4x^4 + 3x^2 - 2$, leading coefficient -1 **29.** $19x^{47} + 112x^5$, leading coefficient 19

31. $\dfrac{3x^9}{4} + \dfrac{x^7}{5} + \dfrac{3x^3}{7} + 1$, leading coefficient $\frac{3}{4}$ **33.** $-2x - 12$, leading coefficient -2

35. $8x^2 - 23$, leading coefficient 8   **37.** $x^3 + 3x^2 + 6x - 14$, leading coefficient 1

39. $-\frac{7}{15}x + \frac{5}{18}$, leading coefficient $-\frac{7}{15}$ **41.** $\frac{4}{5}x^2 - \frac{7}{6}x - \frac{5}{4}$, leading coefficient $\frac{4}{5}$

43. $\frac{2}{3}x^5 - \frac{5}{8}x + \frac{17}{18}$, leading coefficient $\frac{2}{3}$ **45.** $9.56x^4 - 6.63x^2 + 1.24$, leading coefficient 9.56

47. $6x^2 - 4x - 16$ **49.** $x^4 + 9x^3 - x^2 - 3x - 15$ **51.** $15x^2$ **53.** $-3x^3 + 11x^2 + 5x + 8$

55. $3x^2 - 13y^2$ **57.** $\frac{17}{12}x^2 - \frac{5}{8}x + \frac{13}{10}$ **59.** $-\frac{1}{6}x^2 + \frac{5}{12}x + \frac{7}{12}$ **61.** $7.31x^2 - 1.62x - 11.13$

63. $3x^2 - 4x + 8$ **65.** $-10x^2 - 2xy + 12y^2$ **67.** $4x^3 - 2x^2 + 2x - 4$ **69.** $\frac{3}{2}x^2 - \frac{1}{4}x - \frac{7}{8}$

71. $\frac{5}{12}x^2 + \frac{11}{8}x - \frac{1}{15}$ **73.** $6.85x^2 - 2.51x - 2.25$ **75.** $27x - 22$ **77.** $-23x - 22$

79. $-18x^2 - 11x + 24$ **81.** $4x^3 - 2x^2 - 15x + 68$ **83.** $-10x^2 - 13xy + 17y^2$ **85.** $\frac{13}{6}x - \frac{13}{6}$

87. $-4x^2 + \frac{1}{6}x + 3$ **89.** $21.69x - 40.59$

EXERCISES 3.4 (p. 149)

1. $40x^8$ **3.** $-63x^5$ **5.** $-48x^4y^9$ **7.** $40x^8y^{15}$ **9.** $-45x^7y^8z^9$ **11.** $108x^3y^8$ **13.** $10x^2 + 20x$

15. $-24x^4 + 48x^3$ **17.** $16x^3y + 24x^2y^2$ **19.** $-2x^4 - 8x^3 + 8x^2$ **21.** $-4xy^2 - 8x^2y^2 + 20x^2y^3$

23. $6x^2 + 26x + 24$ **25.** $x^2 + 2x - 15$ **27.** $x^2 - 9x + 8$ **29.** $2x^2 - 7x - 15$

31. $4x^2 - 12x + 9$ **33.** $16x^2 + 56x + 49$ **35.** $4x^2 - 14x + 12$ **37.** $36x^2 - 49$

39. $x^2 + 2xy + y^2$ **41.** $2x^2 - 7xy + 3y^2$ **43.** $15x^2 + 22xy + 8y^2$ **45.** $10x^2 + 16xy - 8y^2$

47. $3x^2 - 10xy + 8y^2$ **49.** $-3x^2 + 14xy - 8y^2$ **51.** $6x^2 - 26xy + 8y^2$ **53.** $16x^2 + 40xy + 25y^2$

55. $18x^4 - 27x^2 + 7$ **57.** $10x^3 - 20x^2 + 6x - 12$ **59.** $16x^3 + 80x^2 + 36x$ **61.** $4x^4 - 25$

63. $36x^4 + 12x^2 + 1$ **65.** $15x^3 - 11x^2 - 2x + 8$ **67.** $2x^3 - 5x^2 - 12x + 7$ **69.** $x^3 - 8$

71. $-6x^3 - 26x^2 - 3x + 28$ **73.** $2x^4 + 4x^3 - 7x^2 + 4x - 9$ **75.** $4x^4 - 11x^3 + 18x^2 - 9x$

77. $10x^4 + 4x^3 - 35x^2 + x + 6$ **79.** $16x^4 - 16x^3 + 3x^2 + 8x - 2$ **81.** $12x^3 - 58x^2y + 46xy^2 - 5y^3$

83. $x^3 + 64y^3$ **85.** $4x^3 + 31x^2y + 54xy^2 + 21y^3$ **87.** $3x^3 - 10x^2y + 4xy^2 + 8y^3$

89. $10x^4 + 6x^3 - 22x^2 + 8x - 8$ **91.** $2x^4 - 25x^2 + 20x + 3$ **93.** $3x^4 - 20x^3 - 6x^2 + 27x - 10$

95. $8x^5 + 4x^4 + 12x^3 - 2x^2 - 24x - 10$ **97.** $x^4 - 4x^3 - 12x^2 - 23x + 20$

99. $6x^4 + 22x^3 + 3x^2 - 29x + 5$ **101.** $-3x^4 - 9x^3 - 11x^2 - 14x + 12$

EXERCISES 3.5 (p. 154)

1. $x^2 + 7x + 10$ **3.** $x^2 - 8x - 9$ **5.** $5x^2 - 4x - 12$ **7.** $24x^2 + 20x + 4$ **9.** $6x^2 + x - 15$

11. $15x^2 - 17x + 4$ **13.** $x^2 - 11xy - 18y^2$ **15.** $10x^2 + 17xy + 6y^2$ **17.** $18x^2 - 9xy - 5y^2$

19. $8x^2 - 34xy + 21y^2$ **21.** $x^4 + 2x^2 - 15$ **23.** $x^3 + 6x^2 - 3x - 18$ **25.** $2x^3 - 5x^2 - 12x$

27. $\frac{2}{9}x^2 + \frac{13}{3}x + 20$ **29.** $\frac{3}{16}x^2 - 6x + 45$ **31.** $8.05x^2 + 5.91x - 6.58$ **33.** $13.72x^2 - 30.38x + 13.86$

35. $x^2 - 25$ **37.** $64 - x^2$ **39.** $4x^2 - 36$ **41.** $9x^2 - 25$ **43.** $81x^2 - 1$

45. $x^2 - 9y^2$ **47.** $x^2 - y^2$ **49.** $4x^2 - y^2$ **51.** $16x^2 - 9y^2$ **53.** $81x^2 - 4y^2$ **55.** $\frac{1}{4}x^2 - 25$

57. $1 - \frac{25}{36}x^2$ **59.** $\frac{9}{16}x^2 - \frac{1}{25}$ **61.** $4x^2 - \frac{1}{25}$ **63.** $x^2 - \frac{9}{25}y^2$ **65.** $\frac{1}{4}x^2 - 9y^2$ **67.** $\frac{4}{9}x^2 - \frac{9}{16}y^2$

69. $1.44x^2 - 49$ **71.** $42.25x^2 - 4.41y^2$ **73.** $x^2 + 16x + 64$ **75.** $x^2 - 2x + 1$

77. $4x^2 - 20x + 25$ **79.** $9x^2 + 42x + 49$ **81.** $4x^2 + 40x + 100$ **83.** $x^2 - 4xy + 4y^2$

85. $x^2 + 14xy + 49y^2$ **87.** $16x^2 - 8xy + y^2$ **89.** $9x^2 + 12xy + 4y^2$ **91.** $\frac{1}{25}x^2 - \frac{3}{10}xy + \frac{9}{16}y^2$

93. $\frac{1}{4}x^2 + \frac{2}{3}xy + \frac{4}{9}y^2$ **95.** $1.44x^2 - 7.2x + 9$ **97.** $57.76x^2 + 51.68xy + 11.56y^2$

EXERCISES 3.6 (p. 160)

1. $x + 5$ **3.** $a^2 + 4ab - 8$ **5.** $-p^3 + 2p + 3$ **7.** $4m^3 - 11m + 1$ **9.** $x^3 + 9x^2 - 2x$

11. $16n - 4np + p^2$ **13.** $3r^2 - 4r + 1$ **15.** $-5y^4 + 7y^2 - y$ **17.** $5k^5 + k^4 - 10k^2 + 12k$

19. $q^4 - 4q^3t - 5q^2t^2$ **21.** $\dfrac{2}{y^2} - \dfrac{5}{x^3}$ **23.** $\dfrac{p}{3q} - \dfrac{5}{6} + \dfrac{4q^3}{9p^2}$ **25.** $\dfrac{b^3}{3} - 3b^2 - 2b - \dfrac{4}{5}$

27. $7r^3 + 6r^2t^2 + \dfrac{4rt^3}{3}$ **29.** $2x^2 + \dfrac{9m^3}{4x}$ **31.** $-2y^6 + 3y^4p^2 + 4y^2p^4$ **33.** $\dfrac{9x}{y^3} - \dfrac{6}{y^2} + \dfrac{y}{5x}$

35. $\dfrac{ab^2c^2}{2} + 3a^2b - \dfrac{a^3}{3c}$ **37.** $\dfrac{2n^4p}{m} + 5n^2 + \dfrac{m}{2p}$ **39.** $-\dfrac{3k^2r}{2q^6} + \dfrac{r^3}{4k^3q^3} - \dfrac{8q^2}{kr}$ **41.** $3x - 2 + \dfrac{4}{x + 5}$

43. $5x + 2 - \dfrac{1}{x - 4}$ **45.** $5x - 6$ **47.** $2x + 16 + \dfrac{11}{x - 8}$ **49.** $5x - 1 + \dfrac{2}{2x + 3}$

51. $4x + 1 - \dfrac{7}{5x - 2}$ **53.** $2x + 5$ **55.** $6x + 2 + \dfrac{8}{3x - 1}$ **57.** $x^2 + 4x - 5 - \dfrac{2}{x - 2}$

59. $2x^2 + 5x + 1 + \dfrac{3}{x + 4}$ **61.** $4x^2 - 12x + 3 - \dfrac{5}{x + 3}$ **63.** $x^2 - x + 1$

65. $3x^2 + 2x - 8 + \dfrac{2}{4x - 1}$ **67.** $7x^2 + x + 2 - \dfrac{4}{2x + 5}$ **69.** $x^2 + 5x + 4$

71. $9x^2 + 6x + 4 + \dfrac{10}{3x - 2}$ **73.** $2x - 3 + \dfrac{16}{x^2 + 5x + 2}$ **75.** $4x + 1 + \dfrac{x + 3}{x^2 - 4x - 6}$ **77.** $3x + 5$

79. $2x - 3 - \dfrac{5}{4x^2 + 6x + 1}$ **81.** $3x^2 + x - 4 + \dfrac{2x + 5}{x^2 + 6x - 2}$ **83.** $x^2 - 8x - 4 - \dfrac{6}{2x^2 - 3x + 1}$

85. $2x^2 + 4x - 1 + \dfrac{7x - 5}{3x^2 - 5x - 3}$ **87.** $5x^2 - 6x + 10 + \dfrac{-8x - 15}{x^2 + 5x + 4}$ **89.** $-x^2 + 3x - 4 + \dfrac{5}{2 - 5x}$

91. $2x^2 + 3xy - y^2$ **93.** $3x^2 + x - 3 + \dfrac{2x + 6}{x^2 - 5x + 2}$ **95.** $x^2 + 2x - 1$

97. $2x^2 - 8x + 3 + \dfrac{4x - 12}{x^2 + 5}$ **99.** $4x^2 - 7x - 2$

Chapter 3   REVIEW EXERCISES (p. 163)

1. -49 **3.** 1024 **5.** 81 **7.** k^{13} **9.** y^{24} **11.** $16x^{12}y^{20}$ **13.** x^3 **15.** $36x^{32}y^{20}$ **17.** $64x^9y^{12}$

19. $125x^{15}y^{18}$ **21.** -1 **23.** $\frac{8}{125}$ **25.** $\frac{1}{32}$ **27.** 81 **29.** $\dfrac{1}{x^{18}}$ **31.** $\dfrac{1}{x^7}$ **33.** $\dfrac{4}{3x^2}$ **35.** $\dfrac{y^{25}}{2304}$

37. $\dfrac{128x^{18}y^7}{27}$ **39.** 2.06×10^{-5} **41.** 84 **43.** $17x$, 1st degree, monomial

45. $-3.6x^2y^4 + 2.24xz^3$, 6th degree, binomial   **47.** $-1.95x^2 + 1.6x + 5$; -1.95 **49.** $-\frac{5}{8}x - \frac{7}{6}$; $-\frac{5}{8}$

51. $4x^2 + 7x + 8$ **53.** $5x^3 - 23x^2 + 8x - 6$ **55.** $\frac{6}{5}x^2 + \frac{9}{5}xy - \frac{4}{5}y^2$ **57.** $-x^3 + 6x^2 + 15x + 1$

59. $\frac{7}{6}x^2 - \frac{3}{20}x - \frac{43}{20}$ **61.** $10x^3 - 16x^2 - 42x + 43$ **63.** $15.2x - 1.4$ **65.** $-42x^{14}$

67. $-24x^4 + 40x^2$ **69.** $-8x^3 + 32x^2 - 4x$ **71.** $18x^2 + 33x - 40$ **73.** $40x^2 - 19xy - 14y^2$

75. $3x^2 + 28xy + 60y^2$ **77.** $6x^3 + 13x^2 + 3x + 20$ **79.** $5x^4 + 16x^3 + 12x^2 - 10x - 12$

81. $24x^3 - 58x^2y + 29xy^2 + 7y^3$ **83.** $6x^4 + 9x^3 - x^2 + 35x - .25$ **85.** $x^2 - 25x + 150$

87. $20x^2 + 49x + 9$ **89.** $12x^2 + 80xy - 75y^2$ **91.** $\frac{7}{10}x^2 + \frac{2}{3}x - 10$ **93.** $20.8x^2 - 5.61x - 7.2$

95. $x^2 - 121y^2$ **97.** $4 - \frac{9}{64}x^2$ **99.** $1.69x^2 - 25$ **101.** $16x^2 - 56x + 49$ **103.** $x^2 + \frac{2}{3}x + \frac{1}{9}$

105. $12.25x^2 - 28x + 16$ **107.** $5p^4 + 12p^2 - 3$ **109.** $x^4 + 5x^3 - 2x^2 + 6x$ **111.** $2m - \dfrac{2m^3}{3y}$

113. $\dfrac{8x}{z} + 2yz - \dfrac{4y^2z^3}{3x}$ **115.** $4x + 3$ **117.** $3x^2 + 5x - 5 - \dfrac{9}{x - 4}$ **119.** $2x^2 - x - 6 + \dfrac{4}{3x + 2}$

121. $2x - 7$ **123.** $2x^2 - xy + 4y^2$ **125.** $5x^2 + 2x - 2 + \dfrac{6}{2x^2 + 3}$

Chapter 3   TEST (p. 166)

1. -100 **3.** $\frac{25}{16}$ **5.** $\frac{1}{256}$ **7.** $\frac{4}{x^8}$ **9.** $\frac{3y^6}{2x^8}$ **11.** $\frac{y^7}{16x}$ **13.** 150,000

15. $\frac{6}{5}x^3 - x^2 + \frac{7}{3}x$, 3rd degree, trinomial   **17.** $3x^2 + 5x - 6 + \frac{19}{2x+7}$ **19.** $3x^2 - 7x - 20$

21. $\frac{17}{8}x^2 + \frac{1}{2}xy - \frac{11}{18}y^2$ **23.** $5x^3 - 9x^2 + 9x - 12$ **25.** $\frac{6x^2}{y} + 9 - \frac{5y}{x^2}$ **27.** $\frac{9}{16}x^2 + \frac{9}{5}x + \frac{36}{25}$

TEST YOUR MEMORY: Chapters 1–3 (p. 167)

1. $\frac{11}{12}$ **3.** 10   **5.** -3 **7.** $-.805$ **9.** $-\frac{21}{10}$ **11.** $\{\frac{16}{7}\}$ **13.** $\{0\}$ **15.** $\{\frac{21}{4}\}$
17. $x \geq -6$ **19.** $x > \frac{4}{3}$ **21.** $-5 \leq x \leq 5$

23. $\frac{1}{36}$ **25.** $\frac{64}{27}$ **27.** $\frac{1}{64}$ **29.** $\frac{x^6}{8}$ **31.** $\frac{5x^7}{9y^{15}}$ **33.** $\frac{y}{9x^5}$ **35.** $15x^4y - 18x^3y^2 - 3x^2y^3$

37. $12x^3 + 7x^2 - 13x + 2$ **39.** $2x^2 - x - 4 + \frac{40}{3x+5}$ **41.** $20x^2 - 7xy - 6y^2$ **43.** $-120x^9$

45. $\frac{69}{4}$ feet   **47.** 9×39 kilometers   **49.** 6 39¢ stamps, 8 22¢ stamps

Chapter 4   EXERCISES 4.1 (p. 175)

1. $2 \cdot 3^2$ **3.** $5^2 \cdot 7$ **5.** $2^2 \cdot 5^2$ **7.** $3^2 \cdot 7^2$ **9.** $3 \cdot 5 \cdot 7$ **11.** $2 \cdot 3^2 \cdot 5$ **13.** $3 \cdot 5^2 \cdot 7$
15. $2^2 \cdot 3^2 \cdot 7$ **17.** 18   **19.** 7   **21.** 7   **23.** 15   **25.** $4x^3$ **27.** $9xy^2$ **29.** 6   **31.** x^2
33. $3x^2y$ **35.** $7x^2$ **37.** $8x^2y^3$ **39.** $4(3x - 7y)$ **41.** $5x(3x + 4)$ **43.** $7x^2(2x - 1)$
45. $8x^2y(2x^2 + 3y^2)$ **47.** Prime   **49.** $4abc(9d + 2)$ **51.** $6(2x^2 + 7x + 2)$ **53.** $4(x^2 - 6x - 1)$
55. $x^2(5x + 2y - 3z)$ **57.** $2xy(5x^2 - 3xy - 4)$ **59.** Prime   **61.** $6(-2x^2 + 3y^2 - 4z^2)$
63. $3x^2y(5x - 3y + 7)$ **65.** $4x^2y^2(3y - xy - 2x)$ **67.** $5x^2y(2xy + 3xy^2 + 1)$ **69.** Prime
71. $8x^2y^2(x^2 + xy - 5y^2)$ **73.** $6xyz(-y^2 - 5xy + 3z^2)$ **75.** $9(2x - 9y + 3z + 7)$
77. $2x^2(3x^3 - 4x^2 + 7x - 1)$ **79.** $2xy(4x^2 - 5xy - 6y^2 - 4x^2y^2)$ **81.** $3x^2y(5xy + 2x - 4y + y^2)$
83. $(5a + 3b)(x + 2)$ **85.** $(8x + 5)(x - 3)$ **87.** $(2a - 7b)(x - 1)$ **89.** $(3x - 7)(x + 2)$
91. $(5y + 1)(3y + 4)$ **93.** $4x(x + 2)(x - 6)$ **95.** $3x(3x - 1)(2x - 1)$

EXERCISES 4.2 (p. 180)

1. $(x + 10)(x - 10)$ **3.** $(y + 6)(y - 6)$ **5.** $(t + 1)(t - 1)$ **7.** Prime   **9.** $(x^2 + 9)(x + 3)(x - 3)$
11. $(6 - x)(6 + x)$ **13.** $(11 - y)(11 + y)$ **15.** $(2 - t)(2 + t)$ **17.** Prime   **19.** $(x + y)(x - y)$
21. $(3x + y)(3x - y)$ **23.** $(x + 7y)(x - 7y)$ **25.** $(10x + 3y)(10x - 3y)$ **27.** Prime
29. $(7m + 11n)(7m - 11n)$ **31.** $(x^2 + 9y^2)(x + 3y)(x - 3y)$ **33.** $(4x^2 + 25y^2)(2x + 5y)(2x - 5y)$
35. $5(x + 3)(x - 3)$ **37.** $4(x^2 + 25)$ **39.** $6(x + 10y)(x - 10y)$ **41.** $2(4x + 5y)(4x - 5y)$
43. $3(7x + 2y)(7x - 2y)$ **45.** $2x(x + 5)(x - 5)$ **47.** $9x(x^2 + 9)$ **49.** $xy(2x + 3y)(2x - 3y)$
51. $3xy^3(5x + y)(5x - y)$ **53.** $2x^2y^3(3x + 5y)(3x - 5y)$ **55.** $2y(4x^2 + 9y^2)(2x + 3y)(2x - 3y)$
57. $3x^2y(x^2 + 4y^2)(x + 2y)(x - 2y)$ **59.** $(x + 3)^2$ **61.** $(x + 5)^2$ **63.** $(y - 7)^2$ **65.** $(y - 1)^2$
67. $(2x + 5)^2$ **69.** $(3x + 1)^2$ **71.** $(4x - 3)^2$ **73.** Prime   **75.** $(5x - 2)^2$ **77.** $(x + 2y)^2$
79. $(x + 7y)^2$ **81.** $(x - 6y)^2$ **83.** $(x - 10y)^2$ **85.** $(3x + y)^2$ **87.** $(4x + 5y)^2$ **89.** Prime
91. $(5x - 2y)^2$ **93.** $(4x - 5y)^2$ **95.** $3(x + 2)^2$ **97.** $4(x - 6)^2$ **99.** $3(5x + y)^2$ **101.** $4(2x - 3y)^2$
103. $2x(x - 4)^2$ **105.** $6x^2y(x + 2y)^2$ **107.** $4y(4x^2 + 6xy + 9y^2)$ **109.** $3xy(2x - 3y)^2$
111. $4x^2y^2(4x + y)^2$

EXERCISES 4.3 (p. 186)

1. $(x + 3)(x + 6)$ **3.** $(x - 9)(x - 2)$ **5.** $(x - 1)(x - 1)$ **7.** $(x - 6)(x + 1)$ **9.** Prime
11. $(x - 9)(x + 4)$ **13.** $(x + 12)(x - 1)$ **15.** $(x + 9)(x - 8)$ **17.** $(x - 9)(x + 6)$
19. $(x + 11)(x + 1)$ **21.** $(x - 3)(x - 3)$ **23.** $(x + 2)(x - 1)$ **25.** $(x - 4)(x + 3)$
27. $3(x + 5)(x - 2)$ **29.** $2(x - 3)(x - 4)$ **31.** $4(x - 8)(x - 1)$ **33.** $3(x - 6)(x + 5)$
35. $x^2(x - 6)(x + 3)$ **37.** $4x(x + 8)(x + 1)$ **39.** $2xy(x + 8)(x - 4)$ **41.** $3xy^2(x - 2)(x - 2)$
43. $-(x - 7)(x + 3)$ **45.** $-(x - 6)(x - 2)$ **47.** $-(x + 3)(x + 5)$ **49.** $-(x + 12)(x - 1)$
51. $-2(x + 3)(x + 5)$ **53.** $-3(x - 4)(x - 5)$ **55.** $(x - 8y)(x + 3y)$ **57.** $(x + 8y)(x + y)$
59. $(x + 5y)(x + 9y)$ **61.** $(x + 8y)(x - 5y)$ **63.** $(x - y)(x - y)$ **65.** $(x + 2y)(x + y)$
67. $(x + 8y)(x - y)$ **69.** $(x + 6y)(x - 3y)$ **71.** $(x - 2y)(x - 5y)$ **73.** Prime   **75.** $(x - 9y)(x + 2y)$
77. $(x + 4y)(x - y)$ **79.** $2(x + 6y)(x - 3y)$ **81.** $3(x - 4y)(x - 5y)$ **83.** $4x(x + 2y)(x + 4y)$
85. $2xy(x + 9y)(x - y)$ **87.** $-3(x - 6y)(x + 2y)$ **89.** $-2x(x - 2y)(x - 2y)$

EXERCISES 4.4 (p. 193)

1. $(2x + 1)(x + 7)$ **3.** $(5x + 11)(x + 1)$ **5.** $(7x - 1)(x + 3)$ **7.** $(3x + 7)(x - 1)$ **9.** Prime
11. $(5x - 1)(x - 5)$ **13.** $(3x - 5)(x - 1)$ **15.** $(7x + 5)(x - 1)$ **17.** $(5x - 13)(x + 1)$
19. $(5x - 2)(x - 4)$ **21.** $(7x - 1)(x - 10)$ **23.** $(3x - 4)(2x + 3)$ **25.** Prime
27. $(4x - 3)(2x - 5)$ **29.** $(5x - 2)(5x - 2)$ **31.** $(7x - 6)(2x - 1)$ **33.** $(3x + 5)(3x - 2)$
35. $(6x + 5)(x + 2)$ **37.** $(8x - 7)(x + 2)$ **39.** $2(3x + 1)(2x + 5)$ **41.** $4(3x + 2)(2x - 5)$
43. $3(2x - 3)(2x - 5)$ **45.** $5(2x + 7)(x - 1)$ **47.** $4(3x^2 - 8x + 1)$ **49.** $-(3x - 5)(x - 1)$
51. $-(3x - 2)(2x + 5)$ **53.** $-(5x + 4)(x + 2)$ **55.** $-(3x + 4)(3x - 5)$ **57.** $-2(3x + 2)(4x + 3)$
59. $-5(7x - 1)(2x + 3)$ **61.** $x^2(4x - 5)(2x - 1)$ **63.** $xy(5x + 4))(2x - 3)$ **65.** $(5x + 3y)(x + y)$
67. $(9x - y)(x + 7y)$ **69.** $(7x - 3y)(x - 2y)$ **71.** $(3x + 5y)(2x - y)$ **73.** $(3x - 2y)(3x - 2y)$
75. $(4x + 3y)(2x + 5y)$ **77.** $(9x + y)(x - 4y)$ **79.** $(5x + 2y)(2x - 7y)$ **81.** $-(3x - 4y)(2x - y)$
83. $-(4x - 7y)(3x + y)$ **85.** $2(4x + 5y)(2x - y)$ **87.** $3(7x + 5y)(x + y)$ **89.** $3(2x^2 + 7xy - 6y^2)$
91. $x^2(4x - 3y)(2x - 5y)$ **93.** $2xy(x + 3y)(2x - 7y)$ **95.** $3x^2y(5x + 3y)(x + y)$
97. $-4x(3x - 2y)(5x - 3y)$ **99.** $-2xy^2(4x - 7y)(2x + y)$

EXERCISES 4.5 (p. 199)

1. $(x + 5)(y + 2)$ **3.** $(2x + 7)(y + 4)$ **5.** $(2x + 3)(3y + 2)$ **7.** $(6x + 5)(2y + 7)$
9. $(x + 3)(y - 6)$ **11.** $(3x + 4)(y - 3)$ **13.** $(y + 2)(x - 6)$ **15.** $(2y + 3)(2x - 7)$
17. $(y + 1)(2x + 3)$ **19.** $(5x + 1)(2y - 7)$ **21.** $(y - 1)(x + 5)$ **23.** $(2y - 1)(3x - 4)$
25. $(2x + 3)(y - 5)$ **27.** $(5x - 2)(2y - 3)$ **29.** $(3x + 4y)(a - b)$ **31.** $(2x - 7y)(3a - 2b)$
33. $(x + 4)(y + 8)$ **35.** $(5x + 2)(3y + 4)$ **37.** $(2x + 3)(x - 4y)$ **39.** $(2x - 7)(3x - 2y)$
41. $(4x - 1)(3x - 2y)$ **43.** $(x + 7)(x + 3)(x - 3)$ **45.** $(3x + 7)(x + 1)(x - 1)$
47. $(5x + 8)(2x + 3)(2x - 3)$ **49.** $(3x - 2)(5x + 1)(5x - 1)$ **51.** $(2x + 3)(5x + 2)$
53. $(5x - 4)(x + 2)$ **55.** $(9x - 2)(x - 3)$ **57.** $(2x - 5)(2x + 7)$ **59.** $(2x - 5)(5x - 1)$
61. $(2x + 1)(3x + 10)$ **63.** $(3x - 2)(4x - 3)$ **65.** $(6x + 7)(2x + 1)$ **67.** $(2x + 5)(2x + 5)$
69. $(6x + 7)(x - 1)$ **71.** $(4x + 1)(2x - 3)$ **73.** $(4x + 9)(x - 1)$ **75.** $(5x + 6)(x + 2)$
77. $(3x - 2)(3x - 4)$ **79.** $(3x + 2)(3x - 4)$ **81.** $(8x + 3)(x + 4)$ **83.** $(2x - 3)(2x - 3)$
85. $(5x - 3)(x - 2)$

EXERCISES 4.6 (p. 202)

1. $(x + 3)(x^2 - 3x + 9)$ **3.** $(5 - t)(25 + 5t + t^2)$ **5.** $(r - 6)(r^2 + 6r + 36)$
7. $(x - 2)(x^2 + 2x + 4)$ **9.** $(x + 1)(x^2 - x + 1)$ **11.** $(2x + 5)(4x^2 - 10x + 25)$
13. $(4y - 1)(16y^2 + 4y + 1)$ **15.** $(3 - 4m)(9 + 12m + 16m^2)$ **17.** $(x + 2y)(x^2 - 2xy + 4y^2)$
19. $(5x - 2y)(25x^2 + 10xy + 4y^2)$ **21.** $(5p + q)(25p^2 - 5pq + q^2)$ **23.** $(x - 4y)(x^2 + 4xy + 16y^2)$

25. $(4m + 5n)(16m^2 - 20mn + 25n^2)$ **27.** $2(x + 5)(x^2 - 5x + 25)$ **29.** $3(r - 3)(r^2 + 3r + 9)$
31. $4(2m - 3n)(4m^2 + 6mn + 9n^2)$ **33.** $2(6p + q)(36p^2 - 6pq + q^2)$
35. $2xy(4x - y)(16x^2 + 4xy + y^2)$ **37.** $3xy^2(3x - 5y)(9x^2 + 15xy + 25y^2)$
39. $(2x^2 - y)(4x^4 + 2x^2y + y^2)$ **41.** $(3x + 2y^2)(9x^2 - 6xy^2 + 4y^4)$ **43.** $(2x^3 + y)(4x^6 - 2x^3y + y^2)$
45. $(x^3 - y^2)(x^6 + x^3y^2 + y^4)$ **47.** $3(3x^2 + 4y)(9x^4 - 12x^2y + 16y^2)$
49. $2xy(4x^2 - 5y^2)(16x^4 + 20x^2y^2 + 25y^4)$

EXERCISES 4.7 (p. 203)

1. $(x - 2)(x - 1)$ **3.** $(x - 3)(x + 1)$ **5.** $(2x + 1)(x + 3)$ **7.** $(3x + 1)(3x - 1)$ **9.** $(x + 3)(x + 8)$
11. $(x + 5)^2$ **13.** $(y - 4)(y + 3)$ **15.** $(x + 6)(x - 6)$ **17.** Prime   **19.** $2x(x - 2)$ **21.** $(3x - 7)^2$
23. $(x + 11)(x - 2)$ **25.** $(y - 4)(y - 7)$ **27.** $(2x + 5)(x^2 + 3)$ **29.** $(7 - x)(3 + x)$
31. $(8x + 9)(8x - 9)$ **33.** $(2x - 3)(2x + 5)$ **35.** $(3x - 5)(8x + 1)$
37. $5x(x - 2y)(x + 2y)(x^2 + 4y^2)$ **39.** Prime   **41.** $3x(2x - 7)(x + 2)$ **43.** $(4y + 3)(2y - 3)$
45. $6ab^3(3a^2 - 2b^2 - 1)$ **47.** $(3x - 4)(2x + 3)$ **49.** $(x + 5y)(x + 15y)$ **51.** $(3x + y)(x + 1)(x - 1)$
53. $3x(x + 4)(2x + 1)$ **55.** $(4x^2 - 3)(3x^2 + 5)$ **57.** $5x(x + 5)(x - 5)$ **59.** $(2x + 5)(x + 1)(x - 1)$
61. $(x^2 + 20)(x + 1)(x - 1)$ **63.** $x^2(x + 8y)(x - 5y)$ **65.** $(2x + 3y)(4x^2 - 6xy + 9y^2)$
67. $2x^2(x - 2)(x^2 + 2x + 4)$ **69.** $(x + 2)(x^2 - 2x + 4)(x - 1)(x^2 + x + 1)$

EXERCISES 4.8 (p. 210)

1. $\{-3, -7\}$ **3.** $\{\frac{5}{2}, -12\}$ **5.** $\{\frac{1}{4}\}$ **7.** $\{\frac{4}{3}, -\frac{7}{2}\}$ **9.** $\{0, 2, -3\}$ **11.** $\{-\frac{7}{2}, 3, -\frac{1}{3}\}$
13. $\{-1, 7, -8, 1\}$ **15.** $\{3, 6\}$ **17.** $\{\frac{5}{2}, -3\}$ **19.** $\{0, -3\}$ **21.** $\{-2, 6\}$ **23.** $\{0, \frac{1}{2}\}$ **25.** $\{\frac{1}{2}, \frac{7}{5}\}$
27. $\{-\frac{2}{3}\}$ **29.** $\{-\frac{3}{4}, \frac{7}{2}\}$ **31.** $\{-\frac{1}{2}, -\frac{5}{3}\}$ **33.** $\{-\frac{1}{4}, \frac{1}{4}\}$ **35.** $\{-5, -6\}$ **37.** $\{-\frac{7}{8}, 1\}$ **39.** $\{\frac{1}{3}, \frac{3}{4}\}$
41. $\{\frac{2}{5}, -2\}$ **43.** $\{-7, -2\}$ **45.** $\{8, -7\}$ **47.** $\{\frac{7}{2}, 3\}$ **49.** $\{-\frac{1}{3}, -\frac{2}{3}\}$ **51.** $\{-\frac{7}{3}, \frac{7}{3}\}$ **53.** $\{\frac{1}{7}, 9\}$
55. $\{-2, -\frac{1}{2}\}$ **57.** $\{3, \frac{11}{2}\}$ **59.** $\{4, -\frac{11}{6}\}$ **61.** $\{-2, \frac{3}{4}\}$ **63.** $\{1, 9\}$ **65.** $\{-\frac{4}{3}, -2\}$
67. $\{-7, -1\}$ **69.** $\{0, \frac{9}{2}\}$ **71.** $\{0, -3, 1\}$ **73.** $\{0, -\frac{1}{3}\}$ **75.** $\{0, \frac{3}{2}, \frac{2}{3}\}$ **77.** $\{-2, \frac{5}{2}, \frac{1}{6}\}$
79. $\{\frac{5}{2}, \frac{2}{3}, 9\}$ **81.** $\{2, -2, \frac{5}{2}, -\frac{3}{2}\}$

EXERCISES 4.9 (p. 217)

1. 8 and 9 or -9 and -8 **3.** 10 and 12 or -4 and -2 **5.** 11 and 13 or -9 and -7
7. 6 and 7 or -1 and 0   **9.** -6 and 12   **11.** 4 and 6 or -16 and 26   **13.** $W = 4$ ft and $L = 9$ ft
15. $W = 5\frac{1}{2}$ yd and $L = 14$ yd   **17.** $W = 5$ m and $L = 18$ m   **19.** $b = 12$ ft and $h = 7$ ft
21. $b = 8$ m and $h = 5\frac{1}{2}$ m   **23.** $b = 7$ in. and $h = 15$ in.   **25.** 9 in.   **27.** 7 ft, 24 ft, and 25 ft
29. 3 in., 4 in., and 5 in.   **31.** 3 sec. and 5 sec.   **33.** 10 sec.   **35.** $2\frac{1}{2}$ sec.   **37.** 2 m/sec.   **39.** 5 m

Chapter 4   REVIEW EXERCISES (p. 220)

1. $3 \cdot 7 \cdot 11$ **3.** 14   **5.** $4x^3y^2$ **7.** $3xy^2$ **9.** $6x(x^2 + 2y)$ **11.** $3y(x^3 - 5xy + 2)$
13. $-5x(4x^2 - 5y + 3x)$ **15.** $4x^3y(2x + 3xy + 1)$ **17.** $3xy^2(x^2 + 3xy - y - 4y^2)$
19. $(2x + 5)(3x - y)$ **21.** $(x + 9)(x - 9)$ **23.** $(4x + y)(4x - y)$ **25.** $(6x + 11y)(6x - 11y)$
27. $6(3x + 2y)(3x - 2y)$ **29.** $(y^2 + 16)(y + 4)(y - 4)$ **31.** $4x^2y(2x + 1)(2x - 1)$ **33.** $(y + 6)^2$
35. $(5x + 1)^2$ **37.** $(x - 13y)^2$ **39.** Prime   **41.** $(9x + y)^2$ **43.** $2xy(3x - 2y)^2$
45. $(x + 2)(x + 4)$ **47.** $(x - 2)(x - 5)$ **49.** $(x + 3)(x - 9)$ **51.** $(x + 8)(x - 1)$ **53.** Prime
55. $-(x + 5)(x - 7)$ **57.** $(x + 13)(x - 1)$ **59.** $3(x - 4)(x - 8)$ **61.** $4x^2(x - 6)(x + 2)$
63. $(x - 8y)(x - 3y)$ **65.** $2(x - 10y)(x + 3y)$ **67.** $5x^2y(x - 4y)(x - 5y)$ **69.** $(2x + 3)(x + 1)$
71. $(3x - 2)(x - 1)$ **73.** $(4x - 3)(x - 2)$ **75.** $(6x + 1)(x - 2)$ **77.** Prime
79. $-(2x + 3)(2x - 1)$ **81.** $2(3x + 5)(3x - 2)$ **83.** $x(6x + 5)(x - 2)$ **85.** $3x(2x^2 + x + 3)$
87. $(7x + 2y)(x + 3y)$ **89.** $8(2x + y)(x + 5y)$ **91.** $-2x^3y(3x - y)(2x - 3y)$ **93.** $(3x + 5)(2y + 7)$

95. $(4x - 7)(2y + 5)$ **97.** $(2x^2 + 1)(3x + 7)$ **99.** $(6x + a)(2y + 3)$ **101.** $(x + 3)(x - 3)(5x + 2)$
103. $3x(2x + 9)(y + 4)$ **105.** $(3x + 8)(2x + 5)$ **107.** $(5x + 2)(2x - 1)$ **109.** $(3x + 2y)(2x - 7y)$
111. $(3x + 10y)(2x - 3y)$ **113.** $(x - 5)(x^2 + 5x + 25)$ **115.** $(2m + 3)(4m^2 - 6m + 9)$
117. $4(y - 1)(y^2 + y + 1)$ **119.** $2xy^3(3x + y)(9x^2 - 3xy + y^2)$ **121.** $(4x^2 - y)(16x^4 + 4x^2y + y^2)$
123. $\{-5, 2\}$ **125.** $\{-\frac{5}{2}, \frac{3}{2}, 7\}$ **127.** $\{-\frac{7}{2}, \frac{7}{2}\}$ **129.** $\{5, 8\}$ **131.** $\{-6, 3\}$ **133.** $\{4\}$ **135.** $\{\frac{1}{5}, 4\}$
137. $\{-\frac{10}{3}, \frac{5}{2}\}$ **139.** $\{-5, 0, \frac{5}{2}\}$ **141.** $\{-4, -\frac{5}{2}, 3\}$ **143.** -4 and -2 **145.** -4 and -3
147. $W = 3$ ft and $L = 12$ ft   **149.** $b = 10$ m and $h = 7$ m   **151.** 10 ft and 24 ft   **153.** 6 sec and 8 sec

Chapter 4   TEST (p. 222)

1. $(2x - 5)(2x - 11)$ **3.** $(3m + 11)^2$ **5.** $(x - 6)(x + 4)$ **7.** $2(4x - 1)(4x + 1)$
9. $(3x - 13)(x + 2)$ **11.** $4x^2y^5(2x^2 - 9xy + 5)$ **13.** $(3y - 7)(3y + 7)$ **15.** $3x^2(3x - 1)(3x + 4)$
17. $(y - 11)(y - 3)$ **19.** $2x(x + 5)(2y - 5)$ **21.** $\{\frac{1}{3}, 4\}$ **23.** $\{0, \frac{3}{2}, 3\}$ **25.** $b = 6$ ft and $h = 6$ ft

TEST YOUR MEMORY: Chapters 1–4 (p. 224)

1. $\frac{1}{5}$ **3.** $-\frac{15}{16}$ **5.** $-\frac{1}{4}$ **7.** 1600   **9.** $\{1\}$ **11.** $\{-\frac{17}{6}\}$ **13.** $\{1, 5\}$ **15.** $\{\frac{11}{2}, -3\}$
17. $x \geq \frac{7}{2}$ **19.** $-3 \leq x \leq -1$ **21.** $\frac{3}{5x^{12}}$ **23.** $\frac{x^7}{y^5}$

25. $10x^2 + 20x - 27$ **27.** $-10x^5y^2 - 40x^4y^3 + 15x^4y^5$ **29.** $-\frac{3x^2}{4y} + \frac{5y}{4} - \frac{y^2}{2x}$
31. $33x^2 - 61xy + 10y^2$ **33.** $(3x + 2y)(2x + y)$ **35.** $x^2 - 8$ **37.** $3x(x + 5)(x - 4)$
39. $(7k + 4)(7k - 4)$ **41.** $(x - 4)(x - 10)$ **43.** 22.5 m.p.g.   **45.** 6% account $1500, 9% C.D. $3000
47. 3 and 6   **49.** 6×9 feet

Chapter 5   EXERCISES 5.1 (p. 231)

1. $\frac{2}{3}$ **3.** $\frac{3x^3}{8y^4}$ **5.** $\frac{6}{7x^2y^2}$ **7.** $4xy^3$ **9.** $\frac{1}{2xy^3}$ **11.** $\frac{3xy^2}{2}$ **13.** $\frac{7x^3}{9y^4}$ **15.** $\frac{3x^3}{4y^2}$ **17.** $\frac{2x(x - 4)}{3y(x + 3)}$
19. $\frac{3x(x + 2)}{8y^2}$ **21.** $5x(2x - y)$ **23.** $\frac{1}{2xy(x + 4)}$ **25.** $-\frac{4x}{3y^2}$ **27.** $\frac{x - 4y}{3xy^2(x + 2y)}$ **29.** $\frac{x + 3}{x + 4}$
31. $\frac{2x + 3}{3x + 2}$ **33.** $\frac{x - 1}{2x - 3}$ **35.** $\frac{x + 3}{2x - 5}$ **37.** $\frac{x - 1}{x - 2}$ **39.** $\frac{(x + 3)(x - 4)}{(x - 2)(x + 1)}$ **41.** $-\frac{x + 4}{2x}$
43. $\frac{x + 4}{x + 2}$ **45.** $\frac{1}{x - 5}$ **47.** $\frac{x - 2}{2x}$ **49.** -1 **51.** $\frac{2x + 3}{x - 4}$ **53.** $\frac{-1}{x + 8}$ **55.** $-x + 5$
57. $3x - 2$ **59.** $5xy$ **61.** -3 **63.** $\frac{1}{3x^2}$ **65.** $\frac{3x^2}{2x + 1}$ **67.** $\frac{3(x + 4)}{x(x + 1)}$ **69.** $\frac{2(x - 4)}{x + 3}$
71. $\frac{x - 6}{4xy^2(x + 2)}$ **73.** $\frac{5(x + 2)}{3x(x - 3)}$ **75.** $\frac{-4x(x + 3)}{3(x - 1)}$ **77.** $\frac{2x(x + 2)}{3y(2x - 1)}$ **79.** $\frac{x + 2}{x + 6}$

EXERCISES 5.2 (p. 238)

1. $\frac{6x}{5}$ **3.** $\frac{7}{2x}$ **5.** $\frac{5x}{3y^2}$ **7.** $\frac{4x^2}{9y^2}$ **9.** $\frac{3}{4}$ **11.** $\frac{2x^2y}{5}$ **13.** $\frac{6}{5xy^2}$ **15.** $\frac{1}{6y^2}$ **17.** $\frac{2}{x - 2}$
19. $\frac{2x(x + 1)}{5(x + 3)}$ **21.** $\frac{2x^3}{15y^3}$ **23.** $12x^2y^2$ **25.** $\frac{x^3}{2(2x - 5)}$ **27.** $\frac{3(x - 1)}{2x^2(2x - 3)}$ **29.** $\frac{x + 6}{9x^2}$
31. $\frac{4x^4(x + 2)}{x - 1}$ **33.** $\frac{x - 4}{x - 1}$ **35.** $\frac{x - 2}{x - 6}$ **37.** $\frac{2x - 1}{3x - 2}$ **39.** $\frac{(3x - 1)(x - 3)}{(2x - 1)(x - 1)}$ **41.** $\frac{-(x + 4)}{x - 1}$

43. $\dfrac{1}{(x+6)(x+2)}$ **45.** $(x+3)(x-2)$ **47.** $\dfrac{(x-5)(x+3)}{(x+2)(x+4)}$ **49.** $\dfrac{3x-2}{3x+1}$ **51.** $(x-2)(x+6)$

53. 1 **55.** $\dfrac{x+6}{2x-1}$ **57.** $\dfrac{-(x+1)}{x+4}$ **59.** $\frac{3}{5}$ **61.** $\dfrac{(x+5)(x-2)}{x-1}$ **63.** $x-7$ **65.** $\dfrac{x+2}{(3x-1)(x+5)}$

67. $\dfrac{-(2x+3)(x+1)}{x+6}$ **69.** $\dfrac{2x}{(x+1)(x-2)}$ **71.** $\dfrac{3(x-4)}{x^2}$ **73.** $\dfrac{2x}{x-5}$ **75.** $\frac{2}{5}$ **77.** $\frac{3}{4}$

79. $\dfrac{x^2(x+1)}{3(x-2)}$ **81.** $\dfrac{2(x+7)}{3x(x-9)}$ **83.** $\dfrac{(x+2)(x-1)}{(x-4)(x-2)}$ **85.** $\dfrac{1}{(x-2)(2x+1)}$ **87.** $\dfrac{3x}{2(x-5)}$

89. $\dfrac{2a+b}{2(3a+2b)}$

EXERCISES 5.3 (p. 245)

1. $\dfrac{7x}{3}$ **3.** $\dfrac{3x}{4}$ **5.** $\dfrac{y}{2}$ **7.** $-\dfrac{b}{3}$ **9.** $\dfrac{6}{7x}$ **11.** $\dfrac{5y}{4x^2}$ **13.** $\dfrac{2}{xy}$ **15.** $\dfrac{1}{y^2}$ **17.** $\dfrac{5x+6}{8}$ **19.** $\dfrac{5x+3}{y}$

21. $\dfrac{x+3}{6}$ **23.** $\dfrac{3x-5}{3y}$ **25.** $\dfrac{3}{y}$ **27.** $\dfrac{12x}{3x-2y}$ **29.** 2 **31.** $\dfrac{2x}{2x-5}$ **33.** $\dfrac{x-5}{3x+2}$ **35.** $3x-1$

37. $\dfrac{2x+5}{3x-4}$ **39.** $\dfrac{5}{x-4}$ **41.** $\dfrac{1}{x-2}$ **43.** $\frac{5}{3}$ **45.** $\frac{2}{7}$ **47.** $\dfrac{3x-7}{x^2-x-7}$ **49.** $\dfrac{x^2+16}{x^2-x-12}$

51. $\dfrac{2x-3}{x+5}$ **53.** $\dfrac{1}{x-1}$ **55.** 3 **57.** $\dfrac{2x^2-6}{x^2-4x-32}$ **59.** $\dfrac{2x^2+5x-12}{x^2-5x-6}$ **61.** $\dfrac{2x+5}{x+6}$

63. $\dfrac{3}{x-1}$ **65.** $\dfrac{9}{2x-3}$ **67.** $\dfrac{-2}{5x-4}$ **69.** $\dfrac{2x+13}{5x-2}$ **71.** $\dfrac{3x-3}{2x-3}$ **73.** -1 **75.** $\frac{4}{9}$ **77.** $\dfrac{4}{x-3}$

79. $\dfrac{3y}{2x}$ **81.** $\frac{9}{4}$ **83.** $\dfrac{-4x-3}{2x-5}$ **85.** $3x-5$ **87.** $\dfrac{3x+4}{x-2}$ **89.** $\dfrac{1}{x-6}$

EXERCISES 5.4 (p. 252)

1. $\dfrac{31x}{35}$ **3.** $\dfrac{10x}{9}$ **5.** $\dfrac{4x}{5}$ **7.** $-\dfrac{22x}{45}$ **9.** $\dfrac{4x+15}{6x^2}$ **11.** $\dfrac{4}{3x}$ **13.** $\dfrac{y}{2x}$ **15.** $\dfrac{14xy-20y}{35x^2}$

17. $\dfrac{2y+4x}{xy}$ **19.** $\dfrac{4y^2-9x^2}{6xy}$ **21.** $\dfrac{5y^2+6x}{x^2y^3}$ **23.** $\dfrac{3-5x}{9xy}$ **25.** $\dfrac{6x+13}{(x+3)(2x+1)}$

27. $\dfrac{x-8}{(2x-1)(x-2)}$ **29.** $\dfrac{-5x+41}{(x-5)(x+3)}$ **31.** $\dfrac{6x^2-21x+6}{(x+2)(x-4)}$ **33.** $\dfrac{x^2-3x+8}{(x-4)(x-1)}$

35. $\dfrac{3x^2+5x}{(x-3)(2x+1)}$ **37.** $\dfrac{15x^2+2x+2}{3x(x+1)}$ **39.** $\dfrac{-4x+5}{2x(2x-1)}$ **41.** $\dfrac{10x^2-8x+2}{5x(4x-1)}$

43. $\dfrac{8x^2+36x+27}{8x(4x+3)}$ **45.** $\dfrac{3x^2-4x-16}{(x-5)(x-2)}$ **47.** $\dfrac{x^2+2x-14}{(x-4)(2x-3)}$ **49.** $\dfrac{-x^2+5x-7}{(3x-5)(x-2)}$

51. $\dfrac{13x^2-28x-8}{(2x+3)(3x-7)}$ **53.** $\dfrac{7x}{(x-3)(x+4)}$ **55.** $\dfrac{1}{3x(x+1)}$ **57.** $\dfrac{3x-4}{(x+2)(x-3)}$ **59.** $\dfrac{2x+16}{(x-1)^2(x+5)}$

61. $\dfrac{x+3}{2x(x-3)}$ **63.** $\dfrac{x-1}{(x+6)(x+2)}$ **65.** $\dfrac{x^2+8}{(x+4)^2(x-4)}$ **67.** $\dfrac{x+4}{(x+3)(x+2)}$

69. $\dfrac{2x^2-4x-21}{(x+4)(x+2)(x-5)}$ **71.** 0 **73.** $\dfrac{2x-9}{(x-4)(x-3)}$ **75.** $\dfrac{x-4}{(x-6)(x-1)}$ **77.** $\dfrac{2x+5}{x+2}$

79. $\dfrac{x+6}{2(x+1)}$

EXERCISES 5.5 (p. 259)

1. $\frac{6}{35}$ **3.** $\frac{3}{4}$ **5.** 2 **7.** $\dfrac{y^4}{x}$ **9.** $\dfrac{x^2}{12y^2}$ **11.** $\dfrac{10x^5}{21y^7}$ **13.** $\dfrac{2x^2}{3y^3}$ **15.** $\dfrac{5x^2}{9y^3}$ **17.** $\dfrac{3x}{10}$ **19.** $\dfrac{1}{6xy^3}$

21. $12x^3y$ **23.** $x + 2$ **25.** $\dfrac{x + 3}{2(x - 1)}$ **27.** $\dfrac{1}{3(x + 3)}$ **29.** $\dfrac{x - 3}{7}$ **31.** $\dfrac{1}{3(2x + 3)}$ **33.** $\dfrac{3x^2}{4(3x + 1)}$

35. $\dfrac{4(x - 1)}{5x^2(x + 5)}$ **37.** $\dfrac{2(x - 5)}{5x(x + 3)}$ **39.** $\frac{15}{4}$ **41.** $\dfrac{1}{3(2x - 3)}$ **43.** $\frac{3}{2}$ **45.** $\frac{2}{7}$ **47.** $\frac{7}{16}$ **49.** $\dfrac{3}{x}$ **51.** $\frac{3}{5}$

53. $\dfrac{3x - 2}{2x}$ **55.** $\dfrac{x + 4}{6x}$ **57.** $\dfrac{1}{2x}$ **59.** $\dfrac{25x^2}{x + 5}$ **61.** $\dfrac{4x(2x + 1)}{x - 12}$ **63.** $\dfrac{x - 4}{x - 1}$ **65.** $\dfrac{x + 7}{x + 5}$

67. $\dfrac{x - 3}{x - 5}$ **69.** $\dfrac{12x + 28}{x + 10}$ **71.** $\dfrac{15x - 24}{2x + 7}$ **73.** $\dfrac{x - 1}{2x + 4}$ **75.** $\dfrac{2x + 6}{x + 10}$

EXERCISES 5.6 (p. 265)

1. $\{1\}$ **3.** $\{-\frac{1}{3}\}$ **5.** $\{30\}$ **7.** $\{\frac{6}{5}\}$ **9.** $\{2\}$ **11.** $\{\frac{25}{16}\}$ **13.** $\{-14\}$ **15.** $\{\frac{5}{3}\}$ **17.** $\{\frac{3}{8}\}$
19. $\{-2\}$ **21.** $\{\frac{11}{4}\}$ **23.** $\{-3\}$ **25.** $\{\frac{10}{3}\}$ **27.** $\{\frac{5}{6}\}$ **29.** $\{\frac{1}{2}, 4\}$ **31.** $\{1, -5\}$ **33.** $\{-\frac{4}{7}\}$
35. $\{7, -5\}$ **37.** $\{-7\}$ **39.** $\{0, -20\}$ **41.** $\varnothing$ **43.** $\{6\}$ **45.** $\{\frac{1}{4}\}$ **47.** $\{5\}$ **49.** $\{1\}$ **51.** $\varnothing$
53. $\{3, -3\}$ **55.** $\{-3\}$ **57.** $\varnothing$ **59.** $\{\frac{1}{2}, -5\}$ **61.** $\{-8\}$

EXERCISES 5.7 (p. 270)

1. $\frac{4}{7}$ **3.** $\frac{8}{3}$ **5.** $\frac{1}{6}$ **7.** $\frac{2}{3}$ **9.** $\frac{7}{3}$ **11.** $\frac{5}{2}$ **13.** $\frac{2}{9}$ **15.** $\dfrac{3x}{2}$ **17.** $\dfrac{5x^2}{1}$ **19.** $\frac{4}{3}$ **21.** $\frac{7}{12}$ **23.** $\{4\}$
25. $\{\frac{33}{4}\}$ **27.** $\{-36\}$ **29.** $\{-\frac{9}{8}\}$ **31.** $\{\frac{5}{2}\}$ **33.** $\{\frac{13}{10}\}$ **35.** $\{-\frac{4}{3}\}$ **37.** $\{\frac{7}{6}\}$ **39.** $\{\frac{3}{2}, -4\}$
41. $\{2, 3\}$ **43.** $\{\frac{6}{11}\}$ **45.** $\{\frac{1}{5}\}$ **47.** $\{-1\}$ **49.** $\{-\frac{1}{2}, -3\}$ **51.** $\varnothing$ **53.** $\{6, -1\}$ **55.** \$3.20
57. 6.25 mi   **59.** 11.25 glasses   **61.** $8\frac{2}{3}$ cups   **63.** 6 gal   **65.** 12.5 lb

EXERCISES 5.8 (p. 277)

1. 3   **3.** 5   **5.** $\frac{13}{17}$ **7.** 4, 12 or -6, 2   **9.** 2, 8   **11.** $\frac{1}{2}, \frac{5}{2}$ or $-\frac{5}{3}, \frac{1}{3}$ **13.** $1\frac{1}{3}$ hr (or 1 hr 20 min)
15. 24 min   **17.** 3 hr   **19.** 6 min and 12 min   **21.** 6 hr   **23.** $6\frac{2}{3}$ hr (or 6 hr 40 min)   **25.** 4 mph
27. 6 mph   **29.** 20 mph   **31.** 200 mi   **33.** 125 mph   **35.** 35 mph out and 15 mph back

Chapter 5   REVIEW EXERCISES (p. 280)

1. $\dfrac{3x}{2y^3}$ **3.** $\dfrac{x - 2y^2}{4x^6(2 + y^3)}$ **5.** $\dfrac{-2}{x^2}$ **7.** $\dfrac{x(x + 4)}{3(2x + 3)}$ **9.** $\dfrac{3x - 1}{x - 7}$ **11.** $\dfrac{-1}{3x + 4}$ **13.** $\dfrac{3x(x - 2y)}{3x + y}$

15. $\dfrac{x^2(x - 3)}{2(x + 2)}$ **17.** $\dfrac{4x}{3}$ **19.** $\dfrac{11x^4}{6y^4}$ **21.** -10 **23.** $\dfrac{y}{2x(x - y)}$ **25.** $\dfrac{x + 3}{x - 2}$

27. $-\dfrac{x - 2y}{y - x}$ or $\dfrac{x - 2y}{x - y}$ **29.** $\dfrac{x + 4}{x - 5}$ **31.** $x(3x + 2)$ **33.** $\dfrac{1}{(x + 1)(x - 3)}$ **35.** $(x - 3y)(x + 10y)$

37. $\dfrac{4x}{3}$ **39.** $\dfrac{2x + 10}{5y}$ **41.** 1   **43.** $\dfrac{3}{x - 4}$ **45.** $x + 3$ **47.** $\dfrac{-x^2 + 6x - 13}{x^2 - 2x - 15}$ **49.** 3   **51.** -1

53. $\dfrac{51}{10x}$ **55.** $\dfrac{-13x - 9}{8x(x - 3)}$ **57.** $\dfrac{4x^2 + 7x - 15}{(2x + 1)(x - 3)}$ **59.** $\dfrac{-17x + 6}{x^2 - 4}$ **61.** $\dfrac{2x + 5}{(x + 1)(2x + 3)}$

63. $\dfrac{3x}{(x - 2)(x - 1)}$ **65.** $\dfrac{x^2 - 3x + 11}{(2x - 5)(x - 1)^2}$ **67.** $\dfrac{x - 3}{3(x - 4)}$ **69.** $\dfrac{2x^2y^6}{9}$ **71.** $\dfrac{x + 3}{2}$ **73.** $\dfrac{2x + 1}{75}$

75. $\dfrac{x(x - 5)}{3(x + 3)}$ **77.** $\frac{2}{3}$ **79.** $-\dfrac{5x + 3}{3x}$ **81.** $\dfrac{7x^2}{2x - 7}$ **83.** $-\frac{1}{2}$ **85.** $\{7\}$ **87.** $\{-2\}$ **89.** $\{3\}$

91. $\{-1, 6\}$ **93.** $\{5\}$ **95.** $\{5\}$ **97.** $\frac{4}{3}$ **99.** $\frac{4}{5}$ **101.** $\{-\frac{4}{3}\}$ **103.** $\{4\}$ **105.** $\{1, \frac{5}{2}\}$ **107.** $\{-6\}$
109. \$7.95   **111.** 8   **113.** 12 hr   **115.** 7 mph

Chapter 5   TEST (p. 283)

1. $\dfrac{-5x^2}{x+y}$ **3.** $\dfrac{1}{x+6}$ **5.** $\dfrac{x+4}{(2x+5)(x+3)}$ **7.** $\left\{-\frac{7}{6}\right\}$ **9.** $\frac{7}{11}$ **11.** 6.25 gal

TEST YOUR MEMORY: Chapters 1–5 (p. 285)

1. $\frac{7}{12}$ **3.** $\frac{9}{100}$ **5.** $(x-12)(x-1)$ **7.** $(2x-1)(4x+3y)$ **9.** $1 \le x \le 8$

11. $\{-11\}$ **13.** $\left\{-\frac{10}{3}, 5\right\}$ **15.** $\left\{-\frac{1}{3}\right\}$ **17.** $\left\{-\frac{6}{5}\right\}$ **19.** $\left\{\frac{3}{5}\right\}$ **21.** $\{3, 6\}$ **23.** $\dfrac{x^2}{y^{10}}$ **25.** $\dfrac{3x^2}{x+4}$

27. $\frac{3}{2}$ **29.** $\dfrac{1}{x+1}$ **31.** $-10x^5y - 6x^4y^2 + 8x^3y^5$ **33.** $4x^2 - 2x + 7 + \dfrac{3}{2x-5}$ **35.** $\dfrac{x(x-4)}{(x-2)(x+2)}$

37. x **39.** $\dfrac{2x-7}{x-3}$ **41.** $\dfrac{2x-43}{(x+3)(x-4)}$ **43.** 3 quarters, 13 nickels

45. width = 6 mm, length = 9 mm   **47.** 36 eggs   **49.** 48 min

Chapter 6   EXERCISES 6.1 (p. 292)

1. Yes   **3.** Yes   **5.** No   **7.** Yes   **9.** Yes   **11.** No   **13.** Yes   **15.** No   **17.** No   **19.** Yes
21. Yes   **23.** No   **25.** $(-5, -4), (0, -2), (5, 0), (10, 2)$ **27.** $(-2, 7), \left(0, \frac{7}{2}\right), (2, 0), (6, -7)$
29. $(2, -6), (-1, -1), (-4, 4), (5, -11)$ **31.** $(1, 16), (-2, -2), (-3, -8), (-1, 4)$
33. $(-1, 2), (11, -2), (-7, 4), (8, -1)$ **35.** $\left(-\frac{5}{2}, -\frac{7}{2}\right), \left(\frac{4}{3}, -\frac{22}{3}\right), (-8, 2), \left(-\frac{17}{4}, -\frac{7}{4}\right)$
37. $\left(-\frac{3}{2}, -1\right), \left(4, \frac{3}{8}\right), \left(\frac{1}{2}, -\frac{1}{2}\right), \left(\frac{23}{6}, \frac{1}{3}\right)$ **39.** $(4, 3), (-6, 8), (2, 4), (-10, 10)$
41. $\left(\frac{3}{2}, 6\right), (-1, 16), (1, 8), \left(\frac{17}{5}, -\frac{8}{5}\right)$ **43.** $(2, 0), (10, 5.4), (0, -1.35), (-8, 6.75)$
45. $(3, y)$ where y is any real number, $(3, 4), (3, -2), (3, 11)$
47. $(3, -8), (-5, -8), (0, -8), (x, -8)$ where x is any real number
49. $y = -3x + 8, (-2, 14), (0, 8), (3, -1)$ **51.** $y = 5x + 4, (-4, -16), (0, 4), (1, 9)$
53. $y = -\frac{2}{5}x + 2, (5, 0), (0, 2), (-10, 6)$ **55.** $y = 2x + \frac{5}{3}, \left(-\frac{1}{2}, \frac{2}{3}\right), \left(2, \frac{17}{3}\right), \left(\frac{1}{6}, 2\right)$
57. $y = \frac{4}{7}x + 1, \left(-3, -\frac{5}{7}\right), (7, 5), (14, 9)$ **59.** $y = \frac{4}{3}x - 10, (6, -2), \left(\frac{3}{4}, -9\right), (-3, -14)$
61. $y = -\frac{2}{3}, \left(5, -\frac{2}{3}\right), \left(0, -\frac{2}{3}\right), \left(-2, -\frac{2}{3}\right)$ **63.** $x = -5y + 2, (-3, 1), (2, 0), (7, -1)$
65. $x = 4y - 3, (-11, -2), (-3, 0), (9, 3)$ **67.** $x = -\frac{4}{3}y + 4, (0, 3), (4, 0), (12, -6)$
69. $x = 4y + \frac{5}{2}, \left(3, \frac{1}{8}\right), \left(\frac{13}{2}, 1\right), \left(-\frac{1}{2}, -\frac{3}{4}\right)$ **71.** $x = \frac{9}{4}y + 1, \left(-\frac{5}{4}, -1\right), (-8, -4), (10, 4)$
73. $x = 6y - 8, \left(-4, \frac{2}{3}\right), (4, 2), (-14, -1)$ **75.** $x = \frac{4}{3}, \left(\frac{4}{3}, 8\right), \left(\frac{4}{3}, 0\right), \left(\frac{4}{3}, -\frac{7}{2}\right)$
77. $2L + 2W = 2200; L = 670$ ft when $W = 430$ ft; $W = 250$ ft when $L = 850$ ft
79. $2L + 3W = 2500; L = 725$ ft when $W = 350$ ft; $W = 280$ ft when $L = 830$ ft
81. $C = 0°$ when $F = 32°; F = 212°$ when $C = 100°$

EXERCISES 6.2 (p. 304)

1. 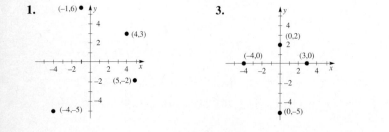 **3.**

5.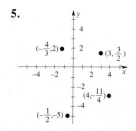

7.

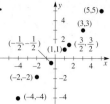

9.

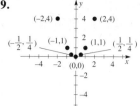

11.

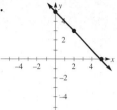

13.

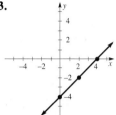

15.

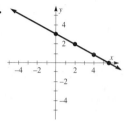

17.

19.

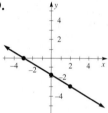

21.

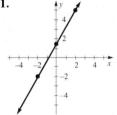

23.

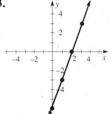

25.

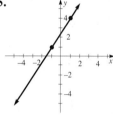

27.

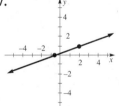

29.

31.

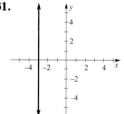

33.

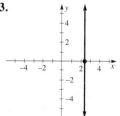

35.

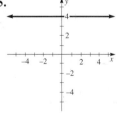

37.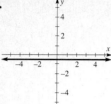

39.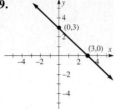

41.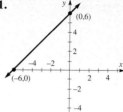

43.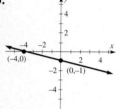

45.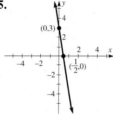

47.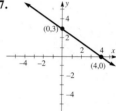

49.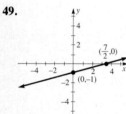

51.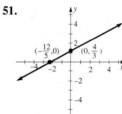

53.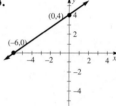

55.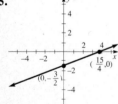

57.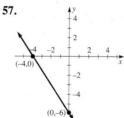

59.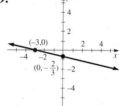

61.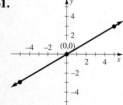

63.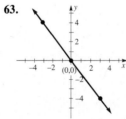

65.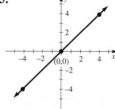

67. 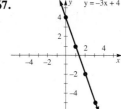 $y = -3x + 4$

69. 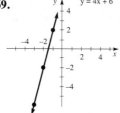 $y = 4x + 6$

71. 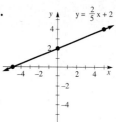 $y = \frac{2}{5}x + 2$

73. 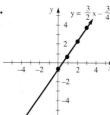 $y = \frac{3}{2}x - \frac{3}{4}$

75. 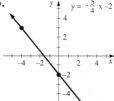 $y = -\frac{5}{4}x - 2$

77. 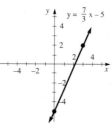 $y = \frac{7}{3}x - 5$

79. 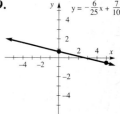 $y = -\frac{6}{25}x + \frac{7}{10}$

EXERCISES 6.3 (p. 311)

1. $\frac{2}{3}$ **3.** 2   **5.** $\frac{4}{3}$ **7.** 0   **9.** Undefined   **11.** -5 **13.** 1.25   **15.** 2   **17.** $\frac{4}{3}$ **19.** $-\frac{7}{5}$
21. $\frac{2}{9}$ **23.** $-\frac{3}{2}$ **25.** 2   **27.** $-\frac{3}{4}$ **29.** $\frac{4}{7}$ **31.** $\frac{4}{3}$ **33.** 0   **35.** Undefined

37.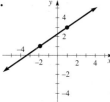

39.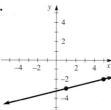

41.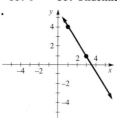

43.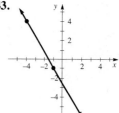

45.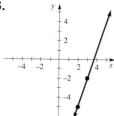

47.

49.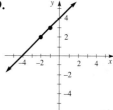

51.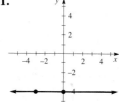

53.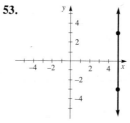

55.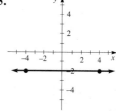

57. $m = \frac{2}{3}$, $b = 2$; slope is the coefficient of x and y-intercept is the constant
59. $m = -3$, $b = 1$; slope is the coefficient of x and y-intercept is the constant

EXERCISES 6.4 (p. 319)

1. $m = 3$, $b = 2$ **3.** $m = -\frac{4}{5}$, $b = 2$ **5.** $m = \frac{7}{3}$, $b = -\frac{5}{3}$ **7.** $m = -\frac{1}{2}$, $b = -\frac{3}{2}$ **9.** $m = 2$, $b = 7$
11. $m = 0$, $b = -\frac{4}{3}$ **13.** $m =$ undefined, no y-intercept; vertical line   **15.** $y = \frac{2}{3}x + 4$
17. $y = -\frac{1}{2}x + \frac{1}{2}$ **19.** $y = -3x - 7$ **21.** $y = \frac{3}{4}x - 1$ **23.** $y = 6$ **25.** $y = 2x + 6$
27. $y = -x - 3$ **29.** $y = -\frac{2}{3}x + \frac{13}{3}$ **31.** $y = 4$ **33.** $y = -\frac{3}{4}x + 3$ **35.** $3x - 5y = -13$
37. $7x + 2y = -3$ **39.** $4x - y = -5$ **41.** $y = 3$ **43.** $x = 4$ **45.** $2x + y = -2$
47. $7x - 2y = 33$ **49.** $y = 3$ **51.** $x = -3$ **53.** $x - y = 3$ **55.** Neither   **57.** Perpendicular
59. Parallel   **61.** Parallel   **63.** Neither   **65.** Perpendicular

EXERCISES 6.5 (p. 325)

1.

3.
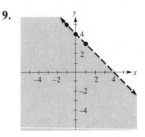

5.
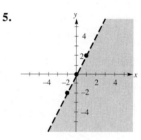

7.
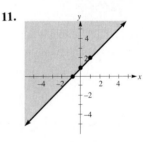

9.
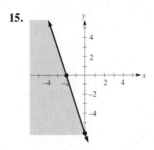

11.
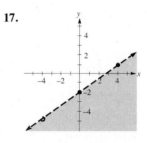

13.
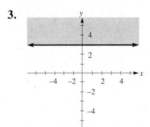

15.

17.

19.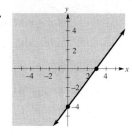

21.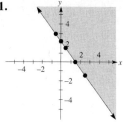

23.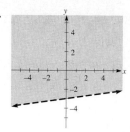

25.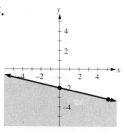

27.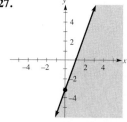

29.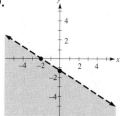

31. $x + y \le 30$
  x = minutes walking
  y = minutes running

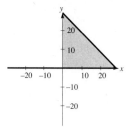

Chapter 6 REVIEW EXERCISES (p. 327)

1. Yes **3.** No **5.** Yes **7.** $(5, 4)$, $(0, -2)$, $\left(\frac{5}{3}, 0\right)$, $(-5, -8)$ **9.** $(1, 3)$, $\left(\frac{7}{3}, 4\right)$, $(-11, -6)$, $\left(-2, \frac{3}{4}\right)$
11. $(4, 1.3)$, $(-10, -7.8)$, $(2, 0)$, $(3, .65)$ **13.** $y = 4x - 7$; $(-2, -15)$, $(0, -7)$, $(5, 13)$
15. $y = -\frac{3}{8}x + 9$, $(8, 6)$, $\left(-4, \frac{21}{2}\right)$, $\left(\frac{8}{3}, 8\right)$ **17.** $x = \frac{5}{3}y + 3$; $(-2, -3)$, $\left(-\frac{1}{3}, -2\right)$, $\left(\frac{14}{3}, 1\right)$
19. $72 = 2W + 2L$; 30 ft; 10 ft

21.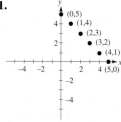

23.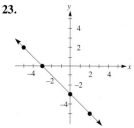

25.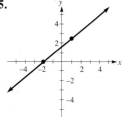

27.

29.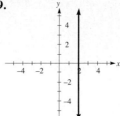

31.

33.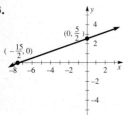

35.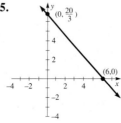

37.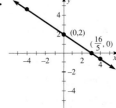

39. $y = 4x - 3$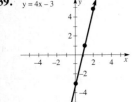

41. $y = -\frac{9}{2}x - 3$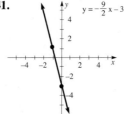

43. -2 **45.** 2 **47.** $-\frac{8}{5}$ **49.** $-\frac{4}{7}$ **51.** Undefined

53.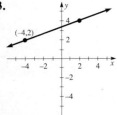

55.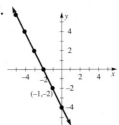

57.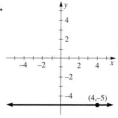

59. $m = \frac{2}{5}$, $b = -2$ **61.** $m = -\frac{1}{2}$, $b = \frac{7}{4}$ **63.** $y = -\frac{1}{3}x + \frac{2}{3}$ **65.** $y = -6$ **67.** $y = \frac{1}{3}x + \frac{16}{3}$
69. $6x - 5y = -20$ **71.** $y = 3$ **73.** $2x + 3y = 6$ **75.** Parallel   **77.** Perpendicular

79.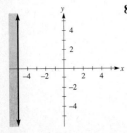

81.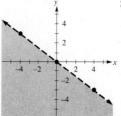

83.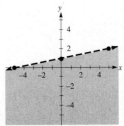

85.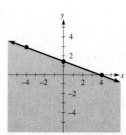

Chapter 6 TEST (p. 329)

1. $(3, 4)$, $\left(0, \frac{8}{5}\right)$, $(-2, 0)$, $\left(-\frac{9}{2}, -2\right)$ **3.** 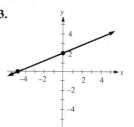   **5.** 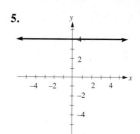   **7.** -5

9. $\frac{7}{3}$ **11.** 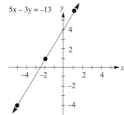   **13.** $x = -6$ **15.** $y = -5$ **17.**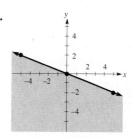

TEST YOUR MEMORY: Chapters 1–6 (p. 331)

1. $-4 < x < 5$ 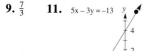   **3.** 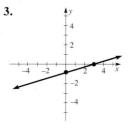   **5.** 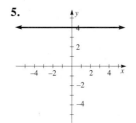   **7.**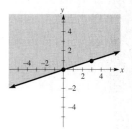

9. $\{0\}$ **11.** $\left\{-\frac{1}{2}, -3\right\}$ **13.** $\left\{-\frac{5}{2}\right\}$ **15.** $\{2\}$ **17.** $\{4\}$ **19.** $x^6 y^9$ **21.** $\dfrac{4x + 3y}{4x - 3y}$ **23.** $\dfrac{x}{x + 2}$

25. $12x^6 y^3 - 42x^3 y^5 - 18x^2 y^6$ **27.** $4x^2 - x - 4 - \dfrac{2}{3x - 2}$ **29.** $\dfrac{2x - 3}{x + 1}$ **31.** $\dfrac{3x - 9y}{4x^2 y}$

33. $\dfrac{4x - 17}{(x + 2)^2(x - 3)}$ **35.** $m = -\frac{6}{5}$ **37.** $m = \frac{2}{5}$ **39.** $3x - 4y = -11$ **41.** $2x + 3y = 2$

43. Perpendicular lines   **45.** 6 quarters, 12 dimes   **47.** 4×8 ft   **49.** 60 hr

Chapter 7 EXERCISES 7.1 (p. 339)

1. Yes   **3.** No   **5.** No   **7.** Yes   **9.** Yes   **11.** $(4, -3)$ **13.** $(2, 2)$ **15.** $(-5, 1)$
17. $(-4, -5)$ **19.** $(-2, 6)$ **21.** $(0, -2)$ **23.** Inconsistent   **25.** $(-3, 3)$ **27.** Dependent
29. $(5, -4)$ **31.** Inconsistent   **33.** $(-2, -5)$ **35.** $(-5, -5)$ **37.** $(3, 1)$ **39.** $(-1, -1)$
41. (a) $\left(\frac{1}{2}, -\frac{2}{3}\right)$ **(b)** Not dependent   **43. (a)** $\left(\frac{3}{4}, -1\right)$, $\left(-\frac{3}{2}, -\frac{1}{4}\right)$ **(b)** Dependent

EXERCISES 7.2 (p. 346)

1. (4, 5)   **3.** $\left(-2, \frac{3}{2}\right)$ **5.** (−6, −1)   **7.** Inconsistent   **9.** (−2, 10)   **11.** $\left(3, -\frac{7}{2}\right)$ **13.** $\left(\frac{2}{3}, -\frac{5}{3}\right)$
15. (−2, 7)   **17.** (3, 1)   **19.** $\left(-\frac{1}{4}, -1\right)$ **21.** (8, 3)   **23.** $\left(5, \frac{1}{3}\right)$ **25.** Dependent   **27.** (−5, −9)
29. (1, −4)   **31.** $\left(-6, -\frac{4}{3}\right)$ **33.** $\left(3, \frac{1}{2}\right)$ **35.** (7, −4)   **37.** (−8, −1)   **39.** Inconsistent   **41.** $\left(\frac{3}{5}, \frac{7}{5}\right)$
43. (4, −2)   **45.** (−12, 3)   **47.** $\left(\frac{3}{2}, \frac{1}{6}\right)$ **49.** $\left(-\frac{5}{2}, \frac{13}{3}\right)$ **51.** (8, −10)   **53.** (−6, −11)
55. Inconsistent

EXERCISES 7.3 (p. 352)

1. (5, −7)   **3.** (9, 2)   **5.** $\left(\frac{3}{2}, -\frac{1}{2}\right)$ **7.** $\left(\frac{2}{3}, -4\right)$ **9.** (−4, 7)   **11.** (−3, −1)   **13.** (−9, 8)
15. Inconsistent   **17.** (3, −3)   **19.** $\left(\frac{2}{5}, -\frac{7}{5}\right)$ **21.** (−2, −5)   **23.** (7, −8)   **25.** (0, −5)
27. Dependent   **29.** $\left(\frac{1}{6}, \frac{2}{3}\right)$ **31.** (12, −9)   **33.** (−8, −6)   **35.** (3, 5)   **37.** (−15, 1)
39. Dependent   **41.** (−9, −3)   **43.** (6, −1)   **45.** (3, 4)   **47.** $\left(\frac{7}{2}, 0\right)$ **49.** Inconsistent
51. (−18, 15)   **53.** (2, −5)

EXERCISES 7.4 (p. 359)

1. 7, 13   **3.** −1, 9   **5.** 25   **7.** $\frac{11}{8}$ **9.** $\frac{2}{5}$ **11.** $\frac{9}{7}$ **13.** 13 nickels, 29 dimes
15. 18 dimes, 52 quarters   **17.** 10 25¢-stamps, 17 3¢-stamps   **19.** 126 adult, 28 children
21. $3500 at 10%, $2500 at 8%   **23.** $1400 at 5%, $3000 at 11%   **25.** $4500 at 15%, $10,000 at 18%
27. 16 oz of Sun Valley, 8 oz of Death Valley   **29.** 16 oz of Ethel's Extract, 4 oz of Ebenezer's Elixir
31. $6.75 for each rose bush, $5.25 for each azalea   **33.** $0.39 for chocolate chip, $0.27 for butter
35. $y = 2x - 5$ **37.** $y = \frac{2}{3}x$

EXERCISES 7.5 (p. 365)

1.

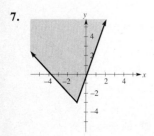

3.

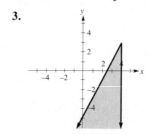

5.

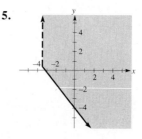

7.

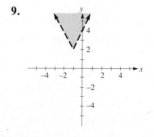

9.

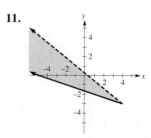

11.

13.
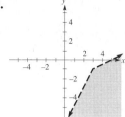

15.
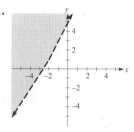

17.
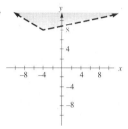

19.
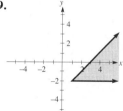

21.
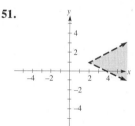

23.
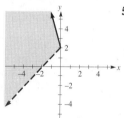

25. $y \geq 3$
$x - y \leq -1$

27. $2x - y > -9$
$4x + 3y < -3$

Chapter 7   REVIEW EXERCISES (p. 369)

1. Yes   **3.** No   **5.** (3, 6)   **7.** (−2, −5)   **9.** (−4, 0)   **11.** Inconsistent   **13.** (−2, −2)
15. (4, 1)   **17.** (5, −8)   **19.** (3, −1)   **21.** Dependent   **23.** $\left(-\frac{3}{2}, 7\right)$ **25.** (6, −10)   **27.** (2, −5)
29. (−1, −4)   **31.** Inconsistent   **33.** (5, 8)   **35.** $\left(\frac{3}{2}, -\frac{5}{2}\right)$ **37.** (−12, 15)   **39.** Dependent
41. −3, 15   **43.** $\frac{13}{7}$ **45.** 6 adult, 42 child   **47.** 6 oz of Ann's, 12 oz of Bennie's

49.
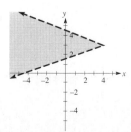

51.
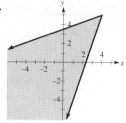

53.
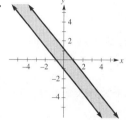

55.
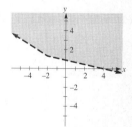

Chapter 7   TEST (p. 371)

1. (4, 5)   **3.** (3, −2)   **5.** (−1, 1)

7.
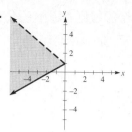

9. $45,000 at 2%, $35,000 at 3%

TEST YOUR MEMORY: Chapters 1–7 (p. 372)

1. 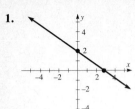   **3.** 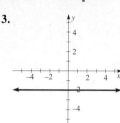   **5.** 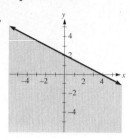   **7.**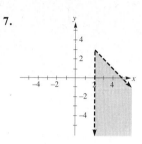

9. $m = -\frac{3}{4}$ **11.** $5x + 2y = -18$ **13.** $x - 3y = -13$ **15.** Neither   **17.** $\{-\frac{6}{7}\}$ **19.** $\{0, 4\}$

21. $\{7\}$ **23.** $\{18\}$ **25.** $\{2, -\frac{1}{4}\}$ **27.** $\frac{y}{2x^7}$ **29.** $\frac{x}{2}$ **31.** $9x^3 + 15x^2 - 26x + 8$ **33.** $\frac{3}{5y}$

35. $\frac{15y + 8x}{6xy}$ **37.** $\frac{x + 27}{(x + 2)(x - 3)}$ **39.** $(3, -1)$ **41.** Dependent   **43.** 8 nickels, 22 dimes

45. $1000 in savings account, $4000 in CD   **47.** 6 m × 8 m   **49.** $\frac{24}{5}$ hr

Chapter 8   EXERCISES 8.1 (p. 381)

1. 8   **3.** -13 **5.** $-\frac{5}{4}$ **7.** $-\frac{5}{11}$ **9.** 5   **11.** $-\frac{11}{12}$ **13.** 1   **15.** $\frac{13}{8}$ **17.** Not a real number
19. 0   **21.** $\frac{5}{2}$ **23.** -9 **25.** Not a real number   **27.** Not a real number   **29.** Not a real number
31. $-\frac{1}{10}$ **33.** 3   **35.** 3.87   **37.** -4.8 **39.** 6.93   **41.** -7.75 **43.** 2   **45.** -5
47. Not a real number   **49.** $-\frac{5}{6}$ **51.** $-\frac{2}{7}$ **53.** 2   **55.** $\frac{1}{5}$ **57.** -5 **59.** $\frac{3}{4}$ **61.** $-\frac{1}{7}$ **63.** 3
65. 6   **67.** $\frac{13}{2}$ **69.** $\frac{5}{3}$ **71.** $-\frac{1}{3}$ **73.** Not a real number   **75.** $\frac{2}{3}$ **77.** $-\frac{1}{4}$ **79.** $-\frac{6}{5}$
81. Not a real number   **83.** $-\frac{4}{5}$ **85.** 3   **87.** -3 **89.** $-\frac{1}{2}$ **91.** 2   **93.** 8   **95.** -4 **97.** 50
99. $x^2 + 5x + 7$ **101.** $125x^6$ **103.** x **105.** xy **107.** $9x$ **109.** $2xy$ **111.** $-5xyz$ **113.** x
115. $3y$ **117.** $-5xy$ **119.** $4pq$ **121.** $x + 1$ **123.** $2y + 3$ **125.** $5x + 2y$

EXERCISES 8.2 (p. 391)

1. $\sqrt{30}$ **3.** $\sqrt{14}$ **5.** 13   **7.** 10   **9.** 6   **11.** $\sqrt{7x}$ **13.** $\sqrt{6xy}$ **15.** $6x$ **17.** $4x$ **19.** x^3
21. $5x^4$ **23.** $8x^5$ **25.** $3\sqrt{2}$ **27.** $2\sqrt{5}$ **29.** $\sqrt{65}$ **31.** $5\sqrt{6}$ **33.** $2\sqrt{10}$ **35.** $3\sqrt{6}$ **37.** $x\sqrt{7}$
39. x^8 **41.** $4x^4$ **43.** $9xy^5$ **45.** $13x^6y^{10}$ **47.** $11\sqrt{x}$ **49.** $x^5\sqrt{x}$ **51.** $8x^2\sqrt{x}$ **53.** $6x\sqrt{2}$
55. $5x\sqrt{3x}$ **57.** $4x^4y^3\sqrt{3}$ **59.** $x^3y^2\sqrt{5y}$ **61.** $4x^4\sqrt{2xy}$ **63.** $\frac{7}{2}$ **65.** $\frac{\sqrt{5}}{8}$ **67.** $\frac{1}{4}$ **69.** $\frac{7}{2}$
71. 5   **73.** $\frac{2\sqrt{6}}{5}$ **75.** $\frac{4\sqrt{2}}{3}$ **77.** $\frac{2\sqrt{2}}{3}$ **79.** $\frac{8x}{7}$ **81.** $\frac{1}{10x}$ **83.** $\frac{4xy^6}{z^2}$ **85.** $\frac{2x\sqrt{6x}}{7}$
87. $\frac{4\sqrt{7x}}{9}$ **89.** $\frac{3\sqrt{2x}}{5y}$ **91.** $\frac{2x^2\sqrt{5}}{3y^2}$ **93.** $\frac{4x\sqrt{2x}}{5y^4}$ **95.** $\frac{5x^3\sqrt{3x}}{4y^3}$ **97.** $3\sqrt[3]{3}$ **99.** $2\sqrt[3]{5}$
101. $5\sqrt[3]{2}$ **103.** $3x\sqrt[3]{5}$ **105.** $3x\sqrt[3]{x^2}$ **107.** $2x\sqrt[3]{7x}$ **109.** $\frac{x}{2}$ **111.** $\frac{5x}{3y}$ **113.** $\frac{x\sqrt[3]{7x}}{4}$ **115.** $\frac{2x\sqrt[3]{2x}}{3y}$

EXERCISES 8.3 (p. 397)

1. $13\sqrt{6}$ **3.** $5\sqrt{7} + 7\sqrt{5}$ **5.** $-2\sqrt{7}$ **7.** $3x\sqrt{5}$ **9.** $-6y\sqrt{11}$ **11.** $\sqrt{2x}$ **13.** 0
15. $12x\sqrt{2y}$ **17.** $6x\sqrt{5x}$ **19.** $8\sqrt{2} + 6\sqrt{6}$ **21.** $3\sqrt{2x} - 6\sqrt{3y}$ **23.** $14x\sqrt{2y} - 4y\sqrt{x}$
25. $6\sqrt{3}$ **27.** $2\sqrt{5}$ **29.** $5\sqrt{3} - 2\sqrt{6}$ **31.** $14\sqrt{7}$ **33.** $-6\sqrt{2}$ **35.** $-8\sqrt{5}$ **37.** $17\sqrt{3}$
39. $-6\sqrt{3} + 3\sqrt{2}$ **41.** $6\sqrt{3} + 14\sqrt{2}$ **43.** $-2\sqrt{5} - 2\sqrt{6}$ **45.** $4\sqrt{5} - 6\sqrt{6}$ **47.** $\sqrt{3}$
49. $10x\sqrt{5}$ **51.** $-7x\sqrt{6}$ **53.** 0   **55.** $-x\sqrt{10}$ **57.** $3\sqrt{5x}$ **59.** $5\sqrt{3x} + 3\sqrt{5x}$ **61.** $9\sqrt{10x}$
63. $-9\sqrt{6x}$ **65.** $14x\sqrt{3x}$ **67.** $-16x\sqrt{2y}$ **69.** $11x\sqrt{6x}$ **71.** $-4y\sqrt{6xy} + 2\sqrt{3xy}$
73. $9\sqrt{2x} + 10x\sqrt{6x}$ **75.** $8x\sqrt{7y} - 4y\sqrt{6x}$ **77.** $11x\sqrt{5} - x\sqrt{7y}$ **79.** $8\sqrt{3xy} - 15\sqrt{xy}$
81. $24xy\sqrt{10y}$ **83.** $8\sqrt[3]{2}$ **85.** $-2\sqrt[3]{3}$ **87.** $x\sqrt[3]{4}$ **89.** $10x\sqrt[3]{3x}$

EXERCISES 8.4 (p. 402)

1. $2\sqrt{35}$　**3.** $18\sqrt{11}$　**5.** $18\sqrt{10}$　**7.** $10\sqrt{21}$　**9.** $30\sqrt{35}$　**11.** $12\sqrt{42}$　**13.** $5\sqrt{21xy}$
15. $3\sqrt{22xy}$　**17.** $35\sqrt{6xy}$　**19.** $16x\sqrt{14y}$　**21.** $48\sqrt{15xy}$　**23.** $210xy$　**25.** $8xy\sqrt{30y}$
27. $36xy\sqrt{10}$　**29.** $6x^2y^2\sqrt{3x}$　**31.** $\sqrt{21}+\sqrt{14}$　**33.** $6\sqrt{30}+40\sqrt{2}$　**35.** $\sqrt{5xy}-x\sqrt{3}$
37. $\sqrt{6}+\sqrt{10}-\sqrt{14}$　**39.** $40\sqrt{3}+15\sqrt{30}-30$　**41.** $6\sqrt{7x}+2\sqrt{2xy}-10x$
43. $\sqrt{35}+\sqrt{15}+\sqrt{14}+\sqrt{6}$　**45.** $\sqrt{30}-\sqrt{10}-3\sqrt{2}+\sqrt{6}$　**47.** $\sqrt{xy}+\sqrt{2x}+\sqrt{3y}+\sqrt{6}$
49. $x\sqrt{6}-\sqrt{10xy}+\sqrt{3xy}-y\sqrt{5}$　**51.** $7+\sqrt{15}$　**53.** $40+11\sqrt{6}$　**55.** $2x-7\sqrt{3x}+15$
57. $5x-4\sqrt{xy}-9y$　**59.** 3　**61.** 3　**63.** $x-9$　**65.** $x-2y$　**67.** $5+2\sqrt{6}$　**69.** $54+18\sqrt{5}$
71. $x-2\sqrt{5x}+5$　**73.** $16x-8\sqrt{x}+1$　**75.** $x+6\sqrt{xy}+9y$　**77.** $x-4\sqrt{x}-4$

EXERCISES 8.5 (p. 412)

1. $\dfrac{\sqrt{6}}{6}$　**3.** $\dfrac{7\sqrt{2}}{2}$　**5.** $\dfrac{5\sqrt{21}}{7}$　**7.** $\dfrac{\sqrt{30}}{5}$　**9.** $\dfrac{\sqrt{6}}{3}$　**11.** $\dfrac{\sqrt{10}}{2}$　**13.** $\dfrac{7\sqrt{3}}{6}$　**15.** $\dfrac{4\sqrt{2}}{7}$　**17.** $\dfrac{3\sqrt{5}}{5}$
19. $\dfrac{3\sqrt{3}}{2}$　**21.** $\dfrac{\sqrt{14}}{4}$　**23.** $\dfrac{4\sqrt{15}}{15}$　**25.** $\dfrac{21}{5}$　**27.** $\dfrac{3\sqrt{10}}{20}$　**29.** $\dfrac{2\sqrt{15}-4\sqrt{10}}{5}$　**31.** $\dfrac{5\sqrt{6}-12}{3}$
33. $\dfrac{\sqrt{2x}}{x}$　**35.** $\dfrac{\sqrt{14x}}{7x}$　**37.** $\dfrac{\sqrt{10xy}}{2y}$　**39.** $\dfrac{\sqrt{55x}}{10x}$　**41.** $\dfrac{3\sqrt{2xy}}{10y}$　**43.** $\dfrac{3\sqrt{55xy}}{22y}$　**45.** $\dfrac{3}{5x^3}$
47. $\dfrac{\sqrt{14x}}{2x^2}$　**49.** $\dfrac{3\sqrt{14x}}{7x^4}$　**51.** $\dfrac{9x^3}{4y^2}$　**53.** $\dfrac{\sqrt{22y}}{2x}$　**55.** $\dfrac{6y\sqrt{10xy}}{5x^2}$　**57.** $\dfrac{11(\sqrt{10}+2)}{6}$　**59.** $\dfrac{3(\sqrt{14}+2)}{2}$
61. $\dfrac{\sqrt{10}-\sqrt{3}}{7}$　**63.** $-\dfrac{2(\sqrt{5}+\sqrt{10})}{5}$　**65.** $5\sqrt{30}+15\sqrt{3}$　**67.** $\dfrac{35-5\sqrt{7}}{6}$
69. $\dfrac{8-\sqrt{10}}{6}$　**71.** $16+5\sqrt{10}$　**73.** $\dfrac{9+2\sqrt{14}}{5}$　**75.** $\dfrac{1-\sqrt{15}}{2}$　**77.** $\dfrac{28+19\sqrt{3}}{23}$
79. $\dfrac{41+7\sqrt{6}}{19}$　**81.** $\dfrac{18\sqrt{14}+12\sqrt{2}+6\sqrt{7}+4}{236}$　**83.** $\dfrac{x+2\sqrt{x}}{x-4}$　**85.** $\dfrac{x\sqrt{2}+5\sqrt{2x}}{x-25}$
87. $\dfrac{x-\sqrt{6x}}{x-6}$　**89.** $\dfrac{5x\sqrt{2}-5\sqrt{7x}}{2x-7}$　**91.** $\dfrac{x-4\sqrt{x}-5}{x-25}$　**93.** $\dfrac{2x+5\sqrt{x}+3}{x-1}$　**95.** $\dfrac{2x-3\sqrt{x}-9}{4x-9}$
97. $\dfrac{2x+5\sqrt{3x}+6}{4x-3}$　**99.** $\dfrac{x-2\sqrt{xy}+y}{x-y}$　**101.** $\dfrac{\sqrt[3]{3}}{3}$　**103.** $\dfrac{\sqrt[3]{90}}{3}$　**105.** $\dfrac{\sqrt[3]{10}}{2x}$　**107.** $\dfrac{\sqrt[3]{20x}}{4y}$

EXERCISES 8.6 (p. 420)

1. $\{9\}$　**3.** $\varnothing$　**5.** $\{\frac{49}{4}\}$　**7.** $\{\frac{1}{50}\}$　**9.** $\{11\}$　**11.** $\{-\frac{1}{2}\}$　**13.** $\varnothing$　**15.** $\{\frac{17}{3}\}$　**17.** $\{-\frac{3}{2}\}$
19. $\{\frac{38}{5}\}$　**21.** $\{\frac{5}{12}\}$　**23.** $\{-\frac{39}{32}\}$　**25.** $\{-1\}$　**27.** $\{\frac{29}{50}\}$　**29.** $\{36\}$　**31.** $\varnothing$　**33.** $\{16\}$　**35.** $\{\frac{49}{9}\}$
37. $\{-5\}$　**39.** $\{15\}$　**41.** $\varnothing$　**43.** $\{\frac{1}{2}\}$　**45.** $\{0\}$　**47.** $\{-2\}$　**49.** $\{6\}$　**51.** $\{2, 3\}$　**53.** $\{\frac{1}{2}, \frac{3}{2}\}$
55. $\{\frac{2}{3}, 2\}$　**57.** $\{\frac{4}{3}\}$　**59.** $\{\frac{8}{3}\}$　**61.** $\{-2, 2\}$　**63.** $\{-3, \frac{1}{2}\}$　**65.** $\{7\}$　**67.** $\{\frac{13}{2}\}$　**69.** $\{2, 3\}$
71. $\{6\}$　**73.** $\{\frac{1}{2}\}$　**75.** $\{\frac{1}{2}\}$　**77.** $\{0\}$　**79.** $\{5\}$　**81.** $\{1\}$　**83.** $\{1, 9\}$　**85.** $\{\frac{1}{4}\}$　**87.** $\{9\}$

Chapter 8　　REVIEW EXERCISES (p. 422)

1. 7　**3.** $-\frac{1}{5}$　**5.** -4　**7.** $\frac{3}{5}$　**9.** Not a real number　**11.** 4　**13.** 21　**15.** 108　**17.** $8x$
19. $-2k$　**21.** 14　**23.** $3y$　**25.** $3\sqrt{7}$　**27.** $7\sqrt{5}$　**29.** $x^5\sqrt{13}$　**31.** $x^7\sqrt{x}$　**33.** $4x^2y^3\sqrt{5y}$
35. $\dfrac{\sqrt{11}}{12}$　**37.** $\frac{3}{7}$　**39.** $\dfrac{11x^2}{y^5}$　**41.** $\dfrac{3\sqrt{10x}}{13y}$　**43.** $2\sqrt[3]{25}$　**45.** $\dfrac{x}{4y}$　**47.** $-5\sqrt{5}$　**49.** $-2x\sqrt{3}$
51. $17x\sqrt{7y}$　**53.** $10\sqrt{2}-5\sqrt{6}$　**55.** $-\sqrt{3}$　**57.** $17\sqrt{7}$　**59.** $39x\sqrt{6}$　**61.** $3\sqrt{7x}+7\sqrt{3x}$
63. $-6x\sqrt{2xy}+2\sqrt{6xy}$　**65.** $5\sqrt[3]{4}$　**67.** $35\sqrt{6}$　**69.** $252\sqrt{10}$　**71.** $12x\sqrt{15}$　**73.** $168x\sqrt{y}$
75. $15xy\sqrt{30}$　**77.** $2\sqrt{21}+\sqrt{70}$　**79.** $28\sqrt{15}-6\sqrt{30}+6\sqrt{14}$　**81.** $\sqrt{30}-7\sqrt{5}+5\sqrt{6}-35$
83. $x\sqrt{10}-5\sqrt{x}+\sqrt{2xy}-\sqrt{5y}$　**85.** $12x+11\sqrt{3x}+6$　**87.** 35　**89.** $23-8\sqrt{7}$
91. $x+10\sqrt{xy}+25y$　**93.** $\dfrac{2\sqrt{7}}{7}$　**95.** $\dfrac{\sqrt{78}}{6}$　**97.** $2\sqrt{2}$　**99.** $\dfrac{3\sqrt{30}}{35}$　**101.** $\dfrac{\sqrt{33x}}{3x}$　**103.** $\dfrac{2\sqrt{6xy}}{9y}$

105. $\dfrac{2\sqrt{65x}}{13x^2}$ **107.** $\dfrac{5(\sqrt{11}+3)}{2}$ **109.** $3\sqrt{2}-2$ **111.** $-3-\sqrt{21}$ **113.** $\dfrac{3x-2\sqrt{x}}{9x-4}$

115. $\dfrac{x-2\sqrt{x}-15}{x-25}$ **117.** $\dfrac{x-\sqrt{5xy}-10y}{x-5y}$ **119.** $\dfrac{\sqrt[3]{10}}{2}$ **121.** $\dfrac{\sqrt[3]{18x}}{9y}$ **123.** $\{23\}$ **125.** $\{\frac{10}{27}\}$

127. $\varnothing$ **129.** $\{3\}$ **131.** $\{2\}$ **133.** $\{7\}$ **135.** $\{1\}$ **137.** $\{1, 7\}$ **139.** $\{4\}$

Chapter 8   TEST (p. 424)

1. $\frac{8}{5}$ **3.** -11 **5.** $3\sqrt{10}$ **7.** $4x^3y^5\sqrt{3y}$ **9.** $\dfrac{3\sqrt[3]{2}}{x^2}$ **11.** $5\sqrt{2xy}$ **13.** $4\sqrt{3}-7\sqrt{5}$

15. $6\sqrt{2xy}+2x\sqrt{5y}-6y\sqrt{x}$ **17.** $x\sqrt{3}-3\sqrt{x}+\sqrt{xy}-\sqrt{3y}$ **19.** $-2\sqrt[3]{3}$

21. $13x\sqrt{5y}+13y\sqrt{5x}$ **23.** $\dfrac{3\sqrt{5}}{5}$ **25.** $\dfrac{5\sqrt{6}}{6}$ **27.** $\dfrac{5x-3\sqrt{x}}{25x-9}$ **29.** $\{13\}$ **31.** $\{9\}$

TEST YOUR MEMORY: Chapters 1–8 (p. 426)

1.   **3.**   **5.** $x+4y=-3$ **7.** $5x-6y=-19$

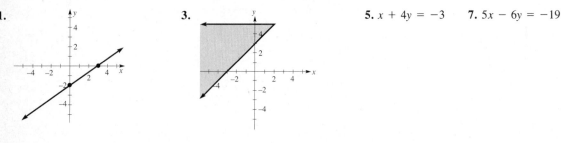

9. $\{\frac{8}{3}\}$ **11.** $\{-\frac{3}{2}\}$ **13.** $\{\frac{9}{2}\}$ **15.** $\{\frac{7}{2}\}$ **17.** $\{5\}$ **19.** $\{0, 3\}$ **21.** $\dfrac{5}{x^{11}y^8}$ **23.** $\frac{7}{10}$ **25.** $8x^2$

27. $6x^4y^4\sqrt{y}$ **29.** 1 **31.** $\dfrac{5x^2-3x-6}{(x-1)(x-3)}$ **33.** $3\sqrt{10xy}$ **35.** $7x-5y$ **37.** $2\sqrt{3}$ **39.** $\dfrac{\sqrt{6}}{3}$

41. $\dfrac{3\sqrt{5}-3}{4}$ **43.** $\left(-\frac{1}{3}, \frac{3}{2}\right)$ **45.** 10 quarters, 13 dimes   **47.** height–4 ft, base–15 ft   **49.** 12 min

Chapter 9   EXERCISES 9.1 (p. 434)

1. $5i$ **3.** $8i$ **5.** $12i$ **7.** $13i$ **9.** $2i\sqrt{5}$ **11.** $3i\sqrt{3}$ **13.** $5i\sqrt{3}$ **15.** $10i\sqrt{2}$ **17.** $\frac{1}{3}i$
19. $\frac{5}{6}i$ **21.** $-7i$ **23.** $-9i$ **25.** $-\frac{1}{4}i$ **27.** $-\frac{7}{2}i$ **29.** $9-i$ **31.** $\frac{3}{4}-\frac{1}{9}i$ **33.** $-3+5i$
35. $5-5i$ **37.** $-2+11i$ **39.** $-\frac{2}{3}-\frac{7}{20}i$ **41.** $12+18i$ **43.** $-2-6i$ **45.** $13-13i$
47. $-13+11i$ **49.** 53 **51.** $-8-6i$ **53.** $-1+3i$ **55.** $\frac{9}{5}-\frac{7}{5}i$ **57.** $\frac{7}{29}+\frac{32}{29}i$ **59.** $3-6i$
61. $5+2i$ **63.** $3+2i$ **65.** $-3+i$ **67.** $-19-3i$ **69.** $1-i$ **71.** $2+\frac{5}{3}i$ **73.** $\frac{4}{5}-\frac{1}{5}i$
75. $-\frac{3}{2}+\frac{1}{2}i$ **77.** $-\frac{3}{2}-\frac{3}{2}i$

EXERCISES 9.2 (p. 442)

1. $\{\pm8\}$ **3.** $\{\pm9\}$ **5.** $\{\pm\sqrt{22}\}$ **7.** $\{\pm2\sqrt{7}\}$ **9.** $\{\pm7i\}$ **11.** $\{\pm11i\}$ **13.** $\{\pm4i\sqrt{2}\}$

15. $\{\pm\frac{5}{2}\}$ **17.** $\left\{\pm\dfrac{\sqrt{7}}{3}\right\}$ **19.** $\left\{\pm\dfrac{2\sqrt{5}}{3}\right\}$ **21.** $\left\{\pm\dfrac{\sqrt{3}}{3}\right\}$ **23.** $\left\{\pm\dfrac{2\sqrt{3}}{3}\right\}$ **25.** $\{\pm\frac{2}{5}i\}$ **27.** $\{\pm\frac{9}{2}i\}$

29. $\{-6, -4\}$ **31.** $\{2, -\frac{4}{3}\}$ **33.** $\{\frac{8}{15}, \frac{2}{15}\}$ **35.** $\{-3\pm\sqrt{38}\}$ **37.** $\{-2\pm4\sqrt{2}\}$ **39.** $\{9\pm7i\}$

41. $\{-1, \frac{5}{3}\}$ **43.** $\{-\frac{11}{5}, \frac{9}{5}\}$ **45.** $\left\{\dfrac{5\pm\sqrt{26}}{2}\right\}$ **47.** $\left\{\dfrac{2\pm2\sqrt{3}}{3}\right\}$ **49.** $\left\{\dfrac{1\pm2\sqrt{3}}{3}\right\}$

51. $\{-2 \pm \sqrt{7}\}$　**53.** $\{\frac{3}{7} \pm \frac{9}{7}i\}$　**55.** $\{2 \pm 5i\}$　**57.** $\{-\frac{5}{2} \pm \frac{1}{2}i\}$　**59.** $\{-\frac{9}{5}, 1\}$　**61.** $\left\{\dfrac{3 \pm \sqrt{7}}{2}\right\}$

63. $\{\frac{3}{2} \pm \frac{1}{2}i\}$　**65.** The number is $3 + \sqrt{30}$ or $3 - \sqrt{30}$　**67.** The number is $\frac{1}{2}$ or $-\frac{11}{2}$

69. Shorter side $= 4$ ft, longer side $= 8$ ft　**71.** Travels 256 ft in 4 sec; hits ground in 7 sec

EXERCISES 9.3 (p. 448)

1. $49; (x + 7)^2$　**3.** $\frac{81}{4}; \left(x + \frac{9}{2}\right)^2$　**5.** $9; (x - 3)^2$　**7.** $\frac{1}{4}; \left(x - \frac{1}{2}\right)^2$　**9.** $\frac{4}{9}; \left(x + \frac{2}{3}\right)^2$　**11.** $\frac{16}{25}; \left(x + \frac{4}{5}\right)^2$

13. $\frac{1}{64}; \left(x + \frac{1}{8}\right)^2$　**15.** $\frac{9}{196}; \left(x + \frac{3}{14}\right)^2$　**17.** $\frac{1}{49}; \left(x - \frac{1}{7}\right)^2$　**19.** $\frac{9}{25}; \left(x - \frac{3}{5}\right)^2$　**21.** $\frac{1}{36}; \left(x - \frac{1}{6}\right)^2$

23. $\frac{49}{16}; \left(x - \frac{7}{4}\right)^2$　**25.** $\{2, 3\}$　**27.** $\{-5, 1\}$　**29.** $\{-4, 4\}$　**31.** $\{1 \pm \sqrt{2}\}$　**33.** $\{-2 \pm \sqrt{3}\}$

35. $\{-4 \pm \sqrt{2}\}$　**37.** $\{5 \pm \sqrt{7}\}$　**39.** $\{2 \pm i\}$　**41.** $\{-4 \pm 3i\}$　**43.** $\{3 \pm \sqrt{6}\}$　**45.** $\{-4 \pm \sqrt{3}\}$

47. $\{\frac{3}{2}, -3\}$　**49.** $\{\frac{2}{3}, 4\}$　**51.** $\{-\frac{1}{2}, \frac{3}{2}\}$　**53.** $\left\{\dfrac{1 \pm \sqrt{2}}{3}\right\}$　**55.** $\left\{\dfrac{2 \pm \sqrt{5}}{2}\right\}$　**57.** $\left\{\dfrac{-3 \pm \sqrt{6}}{2}\right\}$

59. $\left\{\dfrac{-1 \pm 2\sqrt{2}}{2}\right\}$　**61.** $\left\{\dfrac{4 \pm 2\sqrt{3}}{3}\right\}$　**63.** $\left\{\dfrac{-6 \pm \sqrt{5}}{2}\right\}$　**65.** $\{1 \pm \frac{1}{3}i\}$　**67.** $\{-\frac{1}{2} \pm \frac{5}{2}i\}$

69. The number is $-2 + \sqrt{11}$ or $-2 - \sqrt{11}$　**71.** The number is $\dfrac{1 + \sqrt{3}}{2}$ or $\dfrac{1 - \sqrt{3}}{2}$

73. $\dfrac{-1 + \sqrt{41}}{4}$ ft　**75.** After 5 sec

EXERCISES 9.4 (p. 455)

1. $x^2 + 3x - 5 = 0; a = 1, b = 3, c = -5$　**3.** $5x^2 + 4x = 0; a = 5, b = 4, c = 0$

5. $3x^2 + x - 7 = 0; a = 3, b = 1, c = -7$　**7.** $3x^2 + x - 9 = 0; a = 3, b = 1, c = -9$

9. $-x^2 - 2 = 0; a = -1, b = 0, c = -2$　**11.** $5x^2 = 0; a = 5, b = 0, c = 0$

13. $8x^2 + 14x + 1 = 0; a = 8, b = 14, c = 1$　**15.** $x^2 - 8x - 3 = 0; a = 1, b = -8, c = -3$

17. $\{-\frac{1}{3}, 3\}$　**19.** $\{4, -4\}$　**21.** $\{0, \frac{5}{2}\}$　**23.** $\{-1, 4\}$　**25.** $\{-\frac{1}{2}, 3\}$　**27.** $\left\{\pm\dfrac{\sqrt{21}}{7}\right\}$

29. $\left\{\dfrac{-7 \pm \sqrt{33}}{4}\right\}$　**31.** $\left\{\dfrac{-3 \pm 3\sqrt{5}}{2}\right\}$　**33.** $\left\{\dfrac{11 \pm \sqrt{129}}{4}\right\}$　**35.** $\left\{\dfrac{7 \pm \sqrt{65}}{2}\right\}$　**37.** $\{3 \pm \sqrt{3}\}$

39. $\{2 \pm 2\sqrt{2}\}$　**41.** $\left\{\dfrac{-6 \pm \sqrt{5}}{2}\right\}$　**43.** $\left\{\dfrac{2 \pm 5\sqrt{3}}{3}\right\}$　**45.** $\{-5 \pm \sqrt{6}\}$　**47.** $\left\{\dfrac{1 \pm 2\sqrt{3}}{4}\right\}$

49. $\{\pm3\sqrt{2}\}$　**51.** $\left\{\dfrac{-3 \pm 2\sqrt{7}}{2}\right\}$　**53.** $\{\pm\frac{5}{3}i\}$　**55.** $\{1 \pm 3i\}$　**57.** $\{-5 \pm 2i\}$　**59.** $\{\frac{3}{4} \pm \frac{1}{4}i\}$

61. $\{-\frac{1}{2} \pm \frac{3}{4}i\}$　**63.** The number is $2 + \sqrt{7}$ or $2 - \sqrt{7}$　**65.** The number is $-\frac{1}{2} + \frac{3}{2}i$ or $-\frac{1}{2} - \frac{3}{2}i$

67. $\dfrac{-1 + \sqrt{37}}{6}$ ft　**69.** 3 sec

EXERCISES 9.5 (p. 458)

1. $\{\pm\frac{3}{2}\}$　**3.** $\left\{\dfrac{7 \pm \sqrt{41}}{2}\right\}$　**5.** $\left\{\pm\dfrac{\sqrt{10}}{5}\right\}$　**7.** $\{4, -1\}$　**9.** $\{\frac{5}{2} \pm i\}$　**11.** $\{\pm\sqrt{3}\}$　**13.** $\{0, \frac{5}{3}\}$

15. $\{-\frac{5}{2}, 3\}$　**17.** $\{-\frac{7}{6}, \frac{1}{3}\}$　**19.** $\{\pm\frac{1}{2}i\}$　**21.** $\left\{\dfrac{-1 \pm \sqrt{13}}{4}\right\}$　**23.** $\left\{\dfrac{-3 \pm \sqrt{33}}{4}\right\}$　**25.** $\left\{\dfrac{3 \pm \sqrt{3}}{2}\right\}$

27. $\left\{\dfrac{2 \pm \sqrt{2}}{3}\right\}$　**29.** $\{2 \pm \sqrt{13}\}$　**31.** $\{-\frac{8}{3}, -\frac{1}{2}\}$　**33.** $\{-3 \pm 2i\}$　**35.** $\{0, -\frac{1}{2}\}$　**37.** $\{19, -3\}$

39. $\left\{\dfrac{-1 \pm \sqrt{6}}{4}\right\}$　**41.** $\left\{\pm\dfrac{\sqrt{5}}{2}\right\}$　**43.** $\{4 \pm i\}$

EXERCISES 9.6 (p. 465)

1.

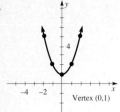

Vertex (0,1)

3.

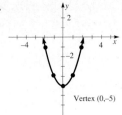

Vertex (0,–5)

5.

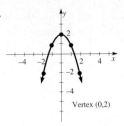

Vertex (0,2)

7.

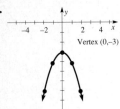

Vertex (0,–3)

9.

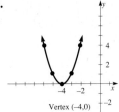

Vertex (–4,0)

11.

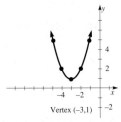

Vertex (–3,1)

13.

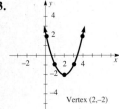

Vertex (2,–2)

15.

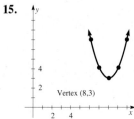

Vertex (8,3)

17.

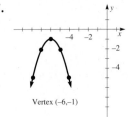

Vertex (–6,–1)

19.

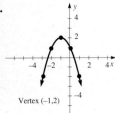

Vertex (–1,2)

21.

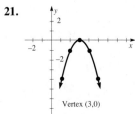

Vertex (3,0)

23.

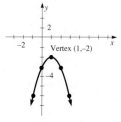

Vertex (1,–2)

25.

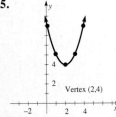

Vertex (2,4)

27.

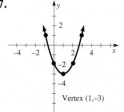

Vertex (1,–3)

29.

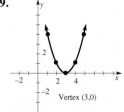

Vertex (3,0)

31.
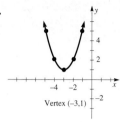
Vertex (–3,1)

33.
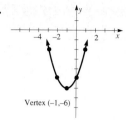
Vertex (–1,–6)

35.
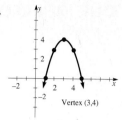
Vertex (3,4)

Chapter 9   REVIEW EXERCISES (p. 468)

1. $9i$ **3.** $\frac{2}{5}i$ **5.** $-3 + 12i$ **7.** $11 + 11i$ **9.** $-20 + 36i$ **11.** $-19 + 26i$ **13.** 85 **15.** $5 - i$

17. $-11 - 3i$ **19.** $-2 + i$ **21.** $\{\pm 4\}$ **23.** $\{\pm 5i\sqrt{2}\}$ **25.** $\left\{\pm\dfrac{2\sqrt{5}}{7}\right\}$ **27.** $\{\pm\frac{5}{3}i\}$

29. $\{-\frac{1}{2}, 2\}$ **31.** $\{-1 \pm 4\sqrt{5}\}$ **33.** $\{-\frac{17}{5}, -1\}$ **35.** $\{\frac{1}{4} \pm \frac{3}{4}i\}$ **37.** $\{\frac{1}{6}, \frac{3}{2}\}$

39. The number is $5 + 2\sqrt{3}$ or $5 - 2\sqrt{3}$ **41.** Shorter side = 6 ft, longer side = 12 ft   **43.** $36; (x - 6)^2$

45. $\frac{9}{49}; \left(x + \frac{3}{7}\right)^2$ **47.** $\{2, 7\}$ **49.** $\{-1 \pm \sqrt{7}\}$ **51.** $\{5 \pm 2\sqrt{6}\}$ **53.** $\left\{\dfrac{3 \pm 2\sqrt{2}}{2}\right\}$

55. The number is $-4 - \sqrt{21}$ or $-4 + \sqrt{21}$ **57.** $\dfrac{-1 + \sqrt{41}}{5}$ ft

59. $4x^2 - 7x = 0; a = 4, b = -7, c = 0$ **61.** $2x^2 - 3x - 19 = 0; a = 2, b = -3, c = -19$

63. $\{\frac{1}{2}, \frac{3}{4}\}$ **65.** $\left\{\pm\dfrac{\sqrt{33}}{3}\right\}$ **67.** $\{5 \pm \sqrt{3}\}$ **69.** $\left\{\dfrac{5}{2} \pm \dfrac{3\sqrt{3}}{2}i\right\}$

71. The number is $-\frac{1}{2} - \frac{1}{2}i$ or $-\frac{1}{2} + \frac{1}{2}i$ **73.** 4 sec   **75.** $\{-2 \pm \frac{1}{2}i\sqrt{10}\}$ **77.** $\{-\frac{9}{2}, 1\}$ **79.** $\{\pm 2\sqrt{2}\}$
81. $\{2, \frac{5}{2}\}$

83.
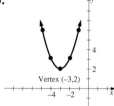
Vertex (–3,2)

85.
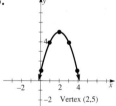
Vertex (2,5)

87.
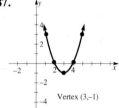
Vertex (3,–1)

Chapter 9   TEST (p. 471)

1. $-1 - 10i$ **3.** $11 - 13i$ **5.** $-9 - 4i$ **7.** $\{6 \pm 2\sqrt{2}\}$ **9.** $\{5 \pm 2\sqrt{3}\}$ **11.** $\{-4 \pm \sqrt{11}\}$
13. $\{-6 \pm \sqrt{43}\}$ **15.** $\{\frac{1}{2}, \frac{5}{4}\}$ **17.** Shorter side = 3 ft, longer side = 12 ft   **19.**

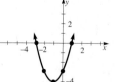
Vertex (–1,–4)

TEST YOUR MEMORY: Chapters 1–9 (p. 472)

1.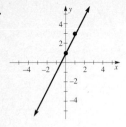

3.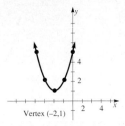

Vertex (−2,1)

5. $y = 4$ **7.** $\{\frac{3}{4}\}$ **9.** $\{0, -\frac{15}{2}\}$ **11.** $\{\frac{20}{3}\}$ **13.** $\{1, 2\}$ **15.** $\{9\}$ **17.** $\{-1, -5\}$ **19.** $\{\frac{1}{3}, -3\}$

21. $64x^{11}y^2$ **23.** -5 **25.** $2x^2y^4\sqrt{5xy}$ **27.** $\dfrac{2x}{x+3}$ **29.** $\dfrac{x+6}{2x(3x+2)}$

31. $30x\sqrt{3} + 20\sqrt{2xy} - 50\sqrt{xy}$ **33.** $5\sqrt{5}$ **35.** $-1 + 9i$ **37.** $-8 - 3i$ **39.** $\dfrac{\sqrt{21xy}}{3x}$

41. Dependent   **43.** $1 \pm \sqrt{7}$ **45.** 25 oz of Andy's; 15 oz of Alan's   **47.** $\frac{5}{2}$ cups   **49.** 9 mph

Appendix I   EXERCISES A.I (p. 478)

1. A function   **3.** Not a function   **5.** A function   **7.** A function   **9.** {All real numbers}
11. {All real numbers}   **13.** {All real numbers except 2}   **15.** {All real numbers except ± 5}
17. {All real numbers except 4 or -2}   **19.** {All real numbers ≥ 3}   **21.** {All real numbers $\geq -\frac{1}{2}$}
23. {All real numbers}   **25.** 10   **27.** -11 **29.** $\frac{11}{2}$ **31.** -2 **33.** $\frac{3}{2}$ **35.** $-\frac{3}{2}$ **37.** -29
39. -6 **41.** -6 **43.** 18

Appendix II   EXERCISES A.II (p. 483)

1. $c = 5$ in   **3.** $a = 15$ ft   **5.** $c = 26$ yd   **7.** $a = 15$ m   **9.** $b = 5$ cm   **11.** $c = 15$ mm
13. $c = 3\sqrt{2}$ m   **15.** $b = \sqrt{7}$ cm   **17.** $c = 3\sqrt{3}$ mm   **19.** $a = 7$ in   **21.** $b = 6$ ft   **23.** $c = 4$ yd
25. 36 ft   **27.** 26 mi

Index